HEAT GENERATION AND TRANSPORT IN THE EARTH

Heat provides the energy that drives almost all geological phenomena and sets the temperature, and hence the rate, at which these phenomena operate. This book provides an up-to-date treatise on heat transport processes. It explains the key physical principles with simple physical arguments and scaling laws that allow quantitative evaluation of heat flux and cooling conditions in a variety of geological settings and systems.

The thermal structure and evolution of magma reservoirs, the crust, the lithosphere and the mantle of the Earth are reviewed within the context of plate tectonics and mantle convection – illustrating how theoretical arguments can be combined with field and laboratory data to arrive at accurate interpretations of geological observations. Recent theoretical advances on free convection in many different configurations and in fluids with complex rheologies are explained, and demonstrations of how past climate changes can be reconstructed from temperature data in deep boreholes are presented. Appendices contain up-to-date information on the thermal properties of rocks and melts, as well as measurements of the surface heat flux and rate of radiogenic heat production in a large number of rocks and terrains.

Heat Generation and Transport in the Earth can be used for advanced courses in geophysics, geodynamics and magmatic processes, and is a valuable reference for researchers in geoscience, environmental science, physics, engineering and fluid dynamics. Electronic figure files, data sets and program codes relating to topics in the book can be downloaded from www.cambridge.org/9780521894883.

CLAUDE JAUPART graduated from the Ecole des Mines de Paris before obtaining a Ph.D. in geophysics from MIT and a doctorat d'Etat at the Universite de Paris 7. He has been associated with the University of Paris Diderot and the Institut de Physique du Globe since 1982, where he served as director between 1999 and 2004 and is currently Professor of Geophysics. Professor Jaupart's research covers diverse aspects of the physics of energy transport in the Earth including volcanic and magmatic systems, continental heat flux and mantle convection. His approach is based on a combination of laboratory experiments in fluid mechanics, field observations and theoretical studies. His contributions have been acknowledged by many distinctions: the Wager prize of the International Association of Volcanology and Chemistry of the Earth's Interior (1993), the silver medal of the CNRS (1995), the

Holweck and the Mergier-Bourdeix prizes of the Academie des Sciences (1995, 1998), the Prestwich medal of the Geological Society of London (1999), and the Holmes medal of the European Geophysical Union (2007).

JEAN-CLAUDE MARESCHAL holds degrees in theoretical physics from the Université Libre de Bruxelles, applied geophysics from the Universite Pierre-et-Marie Curie, Paris and geophysics from Texas A & M University. Following positions at the University of Toronto and Georgia Tech he joined the Université du Québec à Montréal in Canada in 1985, where he is now Professor of Geophysics and teaches geophysics and the physics of the Earth. He was also formerly Director of GEOTOP – the Quebec inter-university network for advanced studies and research in geoscience. Professor Mareschal's research interests include the energy budget and thermal regime of the Earth's lithosphere, the mechanical properties of the continental lithosphere in relation to its formation and evolution, and studies of heat flow at the base of ice sheets to detect signs of climate change. Both authors have worked together on the thermal structure and evolution of cratons and have been mapping the heat flow field of Canada for more than twenty years.

HEAT GENERATION AND TRANSPORT IN THE EARTH

CLAUDE JAUPART

Institut de Physique du Globe de Paris, France
Université Paris-Diderot

JEAN-CLAUDE MARESCHAL

GEOTOP-UQAM-McGill,
University of Quebec,
Montreal, Canada

CAMBRIDGE UNIVERSITY PRESS
Cambridge, New York, Melbourne, Madrid, Cape Town, Singapore,
São Paulo, Delhi, Dubai, Tokyo, Mexico City

Cambridge University Press
The Edinburgh Building, Cambridge CB2 8RU, UK

Published in the United States of America by Cambridge University Press, New York

www.cambridge.org
Information on this title: www.cambridge.org/9780521894883

First published 2011

Printed in the United Kingdom at the University Press, Cambridge

A catalog record for this publication is available from the British Library

Library of Congress Cataloging in Publication data
Jaupart, Claude.
Heat generation and transport in the Earth / Claude Jaupart, Jean-Claude Mareschal.
p. cm.
ISBN 978-0-521-89488-3
1. Terrestrial heat flow. 2. Earth–Internal structure.
I. Mareschal, Jean-Claude, 1945– II. Title.
QE509.J38 2010
551.1–dc22 2010033853

ISBN 978-0-521-89488-3 Hardback

Additional resources for this publication at www.cambridge.org/9780521894883

Contents

Introduction — *page* ix
Credits — xii
1 Historical notes — 1
  1.1 Introduction — 1
  1.2 Kelvin and the age of the Earth — 1
  1.3 The discovery of radioactivity — 3
  1.4 The debate on the cooling mechanism of the Earth — 4
  1.5 Heat flux measurements — 5
  1.6 Energy budget of the Earth — 5
  1.7 Plate tectonics — 6
2 Internal structure of the Earth — 8
  2.1 Introduction — 8
  2.2 Gravity and geodesy — 10
  2.3 Seismology — 15
  2.4 Petrology, mineral physics and seismology: Composition and state of the Earth's interior — 17
  2.5 Lateral variations of seismic structure — 23
  2.6 Core and magnetic field — 28
  2.7 The shallow Earth — 28
  Exercises — 34
3 Basic equations — 35
  3.1 Heat transport mechanisms — 35
  3.2 Definitions. Thermodynamic relationships — 37
  3.3 Conservation of mass — 40
  3.4 Conservation of momentum — 41
  3.5 Energy equation — 42
  3.6 Radial variations of density in the Earth — 46
  3.7 Equations for fluid flow — 47

4 Heat conduction 51
4.1 Heat conduction: Generalities 51
4.2 Steady-state heat equation 53
4.3 Diffusive heat transport: Basic principles 60
4.4 General solutions to the steady-state heat equation 72
4.5 Transient problems 81
4.6 Thermal stresses 95
Exercises 96
5 Heat transport by convection 99
5.1 Isolated heat sources: Plumes and thermals 99
5.2 Rayleigh–Benard convection 111
5.3 Scaling laws for heat flux and velocity in Rayleigh–Benard convection: General theory 123
5.4 Convection in porous media 136
Exercises 145
6 Thermal structure of the oceanic lithosphere 146
6.1 Continental and oceanic heat flow 146
6.2 Cooling models for oceanic heat flux and bathymetry 149
6.3 Hot spots and thermal rejuvenation of the oceanic plates 165
6.4 Other effects of oceanic plate cooling 170
Exercises 175
7 Thermal structure of the continental lithosphere 176
7.1 Continental heat flux 176
7.2 Continental lithosphere in steady state 178
7.3 Long-term transients: Stabilization and secular cooling of the continental lithosphere 194
7.4 Thermal perturbations in compressional orogens 201
7.5 Thermal regime in regions of extension 208
7.6 Passive continental margins. Sedimentary basins 218
7.7 Geophysical constraints on thermal structure 224
Exercises 230
8 Global energy budget. Crust, mantle and core 232
8.1 Thermodynamics of the whole Earth 232
8.2 Heat loss through the ocean floor 239
8.3 Heat loss through continents 244
8.4 Heat loss of the Earth 252
8.5 Radiogenic heat sources in the mantle 253
8.6 Heat flux from the core 257

8.7 Mantle energy budget 259
Exercises 260
9 Mantle convection 261
9.1 Introduction 261
9.2 Elongated convection cells 262
9.3 The impact of continents on convection 266
9.4 Convection with internal heat sources 271
9.5 Temperature-dependent viscosity 278
9.6 Non-Newtonian rheology 284
9.7 Mantle plumes as part of a large convective system 291
9.8 Two scales of convection 296
9.9 Conclusion 299
10 Thermal evolution of the Earth 300
10.1 Initial conditions 300
10.2 Thermal evolution models 307
10.3 Fluctuations of the mantle heat loss 310
10.4 Continental growth and cooling of the Earth 315
10.5 Conclusion 316
11 Magmatic and volcanic systems 317
11.1 A few features of crustal magma reservoirs 317
11.2 Initial conditions: Super-heated magma? 323
11.3 Cooling and crystallization of magma sheets: Conduction 326
11.4 Cooling by convection 342
11.5 Kinetic controls on crystallization 353
11.6 Conclusion 356
12 Environmental problems 357
12.1 The record of past climate in temperature profiles 357
12.2 Ice sheets and glaciers 375
Exercises 379
13 New and old challenges 380
Appendix A A primer on Fourier and Laplace transforms 382
A.1 Impulse response and Green's functions 382
A.2 Fourier series and transforms 392
A.3 Cylindrical symmetry. Hankel transform 401
Appendix B Green's functions 404
B.1 Steady-state heat equation 404
B.2 Transient heat equation 406

Appendix C About measurements 408
C.1 Land heat flow measurements 408
C.2 Oceanic heat flux measurements 413
Appendix D Physical properties 415
D.1 Thermal conductivity 415
Appendix E Heat production 425
E.1 Heat production rate due to uranium, thorium and potassium 425
List of symbols 435
References 437
Index 462

The color plates will be found between pages 84 and 85.

Introduction

The main object of this book is to study the physical processes that have determined, and continue to determine the thermal structure and evolution of the Earth. It has always been recognized that heat provides the energy that drives all the processes internal to the Earth, and that temperature controls the state and the mechanical behavior of the Earth's interior, but few of the available books focus on the thermal structure and evolution of the Earth and its geological systems. We intend to look at geodynamics from this thermal perspective and to provide the tools and data to determine temperatures at depth in a range of natural settings. We also aim at providing basic physical understanding of key aspects of heat transport in the Earth and thorough discussion of processes and physical properties that determine temperature at depth. It is impossible to discuss geological processes and environmental problems without thermal aspects and analysis must rely on sound grasp of physics as well as on knowledge of the parameters and variables at play.

There are many excellent books on the Earth's interior and on the dynamics of the Earth. In order not to forget any of them, and not to offend their authors, we shall not try to list them. Let us simply state that, although it has become in many ways obsolete, Jeffreys' *The Earth* (1959) has set a standard that is difficult to meet. During the past few decades, our understanding of the Earth's internal dynamics has completely changed. This is of course the result of plate tectonics and of the recognition that mantle convection drives the Earth's processes. During the same recent period, the amount of information and data relevant to the Earth's interior has increased. Spectacular new advances have been made because of spatial geodesy, seismic tomography, geochemistry, high pressure experimental geophysics and planetary exploration. Constraints on the Earth's thermal structure include not only heat flux and heat production data, but use information from xenoliths, petrology, mineral physics and seismic tomography. Experiments (physical or numerical) have given a means to explore the physical processes that drive the

Earth. The time when almost all the information concerning the Earth, its structure and evolution could fit in a single book has passed.

We shall present the basic physical principles and methods used to describe the Earth's thermal regime and to provide an up-to-date review of all the recent and not so recent data that have shaped our understanding of the Earth's thermal regime and thermal history. We shall include applications to illustrate the implications for geodynamics, tectonics and the search for natural resources as well as for environmental issues. Additionally, we feel that most of the available textbooks emphasize mathematical and numerical results at the expense of physical understanding and do not foster the physical intuition that would help them address complex problems in Earth sciences.

In this book, we have tried to select the topics that illustrate best our understanding of the Earth's internal dynamics. Analytical and experimental methods are required to investigate the physical models and to compare them with data on the Earth's interior. Neither have we tried to cover exhaustively the methods of geophysical modeling, nor to present a comprehensive review of all the data. The main challenge in Earth sciences has not been to develop methods of ever increasing complexity or to compile an ever increasing amount of data, but to relate theoretical models to the Earth and to infer what the data imply for Earth dynamics. Our choices may reflect our own biases; we hope that they will present a consistent perspective on geodynamics.

As we now believe that we understand fairly well how the Earth works, it may be sobering to remember that, since Kelvin's determination of the age of the Earth in the nineteenth century until the so-called "equality of continental and oceanic heat flow" that was discussed in the 1960s, thermal data have often been misinterpreted. A historical review will not only put our present synthesis in perspective, it will suggest some caution when we claim that we now understand almost everything.

A review of present knowledge on the Earth's interior is a prerequisite to any study of its internal processes. We have access to the physical properties which are determined with variable resolution by geophysical methods. We infer the composition from geophysical, geo and cosmochemical data, and experimental petrophysics. With increasing resolution, geophysical data have shown that the Earth is not spherically symmetric and provide clear evidence that it is not in static equilibrium.

The only direct information on temperature within the Earth comes from surface heat flux measurements. These data also provide an estimate of the total energy budget of the Earth. Extrapolating the surface temperature gradients downward shows inconsistencies in our knowledge of the state of the mantle and core.

Experimental and geophysical data have shown that the Earth can behave as a very viscous fluid on a geologically short time scale (10,000 years). For the estimated viscosity of the mantle, it can not be thermally stable. The conclusion is inescapable that the Earth's mantle and core are cooling by convection. Most of our understanding of convection in the Earth's mantle comes from physical and numerical experiments.

The formation and destruction of oceanic plates is the result of convection in the mantle. The evolution of oceanic plates is easily described by a simple physical model that explains observations and the physical properties of the oceanic lithosphere. The oceanic plate can also witness and record phenomena in the mantle such as the ascent of plumes. The concept of a plate can also be applied in a strict mechanical sense: the oceanic lithosphere responds to loading and bending as an elastic plate whose properties are determined by the temperature. Temperature also determines the fate of the plate that returns into the mantle.

The lithosphere beneath the continents has recorded a long and complex history. The heat flow and thermal regime of the continents are consistent with the presence of thick rigid roots beneath their stable cores, the "cratons." Intra-continental tectonic activity takes place in the form of continental rifts, basins and plateau uplift. Active tectonic belts coincide with collision zones. The syn and post orogenic deformations are controlled by the rheology of the continental lithosphere.

New continental crust is added at the active margins of the continents. An important part of the continental crust was formed early in the Earth's history. The processes that lead to the differentiation of the continental crust from the mantle, and the stabilization of a differentiated crust were also controlled by the temperature regime.

The thermal evolution of the Earth depends on initial conditions, the slowly decreasing internal heat production and internal redistribution of heat sources, and total heat loss. As the Earth cools down, mantle convection and the nature of its surface expression in tectonic activity evolve. On a smaller scale, magma reservoirs and hydrothermal systems have proven to be formidable challenges for the geologist trying to document their mode of operation, and the physicist aiming to establish simple models. Such a wide range of thermal systems and their manifestations have stimulated the interest of scientists from many different disciplines. It has been a motivation for us and we hope to share it with the readers of this book.

This book is intended for a wide variety of readers with diverse backgrounds: advanced undergraduates and graduate students, researchers and professional geoscientists. Our aim was not to write a mathematical treatise, but a quantitative treatment of thermal conduction and convective transport of heat sometimes requires the use of applied mathematics. For the interested readers, we provide a review of some standard techniques in the appendices, but knowledge of these methods is not a prerequisite for reading most chapters. We tried to follow a logical sequence in our presentation, but each chapter can be read independently of the others, and readers are invited to skip sections that are too "technical" and to decide for themselves the order in which they want to read this book.

Credits

All the maps shown in this book were made by the authors with Generic Mapping Tools (GMT) software that was developed for and made available to the geosciences community by Wessel and Smith (1995). Figures were made by the authors with GMT and commercially available software. Data used for the maps comes from several geophysical data repositories. We wish to acknowledge the center for satellite geodesy site of David Sandwell, and the reference earth model site maintained by Gabi Laske and Guy Masters, all at the Scripps Institute of Oceanography. Seismic tomography data were obtained from the CUB2.0 site of Mike Ritzwoller at the University of Colorado and from Susan Van der Lee, at Northwestern University. We are grateful to Maria Richards and David Blackwell at Southern Methodist University for sharing their North American heat flow data base, and to Gabi Laske, and Francis Lucazeau, at Institut de Physique du Globe de Paris, for sharing their data base of world heat flow measurements.

1

Historical notes

1.1 Introduction

The important questions that relate to the Earth's thermal regime and energy budget were raised a long time ago and some are still waiting for a complete answer. These past debates have more than historical interest. Our present understanding of the Earth's dynamics is based on the answers that were given to these questions.

People who live in volcanic areas always had the intuition that temperature increases with depth in the Earth. That it must be so everywhere became clear to scientists and engineers with the development of coal mining and the construction of deep tunnels in the nineteenth century.

Among the many advances in physics during the nineteenth century, development of the theory of heat conduction and of thermodynamics had immediate implications for the understanding of the internal structure and evolution of the Earth. The scarcity of data did not hamper physicists in speculating about the temperature regime inside the Earth.

1.2 Kelvin and the age of the Earth

When Fourier first published *Théorie Analytique de la Chaleur*, the temperature gradient of the Earth was estimated to be $\approx 20\ \mathrm{K\,km^{-1}}$, a value not very different from our present estimates. Fourier analyzed the temperature inside the Earth and concluded that the Earth had retained most of the heat from its formation. This conclusion was the basis for the calculation by Lord Kelvin of the age of the Earth (Thompson, 1862). Kelvin's study triggered a very serious debate between physicists and geologists and has received much attention from the historians. Indeed, it was one of the first examples of the difficult dialogue between physicists and geologists. It had long lasting consequences, not only because for many geologists it discredited the approach of the physicists, but mainly because Kelvin's approach influenced many leading geophysicists during the first half of the twentieth century. Sir Harold Jeffreys, who was the most influential geophysicist of that time, held views on the thermal history of the Earth that were not very different from those of Kelvin. Kelvin's calculation rested on two assumptions, i.e. (1) that the Earth is cooling by conduction, and (2) that there are no sources of heat inside the Earth. All the important questions concerning the Earth's thermal structure and evolution are related to these two assumptions: (1) What is the energy budget of the Earth? (2) What is the mechanism of heat transport inside the Earth? (3) What is the exact amount and distribution of radioactive elements in the continental crust and mantle?

At the time of Kelvin's paper, the Earth's temperature gradient had been estimated to be in the range 20–30 K km^{-1}. Kelvin thought that this data would constrain the age of the Earth. He assumed that the Earth was initially at a uniform temperature of 2000 K and that its surface stayed at constant temperature, $\approx 0\,°C$. Such an initial temperature seems extreme today but it appeared reasonable at the time of Kelvin and is of the correct order of magnitude. Kelvin also assumed that there were no internal heat sources. The only internal heat sources that were known at the time were chemical reactions, or conversion of gravitational potential energy into heat. Chemical reactions were assumed to contribute little because they are not reversible and hence could not go on for a long time. Although Kelvin knew from the mean Earth density and from moment of inertia measurements that density increases inside the Earth, he did not consider gravitational settling of a dense core as a source of energy.

For a conductive half-space, with initial temperature T_0, and with constant surface temperature $T = 0$, Kelvin showed that the surface temperature gradient decreases with time t as $t^{-1/2}$. It is given by

$$\frac{\partial T}{\partial z} = \frac{T_0}{\sqrt{\pi \kappa t}}, \tag{1.1}$$

where κ is the thermal diffusivity of the Earth. Kelvin could thus use this equation to determine how long it would take for the temperature gradient to drop to 20 K km^{-1}. The calculation yields ≈ 100 My. At that time, many geologists were influenced by Hutton's view that there was no beginning or end to geological time ("No vestige of a beginning, no prospect of an end"). Kelvin's result came as a shock and was rapidly challenged. Geologists proposed alternative methods to date the Earth and obtained older ages. They estimated from sedimentation rates and the thickness of sedimentary deposits that the age of the Earth was at least 500 My. It is correct that Kelvin ignored the Earth's radioactivity and convection in his calculation. It is more likely that ignorance of the energy source for the Sun was the main source of Kelvin's error. The only source of solar energy known to Kelvin was the conversion of gravitational potential energy into heat. Dividing the total gravitational potential energy available by the present rate of energy radiation by the sun yields an upper limit for the age of the Sun, which was on the same order as his estimated age of the Earth. This coincidence convinced Kelvin that his calculation was essentially correct.

One should note that in Kelvin's cooling model, the surface heat flow and temperature drop very rapidly and there is no cooling of the deep interior of the Earth: Kelvin's estimate is for the cooling of a shallow boundary layer. It would have made a difference if Kelvin had assumed an isothermal well-mixed "mantle" that cools due to the heat lost through a thin conductive plate < 100 km thick. It takes only a few tens of My for the temperature gradient to drop to 20 K km^{-1}. If applied to oceanic heat flux measurements, Kelvin's method yields a reasonable estimate of the age of the sea floor. Perry (1895a,b,c) showed that the temperature gradient would imply a much greater age if the thermal conductivity increases with depth. This higher thermal conductivity could account for the effect of convection beneath the thin skin of the Earth (England *et al.*, 2007).

Kelvin favored an initial temperature of $\approx$2000 K, close to contemporary estimates of the melting temperature in rocks at room pressure. A much higher estimate of the age of the Earth could have been obtained by including the melting temperature gradient. Assuming an initial temperature of 2000 K at the surface and a melting temperature gradient of 3 K km^{-1}, Jeffreys (1942) obtained an age of 1.6 Gy. Many historical studies have discussed these calculations and the assumptions that went into them. One must remember that Kelvin relied on an estimate of the temperature gradient that had been obtained in continents and could not appreciate the fundamental differences between oceans and continents. After his paper, the debate rapidly focused on internal heat generation by radioactive decay and on the distribution of heat-producing elements within the Earth. The issue of heat transport by convection was raised much later.

1.3 The discovery of radioactivity

Kelvin's assumption that there is no long-lived source of energy in the Earth was soon to be disproved. By 1895, Kelvin was convinced that all the laws of physics had been established and that the end of physics was in sight. The discovery of radioactivity by Becquerel in 1896 was to lead to a revolution in physics and it changed completely our understanding of the Earth's energy budget. Until then, it had been assumed that the Earth is cooling from an initial hot state. The presence of long-lived radioactive elements provided a source of energy that could balance the loss of heat through the Earth's surface. The importance of radioactivity for the Earth's energy budget was soon appreciated and discussed by Strutt (1906), Joly (1909) and Holmes (1915a,b).

The question was raised as to whether the Earth was heating because of radioactivity or whether the heat flow contained also a component due to the cooling of the Earth. Although it is now clear that the Earth must be cooling, the contribution of secular cooling to the energy budget remains very poorly constrained today.

Another consequence of the discovery of radioactivity is that radiogenic heat production of rocks could be compared with the heat flow and used to constrain the composition of the Earth. Strutt (1906) used estimates of radiogenic heat generation and the heat flux at the Earth's surface to conclude that the Earth had a crust that could not be thicker than 60 km. This suggestion was made three years before the confirmation of the existence of the crust by seismology.

Seismology provided the first models of composition of the continental crust. It suggested that the crust was made up of two main layers with seismic velocities consistent with granitic and gabbroic compositions for the upper and lower crust respectively. Granitic rocks that are very enriched in U and Th have an average heat production of $\approx 3\ \mu\mathrm{W\ m^{-3}}$. A granitic composition for the upper 20 km of the Earth's crust was consistent with seismic velocity, but not consistent with heat flux data.

1.4 The debate on the cooling mechanism of the Earth

The other assumption by Kelvin was that in the Earth heat is transported by conduction only. This assumption was not challenged on physical grounds but was questioned by a minority of geoscientists in the wake of the debate following Wegener's continental drift hypothesis. In retrospect, one can see that the most damaging flaw in the various arguments put forward at that time was the lack of reliable heat flux measurements and, more specifically, of measurements at sea. Had Kelvin known that the surface heat flux varies by more than a factor of three over the limited extent of his own country, he would not have been able to advocate a simple cooling model for the whole planet. The large lateral variations of heat flux that occur on Earth provide information on cooling mechanisms and heat sources that are as important as the global average.

Holmes (1931) was among the first to suggest that radioactivity would cause heating of the Earth and that convection was the most efficient mechanism of transporting this heat to the surface. He suggested that the higher heat production in continents would heat the mantle underneath and cause rising convection currents and continental breakup. Pekeris (1935) and Hales (1935) also examined the differences in continental and oceanic thermal regimes. Both authors assumed that the continental mantle was hotter than the oceanic mantle because of the crustal radioactivity. They concluded that these temperature differences would induce large stresses that were probably sufficient to cause convection. Convection in the Earth's mantle was rejected by the vast majority of geophysicists who believed the Earth's mantle could not sustain large deformations, despite the evidence from post-glacial rebound that the mantle can deform on a 10,000 years time scale.

1.5 Heat flux measurements

Few reliable heat flux estimates had been obtained before 1939, although it was becoming increasingly clear that heat flow data would provide constraints on crustal composition. Heat flow measurements require simultaneous determination of the temperature gradient in the Earth and of the thermal conductivity. Anderson (1934) made the first estimates of heat flow in England from temperature gradients measured in boreholes and the thermal conductivity of the main rock sections. From values in seven drill-holes, he concluded that the average heat loss of the Earth was 63 mW m^{-2}. In two papers reporting on measurements in England and South Africa, Benfield (1939) and Bullard (1939) established the standard procedure for measuring continental heat flux. Some ten years later, following the first determinations by Petterson (1949), Bullard *et al.* (1956) took the initiative of developing a program of oceanic heat flux measurements. He developed a probe that measures and records temperature and the thermal diffusivity of sea-floor sediments. Bullard had conjectured that heat flux would be lower in the oceans because of the absence of a radioactive crust. The first heat flux measurements suggested an approximate equality of continental and oceanic heat flow. This "equality" was interpreted as suggesting that the continental crust had differentiated from the underlying mantle, now depleted in radioelements, while the oceanic crust rested on a mantle that had retained its heat producing elements. The implication was that the oceans and continents were fixed relative to the mantle. The so-called equality of oceanic and continental heat flux was used as an argument against continental drift even after oceanic heat flux and bathymetry were explained by the cooling plate model. The inconsistency of the argument had not escaped Bullard (1962) who pointed out that heat originating from the lower mantle can not be brought to the surface of the Earth by conduction in the time available.

1.6 Energy budget of the Earth

Kelvin's basic assumption was that the Earth is cooling from an initially hot state. Following the discovery of radioactivity, the flow of heat out of the Earth no longer required the Earth to be cooling. The Earth's energy budget could not be determined as long as the total amount of radioelements in the Earth was not known. Holmes (1931) argued that, with continental heat flux low relative to heat production, the Earth might even warm up. He also proposed that the Earth's surface gets rejuvenated and shaped by convection currents. Following Jeffreys (1921), most of the opponents to convection in the mantle argued for cooling of the Earth because thermal contraction was their favored mechanism of mountain building. The mechanism of heat transfer and the energy budget are two independent issues, however, as convection can occur in both cooling and heating planets. Ironically, that part of Earth's topography which lies below sea level is indeed due to thermal contraction, but we know now that it is one of the most prominent manifestations of convection. Once again, the different processes that affect continents and oceans confused the best physicists of their time.

Urey (1964) noted that if the Earth had a composition identical to that of the chondritic meteorites, its heat production would be equal to the heat loss (then estimated at 30 TW), suggesting that the Earth's energy budget could be in equilibrium. However, Birch (1965) and Wasserburg *et al.* (1964) noted that the Earth is depleted in K relative to the chondrites and that a K-depleted chondritic Earth could account only for a fraction of the Earth's energy budget. The value of the ratio of total heat production to total heat loss of the Earth, the *Urey* ratio, remains controversial today with estimates ranging between 0.2 and 0.8.

The analysis of Pb isotopes by Patterson (1956) yielded an estimate of 4.55 Gy for the age of the Earth. Calculating the thermal history of a conducting Earth requires knowledge of the distribution of heat sources and of variations in thermal conductivity within the Earth, as well as the initial and boundary conditions. Only the boundary conditions are known. With improved understanding of heat production and the physical properties of the interior of the Earth, it became feasible to investigate conductive thermal evolution models for the interior of the Earth. Jacobs and Allan (1956) and MacDonald (1959), had to introduce questionable initial conditions in their calculations to obtain a thermal history consistent with the few constraints on temperature in the interior of the Earth. The thermal models of MacDonald (1959) required that radiation be the dominant mechanism of heat transfer in the Earth's mantle, which was later ruled out by experimental data.

1.7 Plate tectonics

The failure of conductive thermal evolution models of the Earth was not an important factor in the establishment of plate tectonics. On the contrary, it is the geophysical and geological evidence that led to recognition that the motion of tectonic plates accompanies mantle convection and cooling of the Earth's mantle.

As plate tectonics was emerging, thermal evolution models successfully explained the thermal regime of the oceanic lithosphere and could address that of the Earth's mantle. The success of the cooling plate model to explain variations in sea-floor heat flux and bathymetry was one of the first examples of the successful application of physical models in geology. The discrepancy between heat flux observations and model predictions led to discovery of hydrothermal circulation near the mid-oceanic ridges and to an understanding of the physical processes that control it.

On a different scale, although the actual numbers have not changed by orders of magnitude, the energy budget of the Earth is much better quantified today than it was 50 years ago. Not only is the total heat loss estimated with greater precision than before, but other terms that enter into the budget have been identified and estimated.

Kelvin's model of a cooling Earth raised two questions: what is the cooling mechanism? what are the sources of energy? Today, there can be no doubt that mantle convection is the dominant mechanism of heat transport in the Earth, even though many details of how it operates remain elusive. On the other hand, uncertainties about the different terms that enter the energy budget remain incapacitating. This is only one among many open questions about the Earth's thermal history and convection regime. This short historical account illustrates that significant advances have been achieved through both theory and observation. With no theory, the effect of time, a key variable, cannot be accounted for properly. With insufficient data, theoretical assumptions cannot be tested convincingly and calculations may run on empty. Kelvin would probably have thought differently had he realized how variable the heat flux and temperature gradient are at the Earth's surface. We shall see in this book that convection theory must still be considered as being in a development stage and that some critical data are still missing. This statement is valid for both the large-scale question of secular cooling of the Earth, intermediate-scale tectonic problems which require good control of crustal rheology and hence temperature, as well as small-scale issues on the behavior of magma bodies and their effects on crustal processes.

2

Internal structure of the Earth

Objectives of this chapter

The Earth can be compared to a big thermal engine: its internal heat provides the energy that drives all geodynamic processes and its long term evolution is governed by cooling. The total energy of the Earth depends on its internal structure and composition. How energy is transported in the Earth depends on its physical properties, which are controlled by the thermal structure. Here, we review some basic geophysical information about the present state of the Earth's interior and show how it is related to the thermal regime and the energy budget of the Earth. This chapter is not intended as a comprehensive description of the Earth and its main units and is focused on aspects that are most relevant to heat generation and transport.

2.1 Introduction

This chapter is focused on the silicate Earth which is made of a thin crust over a thick mantle, lying above a central metallic core. For the sake of brevity, we do not explain plate tectonics and assume familiarity with some of its basic premises and terminology. Mid-ocean ridges are zones of shallow sea floor where volcanic eruptions and earthquakes occur frequently. The sea floor is formed out of the mantle there, moves horizontally, and eventually returns into the mantle through deep trenches in a process called *subduction*. The implicit assumption is that the sea floor does not deform as it moves away from a ridge so that the velocity field at the Earth's surface can be accounted for by the relative motions of a small number of rigid plates. Their velocities have been determined by many different methods, from geophysical and geological techniques tracking displacements on time scales of several million years to land-based or satellite-based laser ranging on time scales of a few years. Velocity values vary by about one order of magnitude between ≈ 1.5 and 15 cm y^{-1}. We know a lot about the surface of the Earth for obvious reasons and the challenge we face is to link surface activity to the deep mantle motions and forces. On Earth, the main driving force is buoyancy, which involves large volumes and hence large depth extents. Buoyancy forces maintain movement in a viscous material through a process called *convection*. Convection also transports heat and may be contrasted to *conduction*, a process that transports heat through a motionless material. We shall often refer to the lithosphere, which has several different definitions. We shall discuss this in detail in various parts of the book. For the moment, we will use the term lithosphere to denote the rigid outer shell of the Earth that moves coherently in response to convective forces. The energy that must be expended to sustain deep mantle motions as well as surface deformation can only be drawn from the interior of the Earth. Thus, solutions to the vast majority of geological problems must ultimately be sought at depth, in the mantle. Due to current limitations, neither theory nor observation can on their own provide the required answers and must be used in combination. This will be a recurrent theme of this book.

The general method used in geophysics is to determine an average spherically symmetric structure described by radial profiles of the relevant properties and variables, and to seek lateral deviations from this gross structure, called *anomalies*. We shall see, however, that even radial profiles provide useful information on dynamics and convection. By 1940, the gross structure of the Earth had been worked out. There is surprisingly little difference between that structure, revealed by Jeffreys and Bullen in 1940, and recent Earth reference models like PREM or IASPEI91 (Dziewonski and Anderson, 1981; Kennett and Engdahl, 1991). Today, these reference models serve a completely different purpose: they provide the standard from which departures from spherical symmetry are determined. This change in objective reflects the evolution from a static perspective of the internal structure of the Earth into a dynamic one.

The Earth and meteorites have approximately the same age of 4.55 Gy. It is thought that Earth and the planets accreted very rapidly from planetesimals that formed in the nebular cloud. In the Earth, the differentiation between the metallic core and the silicate mantle took place in <100 My, during the accretion. It is also believed that, at the end of the accretion, the Earth was impacted by a Mars-sized body and the ejecta from this collision formed the Moon and may have re-homogenized the Earth. After impact, the Earth had reached a mass close to present, and the core possibly differentiated for a second time. The oldest preserved crust dates to 4.18 Gy, but older ages from recycled minerals (zircons) indicate that some crust had already formed at 4.4 Gy, i.e. very soon after the Moon forming impact event. There are also clear indications from extinct isotopes in very old gneisses that crustal differentiation started very early in Earth history. The heat released during accretion, core formation and the impact event determined the initial thermal state of the Earth. The early stabilization of some crust provided some constraint on the cooling of the young Earth.

Most of the recent progress in geodynamics has followed technological advances that provide more and better measurements, in particular the development of space geodesy, and the capability to process large amounts of data. Satellite measurements provide very precise and high resolution data sets for sea-floor bathymetry and land topography, continuous data sets for potential fields and high resolution local images. It is now possible to measure directly deformations of the Earth surface over wide areas thanks to satellites. Global and regional seismic networks provide three-dimensional images of the mantle. Gravity anomalies provide information on departure from spherical symmetry. Finally, theory, laboratory and numerical experiments allow us to investigate the underlying physical processes.

The crust is much thicker under continents than under oceans. More subtle differences between continents and oceans extend to great depth (300 km) and involve lateral variations of both composition and temperature. The nature of these differences and their implications for geodynamics are beginning to be understood in part because sufficient heat flux and heat production data are now available.

2.2 Gravity and geodesy

The shape of the Earth and its gravity field depend on internal density structure and rheology. The large free-air gravity anomalies that are observed indicate the presence of lateral variations in density due to temperature and composition. These anomalies are associated with large buoyancy forces that drive convective motions.

2.2.1 Moment of inertia, angular momentum and energy of rotation

The moment of inertia of a body relative to an axis is

$$I = \int_V \rho r^2 dV, \tag{2.1}$$

where r is the distance to the axis. The moment of inertia is measured in kg m^2. For a sphere of mass M with uniform density and radius R, the moment of inertia $I_H = 0.4\ MR^2$. For the Earth, the polar moment of inertia (i.e. relative to the axis of rotation) $C = 0.33\ MR^2$, which is less than I_H. Using equation 2.1, one can deduce that density increases toward the Earth's center. This had been noted in the early nineteenth century and led to the discovery of a core denser than the mantle. The moment of inertia is a powerful constraint on the radial density distribution of the Earth. For example, the density distribution that was first derived from seismic velocity through the Adams–Williamson equation (equation 3.48) failed to satisfy this constraint. The moment of inertia is important because it appears in the conservation laws for a rotating body. Without a torque exerted by external forces, the angular momentum and the rotational energy are conserved. If the rotation vector $\vec{\omega}$ is parallel to one of the principal axes of inertia with moment C, the angular momentum is $\vec{J} = C\vec{\omega}$ and the energy of rotation is $C\omega^2/2$. Changes in the distribution of mass within the Earth and in the shape of the Earth induce modifications of the moment of inertia and hence of the rotation. The present rotational energy of the Earth is $\approx 2.1 \times 10^{29}$ J.

2.2.2 Gravitational potential energy of a self-gravitating sphere

Any mass distribution has a self potential energy, i.e. the potential energy of each mass in the gravity field of the others:

$$E = \int_V \rho(\vec{r})U(\vec{r})dV = \int_V \rho(\vec{r})G \int_{V'} \frac{\rho(\vec{r}')}{|\vec{r} - \vec{r}'|} dV', \tag{2.2}$$

where $\vec{r} \neq \vec{r}'$, U is the gravitational potential and G the gravitational constant. For a spherically symmetric mass distribution, this gives

$$E = 16\pi^2 G \int_0^R \rho(r) r dr \int_0^r \rho(r') r'^2 dr'. \tag{2.3}$$

The potential energy of a sphere with uniform density ρ_a is

$$E_g = \frac{16}{15}\pi^2 G \rho_a^2 R^5 = \frac{3GM^2}{5R}. \tag{2.4}$$

For a sphere with the same radius and mass as Earth, this energy is $\approx 2.2 \times 10^{32}$ J. We shall see later that this very large amount of energy is not available to drive internal motions.

2.2.3 Shape of the Earth

Equilibrium in a self-gravitating fluid requires a balance between gravity, inertia (i.e. the centrifugal force due to Earth rotation), and pressure gradients. Therefore, gravity equipotential surfaces must also be isobars. The present shape of the Earth is approximately that of an ellipsoid of revolution with equatorial radius a about 20 km longer than the polar radius c. This is reflected in the Earth's reference gravitational field, which departs from that of a spherical body and varies with latitude. The flattening of the Earth, $f = (a - c)/a$, is $\approx 1/300$ and can be calculated for a rotating fluid in mechanical equilibrium. There is a small difference between the observed and theoretical values. On the one hand, this difference is small and shows that the Earth's shape is very close to the figure of equilibrium for a fluid at the present rotation rate. On the other hand, this difference provides information on re-adjustment mechanisms to changes in rotation rate.

The rotation rate of the Earth varies on different time scales. On a very long time scale, the Earth's rotation is slowing down because of tidal friction. As a result of the short time lag between the tide and the lunar gravitational pull, angular momentum is transferred from the Earth to the Moon, and the Moon is slowly moving away. This phenomenon is quite significant as the Moon is presently recessing at a rate of ≈ 3 cm y^{-1}. Evidence from growth lines in corals suggests that the rotation rate of the Earth has slowed down by 15% in the past 400 My. The important deductions are that on a geological time scale, the shape of the Earth has adjusted to its rotation rate and thus the Earth behaves like a fluid. On shorter time scales, the moment of inertia of the Earth has changed during the ice ages, when large ice sheets formed near the poles and the moment of inertia decreased. After the melting of the ice sheets, the increase in moment of inertia resulted in slowing down of the Earth's rotation. This re-adjustment is on a geologically short time scale (10,000 y). Incomplete post-glacial re-adjustment can account for the small difference between observed and theoretical flattening. The relaxation time for this re-adjustment yields a constraint on the viscosity of the Earth's mantle.

2.2.4 Gravity anomalies. Isostasy

The gravity field has been determined with high precision and resolution better than 100 km over the entire Earth by differential tracking of two satellites on the same orbit. The mean sea surface has also been measured very accurately by satellite altimetry. The surface of the Earth and its gravity field do not coincide with those of an ellipsoid since density does not depend only on radial distance r and varies laterally. The surface that best fits the observed mean sea level defines the reference ellipsoid with a flattening of $\approx 1/300$. The observed mean sea level defines the geoid. Geoid anomalies represent the differences between the observed sea level and the reference ellipsoid. They are small (< 100 m), indicating that the Earth does not support very large density differences. At the scale $\gtrapprox 500$ km, free-air gravity anomalies are small over continents and oceans, and usually show no correlation with elevation, even in recent mountain belts. This supports the concept of isostatic compensation. In the oceans, it will be seen that variations in bathymetry are the consequence of thermal contraction due to cooling of the mantle. On land, isostasy seems to hold regardless of the age or erosion level, which suggests that adjustments take place continuously. The preferred model to explain isostatic compensation implies flexure under loading of an elastic plate over

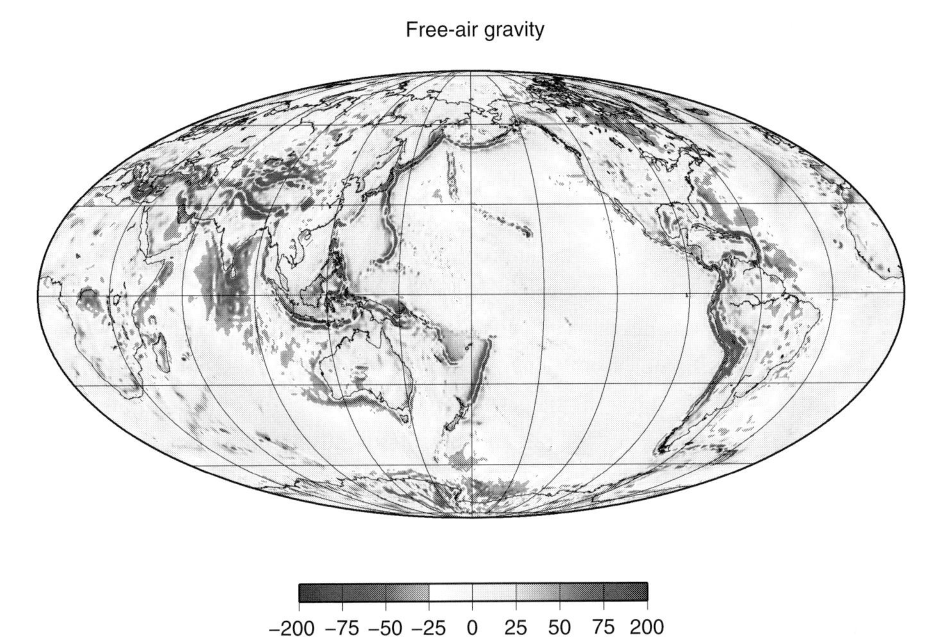

Figure 2.1. Free-air gravity map based on model GGM02 including data from the GRACE satellite mission (Tapley *et al.*, 2005). Note that the major gravity anomalies are in narrow belts. See plate section for color version.

a viscous substrate. An "effective elastic thickness" can be determined from gravity and topography data, but it is not the thickness of the mechanical lithosphere. As we shall discuss further, temperature is a key controlling parameter on the strength of the rocks and hence on the effective elastic thickness and on the thickness of the lithosphere. On the global free-air gravity map (Figure 2.1), large positive gravity anomalies are observed near deep sea trenches. These anomalies imply negative buoyancy that forces the subducted slabs to sink deeper into the mantle. One should not, however, conclude that such regions are in a state of dynamical disequilibrium. In the Earth, inertial forces are negligible and buoyancy forces are equilibrated by viscous forces.

2.2.5 Post-glacial rebound. Viscosity of the mantle

During the last glacial period (10–40,000 years ago), the Earth's surface was depressed under the load of ice sheets up to 4 km thick that covered most of Canada and Scandinavia. After the glacial retreat and removal of the load, re-adjustment produced slow uplift of the Earth's surface. The return to equilibrium was not instantaneous because it involved mantle deformation. The relaxation time of this re-adjustment is directly related to the viscosity of the mantle. This *post-glacial rebound* is not complete yet; Fennoscandia and the region around Hudson Bay, Canada, are still slowly emerging. The rates of uplift are on the order of 1 cm y^{-1} in Fennoscandia today and were larger when rebound started. These velocities are comparable to those of convective motions in the mantle.

The glacial loads are very wide and can be decomposed into components with different wavelengths, hence permitting a depth sounding of mantle viscosity: the long wavelength components contain more information about the deep mantle, the short wavelength ones about the shallow mantle. Analysis of rebound data includes vertical layering of the viscosity, the spherical geometry of the Earth, the redistribution of the load over the oceans, and models for time-dependent ice sheet thickness (McConnell, 1965; Peltier, 1974, 1998). The estimate of Haskell (1935) that mantle viscosity is on the order of 10^{20} Pa s has essentially survived. This value

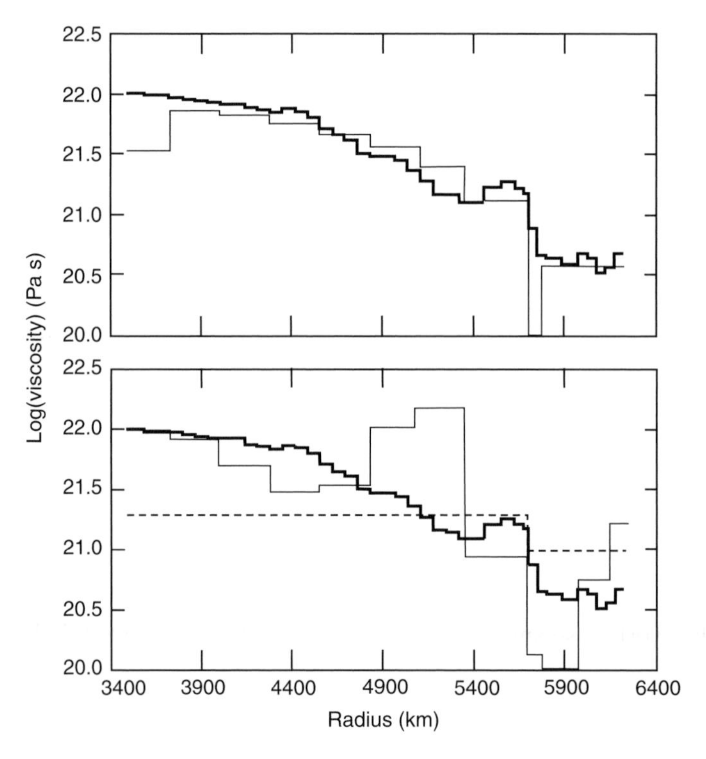

Figure 2.2. Mantle viscosity profiles obtained by inversion of post-glacial rebound data (top panel), or post-glacial rebound and geoid data (bottom panel). Thick line: Peltier (1998); thin line: Mitrovica and Forte (1997); dashed line: two-layer approximation.

is close to those suggested by laboratory experiments on mantle rocks and minerals at high temperature.

The mantle viscosity profile derived from inversion of post-glacial rebound data shows a viscosity $\approx 3 \times 10^{20}$ Pa s in the upper mantle and $\approx 10^{25}$ Pa s in the lower mantle, i.e. below the 670 km transition zone (Figure 2.2). Although temperature and to a lesser extent pressure are controlling factors on the viscosity, their variations with depth can hardly account for some mantle viscosity profiles that have been proposed. Other factors including phase changes and the presence of volatiles may play a role.

The post-glacial rebound studies are compatible with a Newtonian (i.e. linear) mantle viscosity. In the physical conditions of mantle convection, however, mantle minerals are thought to deform through the migration of dislocations, a mechanism that is inherently non-Newtonian. For such materials, viscosity depends on the rate of deformation, and hence on the characteristic time scale of the driving mechanism. As noted above, mantle convection and post-glacial rebound are associated with similar velocity values, and hence similar deformation rates in the shallow mantle. Deep in the mantle however, deformation rates due to post-glacial rebound are small. Constraints other than post-glacial rebound must be introduced to determine the viscosity for convection in the entire mantle.

2.3 Seismology

Seismology has been the main source of information on the internal structure of the Earth. The boundaries of the tectonic plates and their relative motions are revealed by the geographic distribution of seismicity and the focal mechanisms of earthquakes. The velocity of seismic waves depends on elastic parameters and on the density, which vary as a function of pressure, temperature and composition. Major discontinuities have been inferred from observations of travel time of body waves versus distance and shadow zones as well as from reflections and conversions between P and S waves.

2.3.1 Seismicity

Most of the world's seismic activity takes place in very narrow belts. It is now understood that these tectonically active zones mark the boundaries of tectonic plates. In continental collision zones, seismicity affects very wide regions. Diffuse seismicity also affects large areas of extension in the continents, for example in the Basin-and-Range geological province, western USA. Except in zones of plate convergence, seismicity is shallow (<30 km). Below the seismogenic layer, continuous rock deformation by steady-state creep requires lower stress than motion along faults. The thickness of the seismogenic layer, i.e. the depth below which steady-state creep is the dominant deformation mechanism, depends on temperature and varies amongst geological provinces. Zones of plate convergence host the deepest and most powerful earthquakes. Deep earthquakes are confined to thin dipping layers, the Benioff–Wadati zones, and do not occur below 650 km.

2.3.2 *Reference Earth model*

The two largest seismic discontinuities separate the crust from the mantle, and the mantle from the core. They correspond to changes in composition but between the core and mantle there is also a change of physical state between the "solid" mantle and the "liquid" core. At shallow depths, there are important horizontal differences in seismic structure: beneath the oceans, the crust is uniformly thin (5–6 km) while it is thicker and more variable (20–80 km) beneath the continents. Radial models of seismic velocities and density are shown in Figure 2.3.

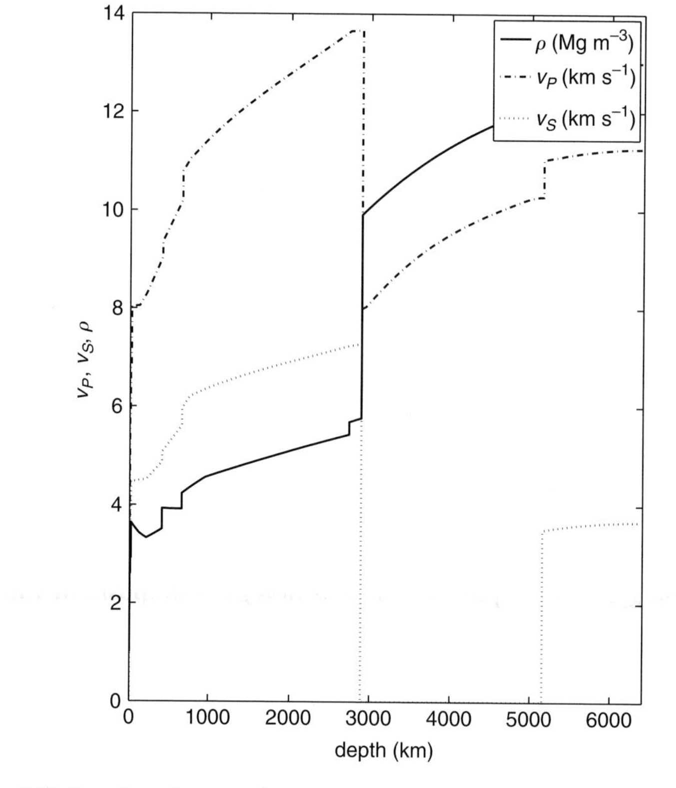

Figure 2.3. Density, shear and compressional wave velocities for the AK135 reference Earth model (Kennett and Engdahl, 1991).

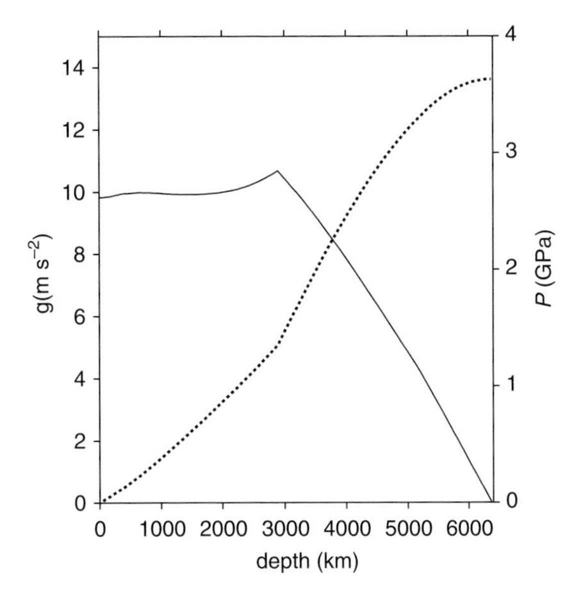

Figure 2.4. Variations in g and pressure through the mantle and core for the Earth reference model. Solid line: g; Dashed line: P.

The seismic reference models show an upper mantle with many distinctive features and a lower mantle with much smoother velocity variations. The fine structure in the Earth's upper mantle appears through several changes of the seismic velocity gradient which include a low velocity zone, followed by an increase in velocity and two discontinuities at 410 and 660 km depth. Study of the amplitude of seismic waves across these regions shows that there is a very strong gradient over a region < 20 km thick. At the base of the mantle there is a transition zone identified as D″, which now appears to be extremely heterogeneous.

Radial density variations from the reference model are useful to determine how gravity and pressure increase in the Earth's mantle and core (Figure 2.4). It is noteworthy that the acceleration due to gravity is almost constant throughout the mantle.

2.4 Petrology, mineral physics and seismology: Composition and state of the Earth's interior

Petrological and mineral physics studies have made possible the interpretation of seismic data in terms of composition and physical state. For a given bulk composition, variations in density with depth can be calculated. This requires accurate knowledge of elastic properties as a function of pressure and temperature, and can be done quite accurately in the upper mantle. Uncertainties are larger in the lower mantle and it is not clear that the lower mantle is chemically homogeneous. A widely used model composition for the Earth's mantle is pyrolite. Partial melting of the mantle generates basaltic magmas and solid residues (peridotite) that can be sampled in many locations. The pyrolite composition can be determined by mixing peridotite and basalt in appropriate proportions or through models of Earth formation out of primordial material sampled in meteorites.

The correspondence between seismic velocity discontinuities in the upper mantle and solid–solid phase transitions was first proposed by Bernal (1936) and confirmed with high pressure experiments by Ringwood (1969). The transition zone represents three important phase transitions in the olivine from α to β to γ phases followed by the breakdown of olivine into perovskite and magnesiowustite at a depth of 660 km. Some questions remain concerning these phase transition zones because seismic data require them to be thinner than indicated by mineral physics experiments. At the high pressures of the lowermost mantle, the perovskite structure probably breaks down into a post-perovskite phase.

The identification of Fe as the main component in the core was inferred from cosmochemical data on the abundance of elements in the solar system. Measurements of the hydrodynamic velocity of minerals at high pressure by Birch (1964) demonstrated that the core could not be a phase transformation of silicate minerals but that it must be made up of liquid metallic alloy: mainly Fe and Ni with some additional lighter elements such as oxygen and sulfur. The surface separating the fluid outer core from the mantle – the core–mantle boundary (CMB) – is characterized by a density discontinuity with a large jump > 5000 kg m^{-3}. The core viscosity is very low, possibly as low as that of water at room temperature.

As the Earth cools, the core solidifies growing an inner core whose radius is currently about 1200 km. The inner core is made of a solid Fe–Ni alloy and may be partially molten. The surface separating the solid inner core from the liquid outer core is the inner core boundary (ICB). Growth of the inner core proceeds at a rate which is dictated by the cooling of the core and hence by the amount of heat lost to the lower mantle.

2.4.1 Constraints on the thermal structure of the Earth

The radial profiles of seismic velocities and density allow temperature determinations at the depths of major phase changes as well as constraints on the radial temperature profile. Unfortunately, the mantle may be compositionally heterogeneous which makes detailed interpretation of seismic data very difficult. The existence of some compositional stratification, even if it is minor by chemical standards, would have major consequences for the convection regime and hence for the rate of heat loss of the Earth. Because of the small coefficient of thermal expansion of mantle rocks, even subtle changes of composition may have a significant impact on density differences and hence on buoyancy forces.

We shall see that the occurrence of thermal convection in a layer is marked by two characteristic features. One is the existence of large temperature gradients in thermal boundary layers at the top and bottom. The other is a well-mixed interior with a radial temperature profile that depends on the energy source driving the motions. Mantle convection is driven by secular cooling of the Earth, radiogenic heat production in mantle rocks and heating from the core. The exact balance between these different sources is not well known and remains one of the major stumbling blocks in studies of Earth's evolution through geological time. Here, we review what can be inferred on thermal structure from geophysical observables.

Absolute temperature determinations

The potential temperature of the shallow oceanic mantle may be calculated from the composition of mid-ocean ridge basalts which have not been affected by fractional crystallization and from heat flux data, as will be explained in Chapter 6. Two such independent determinations are included in Table 2.1 and are in very good agreement with one another. At greater depth, mantle temperatures can be determined at the depths of seismic discontinuities. The pressure at each discontinuity can be determined from its depth and hence the temperature can be calculated from the phase equilibrium diagram in (P,T) space. This method requires specification of the mantle composition, which is usually taken to be pyrolite (Ringwood, 1962). Well-defined discontinuities at depths of 410 and 660 km have been linked to the olivine–wadsleyite transition and to the dissociation of spinel to perovskite and magnesowustite, the so-called "post-spinel" transition. Other seismic discontinuities have been identified, notably at a depth of about 500 km. Some of these are not detected everywhere and seem to have a regional character, and their interpretation is still tentative. Recent estimates of temperature at these discontinuities are listed in Table 2.1.

The other major discontinuity, the core–mantle boundary, is a chemical boundary. Its temperature can be calculated from the core side to be 4080 ± 130 K (Table 2.1). At the high pressures that prevail in the lower mantle, the perovskite structure may break down and change to post-perovskite if temperatures are high enough (Murakami *et al.*, 2004; Oganov and Ono, 2004). The temperature of this phase change for the core–mantle boundary pressure has been estimated to be in the range 3800–4200 K (Hirose *et al.*, 2006; Hirose, 2006). This is below or at the lower end of the range of temperatures at the base of the mantle, implying that the phase change is likely to occur. It has been associated with weak seismic discontinuities near the core–mantle boundary. These discontinuities are not detected everywhere (Hernlund *et al.*, 2005), however, for reasons that will be discussed below.

Table 2.1. *Constraints on the mantle geotherm*

Boundary	Depth, km	Temperature, K	Reference
Potential temperature	0	1590 †	McKenzie *et al.* (2005)
MORB generation	50	1590–1750 ‡	Kinzler and Grove (1992)
Olivine–wadsleyite	410	1760 ± 45	Katsura *et al.* (2004)
Post-spinel	660	1870 ± 50	Katsura *et al.* (2003, 2004)
Core–mantle	2900	4080 ± 130	Alfé *et al.* (2002) Labrosse (2003) Jaupart *et al.* (2007)

† Temperature of the mantle isentrope at atmospheric pressure deduced from a cooling model of the oceanic lithosphere that accounts for heat flux and bathymetry data.
‡ indicates true range of temperatures in the shallow mantle.

The four "anchor" points for the geotherm that are listed in Table 2.1 show that temperature increases with depth in the Earth's mantle and that the radial temperature gradient varies by large amounts. To determine the full temperature profile, one cannot rely on such a meager data set and must introduce additional assumptions. The common practice is to assume that the temperature follows an isentrope, along which the entropy of a parcel moving up or down is conserved (Chapter 3). This is often referred to as the mantle "adiabat", because it is such that no heat is exchanged in the process, but this terminology is not strictly correct, as discussed in Chapter 3. By definition, an isentropic temperature profile only makes sense if there are internal motions within the mantle. It is a theoretical profile that may not be achieved exactly. Such a profile must be considered only as a convenient reference accounting for the effect of pressure on temperature. Figure 2.5 shows one recent model isentrope derived by Katsura *et al.* (2004), which is only accurate to $\approx \pm 50$ K due to experimental uncertainties.

To connect the isentrope of Figure 2.5 to temperatures at the top and base of the mantle, two boundary layers with large temperature differences are required. From Table 2.1, temperature differences of ≈ 1300 and 2000 K are inferred for the upper and lower thermal boundary layers, respectively. One could in principle combine these values with estimates of the boundary layer thicknesses to calculate the conductive heat flux, but we shall see that the procedure is not reliable for a host of reasons.

Radial density and temperature profiles

For the time being, we take the reference temperature profile to be an isentrope, such that, as will be demonstrated later,

$$\frac{dT_S}{dr} = -\frac{\alpha g T_S}{C_p}, \tag{2.5}$$

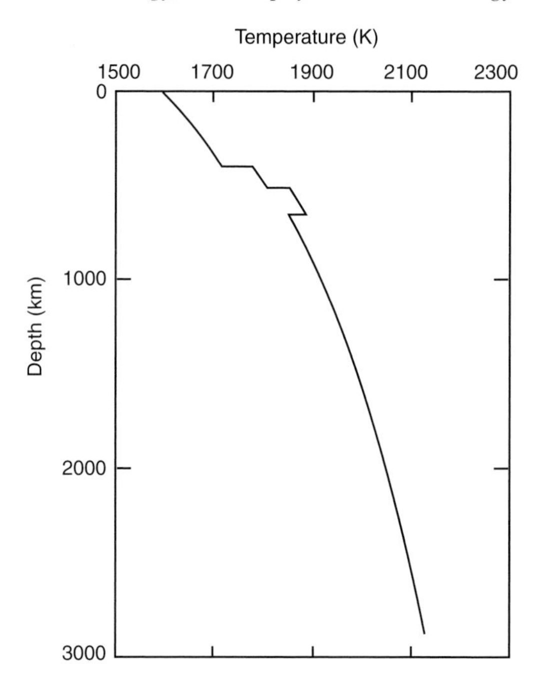

Figure 2.5. Isentropic temperature profile through the mantle derived from mineral physical properties and phase change characteristics from (Katsura *et al.*, 2004). Phase change boundaries are associated with temperature jumps and seismic discontinuities. Phase change temperatures are accurate to $\approx \pm 50$ K.

where r is the radial distance, α is the coefficient of thermal expansion, g is the acceleration due to gravity, and C_p is the heat capacity per unit mass at constant pressure. Along the isentrope, density changes due to pressure and temperature variations according to

$$\frac{d\rho_S}{dP} = \frac{\rho_S}{K_S}, \tag{2.6}$$

where K_S is the isentropic bulk modulus. Using these equations, one can calculate the isentrope and the corresponding density profile for a model mantle composition such as pyrolite for example, and compare the model profile to geophysical data. Conversely, one can detect departures of the isentrope from the geophysical data. Current data on physical properties allow reliable calculations in the upper mantle. The isentrope that passes through the phase change at a depth of 410 km has a temperature of 1600 ± 50 K at atmospheric pressure (Katsura *et al.*, 2004), what is called the potential temperature. This is within the range of values deduced independently from the compositions of mid-ocean ridge basalts and oceanic heat flux data (Table 2.1). This indicates that the geotherm is close to an isentrope in the upper mantle.

Over the large depth range of the lower mantle, one may detect departures from an isentrope. From the seismic velocities and density profiles (Figure 2.3), one can derive an equivalent "sound" velocity, V_Φ, such that

$$V_\Phi^2 = V_P^2 - \frac{4}{3}V_S^2 = \frac{K_S}{\rho_S}, \qquad (2.7)$$

where ρ_S is the reference density for an isentropic process. Using the equation of state (2.6), this can be rewritten as follows:

$$V_\Phi^2 = \left(\frac{d\rho_S}{dP}\right)^{-1}. \qquad (2.8)$$

Thus, seismic velocity data allow determination of the values of $d\rho_S/dP$ along an isentrope. One can also independently derive the actual pressure derivative of density in the Earth:

$$\frac{d\rho_E}{dP} = \frac{d\rho_E}{dr}\left(\frac{dP}{dr}\right)^{-1} \qquad (2.9)$$

where ρ_E is the true value of the mantle density, as opposed to the hypothetical isentropic value ρ_S. Departures from the isentrope are determined by the Bullen parameter (Bullen, 1963):

$$\eta = V_\Phi^2 \frac{d\rho_E}{dr}\left(\frac{dP}{dr}\right)^{-1} = 1 + V_\Phi^2 \frac{\alpha}{\rho g}\left[\frac{dT_E}{dr} + \frac{\alpha g T_E}{C_p}\right], \qquad (2.10)$$

where T_E is the true mantle temperature. From equation 2.5, the Bullen parameter is equal to 1 along an isentrope. Observations show that $\eta = 1 \pm 0.02$ on average and that it varies with radial distance (Figure 2.6). In regions where $\eta < 1$, the magnitude of the mantle temperature gradient is larger than that of the isentrope (remember that the radial gradient is negative) and the opposite situation holds when $\eta > 1$.

We shall see later, in the chapter on mantle convection, that one does not expect the mantle to lie along an isentrope indeed, but that observed departures from an isentrope are not consistent with the thermal structure of a homogeneous convecting system. This provides important information on mantle convection as well as on the structure and composition of the Earth's mantle.

2.4.2 Attenuation: Q factor

Attenuation of seismic waves occurs because of imperfect elastic behavior, which causes the dissipation of seismic energy. The quality factor Q is a measure of the

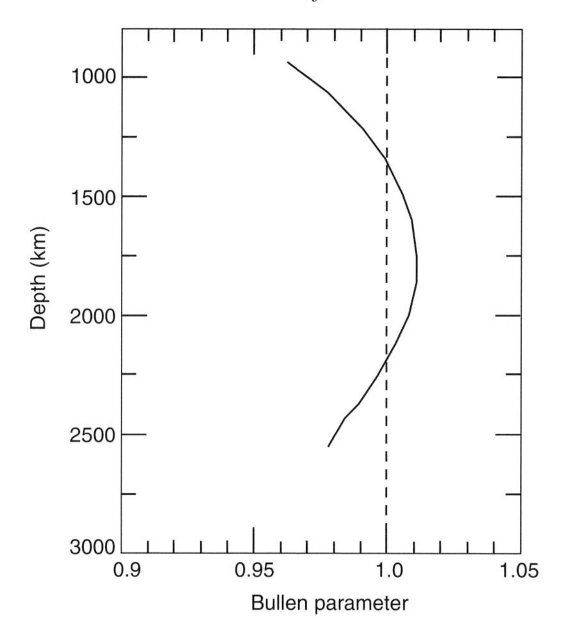

Figure 2.6. Bullen parameter η (equation 2.10) as a function of depth in the Earth's mantle, from the AK135 seismological model of Kennett and Engdahl (1991). A value of 1 for the Bullen parameter indicates that the mantle lies along an isentrope. Deviations from an isentropic temperature profile provide information on mantle structure and convection.

medium capacity to transmit seismic energy without attenuation: it is inversely proportional to the fraction of energy lost per cycle and is infinite for non-viscous materials. For body waves (P and S), quality factors vary between $\approx$200 in the upper mantle to more than 1000 in the lower mantle (Figure 2.7).

At lower than seismic frequencies, an-elasticity of the mantle can be inferred from the decay of the Chandler wobble,[1] or from the phase of the solid Earth tide. The very small phase lag between the solid Earth tide and the tidal potential shows that Q must be large at these frequencies.

2.5 Lateral variations of seismic structure

The data discussed so far allow construction of an accurate reference model for the Earth's mantle, i.e. a model for the radial variations of physical properties and temperature. In a convecting mantle, however, motions are driven by lateral temperature differences which must induce lateral deviations from reference models.

[1] The Chandler wobble refers to movements of the rotation axis relative to the Earth surface with a period of $\approx$433 days.

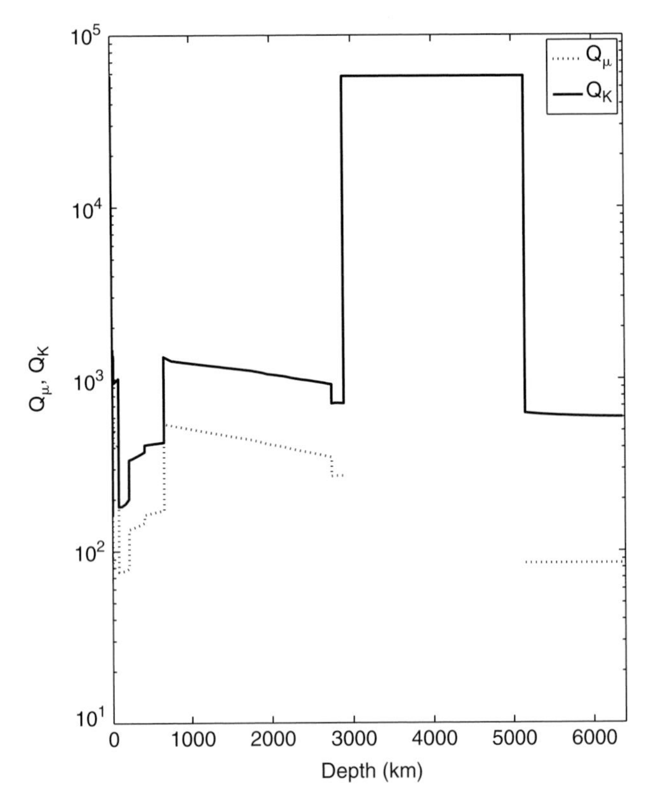

Figure 2.7. Radial variation of the quality factors associated with bulk Q_k and Q_μ shear moduli from reference model AK135 (Kennett and Engdahl, 1991).

2.5.1 Topography of the main seismic discontinuities

Prominent seismic discontinuities have been used to determine mantle temperatures at depths of 410 and 660 km. In reality, these discontinuities are not found at the same depths everywhere. For example, the depth to the olivine–wadsleyite phase change is only 410 km on average and varies by as much as $\pm$ 30 km (Chambers *et al.*, 2005). The slope of the phase change boundary, the Clapeyron slope dP_p/dT_p, is known to be 4 MPa K^{-1} with good accuracy (Katsura *et al.*, 2004). Thus, if there are no lateral variations of composition in that part of the mantle, one deduces that temperature varies laterally by $\approx \pm 300$ K at a depth of 410 km. For the other prominent transition, which lies at a depth of 660 km on average, vagaries of the seismic discontinuity defy simple explanations (Chambers *et al.*, 2005). The detailed mapping of mantle seismic discontinuities is being pursued very actively at the moment and appears to indicate that temperature cannot be the only control variable.

The perovskite to post-perovskite phase change is expected to occur near the base of the mantle, according to Hirose *et al.* (2006). Hernlund *et al.* (2005) argued

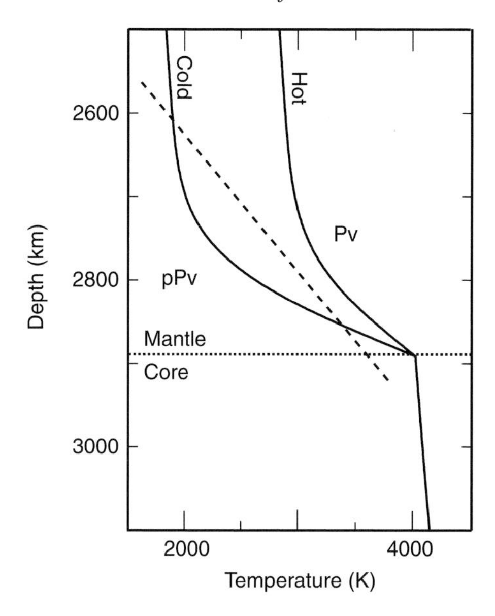

Figure 2.8. Sketch of temperatures in the lowermost mantle, from Hernlund *et al.* (2005). The dashed curve stands for the perovskite to post-perovskite phase change boundary. Plain curves indicate two profiles in cold and hot regions at the base of a mantle associated with subducted material and a mantle plume, respectively. Values for the temperature at the core–mantle boundary and for the post-perovskite phase change are only known to $\approx \pm 200$ K.

that the phase change boundary should be crossed twice in cold regions and should not occur in hot ones (Figure 2.8), which seems to be consistent with seismic observations. This provides an extreme example of lateral variations of the depth to a phase boundary in the Earth's mantle. In hot regions near the core–mantle boundary, the post-perovskite phase change just does not occur and hence induces no seismic discontinuity. Lateral variations of the depth of this discontinuity imply lateral temperature variations of $\approx$1500 K (Hutko *et al.*, 2006; Hirose, 2006). This is close to the total temperature difference across the thermal boundary layer at the base of the mantle and is consistent with the convective disruption of this boundary layer.

2.5.2 *Seismic tomography*

Several types of seismic data are being used to image the Earth's three-dimensional structure. Tomography models are being sought actively by different research teams and are in a state of flux, with each year witnessing new models and improvements to old ones. Several robust features have emerged that are shared between all the global tomography models:

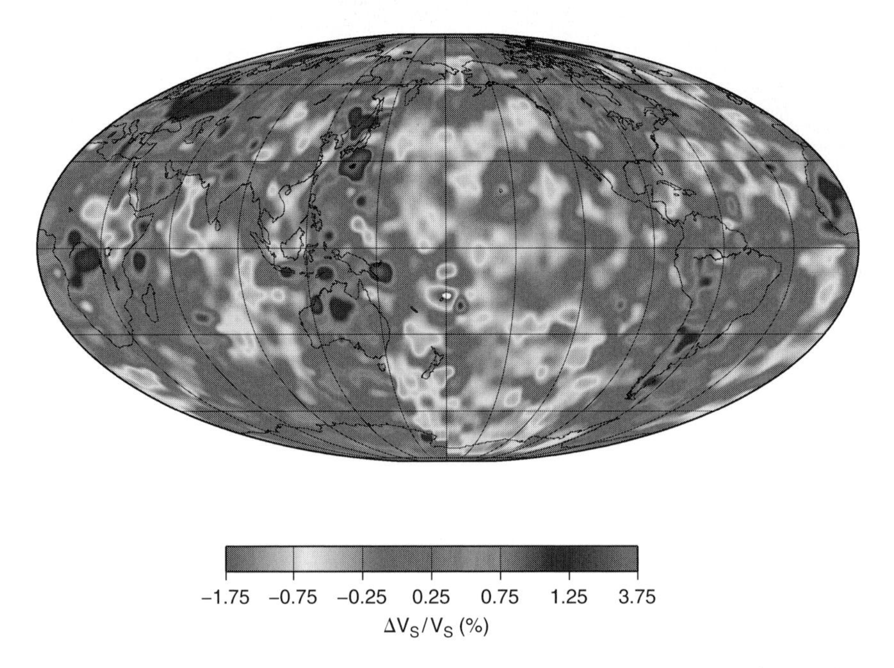

Figure 2.9. Shear wave velocity anomalies at 350 km depth. The anomalies relative to the mean shear wave velocity at that depth are given in %. By convention, the color scale is inverted as negative velocity anomalies are interpreted as hot. The seismic model is that of Grand (1997). See plate section for color version.

(1) the upper mantle is very heterogeneous and the anomalies can be correlated with geological features. The roots beneath the cratons can be followed as deep as 350 km (Figure 2.9);
(2) some subducting slabs can be followed below the transition zone, deep into the lower mantle (Figure 2.10);
(3) the deepest part of the mantle is also extremely heterogeneous: the D″ layer at the base of the mantle is perhaps the most complicated region of the mantle (Figure 2.11).

The differences in shear wave velocity within D″ require temperature variations comparable to those inferred from the post-perovskite phase transition. In the upper mantle, subducting slabs are fast and exhibit much diversity. Their dips are highly variable and sometimes vary down a single slab; some slabs penetrate in the lower mantle whereas others stagnate in the transition zone. These structures are of particular interest because they are the downwellings of mantle convection.

Linear chains of volcanic islands on the sea floor have often been explained by the activity of focused upwellings called plumes rising from great depths in the mantle. A few plume structures have been followed to great depth by seismic tomography but the majority of them remain elusive.

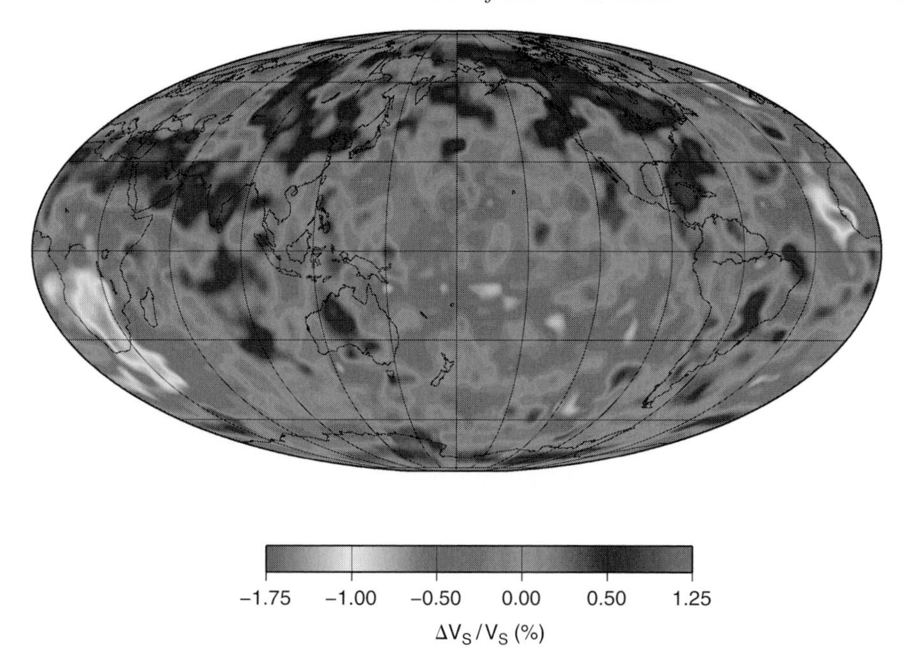

Figure 2.10. Shear wave velocity anomalies at 1810 km depth (Grand, 1997). Note the reduced scale relative to Figure 2.9. See plate section for color version.

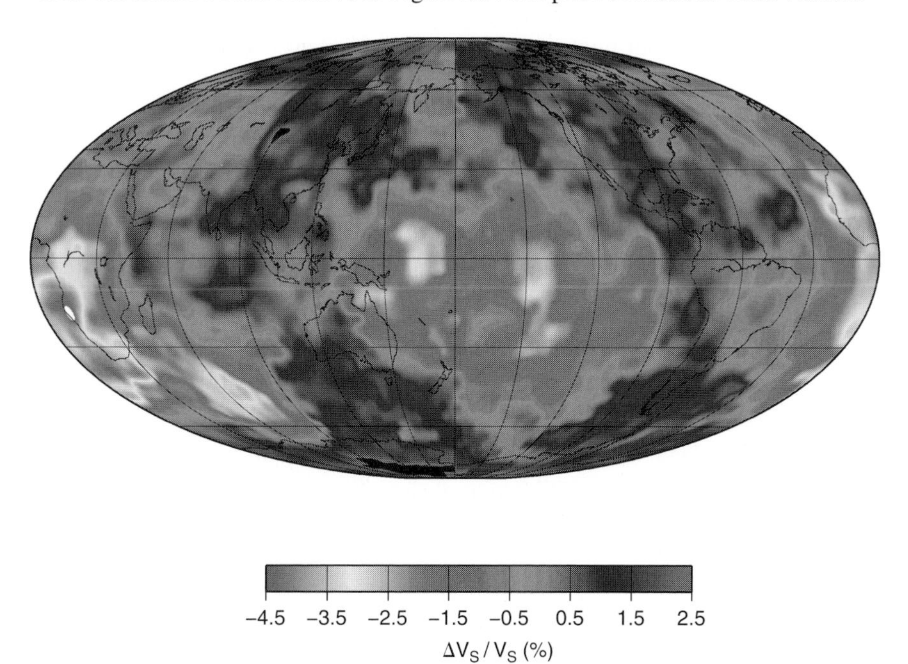

Figure 2.11. Shear wave velocity anomalies in the D″ region at 2800 km depth (Grand, 1997). Anomalies are enhanced relative to the lower mantle. See plate section for color version.

2.5.3 Anisotropy

The main mantle minerals have anisotropic crystallographic structures and generate bulk anisotropy of elastic properties and hence seismic wave speeds if they are aligned by a deviatoric stress field, such as that in a flow. Anisotropy is detected by the azimuthal dependence of the velocity of compressional waves, or by variation with polarization of shear waves. In the "cold lithosphere", seismic anisotropy is often fossil and a remnant of the stress field at the time the lithosphere cooled down. In the mantle below, the anisotropy is due to the present mantle flow.

2.6 Core and magnetic field

The Earth has a strong magnetic field of which the main component is a geocentric dipole. The magnetic field varies on many spatial and temporal scales. It is generated by convection currents in the liquid and electrically conductive outer core by a self-exciting dynamo mechanism.

The dipolar field varies in strength and direction but its long-term average is parallel to the axis of rotation of the Earth. It experiences reversals of dipole polarity. Multipolar components also vary on geologically short time scales. Several sources are available to provide the energy needed to maintain the magnetic field: cooling of the core, latent heat of freezing of the inner core and gravitational settling of the inner core. Their relative contributions to the core energy budget are still very much in debate.

The magnetic field and its origin by dynamo action in the outer core provides two different types of information on the energy budget of the Earth. One is the amount of energy required to drive the dynamo and to sustain the intensity of the magnetic field at its present value. The other information is that changes in the dynamic regime and efficiency of the dynamo probably occurred when the inner core started to grow. In some models, for example, a magnetic field with an intensity similar to today's field can only be achieved through the differentiation and growth of an inner core. Evidence in the paleomagnetic record for change in strength of the magnetic field can be used to date the emergence of an inner core and to tighten estimates of the present temperature and cooling rate of the core.

2.7 The shallow Earth

Large deviations from seismic reference models occur at shallow depths in association with continents and oceans. These can be observed, for example, in the delays in arrivals of seismic shear waves. The calculated travel times of seismic shear waves across the region from 60 to 300 km depth vary by ± 3.5 s (Figure 2.12). They

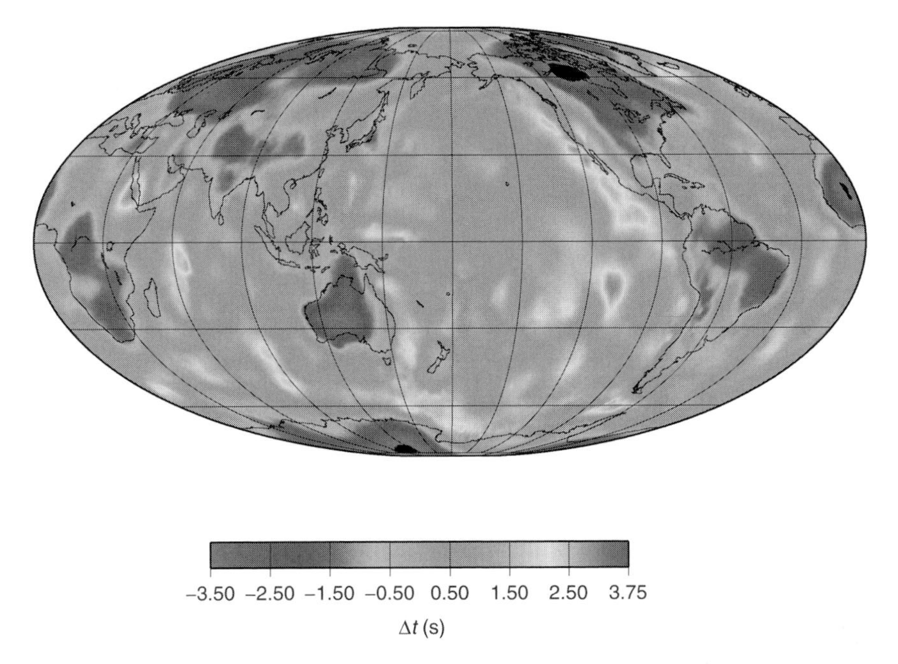

Figure 2.12. Shear wave travel time anomalies across the 60 to 300 km region. The travel times are calculated from the CUB2.0 model of Shapiro and Ritzwoller (2002), and the anomalies are relative to the world average. The value of 60 km for the upper limit is selected to exclude the crust which is not very well resolved by the seismic model. See plate section for color version.

record differences in structure and composition within both oceanic and continental lithospheres, and between them. The concept of a lithosphere has been around for several decades and has been used with different meanings by different authors. It has evolved considerably as more information about the shallow constitution of the Earth became available but also, more importantly, as dynamical models for geological processes have improved. It was first introduced to explain isostasy in continents. The seismic lithosphere was initially defined as a static unit lying above a low velocity zone for the shear waves (e.g. Anderson and Sammis, 1970), which was itself associated with partial melting and weak mechanical behavior. It was later associated with the cold upper part of the mantle where heat is transported by conduction, as opposed to the deeper mantle where convection is the dominant heat transfer mechanism. Studies of flexure and deformation due to topographic loads such as volcanic edifices and mountain belts led to yet another definition of the lithosphere, such that it sustains elastic stresses to support these loads. It is now clear that the lithosphere must also be regarded as a chemical entity with a root that is made of chemically depleted and dehydrated mantle. Electrical conductivity profiles provide yet another means of determining lithosphere thickness, as conductivity is sensitive to both temperature and the water content of mantle rocks. Analysis of data from the Archean Slave Province, Canada, and the northeastern Pacific shows that old continental lithosphere contains less water than the oceanic mantle in the depth range 150–250 km (Hirth *et al.*, 2000).

2.7.1 The seismic lithosphere

On a world-wide scale, with very few exceptions, old continents (cratons) are systematically associated with fast regions extending to depths of about 200–300 km. The lithosphere preserves compositional heterogeneities in contrast to the well mixed asthenosphere. Thus, its thickness can be defined by seismology as the depth where small-scale lateral heterogeneities disappear. Discontinuities in the depth range 80–240 km have been observed by seismic reflection, refraction experiments and by teleseismic body wave studies. They cannot be explained by thermal effects but could be due to variations in water content.

Independent information may come from the vertical distribution of seismic anisotropy. One may expect different anisotropy characteristics and orientation depending on depth (Gung *et al.*, 2003). Below the base of the lithosphere, anisotropy is due to convective shear stresses and should be aligned with the direction of plate motion. Within the lithosphere, anisotropy probably reflects fabrics inherited from past tectonic events (Silver, 1996). The depth at which the anisotropy characteristics are correlated with plate velocity may be interpreted as the base of the lithosphere.

2.7.2 Oceanic crust

The rate of sea-floor formation varies between spreading centers, and there is no reason to believe that it has been constant through time. One can not just extrapolate present rates to date old sea floor. The linear marine magnetic anomalies, that have confirmed the sea floor spreading hypothesis on young sea floor, represent "isochrons" that can be correlated with the chronometry of geomagnetic reversals and can be used to date the age of the sea floor. The correlation between sea-floor ages (Figure 2.14) and bathymetry (Figure 2.15) is not a coincidence. As will be explained later, the relationship between bathymetry and sea-floor age provides a test and a constraint for the cooling models of the sea floor.

2.7.3 Continental crust

The continental crust plays a major role because its low density makes the continents buoyant. Also, the crust contains large amounts of heat-producing elements such as

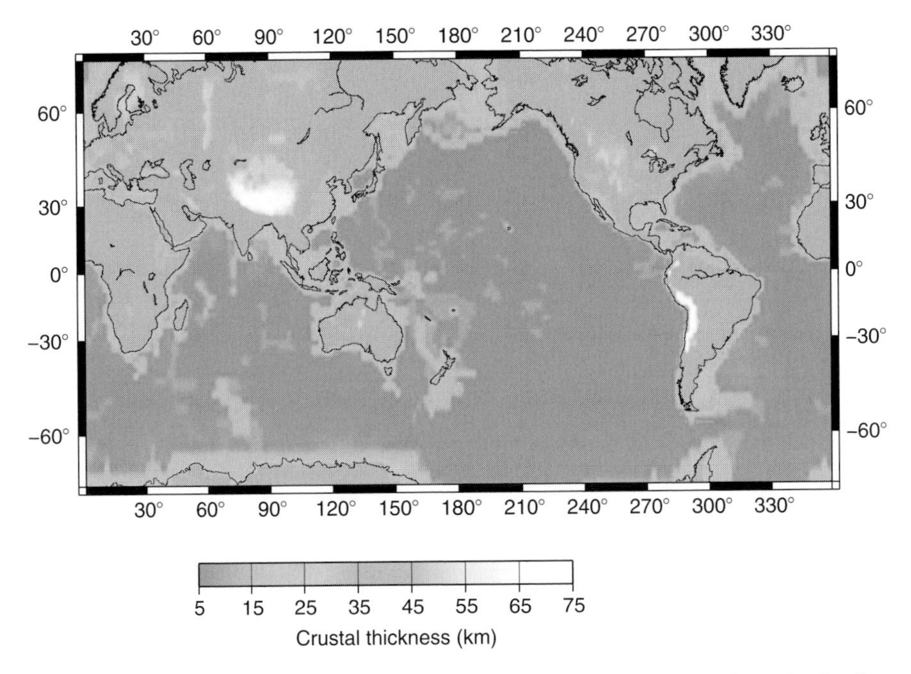

Figure 2.13. Crustal thickness variations from the global compilation of seismic crustal thickness data, CRUST2.0 by Mooney *et al.* (1998). See plate section for color version.

uranium, thorium and potassium. It is important not only for the thermal structure and hence the rheology of continents, but also for the very energy budget of the planet. It stores large amounts of heat sources at shallow levels, where they are not available to drive mantle convection.

On geological time scales, the area, volume and surface distribution of continental crust have changed, with two consequences for the Earth's thermal regime. The long-term process of continental growth has implied a gradual depletion of radioactive heat sources in the mantle and a change of boundary conditions at the Earth's surface. Another consequence is that the cycle of continental assembly and breakup, which occurs on the time scale of a few hundred million years, generates important changes of flow pattern at the surface and, by way of consequence, at depth in the mantle.

From the perspective of global geophysical models, the continental crust is a mosaic of different blocks that follows a large-scale geological pattern, and hence is responsible for large-scale lateral variations in temperature and seismic travel times. The continental crust is always thicker than the oceanic one. Crustal thickness is very variable beneath the continents and almost uniform beneath the oceans (Figure 2.13). The global elevation topography map shows the obvious dichotomy

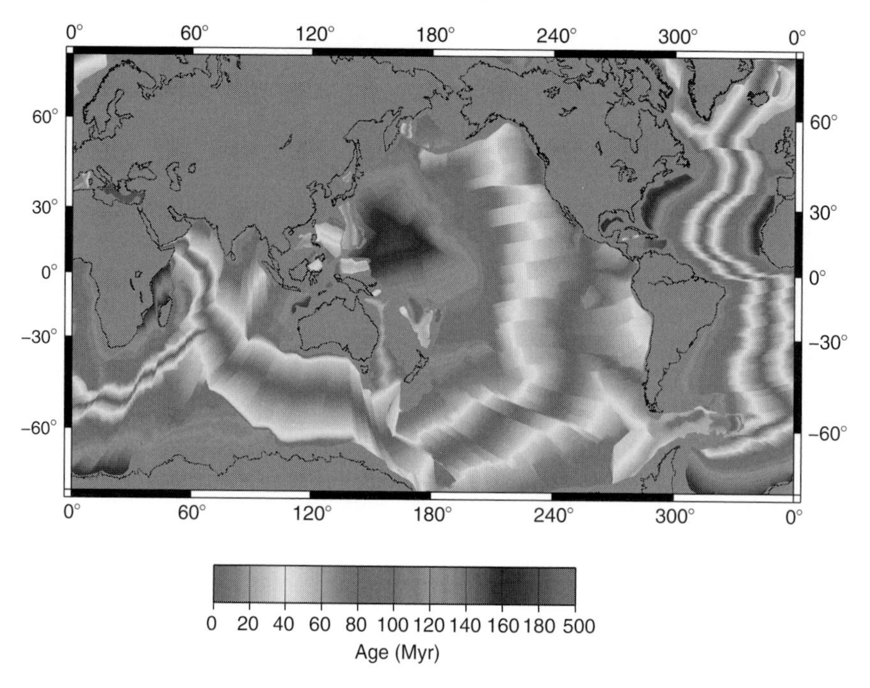

Figure 2.14. Map of the ages of the sea floor from Müller *et al.* (2008). See plate section for color version.

between oceans and continents (Figure 2.15). More importantly, the comparison between crustal thickness and elevation demonstrates that isostasy is achieved by different processes beneath the continents and the oceans. In the continents, elevation is correlated with crustal thickness, but it is not so beneath the oceans.

2.7.4 Melting and melt transport

In the Earth, convective motions are associated with melting and volcanic activity. One surprising feature is that magmas get generated in both upwellings and downwellings through different mechanisms. Decompression melting is responsible for the largest fraction of the total magma output of the planet, whereas heating and dehydration of subducting slabs generate a small, but significant, amount of magma. The emplacement and cooling of magmas at the Earth's surface or at shallow crustal levels is one of the heat loss mechanisms operating on Earth but it has been shown that it is a minor one and that it is overwhelmed by cooling of the oceanic plates.

Locally, however, magmatic activity represents large inputs of energy into the crust and must be accounted for in thermal models. The focusing of magma transport leads to the formation of large magma reservoirs that are of considerable interest in

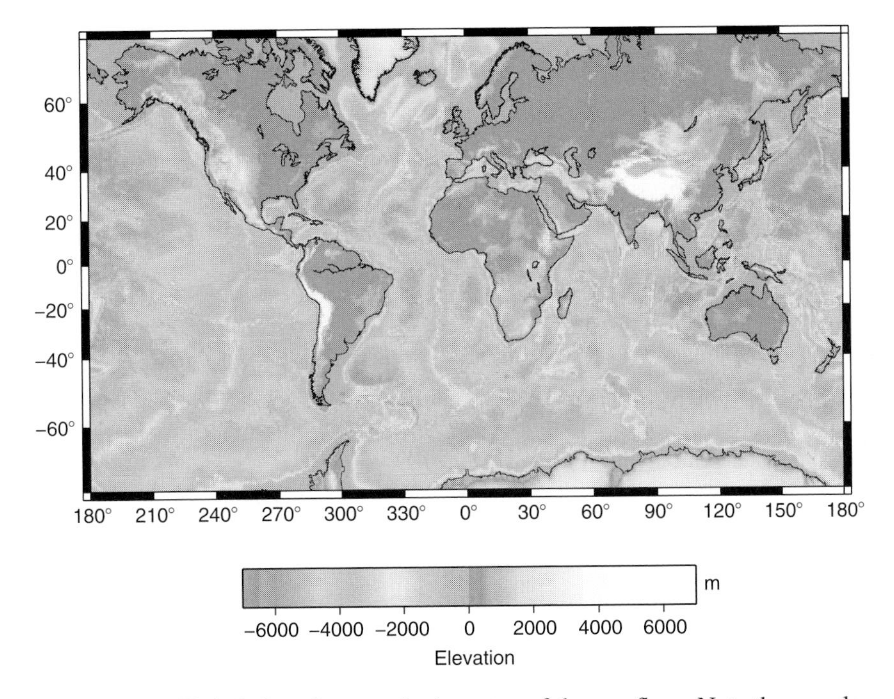

Figure 2.15. Global elevation map bathymetry of the sea-floor. Note the correlation of continental elevation with crustal thickness and sea-floor bathymetry with age. See plate section for color version.

Table 2.2. *Four large mafic igneous intrusions*

Name	Age (My)	Area (km^2)	Average thickness (km)
Stillwater, Montana, USA	2740	4,400	7
Bushveld, South Africa	2060	66,000	7–8 †
Duluth, Minnesota, USA	1096–1107 §	5,000	≈ 4 ‡
Dufek, Antarctica	182	6,600	7

† The intrusion is made of two different parts with different thicknesses.
§ Igneous activity occurred in several pulses over about 11 My.
‡ The intrusion is funnel-shaped and reaches at least 15 km thickness in its central part.

themselves. Table 2.2 lists the characteristics of four large mafic igneous complexes found on Earth. The formation and cooling of such complexes involve a number of physical processes, including convection and the settling or flotation of crystals. The end result, documented in considerable detail through structural, geochemical and petrological studies, is a complicated series of igneous layers with different compositions and minerals as well as large-scale chemical trends. These igneous bodies provide insights into the complex interactions between cooling, crystallization and chemical differentiation, and may serve as templates for the even more formidable problems posed by the evolution of the liquid core, and the magma ocean that probably existed in the early Earth.

Exercises

2.1 Calculate the moment of inertia of a sphere made of a core with density ρ_c surrounded by a mantle with density ρ_m, the radius of the sphere is R. The radius of the core $a \times R$. Find the average density ρ_a in terms of ρ_m, ρ_c, and a. Express the reduced moment of inertia in a function of a and $\delta = \rho_m/\rho_a$. If $\rho_a = 5500$ kg m^{-3}, $\rho_m = 4000$ kg m^{-3}, and the reduced moment is 0.35, calculate a/R and ρ_c.

2.2 How does the gravitational potential energy of a homogeneous sphere scale to its radius?

2.3 During the glacial ages, a large volume of sea water is stored in ice caps near the poles where they contribute very little to the moment of inertia. After deglaciation, the water is redistributed over the entire Earth's surface and the moment of inertia increases. As a crude approximation for this change in moment of inertia, determine the moment of inertia of a 100 m thick layer of water on a sphere of radius 6378 km. Compare with the polar moment of inertia of the Earth and estimate the change in rotation rate after melting of the glaciers.

2.4 Determine the gravitational potential energy of a sphere made of two concentric layers with density ρ_c and ρ_m, radius of the sphere R and radius of the core $a \times R$. Compare with the potential energy of the same mass uniformly distributed in the sphere. Estimate the energy released during differentiation of the Earth's core.

2.5 Consider a sphere of mass M growing at a constant rate $dM/dt = \alpha$ by accretion. Assume that the masses that are accreted to the sphere come from ∞ and that all the gravitational potential energy is converted into heat and redistributed in the sphere. Calculate how the total energy and the temperature of the sphere change with time. As temperature increases, the sphere loses energy from its surface by black body radiation. The total flux of energy at the surface is $A\sigma T^4$, where A is the surface area of the sphere and $\sigma = 5.67 \times 10^{-8}$ W m^{-2} K^{-4} (See Section 3.1.2). Write the differential equation that determines how the temperature varies as a function of time.

3

Basic equations

Objectives of this chapter

We shall first review the mechanisms of heat transport and some basic thermodynamic considerations that are important for understanding thermal processes in the Earth, and the relationships between physical properties and the thermal regime. We shall also review the fundamental equations of conservation of energy and momentum.

3.1 Heat transport mechanisms

There are three basic mechanisms of heat transport: radiation, convection and conduction. Radiation is the transport of energy by electromagnetic waves; it is the only possible mechanism of heat transport in a vacuum. Convection (or advection) refers to the transport of energy by matter in movement. When energy is transferred from one part to another in a continuous medium without mass transport, it is said to be conducted. All three mechanisms are important in different parts of the Earth. The light that carries the energy from the sun to the Earth's surface is electromagnetic radiation, in the visible part of the electromagnetic spectrum; it is radiated back in the infra-red. In the oceans and atmosphere, convection is the dominant mechanism of heat transport. Within the solid Earth, all three mechanisms play a role at different depths: in the outer shell of the Earth, conduction dominates, but locally, one must include hydrothermal convection generated in porous and permeable rocks and magma ascent. Hydrothermal circulation develops in fractures and pores, which get closed by the confining pressure deeper than 10 km. This mechanism is of small importance at the lithospheric scale but plays a crucial role in shallow environments where heat flux measurements are made. Conduction dominates over regions that are cold and cannot be deformed over geological time scales. At sufficient depth, the temperature is high enough for rocks to deform at significant rates (on the geologic time scale) and convective heat transfer dominates. One can see here the strong link between thermal and mechanical processes. Electromagnetic waves are attenuated over very short distances in rocks, but at very high temperatures, electromagnetic waves are re-emitted, and radiation may play a significant role.

3.1.1 Heat conduction: Fourier's law

The flow of heat from one part of a body to another is a vector, $\mathbf{q}$. The rate at which heat flows across a surface S is

$$\frac{dQ}{dt} = \int_S \mathbf{q} \cdot \mathbf{n} dS, \tag{3.1}$$

where $\mathbf{n}$ is the unit vector normal to the surface. The heat flux is thus measured in W m^{-2}.

In a solid body, heat transport is in large part effected by lattice vibrations. Fourier's law states that the heat flux is proportional to the temperature gradient. For an isotropic medium, Fourier's law states that

$$\mathbf{q} = -\lambda \nabla T, \tag{3.2}$$

where λ is the thermal conductivity, measured in W m^{-1} K^{-1}. The negative sign in equation 3.2 is a consequence of the second law of thermodynamics which requires that heat flows from hot to cold (in the direction opposite to that of the gradient).

More generally, if the medium is not isotropic, the thermal conductivity must be defined as a second-order tensor and Fourier's law takes the form:

$$q_i = -\lambda_{ij} \frac{\partial T}{\partial x_j} \quad i,j = 1,2,3 \tag{3.3}$$

with the summation convention and where λ_{ij} is the thermal conductivity tensor (Nye, 1985). There is no simple physical argument that the thermal conductivity tensor is symmetric. The symmetry can be demonstrated experimentally (Lovett, 1999) and it follows from thermodynamical arguments based on Onsager's principle (Nye, 1985). Because the tensor is symmetric, it is diagonal in the principal coordinate system. Along the three principal axes, the heat flow is parallel to the temperature gradient. On the rock scale, the anisotropy of thermal conductivity is the result of foliation and usually involves only two distinct principal values (in the direction perpendicular to the foliation and in the plane of foliation). At the scale of meters to hundreds of meters, a rock assemblage, such as a series of sedimentary layers, may also be treated as an anisotropic medium.

Thermal conductivity depends on rock composition and increases with quartz content. One can estimate the conductivity of a rock by treating it as a mixture of different minerals (see Appendix D). Amongst the most common minerals, the smallest and largest values of thermal conductivity are achieved in micas and quartz respectively. Typical values of thermal conductivity in crystalline rocks are $\approx 2\text{–}3$ W m^{-1} K^{-1}. The thermal conductivity is not constant but decreases with increasing temperature. More information and references on thermal conductivity can be found in Appendix D.

3.1.2 Radiation

In empty space, energy is transported by electromagnetic radiation. The flow of energy depends on the frequency spectrum of the electromagnetic waves, which in turns depend on the temperature of the source. The flux of energy Φ radiated at the surface of a "black body" depends only on temperature and is given by the Stefan–Boltzman law:

$$\Phi = \sigma T^4, \tag{3.4}$$

where the Stefan constant $\sigma = 5.67 \times 10^{-8}$ W m^{-2} K^{-4}, and T is the absolute temperature. Electromagnetic waves can also transport energy in continuous media. In this case, the radiation is rapidly attenuated but the medium re-emits energy if the temperature is sufficiently high. In a temperature gradient, the net energy flux will be $\propto T^3 \nabla T$, which has the same form as the flux for lattice conduction. One can therefore define a radiative component of conductivity, λ_r, such that the radiative heat flux is:

$$\mathbf{q}_r = -\lambda_r \nabla T. \tag{3.5}$$

The grey body law states that:

$$\lambda_r = \frac{16}{3} \frac{n^2}{\epsilon} \sigma T^3, \tag{3.6}$$

where n is the refractive index and ϵ is the opacity. In the following, radiative heat transport will be accounted for in this manner and will be lumped together with lattice conduction.

3.2 Definitions. Thermodynamic relationships

Physical properties of Earth materials are inter-related and enter in the equation of state that relates density, pressure and temperature. These definitions are useful to relate physical properties to observations, mostly from seismic data.

3.2.1 Definitions

The volume thermal expansion coefficient α relates how density varies with temperature:

$$\alpha = \frac{-1}{\rho}\left(\frac{\partial \rho}{\partial T}\right)_P = \frac{1}{V}\left(\frac{\partial V}{\partial T}\right)_P, \tag{3.7}$$

where ρ is density and $V(=1/\rho)$ specific volume. This coefficient α depends only weakly on temperature, and can be considered constant over a wide range of temperatures. In the upper mantle $\alpha \approx 3 \times 10^{-5}\ \mathrm{K}^{-1}$.

The bulk modulus relates density variations to pressure. It is an elastic parameter that can be derived from seismic velocities. At constant temperature, the isothermal bulk modulus K_T is

$$\frac{1}{K_T} = \frac{1}{\rho}\left(\frac{\partial \rho}{\partial P}\right)_T. \tag{3.8}$$

It is different from the isentropic bulk modulus K_S:

$$\frac{1}{K_S} = \frac{1}{\rho}\left(\frac{\partial \rho}{\partial P}\right)_S. \tag{3.9}$$

Changes in volume accompanying the passage of a seismic wave are quasi-instantaneous and involve no exchange of energy. The seismic compressional wave speeds are associated with the isentropic K_S. The bulk modulus of mantle rock is on the order of 10^{11} Pa (100 GPa).

The hydrodynamic velocity or "equivalent" sound velocity is derived from seismic velocities:

$$V_\phi^2 = \frac{K_S}{\rho} = V_P^2 - \frac{4}{3}V_S^2 = \left(\frac{\partial P}{\partial \rho}\right)_S. \tag{3.10}$$

From the first law of thermodynamics, for a reversible transformation, the change in internal energy U is the sum of the heat stored in the system Q, and of the work done on the system:

$$dU = dQ - PdV = TdS - PdV. \tag{3.11}$$

The specific heat relates the internal energy U to changes in temperature. At constant volume, C_V is:

$$C_V = \left(\frac{\partial U}{\partial T}\right)_V = T\left(\frac{\partial S}{\partial T}\right)_V. \tag{3.12}$$

The specific heat at constant pressure, which appears in the heat equation, is defined from the enthalpy $H = U + PV$:

$$C_P = \left(\frac{\partial H}{\partial T}\right)_P = T\left(\frac{\partial S}{\partial T}\right)_P. \tag{3.13}$$

There is a relationship between C_V and C_P involving the other physical properties defined above:

$$C_P - C_V = T\frac{\alpha^2}{\rho K_T}, \tag{3.14}$$

this implies that $C_P > C_V$.

We shall be concerned with C_P which enters the heat equation. For water $C_P = 4180$ J kg^{-1} K^{-1}; it is much larger than in most crustal rocks, where C_P ranges between 800 and 1100 J kg^{-1} K^{-1}. In the mantle $C_P \approx 1200$ J kg^{-1} K^{-1}. The Gruneisen parameter γ was introduced in microphysics to relate lattice vibrations to temperature. It has a macroscopic definition involving parameters defined above:

$$\gamma = \frac{\rho}{T}\left(\frac{\partial T}{\partial \rho}\right)_s = \frac{\alpha K_T}{\rho C_V} = \frac{\alpha K_S}{\rho C_P}. \tag{3.15}$$

The Gruneisen parameter ranges between 0.5 and 1 in the mantle.

3.2.2 Latent heat. Clausius Clapeyron equation

When two phases of a single chemical component are in equilibrium, the number of free thermodynamic variables is one; any variation in pressure must be accompanied by a variation in temperature to maintain the conditions of equilibrium. Also, there is a difference in entropy ΔS and specific volume ΔV between the two phases. Transformation from one phase to another is thus accompanied by the release or absorption of heat $L = T\Delta S$. For instance, the entropy of water is greater than that of ice, melting of ice thus requires that heat be provided to the system. The amount of heat absorbed depends on the phase equilibrium conditions:

$$\frac{L}{T} = \Delta S = \Delta V\frac{dP}{dT}. \tag{3.16}$$

In the case of water, the liquid has higher entropy and lower specific volume (it is denser) than ice. It thus implies that $dP/dT < 0$ (i.e. increasing the pressure lowers the melting point of ice). For the water–ice system $L = 334$ kJ kg^{-1} and $\Delta V = 9 \times 10^{-5}$ m^3 kg^{-1}, $dP/dT = -13.5$ MPa K^{-1}. At the base of a 2 km thick ice sheet, the melting temperature decreases by 1.5 K.

For rocks, the latent heat of melting $L \approx 300$ kJ kg^{-1}, i.e. it is in the same range as that of ice. Values of C_P are very different for rocks and water: the energy needed to melt ice is equivalent to that required to raise the temperature of water by 80 K; the energy for melting a rock is equivalent to a temperature increase of ≈ 300 K in the rock.

3.3 Conservation of mass

The equation of state of a material specifies how its density varies as a function of the main thermodynamic state variables, pressure and temperature: $\rho = \rho(P, T)$. Some systems involve different types of materials, which may or may not mix, or material with smooth variations of composition, such as seawater, where the concentration of salt changes due to evaporation and the input of fresh water from rain or continental drainage networks. For those, density must be also specified as a function of composition. In a dynamic system, material moves and deforms, so that composition, pressure and temperature all vary as a function of position and time. In this case, it is more practical to write density as a function of time and spatial coordinates, $\rho = \rho(x, y, z, t)$. The same can be said for other properties and variables, and we shall work with both types of representation.

Conservation of mass implies that

$$\frac{\partial \rho}{\partial t} + \nabla \cdot (\rho \mathbf{v}) = 0, \tag{3.17}$$

where $\mathbf{v}$ is the velocity vector. With this equation, one can track how density changes act to modify velocity, for example how expansion induces acceleration. In order to write thermodynamic principles, one must work with material points and this form of the equation is not of practical use. A material point moves with the fluid and its density may therefore vary as a function of time and space variables. In a frame of reference that moves with the parcel, however, density can only vary as a function of time, and this is what we will be interested in when we introduce thermodynamics. For a material point located at point M with coordinates $x_M(t), y_M(t)$ and $z_M(t)$, we must deal with a "tagged" density, noted ρ_M, such that $\rho_M = \rho(x_M, y_M, z_M, t)$. The time-derivative of that function is obtained by standard derivation formulae:

$$\frac{d\rho_M}{dt} = \left(\frac{\partial \rho}{\partial t}\right)_M + \frac{\partial \rho}{\partial x}\frac{dx_M}{dt} + \frac{\partial \rho}{\partial z}\frac{dy_M}{dt} + \frac{\partial \rho}{\partial z}\frac{dz_M}{dt}. \tag{3.18}$$

This can be simplified by noting that, by the definition of velocity, $dx_M/dt = v_x, dy_M/dt = v_y$ and $dz_M/dt = v_z$, so that one has :

$$\frac{d\rho_M}{dt} = \left(\frac{\partial \rho}{\partial t}\right)_M + \nabla \rho \cdot \mathbf{v} \equiv \frac{D\rho}{Dt} \tag{3.19}$$

where, for simplicity, we have introduced a new notation, D/Dt, which is called the *material derivative*. This derivative can be used for all variables, of course. We can introduce it into the mass conservation equation, which takes the following form:

$$\frac{D\rho}{Dt} + \rho \nabla \cdot \mathbf{v} = 0. \tag{3.20}$$

This equation specifies how the density of the material point changes. For comparison with usual thermodynamics, one can introduce the *specific volume* V. The mass conservation equation can then be written in yet another form:

$$\nabla \cdot \mathbf{v} = -\frac{1}{\rho}\frac{D\rho}{Dt} = \frac{1}{V}\frac{DV}{Dt}, \tag{3.21}$$

which states that the divergence of velocity is equal to the rate of expansion or contraction (depending on the sign).

3.4 Conservation of momentum

The fundamental principle of dynamics states that acceleration is equal to the sum of the applied forces. We consider weight in a gravity field as the only body force and introduce the stress tensor $\mathbf{\Pi}$ to describe the internal forces. These forces between two parts of the continuum act on the surface that separates them. For an arbitrary elementary surface within the material, stress describes the force per unit area. The internal force acting on a volume inside the continuum is the integral of the surface forces:

$$\mathbf{f} = -\oint_S \mathbf{\Pi} \cdot \mathbf{n} dS. \tag{3.22}$$

The stress tensor is decomposed into an isotropic diagonal part, which defines *pressure* p, and the deviatoric stress, $\boldsymbol{\sigma}$, a tensor with zero trace:

$$\mathbf{\Pi} = p\mathbf{I} + \boldsymbol{\sigma}, \tag{3.23}$$

where $\mathbf{I}$ is the identity tensor. By convention, this stress tensor refers to the force that is applied on material located on the positive side of the surface.[1] Conservation of momentum is written as follows:

$$\rho \frac{D\mathbf{v}}{Dt} = \rho \left(\frac{\partial \rho}{\partial t} + \mathbf{v} \cdot \nabla \mathbf{v} \right) = -\nabla p - \nabla \cdot \boldsymbol{\sigma} + \rho \mathbf{g}, \tag{3.24}$$

where the negative sign on the second term on the right-hand side follows from the sign convention on the stress transfer.

[1] The opposite sign convention can be found in many books.

In addition to this general momentum conservation equation, we shall need a constitutive equation that defines the rheology, i.e. that relates the stress to the deformation and/or deformation rate.

3.4.1 Equilibrium condition

In a continuum at rest with no shear stress, we have the condition

$$-\nabla p + \rho \mathbf{g} = 0. \tag{3.25}$$

This condition defines hydrostatic or lithostatic (in the solid Earth) equilibrium.

3.5 Energy equation

3.5.1 General form

We are interested in fluid motions in a planet where density varies by large amounts due to pressure variations and develop thermodynamics where quantities are defined per unit mass. The first law of thermodynamics (Bird *et al.*, 1960) specifies how the total energy of a system changes. The total energy is the sum of internal energy u and kinetic energy e_c:

$$\rho \frac{D(u + e_c)}{Dt} = -\nabla \cdot \boldsymbol{q} - \nabla \cdot (pv) - \nabla \cdot [\boldsymbol{\sigma} \cdot v] + H + \phi + \rho \mathbf{g} \cdot \mathbf{v}, \tag{3.26}$$

where $\mathbf{q}$ is the conductive heat flux, H the rate of internal heat generation (due to radioactive decay) and ϕ collects work done by external sources (such as tidal dissipation in the Earth). From the momentum equation, we get

$$\rho \frac{De_c}{Dt} = \rho \frac{D\left(\frac{v^2}{2}\right)}{Dt} = -\nabla p \cdot \mathbf{v} - \mathbf{v} \cdot \nabla \cdot \boldsymbol{\sigma} + \rho \mathbf{g} \cdot \mathbf{v}. \tag{3.27}$$

Subtracting this from the total energy balance leads to an equation for the internal energy,

$$\rho \frac{Du}{Dt} = -\nabla \cdot \mathbf{q} + H + \psi + \phi - p \nabla \cdot \mathbf{v}, \tag{3.28}$$

where ψ stands for the work of the deviatoric stresses.

$$\psi = -\nabla \cdot [\boldsymbol{\sigma} \cdot \mathbf{v}] + \mathbf{v} \cdot \nabla \cdot \boldsymbol{\sigma} = -\boldsymbol{\sigma} : \nabla \mathbf{v}. \tag{3.29}$$

In a viscous fluid, this term corresponds to shear dissipation. Equation (3.28) is thus the usual statement that changes of internal energy u are due to heat gains or losses (which are broken into four contributions) and to the work of pressure (the last term on the right).

The internal energy can be broken down into two components, heat content and strain energy: $u = u_T + u_D$. The former will be written as a function of thermodynamic state variables, such that $du_T = Tds$, where s is the entropy per unit mass, and the other one is the energy that gets released as the material contracts in its own gravity field. We shall dwell on this when we deal with secular cooling of the Earth.

3.5.2 Specialized form for no net contraction or expansion

We take $u \equiv u_T$ and do not account for external source terms, so that ϕ in equation 3.26 is set to zero. Introducing variables of state, we write,

$$\rho \frac{Du_T}{Dt} = \rho T \frac{Ds}{Dt} = \rho C_p \frac{DT}{Dt} - \alpha T \frac{Dp}{Dt}, \tag{3.30}$$

where s, the entropy per unit mass, has been expressed as a function of temperature and pressure and α is the coefficient of thermal expansion.

From (3.28), we deduce that,

$$\rho C_p \frac{DT}{Dt} - \alpha T \frac{Dp}{Dt} = -\nabla \cdot \mathbf{q} + H + \psi. \tag{3.31}$$

Isentropes and adiabats

We introduce two different reference temperature profiles. The equation for the entropy per unit mass, s, is,

$$\rho T \frac{Ds}{Dt} = \rho C_p \frac{DT}{Dt} - \alpha T \frac{Dp}{Dt} \tag{3.32}$$

$$= -\nabla \cdot \mathbf{q} + \psi + H. \tag{3.33}$$

This shows that entropy is not conserved due to irreversible dissipation and radioactive decay. Density changes due to temperature have a small impact on pressure, and dynamic pressure variations are small compared to the hydrostatic pressure. Thus,

$$\frac{Dp}{Dt} \approx -\rho g w, \tag{3.34}$$

where w is the radial velocity component in a spherical system, or the vertical one in a Cartesian system.

From these equations, we may deduce the isentropic temperature profile, such that $Ds/Dt = 0$. In the interior of the convecting system, far from the upper and lower boundaries, the dominant velocity component is along the vertical. Assuming steady state and using equation (3.32):

$$\rho C_p w \frac{dT_S}{dr} = -\alpha T_S \rho g w, \tag{3.35}$$

where T_S stands for the isentropic temperature profile. Thus,

$$\frac{dT_S}{dr} = -\frac{\alpha g}{C_p} T_S = -\frac{g\gamma\rho}{K_S} T_S = -\frac{g\gamma}{V_\phi^2} T_S, \tag{3.36}$$

where γ is the Gruneisen ratio (equation 3.15) and here $V_\phi^2 = K_S/\rho$ is hydrodynamic velocity. The isentropic temperature gradient derived above provides a convenient *reference* profile which illustrates the role of compressibility. However, it is a poor approximation for the temperature path followed by a rising (or sinking) material point. We may consider for simplicity that this material point does not exchange heat with its surroundings. In this case, we set $\mathbf{q} = 0$, use the same approximation as before for pressure and obtain:

$$\rho C_p \frac{DT}{Dt} \approx \rho C_p \left(\frac{\partial T}{\partial t} + w \frac{\partial T}{\partial r} \right) = -(\alpha T) \rho g w + H + \psi. \tag{3.37}$$

The material point temperature changes due to radiogenic heat production and dissipation as well as due to the work of pressure forces. This may be recast as follows:

$$\frac{\partial T}{\partial r} = -\frac{\alpha g}{C_p} T + \frac{1}{w} \left(\frac{\psi + H}{\rho C_p} - \frac{\partial T}{\partial t} \right), \tag{3.38}$$

where one should note that secular cooling acts in the same direction as internal heat production. This radial temperature gradient may be called an *adiabat*. In the Earth, it differs from the isentrope by about 30%.

Approximate equations for convection in a compressible material

Density variations due to temperature are small because of the small value of the thermal expansion coefficient:

$$\left(\frac{\Delta\rho}{\rho} \right)_P = -\alpha \Delta T \approx 3\% \text{ for } \Delta T = 10^3\,\text{K}. \tag{3.39}$$

Such density changes have little impact on the fluxes of mass and energy, but they are responsible for the buoyancy forces that drive convection. As a consequence, a great simplification to the governing equations is to neglect density variations due to temperature in all terms except the body force term. This is called the *Boussinesq–Oberbeck* approximation and has been evaluated in great detail. It leads to little error and will be adopted here. One must also deal with density variations due to pressure which are negligible on the scale of a magma reservoir but which are large in the Earth's mantle. A comprehensive treatment of the energy equation may be found in Jarvis and McKenzie (1980) and Schubert *et al.* (2001) and we restrict ourselves to a few salient points. In steady state, the heat equation can be written as follows:

$$\rho C_p \left[\mathbf{v} \cdot \left(\nabla T - \frac{\alpha T}{\rho C_p} \nabla p \right) \right] = \nabla \cdot (\lambda \nabla T) + H + \psi, \tag{3.40}$$

where we have specified that the heat flux $\mathbf{q}$ obeys Fourier's conduction law. Viscous stresses and buoyancy forces are small compared to the gravity body force and induce only small perturbations to the gravity and pressure fields (in all rigour, this should be assessed carefully using a perturbation analysis, as was done by Jarvis and McKenzie (1980). Thus, in the heat balance equation, we may approximate the pressure field by that of a system in hydrostatic equilibrium, i.e. $\nabla p \approx \rho \mathbf{g}$ and obtain,

$$\rho C_p \left[\mathbf{v} \cdot \left(\nabla T - \frac{\alpha T}{\rho C_p} \nabla p \right) \right] \approx \rho C_p \left[\mathbf{v} \cdot \left(\nabla T - \frac{\alpha T}{C_p} \mathbf{g} \right) \right], \tag{3.41}$$

where we may identify a term that looks like the definition of the isentrope. We split temperature into the average isentropic field T_S and a deviation: $T = T_S + T^*$. By definition, $\nabla T_S = (\alpha T_S / C_p)\mathbf{g}$. Deviations from the isentrope are only important in the thin thermal boundary layers that exist at the top and bottom of a convecting layer, where the dominant heat transport mechanism is conduction. Thus, one may write that $(\alpha T / C_p)\mathbf{g} \approx (\alpha T_S / C_p)\mathbf{g} = \nabla T_S$, and hence:

$$\rho C_p \left[\mathbf{v} \cdot \left(\nabla T - \frac{\alpha T}{\rho C_p} \nabla p \right) \right] \approx \rho C_p \left[\mathbf{v} \cdot \nabla (T - T_S) \right]. \tag{3.42}$$

Thus, using the deviation from the isentrope $T^* = T - T_S$ in the heat balance equation and neglecting heat that is conducted along the isentrope, one reduces the system to:

$$\nabla \cdot \mathbf{v} = 0 \tag{3.43}$$

$$\rho C_p \frac{DT^*}{Dt} = \nabla \cdot (\lambda \nabla T^*) + H + \psi \tag{3.44}$$

$$\rho \frac{D\mathbf{v}}{Dt} = -\nabla p - \nabla \cdot \boldsymbol{\sigma} + \rho \mathbf{g}, \tag{3.45}$$

where ρ depends on pressure only. Save for this pressure dependence, these are the Boussinesq equations for an incompressible fluid.

The physical framework provided by equations 3.43–3.45 has been explored comprehensively in laboratory experiments, numerical calculations and theoretical scaling arguments. When applying it to convection in the Earth's mantle, one must remember that it is only an approximation and also that one must allow for the isentropic reference temperature profile. For example, vigourous convection at high Rayleigh number in a laboratory tank is such that the well-mixed interior is almost isothermal. Depending on the mode of heating, slight deviations from a uniform interior temperature are observed, however, which shed light on significant aspects of convection. Transposed to the Earth's mantle, such deviations must be understood as departures from the isentrope. For example, one should associate the negative temperature gradient of internally heated convection to a sub-isentropic gradient. In quantitative thermal studies, one must account for the heat flux conducted along the isentrope. In the Earth's mantle, the isentrope is $dT_s/dr \approx -0.5$ K km^{-1}, corresponding to a heat flux of ≈ 2 mW m^{-2}. This is small compared to the average surface heat flux on Earth, which is ≈ 80 mW m^{-2}. It is not small, however, compared to the small values of the heat flux at the base of an Archean continental lithosphere (≈ 15 mW m^{-2}).

3.6 Radial variations of density in the Earth

The seismic P and S wave velocities can be used to determine the variation of density with pressure in a homogeneous shell inside the Earth. For isentropic compression:

$$\frac{dP}{dr} = -g(r)\rho \tag{3.46}$$

$$\left(\frac{\partial P}{\partial \rho}\right)_S = \frac{K_S}{\rho}. \tag{3.47}$$

One deduces from these an equation for the radial density gradient, called the Adams–Williamson equation (Birch, 1952):

$$\frac{d\rho}{dr} = \frac{d\rho}{dP}\frac{dP}{dr} = \frac{-g(r)\rho^2}{K_S} = \frac{-g(r)\rho}{V_\phi^2}. \tag{3.48}$$

This equation is specific to isentropic compression and a more general form for the change of density is (Birch, 1952):

$$\frac{d\rho}{dr} = \left(\frac{\partial \rho}{\partial P}\right)_T \frac{dP}{dr} + \left(\frac{\partial \rho}{\partial T}\right)_P \frac{dT}{dr} \tag{3.49}$$

$$= \frac{-g\rho^2}{K_T} - \alpha\rho\left(\frac{dT}{dr} - \left(\frac{\partial T}{\partial r}\right)_S\right), \tag{3.50}$$

which introduces the deviation from an isentrope.

3.7 Equations for fluid flow

3.7.1 Viscosity

Experiments show that forces must be applied to maintain velocity differences in a fluid. For example, if a viscous fluid is present between two plates moving with a constant relative velocity, a tangential stress σ applied to the upper plate is proportional to the velocity difference Δv, and inversely proportional to the distance between the plates d. The proportionality constant is the viscosity μ, which is a physical property characteristic of the fluid:

$$\sigma = \mu \frac{\Delta v}{d}. \tag{3.51}$$

The viscosity is measured in Pa s and is the physical property that varies by the largest amount amongst natural materials. It is 10^{-3} Pa s for water and $\approx 10^{20}$ to $\approx 10^{22}$ Pa s for the upper and lower mantles, respectively (see Section 2.2.5). The viscosity of magmas depends on the composition in major elements as well as on the concentration of dissolved volatiles, and is in a range of 1–10^{10} Pa s in most cases. The viscosity of the outer core is believed to be ≈ 0.01 Pa s (see Appendix D.1.6 for typical values of viscosity).

The deviatoric stress tensor is

$$\boldsymbol{\sigma} = -\mu \left(\nabla \mathbf{v} + \nabla \mathbf{v}^T - \frac{2}{3} (\nabla \cdot \mathbf{v}) \mathbf{I} \right), \tag{3.52}$$

where the ^{T} symbol stands for the transpose operation. This equation is the constitutive equation for a Newtonian (linear) fluid. In a compressible fluid, the dynamic pressure p may not be equal to the thermodynamic pressure, which corresponds to an equilibrium relaxed state. In this case, one must introduce a second viscosity coefficient. We shall not deal with such situations here and will only consider incompressible fluids.

3.7.2 Navier–Stokes equations

For an incompressible fluid, $\nabla \cdot \mathbf{v} = 0$, the momentum conservation equation simplifies to:

$$\rho \left[\frac{\partial \mathbf{v}}{\partial t} + \mathbf{v} \cdot \nabla \mathbf{v} \right] = -\nabla P + \mu \nabla^2 \mathbf{v} + \rho \mathbf{g}. \tag{3.53}$$

This is the Navier–Stokes equation, which must be solved with the incompressibility condition. It is also written as:

$$\frac{\partial \mathbf{v}}{\partial t} + \mathbf{v} \cdot \nabla \mathbf{v} = -\frac{\nabla P}{\rho} + \nu \nabla^2 \mathbf{v} + \mathbf{g}, \tag{3.54}$$

where $\nu = \mu/\rho$ is the kinematic viscosity. Note that the units for ν ($m^2\ s^{-1}$) are the same as the units for the thermal diffusivity ($\kappa = \lambda/\rho C_P$) in the heat equation. This is not coincidental and the Navier–Stokes equation is similar to the heat equation because it describes the diffusion of momentum in a fluid. To illustrate this, one may consider fluid that is injected at a finite velocity on top of a layer of stagnant fluid at rest. The velocity difference between the two fluid regions induces friction at the interface which sets the lower fluid in motion. A velocity gradient develops through the lower fluid layer in a finite amount of time. Thus, momentum is being diffused away by friction. We shall illustrate this with thermal plumes. A key difference between equation 3.54 and the heat equation is that the advection term in the Navier–Stokes equation is non-linear. The non-linearity of the Navier–Stokes equation is fundamental for most problems in fluid mechanics, except in the solid Earth.

3.7.3 Reynolds number

The Navier–Stokes equation is complicated because it involves many different terms and it is useful to evaluate conditions for which some of these terms are negligible in comparison with the others. In a first, and general, step, one considers the two terms that account for effects that resist motion: inertia and viscous shear. Depending on which one dominates over the other, the flow regimes and the governing equations are completely different. In order to determine which regime prevails, one must estimate the magnitudes of the two corresponding terms in the momentum balance. Let us consider that the flow involves a characteristic length scale L and velocity U. We can change the variables in equation 3.54 in order to use the characteristic dimensions:

$$
\begin{aligned}
v' &= \frac{v}{U} \\
x' &= \frac{x}{L} \\
t' &= \frac{tU}{L}.
\end{aligned}
\tag{3.55}
$$

For pressure, one may use two different scales depending on the flow regime. An inertial pressure scale may be obtained by a balance between pressure and inertial forces (on the left-hand side of the Navier–Stokes equation). Similarly, a viscous pressure scale emerges from a balance between pressure gradients and viscous forces. These two pressure scales are:

$$
\Delta P_i = \rho U^2, \quad \Delta P_v = \mu \frac{U}{L}, \tag{3.56}
$$

where indices i and v stand for inertial and viscous. We arbitrarily choose the inertial scale, so that dimensionless pressure is

$$P' = \frac{P - P_0}{\rho U^2}, \tag{3.57}$$

where P_0 is the hydrostatic pressure, i.e. $\nabla P_0 = \rho \mathbf{g}$. Introducing these variables, we get the Navier–Stokes equation in dimensionless form:

$$\frac{\partial \mathbf{v}'}{\partial t'} + \mathbf{v}' \cdot \nabla' \mathbf{v}' = \frac{1}{\mathrm{Re}} \nabla'^2 \mathbf{v}' - \nabla P', \tag{3.58}$$

where $\mathrm{Re} = LU/\nu = \rho LU/\mu$ is the *Reynolds number*.

Physically, the *Reynolds number* can be understood as the ratio of the scale of the inertial forces (the left-hand side of the Navier–Stokes equation) to the friction forces, the first term in the right-hand side. The pressure scale is adjusted to eliminate the hydrostatic pressure.

In general, when the Reynolds number is large, the inertial forces are important, and the non-linear terms in Navier–Stokes dominate the behavior, leading to *turbulence*. In the other limit, such that the Reynolds number is small, friction forces must be accounted for. One usually associates this regime with *laminar* flow conditions, such that flow lines are quasi-parallel, no mixing occurs between neighboring fluid regions and the flow field is reversible, i.e. one recovers an initial configuration by inverting the driving force for the flow. A small Re is a necessary condition for laminar flow, but it is not sufficient. For example, it is not a sufficient condition for thermal convection to be organized in stationary cells. For flow at small Reynolds numbers, one must use the viscous pressure scale instead of the inertial one. It is instructive to calculate the ratio between the two pressure scales:

$$\frac{\Delta P_i}{\Delta P_v} = \frac{\rho U^2}{\mu U/L} = \frac{\rho UL}{\mu}, \tag{3.59}$$

where we recognize the Reynolds number. Thus, $\mathrm{Re} \gg 1$ implies that $\Delta P_i \gg \Delta P_v$, and vice versa. In other words, one must select the pressure scale that is largest in order to obtain the appropriate dimensionless governing equations.

This analysis relies on a velocity scale U that has been prescribed. For convection, as we shall see, this is not so and the control variable is, for example, the temperature difference that is imposed across the fluid layer. In this case, the Reynolds number must be solved for. For mantle convection, however, one may measure the velocity and hence one may estimate the Reynolds number directly. The mantle kinematic viscosity is on the order of 10^{18} m^2 s^{-1}, velocities are on the order of 10^{-9} m s^{-1}, and the characteristic length scale is always less than the radius of the Earth, 6.4×10^6 m, so that $\mathrm{Re} \sim 10^{21} \gg 1$, implying that inertial forces can be neglected. This does not hold in the outer core and in basaltic magma reservoirs, as we shall see later.

3.7.4 Other dimensionless numbers

The *Rossby number* is the ratio of inertial to Coriolis forces:

$$\mathrm{Ro} = \frac{U}{L\Omega}, \tag{3.60}$$

with Ω the frequency of rotation of the Earth. When $\mathrm{Ro} > 1$, the Coriolis force will cause only small perturbations to the flow.

For convection in the Earth's mantle, the Reynolds number is small and inertia is negligible. In this case, it is more appropriate to compare the Coriolis and frictional forces. The relevant dimensionless number is the *Taylor number*:

$$\mathrm{Ta} = \mathrm{Re}\mathrm{Ro}^{-1} = \frac{\Omega L^2}{\nu}. \tag{3.61}$$

For the Earth's mantle, $\mathrm{Ta} \ll 1$ and one can neglect the Coriolis force.

The *Prandtl number* is the ratio of viscous to thermal diffusion:

$$\mathrm{Pr} = \frac{\nu}{\kappa}. \tag{3.62}$$

When $\mathrm{Pr} > 1$, momentum is diffused away more rapidly than heat, which is true for the materials that we will be dealing with here, magmas and the Earth's mantle. Because of the large differences in viscosity that exist, however, the Prandtl number varies by large amounts with important consequences for the dynamics of convection. In contrast to the other dimensionless numbers that we shall be dealing with, the Prandtl number depends neither on the physical dimensions of the system nor on the boundary conditions: it is an intrinsic property of the material.

The *Péclet number* measures the ratio of convective to conductive heat transfer. By comparing the convective to conductive terms in the heat equation, we obtain:

$$\mathrm{Pe} = \frac{LU}{\kappa}, \tag{3.63}$$

which can be understood as the "thermal" equivalent of the Reynolds number. One must note that:

$$\mathrm{Re} = \frac{LU}{\nu} = \mathrm{Pe}\frac{\nu}{\kappa} = \mathrm{Pe}\mathrm{Pr}. \tag{3.64}$$

4
Heat conduction

Objectives of this chapter

The chapter is divided in two different parts. The first part illustrates key physical principles of heat transport by diffusion and sketches the relevant mathematical techniques for obtaining solutions to the heat equation. We shall draw a parallel between the physics of heat diffusion, expressed as scaling laws, and the form of the mathematical solutions. Three main mathematical methods are used: separation of variables, similarity solutions and Fourier transforms. We then develop a number of solutions using more involved, but standard, techniques such as Fourier and Laplace transforms and Green's functions. These techniques are powerful because they can be applied to almost all problems provided that they are linear, whereas the others (separation of variables, similarity solutions) can only be used in specific cases. In a few cases, the same problem will be solved by different methods in order to illustrate the advantages and drawbacks of each method.

4.1 Heat conduction: Generalities

For a solid or a motionless fluid, the energy conservation equation (3.27) becomes the heat conduction equation:

$$\rho C_p \frac{\partial T}{\partial t} = -\nabla \cdot \mathbf{q} + H = \nabla(\lambda \nabla T) + H, \tag{4.1}$$

where the conductivity λ includes a radiative component. If physical properties are constant, equation 4.1 becomes

$$\frac{\partial T}{\partial t} = \kappa \nabla^2 T + \frac{H}{\rho C_p}, \tag{4.2}$$

where $\kappa = \lambda/(\rho C_p)$ is the thermal diffusivity. The units for diffusivity are $m^2\ s^{-1}$. The thermal diffusivity of rocks varies little and is of the order of ($\approx 10^{-6}\ m^2\ s^{-1}$).

For geological scale problems, it is practical to use the following values for κ: $31.5\ \mathrm{m}^2\ \mathrm{y}^{-1}$ or $31.5\ \mathrm{km}^2\ \mathrm{My}^{-1}$. The solution of equation 4.2 needs both initial and boundary conditions to be specified. In order to determine how temperature varies with time in a region of space, it is necessary to know what the temperature was initially in that region and what are the conditions that determine the flow of heat on the boundaries of that region.

In practice, we shall include the mechanical or potential energy converted into heat in the heat production density. In the Earth, the main source of heat production is the radioactive decay of long-lived isotopes (^{235}U,^{238}U,^{40}K,^{232}Th). The concentration of radioactive elements varies much with the chemical composition of the rocks, and tend to be concentrated in the uppermost part of the continental crust (see Chapter 7 and Appendix E). Other sources are the release of latent heat during phase transformations and the conversion of gravitational potential energy, or mechanical energy into heat by friction or viscous dissipation.

4.1.1 Time and distance scales

Without solving equation 4.2, it is interesting to note that it implies a relationship between length and time scales. If l is the length scale and τ the time scale, $\tau = l^2/\kappa$ or $l = \sqrt{\kappa\tau}$.

This relationship can be used to get a rough estimate of the distance reached by transient temperature variations, or the time required for temperature perturbations to propagate a given distance. For instance, for a standard value of thermal diffusivity of $10^{-6}\ \mathrm{m}^2\ \mathrm{s}^{-1}$ the depth scale of the daily temperature cycle ≈ 0.3 m. For the annual cycle, it is ≈ 5.5 m.

Using these relationships, the characteristic time of cooling for a dyke 1 m thick is 10^6 s ≈ 15 days; for a kilometric intrusion, it is $\approx 30{,}000$ years, for a body the size of the Earth, it is >100 Gy, several times the age of the Universe. These scales are useful in obtaining order-of-magnitude estimates, but the precise determination of decay time or propagation of thermal perturbations depend on boundary conditions and geometry and require full solution of the heat equation (4.2).

4.1.2 Superposition of solutions

When the thermal conductivity does not depend on temperature, the heat equation (4.2) is linear which allows us to decompose complex problems into elementary ones and to superpose their solutions. Solutions can be superposed provided boundary and initial conditions are combined. For instance the time-dependent heat equation with source terms, initial and boundary conditions, can be decomposed into the transient equation without sources, and the steady-state equation with sources.

This superposition principle will be used in many applications and allows us, for instance, to calculate separately a steady-state solution and the transient relaxation of the initial conditions. For the steady-state equation, we can decompose a solution and treat separately internal heat generation and heating or cooling through the boundaries.

Consider for example the simple problem of the steady-state temperature distribution in the Earth's crust. If crustal radiogenic heat production is uniformly distributed and the heat flow at the Moho discontinuity is also uniform, heat transfer occurs in the vertical z direction only and temperature obeys the following equation:

$$\lambda \frac{d^2T}{dz^2} + H = 0, \tag{4.3}$$

where z is positive downward, with the following boundary conditions:

$$T(0) = T_o \text{ at the Earth's surface}; \; -\lambda \frac{dT}{dz}(h) = Q_M \text{ at the Moho}, \tag{4.4}$$

where Q_M is negative since heat is supplied to the crust by the underlying mantle. The solution is

$$T(z) = T_o - \frac{Q_M}{\lambda} z + \frac{H}{2\lambda} z(2h - z). \tag{4.5}$$

This can be understood as the superposition of solutions for three problems involving different boundary conditions and heat production values:

$$(1)\, T(0) = T_o, \quad H = 0, \quad -\lambda \frac{dT}{dz}(h) = 0 \tag{4.6}$$

$$(2)\, T(0) = 0, \quad H = 0, \quad -\lambda \frac{dT}{dz}(h) = Q_M \tag{4.7}$$

$$(3)\, T(0) = 0, \quad H \neq 0, \quad -\lambda \frac{dT}{dz}(h) = 0. \tag{4.8}$$

We shall use this principle to treat separately different effects that are active simultaneously.

4.2 Steady-state heat equation

4.2.1 Poisson's and Laplace's equations

If the temperature does not change with time, the heat conduction equation becomes

$$\nabla \cdot \vec{\mathbf{q}} = -\nabla \cdot \lambda \nabla T = H. \tag{4.9}$$

It states that, in steady state, the flux of heat across a closed surface is equal to the heat generated in the volume within that surface. If thermal conductivity λ is constant, the equation becomes

$$\nabla^2 T = -\frac{H}{\lambda}. \tag{4.10}$$

This equation, known as Poisson's equation, is a standard equation in potential theory. If there are no heat sources, the homogeneous Poisson's equation is known as Laplace's equation:

$$\nabla^2 T = 0. \tag{4.11}$$

In steady state, therefore, temperature is a potential field. Temperature is analogous to the gravity potential, and heat flux corresponds to the acceleration of gravity. One may thus use many results from potential theory, as will be shown below. The main difference between temperature and gravity potential lies in the boundary conditions. In gravity studies, the convention is to assume that the gravity potential vanishes at infinity because it is solution of Laplace's equation in free space. In contrast, the heat equation does not apply in free space but only within a material body. For the heat equation, usually the temperature or the heat flux is specified on the boundary surrounding the domain of interest.

Equations 4.10 and 4.11 hold only within a region where thermal conductivity does not vary. On the surface separating two regions with different thermal conductivities, the temperature and the normal component of the heat flux must be continuous. The normal component of the temperature gradient is therefore discontinuous:

$$\lambda_1 \frac{\partial T_1}{\partial n} = \lambda_2 \frac{\partial T_2}{\partial n}, \tag{4.12}$$

where n is the direction normal to the surface separating two regions with thermal conductivity λ_1 and λ_2.

4.2.2 Steady-state heat equation in one dimension

When temperature (and thermal properties) only vary in one direction, the heat equation is an ordinary differential equation:

$$\frac{d}{dz}\left(\lambda \frac{dT}{dz}\right) = -H. \tag{4.13}$$

This equation can be solved within a layer $a < z < b$, provided conditions are specified at both limits, or in a half-space $z > 0$ if both temperature and its gradient (or heat flux) are specified at the boundary $z = 0$. This equation will be used to estimate the geotherm (i.e. how temperature varies with depth) in the lithosphere of stable continents. Denoting the vertical heat flux by q, the heat equation can also be written as

$$-\frac{dq}{dz} + H = 0. \tag{4.14}$$

Note that the heat flux is negative.

For a layer $a < z < b$ with constant heat production, we may integrate to obtain

$$q(a) = q(b) + \int_a^b H dz = q(b) - H(b-a), \tag{4.15}$$

showing that the heat flux at the top of a radiogenic layer is that at the base augmented by heat production within the layer.

4.2.3 Steady-state heat equation with sources in a half-space

For thermal studies of the Earth's crust, one often deals with a heterogeneous distribution of heat production, with, for example, enriched granite bodies embedded in depleted gneissic or sedimentary rocks. In this case, one can obtain the temperature distribution by considering a sum of heat sources. In contrast to gravity studies for example, the temperature field must be solved for a half-space with fixed temperature at the boundary. The boundary temperature can be set to zero without loss of generality (see the superposition principle) and the boundary condition can be satisfied by considering heat sources in a mirror arrangement on the other side of the boundary.

Point source in full space

We first solve for the temperature field due to a point heat source in infinite space. The source is generating heat at a steady rate H and is located at the origin. The temperature field depends only on the radial distance to the source, r. In steady state, the amount of heat flowing across a sphere of any radius r is H:

$$4\pi r^2 q(r) = H, \tag{4.16}$$

where $q(r) = -\lambda dT/dr$ is the radial heat flux at distance r. From this, assuming that the temperature is 0 at ∞, we obtain by integration

$$T(r) = \frac{H}{4\pi \lambda r}. \tag{4.17}$$

Note that for uniformly distributed heat sources within a sphere, or for any spherically symmetric distribution of heat source, the heat flux outside the sphere is identical to that of all the sources concentrated at the center of the sphere, and is independent of the radial variation of the heat sources.

The general solution of the steady-state heat equation with sources (Poisson's equation 4.10) for a point source of strength one located at (x', y', z') is thus

$$T(x,y,z) = \frac{1}{4\pi\lambda\left[(x-x')^2+(y-y')^2+(z-z')^2\right]^{1/2}}. \tag{4.18}$$

This is the solution for a source $\delta(x-x')\delta(y-y')\delta(z-z')$, where $\delta(x)$ is Dirac's delta function (equation A.1). Equation 4.18 defines the full space Green's function (see Appendix B). We can use it to calculate the steady-state temperature distribution for any heat source distribution $h(x,y,z)$:

$$T(x,y,z) = \frac{1}{4\pi\lambda}\int_V \frac{h(x',y',z')\,dx'dy'dz'}{\left[(x-x')^2+(y-y')^2+(z-z')^2\right]^{1/2}}. \tag{4.19}$$

Solutions for the half-space

The solution for a point source in the half-space $z > 0$ with $T(x,y,z=0) = 0$ is obtained by the method of images, used for many problems in potential field theory. The solution is forced to vanish on the plane $z = 0$ by adding the effect of a point sink of heat (i.e. a negative source) at the mirror image point of the source with respect to the plane $z = 0$. We thus get the two half-space Green's functions:

$$T(x,y,z) = \frac{1}{4\pi\lambda}\left(\frac{1}{\left[(x-x')^2+(y-y')^2+(z-z')^2\right]^{1/2}} \mp \frac{1}{\left[(x-x')^2+(y-y')^2+(z+z')^2\right]^{1/2}}\right), \tag{4.20}$$

The heat flux across the plane $z = 0$ can be forced to 0, by adding the image of the source with respect to the plane $z = 0$.

Solutions for 2D problems

For 2D problems in (x,z), one can use an infinite line source in the y direction. The line source solution obtained by integrating point sources has a logarithmic singularity. We can follow the same reasoning as above using cylindrical coordinates. For a line heat source of strength H in full space located along an axis perpendicular to the plane (x,z), temperature depends only on the radial distance to the axis, r. The radial heat flux q_r and temperature can be calculated:

$$q_r = \frac{H}{2\pi r}$$
$$T(r) = \frac{1}{2\pi\lambda}\log(1/r). \tag{4.21}$$

Considering now the half-plane $z > 0$ with a point source located at (x',z') and $T = 0$ at $z = 0$, the solution obtained with the method of images is

$$T(x,z) = \frac{H}{2\pi\lambda}\left(\log\left[(x-x')^2 + (z+z')^2\right] - \log\left[(x-x')^2 + (z-z')^2\right]\right). \tag{4.22}$$

This can be integrated to obtain the temperature field due to heat production in a body of arbitrary shape.

From these solutions, one can derive compact solutions for the surface heat flux (at $z = 0$). For a line source located at $(0,a)$, the vertical heat flux is

$$q_z = \frac{ha}{\pi r^2}, \tag{4.23}$$

where $r = \sqrt{x^2 + a^2}$. The heat flux due to a infinite cylinder of radius b with heat production H at depth a below the origin is

$$q_z = \frac{Hb^2 a}{(a^2 + x^2)}. \tag{4.24}$$

For heat sources with density h concentrated on the half-plane $x < 0$ at depth $z = a$, the heat flux across $z = 0$ is

$$q_z = h\left(\frac{1}{2} - \frac{1}{\pi}\tan^{-1}(x/a)\right). \tag{4.25}$$

This shows that the heat flux varies over a characteristic distance a on either side of the discontinuity. For problems of this kind, one can use the vast literature on gravity anomalies. As mentioned above, the vertical heat flux is analogous to the vertical component of the gravity anomaly field with one caveat: the vertical heat flux at the surface $z = 0$ is double that of the corresponding gravity source. For two-dimensional bodies with polygonal shapes, the heat flux can be calculated using the formulas of Talwani *et al.* (1959).

4.2.4 Steady-state heat equation for the sphere

In a sphere, assuming spherical symmetry of the temperature (i.e. the temperature varies only with distance to the center of the sphere, r), Poisson's equation becomes

$$\frac{1}{r^2}\frac{d}{dr}\left(r^2\frac{dT}{dr}\right) = -\frac{H}{\lambda}. \tag{4.26}$$

The general solution of this equation is:

$$T(r) = \frac{-Hr^2}{6\lambda} - \frac{2C_1}{r^3} + C_2, \tag{4.27}$$

where C_1 and C_2 are two constants determined by the boundary conditions.

Sphere with uniform heat production

For the temperature to be finite at the center of the sphere, we must have $C_1 = 0$. If the temperature is 0 at $r = a$, then

$$T(r) = \frac{H}{6\lambda}(a^2 - r^2). \tag{4.28}$$

The surface heat flux is $q_a = Ha/3\lambda$, the total flux on the surface of the sphere is $4\pi a^3 H/3$ and the temperature at the center is $Ha^2/6\lambda$.

Spherical shell with constant heat production

Within the shell, $b < r < a$, heat production is constant. If $T(r = a) = 0$, and $dT/dr(r = b) = 0$, the temperature in the shell is

$$T(r) = \frac{H}{6\lambda}(a^2 - r^2) + \frac{Hb^5}{9\lambda}\left(\frac{1}{a^3} - \frac{1}{r^3}\right). \tag{4.29}$$

In the central part of the sphere ($r < b$), the temperature is constant.

4.2.5 Variations in thermal conductivity

Rocks are poly-crystalline assemblages and geological environments are often made of juxtaposed rocks. Thermal conductivity values vary on many different scales, from those of individual minerals to those of geological formations within a province. Thus, rather than solving for temperature in each and every individual constituent, one deals with a restricted set of materials with average thermal properties.

Layered system

Consider first two layers of thickness h_1 and h_2 with conductivities λ_1 and λ_2 that are heated from below with heat flux Q. The temperature drops across the two layers are ΔT_1 and ΔT_2, respectively. In steady state, the heat flux through each layer is Q, so that

$$Q = -\lambda_i \frac{\Delta T_i}{h_i}. \tag{4.30}$$

The average medium has thickness $h = h_1 + h_2$ and temperature drop ΔT. Thus, its conductivity $\bar{\lambda}$ is such that $Q = -\bar{\lambda}\Delta T/h$. Writing that $\Delta T = \Delta T_1 + \Delta T_2$, one gets,

$$\frac{1}{\bar{\lambda}} = \frac{h_1/h}{\lambda_1} + \frac{h_2/h}{\lambda_2}. \tag{4.31}$$

This can be generalized for an arbitrary vertical distribution of conductivity defined by some function $\lambda(z)$:

$$\begin{aligned} q(z) &= -\lambda(z)\frac{dT}{dz} \\ T(h) - T(0) &= -q\int_0^h \frac{dz}{\lambda(z)}, \end{aligned} \tag{4.32}$$

so that

$$\frac{1}{\bar{\lambda}} = \frac{1}{h}\int_0^h \frac{dz}{\lambda(z)}. \tag{4.33}$$

It is useful to rewrite equation 4.31 using the volume fraction of each material in the whole layer. In that example, the volume fraction of material (i) is $(\phi_i) = h_i/h$, and the equivalent conductivity is given by

$$\frac{1}{\bar{\lambda}} = \sum_i \frac{(\phi_i)}{\lambda_i}. \tag{4.34}$$

This result is valid if materials (1) and (2) are in fact distributed in any number of intercalated layers with arbitrary thicknesses. This can be demonstrated using equation 4.33. For a stratified system, therefore, the equivalent thermal conductivity does not depend on the details of the layering and only depends on the volume fraction of each material in the bulk.

For a layered system, equation 4.33 is used to define the thermal resistance to a given depth h:

$$R(h) = \int_0^h \frac{dz}{\lambda(z)} \tag{4.35}$$

such that, without heat sources, the temperature difference across a layer is equal to the thermal resistance times the heat flux.

Bounds on the average thermal conductivity of a mixture

We now consider a layer of thickness h made of individual blocks of different materials. Each block with conductivity λ_i has area A_i and the layer covers a total area A. Applying a temperature difference ΔT between the top and bottom of the layer, each block carries heat flux $q_i = -\lambda_i \Delta T/h$. The total heat flux $Q = \sum_i A_i q_i = -\sum_i A_i \lambda_i \Delta T/h$. The average heat flux across the layer is therefore the weighted average of the individual heat fluxes through each individual strip, such that $\bar{q} = \sum_i (A_i/A) q_i$. The equivalent conductivity of the layer is therefore

$$\bar{\lambda} = \sum_i \frac{A_i}{A} \lambda_i = \sum_i \phi_i \lambda_i, \tag{4.36}$$

where ϕ_i is the volume fraction of material with conductivity λ_i. It is obvious that this result is valid for any distribution of blocks, with, for example, material (1) spread over any number of units of different widths.

Equations 4.34 and 4.36 correspond to the series and parallel formulae for electrical resistances. Their usefulness is not restricted to such particular geometrical arrangements as they provide absolute bounds for the average conductivity of any material mixture. The average of the two bounds is within a few percent of the true average conductivity in most cases and has often been used. More restrictive bounds for the effective thermal conductivity of mixtures have been derived and are discussed in Appendix D.

4.3 Diffusive heat transport: Basic principles

Here, we illustrate how heat propagates by diffusion. We show how one can derive scaling laws for temperature and how one can deduce from those the mathematical form of solutions to the heat equation. We shall also discuss the key influence of the boundary condition. All problems discussed here involve the same heat equation, and hence the different behaviors and solutions are specified by the boundary and initial conditions. Specifying a boundary condition is a key step in formulating a problem and one should pay attention to its relevance to the true physical situation that is dealt with. In many cases, setting a boundary condition glosses over the complex physics of heat transport between two systems on either side of that boundary.

4.3.1 Heating a half-space from its surface

Here, we consider a change of thermal conditions on $z = 0$, the upper boundary of a conductive half-space.

Sudden change in temperature

The region $z > 0$ is initially at a uniform temperature T_o and the surface is set at temperature $T_o + \Delta T$ at time $t = 0$. The problem can be solved for $T_o = 0$ without loss of generality. In this problem, there are no externally defined length scales or time scales. However, we can guess by dimensional argument that the thickness δ that has been heated after time t is related to time through $\delta \propto \sqrt{\kappa t}$. To demonstrate this, we seek an approximate solution to the heat equation using what is called the "integral" method. This method relies on the bulk heat balance over the whole material volume and on a "trial" function for the temperature profile $T(z,t)$:

$$\frac{d}{dt}\left(\int_0^{\infty} \rho C_p T dz\right) = q(0,t) = -\lambda \frac{\partial T}{\partial z}(0,t). \tag{4.37}$$

At time t, material is heated over thickness $\delta(t)$ such that temperature varies from ΔT at $z = 0$ to 0 at $z = \delta$. The simplest profile that meets these requirements is obviously

$$\begin{aligned} &T(z,t) = \Delta T\left(1 - \frac{z}{\delta}\right), \text{ for } 0 \leq z \leq \delta \\ &T(z,t) = 0 \text{ for } z \geq \delta. \end{aligned} \tag{4.38}$$

Using this, the bulk heat balance leads to

$$\frac{d}{dt}\left(\frac{1}{2}\rho C_p \Delta T \delta\right) = \lambda \frac{\Delta T}{\delta}, \tag{4.39}$$

which yields a differential equation for $\delta(t)$:

$$\delta \frac{d\delta}{dt} = 2\kappa, \tag{4.40}$$

so that $\delta = 2\sqrt{\kappa t}$. We shall see that this result is only accurate to about 10%, due to the rather simplistic "trial" temperature profile. One can obtain a better approximation using a more realistic profile, using for example a second-degree polynomial. For our present purposes, we note that this simple result illustrates one basic property of heat diffusion: heat does not propagate at a constant velocity but at a rate which decreases as $t^{-1/2}$ instead. $\delta = 2\sqrt{\kappa t}$ has been called the "diffusion length" and shows up in many different situations, as will be illustrated below.

We now seek an exact solution to the same problem, which can be obtained by reducing the problem to a single variable. This problem involves no external length scale and time scale and is such that distance and time are linked to one another. Thus, we seek a similarity solution in terms of variable $\eta = f(z,t)$ which depends on both z and t, so that $T(z,t) = \Delta T \theta(\eta)$. Dimensional analysis suggests using the variable $\eta = z/\delta = z/2\sqrt{\kappa t}$. The temperature profile used in the integral method had already been written in this form.

Substituting into the heat equation, one finds,

$$-2\eta\dot{\theta} = \ddot{\theta}, \tag{4.41}$$

with the following boundary conditions:

$$\theta(0) = 1 \text{ and } \lim_{\eta\to\infty} \theta = 0. \tag{4.42}$$

The well-known solution is

$$\theta = \operatorname{erfc}(\eta) = 1 - \operatorname{erf}(\eta) = 1 - \frac{2}{\sqrt{\pi}} \int_0^\eta \exp(-u^2) du, \tag{4.43}$$

which introduces the "error function" $\operatorname{erf}(\eta)$ and "complementary error function" $\operatorname{erfc}(\eta) = 1 - \operatorname{erf}(\eta)$. These functions derive their names from their extensive use in Gaussian statistics.

The solution is displayed in Figure 4.1. Note that all the profiles are similar one to another, with depth scaling to $t^{1/2}$. This equation was used by Kelvin when he calculated the age of the Earth. It has been used in many other applications. As will be seen later, it can be used for calculating the cooling of the oceanic lithosphere. The correction to heat flux measurements in regions that were glaciated is based on this equation (see Appendix C on measurements). The same equation is also used to determine recent changes in ground surface temperature from perturbations of the shallow part of the temperature profile measured in boreholes. This topic is covered in detail in Chapter 12.

One can compare the exact solution with the approximate one derived from the integral method:

$$-\lambda \frac{\partial T}{\partial z}(0,t) = \lambda \frac{\Delta T}{2\sqrt{\kappa t}}, \text{ for the integral method} \tag{4.44}$$

$$= \lambda \frac{\Delta T}{\sqrt{\pi \kappa t}}, \text{ for the exact solution.} \tag{4.45}$$

Two important points are illustrated here. The "integral method" solution does lead to a solution with the correct dimensions and provides a reasonable approximation

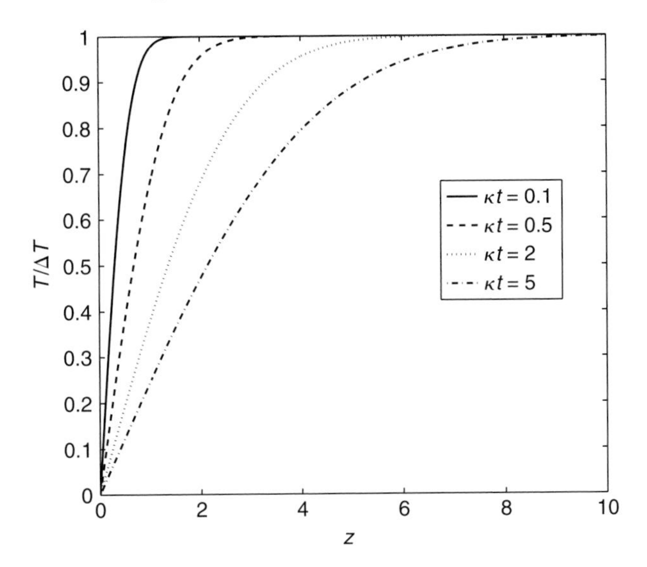

Figure 4.1. Temperature profiles in a half-space at different times after it starts cooling. Initially the temperature is uniformly ΔT.

to the exact solution. In the present problem, the exact solution is quite straightforward, and the advantage of the "integral method" may seem slight. In more complicated problems, however, the approximate method can be very handy. This problem can also be solved using the Laplace transform technique, which will be abundantly discussed in the "advanced" section below.

Contact between two materials with different thermal properties

We have described above the sudden heating of a half-space and have assumed that the surface temperature can be set instantaneously at some value $T_o + \Delta T$. In reality, heat is extracted out of some other system and one should solve for the exchange of heat between two media. In this section, we consider that heat is brought from an overlying layer by diffusion. One geological example is the condition at the base of an ice sheet over a continent.

Specifically, we consider two media with different properties lying on either side of the plane $z = 0$. Regions (1) and (2) with properties λ_1, ρ_1, C_1 and λ_2, ρ_2, C_2, respectively, extend in the $z > 0$ and $z < 0$ directions respectively. The two regions are initially at different temperatures, such that,

$$T_1(z,t) = T_{01} \quad \text{for z} > 0; \quad T_2(z,t) = T_{02} \quad \text{for z} < 0. \tag{4.46}$$

The full solution is developed in the next paragraph but a simple argument allows calculation of the contact temperature. Over time t, the two regions develop thermal boundary layers as heat is exchanged between them. We know that heat propagates over a distance $\delta \approx 2\sqrt{\kappa_n t}$ in both regions, where $n = 1, 2$, depending on the layer. The contact temperature T_i must be such that the two boundary layers have the same total energy before and after they equilibrate:

$$\rho_1 C_1 2\sqrt{\kappa_1 t}\left(\frac{T_i + T_{01}}{2}\right) + \rho_2 C_2 2\sqrt{\kappa_2 t}\left(\frac{T_i + T_{02}}{2}\right) \\ = \rho_1 C_1 2\sqrt{\kappa_1 t}\, T_{01} + \rho_2 C_2 2\sqrt{\kappa_2 t}\, T_{02}. \tag{4.47}$$

We note that t cancels out of this equation and hence the contact temperature is constant. From equation 4.47, we obtain,

$$T_i = \frac{\sqrt{\lambda_1 \rho_1 C_1}\, T_{01} + \sqrt{\lambda_2 \rho_2 C_2}\, T_{02}}{\sqrt{\lambda_1 \rho_1 C_1} + \sqrt{\lambda_2 \rho_2 C_2}}. \tag{4.48}$$

From this, we deduce that $T_i \approx T_{01}$ if $\sqrt{\lambda_1} \gg \sqrt{\lambda_2}$, illustrating the common-sense principle that the most conductive material imposes its temperature. In most cases, however, one has $\lambda_1 \rho_1 C_1 \approx \lambda_2 \rho_2 C_2$, and $T_i = (T_{01} + T_{02})/2$. This is the situation for the contact between rock and ice at the base of a glacier.

To derive the complete solution, we seek solutions in both regions in the form $T_n(z,t) = A_n + B_n \text{erf}\,(z/2\sqrt{\kappa_n t}\,)$ with the appropriate subscript in each region. These two functions are solutions to the heat equation in their respective regions and the constants are obtained from the boundary conditions (temperature and heat flux are continuous at the interface $z = 0$) and from the initial conditions defined above. One finds indeed that the interface temperature is given by equation 4.48 and that temperature in the two regions is given by,

$$T_1(z,t) = T_i + \frac{\sqrt{\rho_2 C_2 \lambda_2}}{\sqrt{\rho_1 C_1 \lambda_1} + \sqrt{\rho_2 C_2 \lambda_2}}(T_{01} - T_{02})\text{erf}\left(\frac{z}{2\sqrt{\kappa_1 t}}\right), \quad z > 0 \\ T_2(z,t) = T_i + \frac{\sqrt{\rho_1 C_1 \lambda_1}}{\sqrt{\rho_1 C_1 \lambda_1} + \sqrt{\rho_2 C_2 \lambda_2}}(T_{02} - T_{01})\text{erf}\left(\frac{z}{2\sqrt{\kappa_2 t}}\right), \quad z < 0. \tag{4.49}$$

The air–ground interface

When very hot material is deposited at the surface of the Earth (for instance a lava flow), its surface temperature does not equilibrate instantly with air temperature. Energy is lost at the surface by radiation. The radiative heat flux is determined by Stefan–Boltzmann law that the flux is proportional to the fourth power of thermodynamic temperature (equation 3.4). The Earth's surface also absorbs energy received from solar radiation, this flux of energy corresponds to a black body temperature T_0. Averaged over the daily cycle, this temperature is slightly less than that of the air. The net heat loss at the Earth's surface is thus

$$\Phi = \sigma \times (T^4 - T_0^4), \tag{4.50}$$

where T is the surface temperature and T_0 is the black body temperature of the incoming radiation. This flux must be balanced by the conductive heat flux inside the hot layer. The surface temperature will adjust to the heat loss as follows:

$$T(t) = T(t=0) + \int_0^t \Phi(t')G_q(t-t')dt'. \tag{4.51}$$

The specific form of the function G_q will be obtained later. This equation appears as a convolution, but it is a non-linear integral equation that must be solved numerically. Neglecting other effects, such as the latent heat released by the freezing lava, one finds that $T(t) \rightarrow T_0$ in a fairly short time (typically less than a day for a lava at 1500 K).

Temperature pulse on a plane

At time $t = 0$, a finite amount of energy ΔE is released per unit area at the surface of the region $z = 0$, where the initial temperature is 0. Thereafter, no heat is brought or lost through that boundary. At time t, material has been heated over distance $\delta \propto \sqrt{\kappa t}$, so that the temperature scale Θ is such that:

$$\begin{aligned} \rho C_p \Theta \delta &\propto \Delta E \\ \Theta &\propto \frac{\Delta E}{\rho C_p \sqrt{\kappa t}}. \end{aligned} \tag{4.52}$$

This specifies, to an unknown constant of order one, how the surface temperature evolves. To determine the full temperature distribution $T(z,t)$, we seek a similarity solution and we know that the similarity variable must again be $\eta = z/\left(2\sqrt{\kappa t}\right)$. From the argument above, we may write temperature as follows:

$$T(z,t) = \frac{\Delta E}{\rho C_p \sqrt{\kappa t}} f\left(\frac{z}{2\sqrt{\kappa t}}\right), \tag{4.53}$$

where f is the function to be determined. Substituting into the heat equation and requiring that

$$\begin{aligned} \int_0^\infty \rho C_p T(z,t)dz &= \Delta E \\ \frac{\partial T(0,t)}{\partial z} &= 0, \end{aligned} \tag{4.54}$$

we find,

$$T(z,t) = \frac{\Delta E}{\rho C_p \sqrt{\pi \kappa t}} \exp\left(-\frac{z^2}{4\kappa t}\right). \tag{4.55}$$

This equation defines the Green's function for a heat source on the plane $z = 0$. The solution for any time varying heat source on the plane $z = 0$ can be obtained by superposition of solutions for each time. It gives a convolution (see Appendix B):

$$T(z,t) = \int_0^t \frac{\frac{\Delta E(t')}{\rho C_p}}{\sqrt{\pi \kappa (t-t')}} \exp\left(-\frac{z^2}{4\kappa (t-t')}\right) dt'. \tag{4.56}$$

Solution 4.55 illustrates one serious physical problem raised by the heat equation. For any time $\tau > 0, T(z,\tau) > 0$, regardless of how large is z; in other words, it implies that heat propagates with ∞ velocity (e.g. Sommerfeld, 1949). Although the heat conduction equation accounts perfectly for the diffusion of heat, it is not consistent with basic physical principles. Several ways have been proposed to resolve this inconsistency, but they result in corrections that are too small to be observed.

Temperature pulse at the surface of the half-space

At time $t = 0$, the surface of the region $z > 0$, where the initial temperature is 0, is brought to temperature ΔT, thereafter, it is kept at temperature 0. The problem can be treated like the previous one, but the surface boundary condition for $t > 0$ is now,

$$T(0,t) = 0 \tag{4.57}$$

We find,

$$T(z,t) = \frac{\Delta T}{\sqrt{\pi \kappa t^3}} \exp\left(-\frac{z^2}{4\kappa t}\right). \tag{4.58}$$

This function is the Green's function for the temperature boundary condition, i.e. the response for a temperature pulse $\Delta T \delta(t)$ at the surface $z = 0$ (Figure 4.2). It can be used to determine the temperature in the region $z > 0$ for any temperature variation $\Delta T(t)$ on the surface $z = 0$. The temperature in the region $z > 0$ is obtained by "convolution" of the Green's function with the time-varying surface temperature $\Delta T(t)$:

$$T(z,t) = \int_0^t \frac{\Delta T(t') z}{\sqrt{\pi \kappa (t-t')^3}} \exp\left(-\frac{z^2}{4\kappa (t-t')}\right) dt'. \tag{4.59}$$

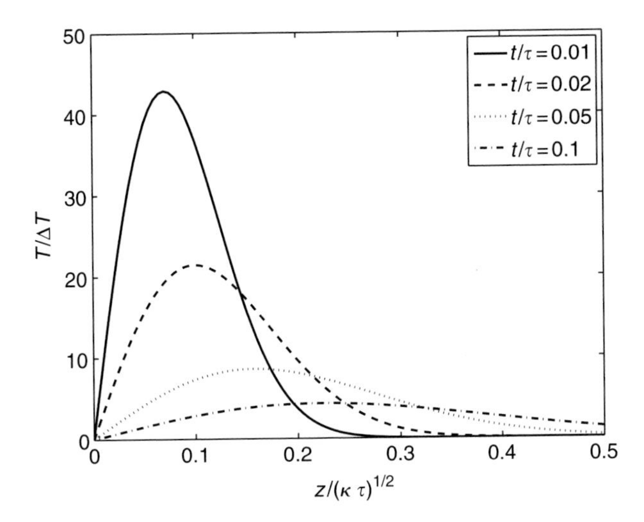

Figure 4.2. Propagation of a temperature pulse from the surface of a half-space. The pulse widens as it moves downward.

Fixed heat flux at the boundary

Starting with an isothermal region, heat flux Q is applied at the surface $z = 0$ starting at time $t = 0$ and is sustained. The corresponding boundary condition is,

$$\lambda \frac{\partial T}{\partial z}(0,t) = Q \text{ for } t \geq 0. \tag{4.60}$$

Using the same argument as before, we can obtain a scale for the temperature at the boundary Θ:

$$\rho C_p \Theta \sqrt{\kappa t} \propto Qt, \tag{4.61}$$

which specifies how the surface temperature $T(0,t)$ evolves. The full solution can then be obtained using the same method as before,

$$T(z,t) = \frac{Q}{\lambda}\left(2\sqrt{\frac{\kappa t}{\pi}}\exp\left(-\frac{z^2}{4\kappa t}\right) - z\,\mathrm{erfc}\left(\frac{z}{2\sqrt{\kappa t}}\right)\right), \tag{4.62}$$

where erfc stands for the the complementary error function which has already been defined.

This solution can be obtained by many other methods. One of them is to consider that the heat flux corresponds to the release of $dE = Qdt'$ between times t' and $t' + dt'$, and to calculate the convolution (4.56) for a pulse of energy:

$$T(z,t) = \int_0^t \frac{Q}{\rho C_p \sqrt{\pi \kappa (t-t')}} \exp\left(-\frac{z^2}{4\kappa (t-t')}\right) dt'. \tag{4.63}$$

A simpler method is to solve directly for the flux of heat, $q = -\lambda \partial T/\partial z$. If we differentiate the heat equation with respect to z, we see that the heat flux q obeys the same equation,

$$\frac{\partial q}{\partial t} = \kappa \frac{\partial^2 q}{\partial z^2}. \tag{4.64}$$

For this problem, the boundary and initial conditions are,

$$q(0,t) = Q; \; q(z,0) = 0, \tag{4.65}$$

and the solution is the same as that for temperature with a step-change at the surface at time $t = 0$ (equation 4.43):

$$q(z,t) = Q \, \mathrm{erfc}\left(\frac{z}{2\sqrt{\kappa t}}\right). \tag{4.66}$$

One then integrates this equation with respect to z to find the solution (4.62) as above.

The methods outlined above can be generalized for a surface heat flux that varies as a function of time, i.e. $Q(t)$. For power-law functions of time, i.e. $Q(t) \propto t^{\alpha}$, for example, one readily finds that $T(0,t) \propto t^{\alpha+1/2}$.

4.3.2 Point source

Heat pulse

A finite amount of energy ΔE is released at point O in a very large region. Heat propagates away from the point source and leads to elevated temperatures in a sphere of radius $R(t)$ at time t. We know from the characteristics of diffusion that $R \propto \sqrt{\kappa t}$ and hence that temperature scales with Θ such that,

$$\rho C_p \Theta R^3 = \rho C_P \Theta \, (\kappa t)^{3/2} = \Delta E, \tag{4.67}$$

which shows for example that the temperature at the source decays as $t^{-3/2}$. From this, we seek a similarity solution to the heat equation in the form,

$$T(r,t) = \frac{\Delta E}{\rho C_p} \frac{1}{(\kappa t)^{3/2}} f\left(\frac{r}{2\sqrt{\kappa t}}\right), \tag{4.68}$$

where f is some function to be determined. Substitution into the heat equation leads to,

$$T(r,t) = \frac{\Delta E}{\rho C_p} \frac{1}{8\,(\pi \kappa t)^{3/2}} \exp\left(-\frac{r^2}{4\kappa t}\right). \tag{4.69}$$

It is straightforward to verify that energy is conserved at all times as,

$$\Delta E = \rho C_p 4\pi \int_0^\infty T(r,t) r^2 dr. \tag{4.70}$$

This solution corresponds to a temperature pulse or Dirac delta function as:

$$\begin{aligned} &\lim_{t\to 0} T(r,t) = 0, \quad r \neq 0 \\ &4\pi \int r^2 T(r,t) dr = 1. \end{aligned} \tag{4.71}$$

Steady heat release

Heat is now released at constant rate P at point O. The solution can be obtained by integrating the previous solution, i.e. by considering that an amount of heat equal to $dE = Pdt'$ is released between times t' and $t' + dt'$. Thus, at time t, temperature is,

$$\begin{aligned} T(r,t) &= \int_0^t \frac{P}{8\rho C_p [\pi\kappa(t-t')]^{3/2}} \exp\left(\frac{-r^2}{4\kappa(t-t')}\right) dt' \\ T(r,t) &= \frac{P}{4\pi\lambda r} \operatorname{erfc}\left(\frac{r}{2\sqrt{\kappa t}}\right). \end{aligned} \tag{4.72}$$

At any time t, we must have

$$4\pi\rho C_p \int_0^\infty r^2 T(r,t) dr = Pt. \tag{4.73}$$

If we let $t \to \infty$, we obtain the steady-state temperature distribution for a constant heat source. Using the property that $\operatorname{erfc}(\eta) \to 1$ as $\eta \to 0$, we obtain,

$$T(r) = \frac{P}{4\pi\lambda r}, \tag{4.74}$$

which has already been derived above (equation 4.17) using a different, and simpler, method.

4.3.3 Damping of thermal fluctuations

Heat diffusion acts to smooth out temperature differences with two consequences: spatial variations are progressively erased and temporal variations that are generated at some boundary get damped away from that boundary. The key point is that the magnitude of damping depends on distance and time: short-wavelength fluctuations are rapidly smoothed out and high-frequency time variations are not detectable even at short distances. We specify below what is meant by "rapid" and "short distances".

Spatial variations

We consider for example a very long cylindrical bar which is thermally insulated, so that heat can propagate only along the axis. The initial temperature distribution is known and for $t > 0$ will evolve by diffusion. If the length of the bar is much longer than its radius, the initial temperature distribution depends on distance only. For this problem, the heat equation can be solved in one dimension over a domain that extends to $z \to \pm\infty$.

We must find a solution to the 1D heat equation,

$$\frac{\partial T}{\partial t} = \kappa \frac{\partial^2 T}{\partial z^2}. \tag{4.75}$$

Let us consider the initial condition:

$$T(z,t=0) = T_0 \cos\{kz\} = T_0 \cos\left\{\frac{2\pi z}{\lambda}\right\}, \tag{4.76}$$

where k and λ are the wave-number and wavelength of the initial temperature fluctuation, respectively. This may be taken as one component of the Fourier transform of a more complicated temperature distribution (see Appendix A, for a more extensive development). The initial condition introduces a length scale, λ, and a time scale $\tau = \lambda^2/\kappa$. We may therefore surmise that fluctuations over wavelength λ decay to negligible values over a time which is long compared to τ.

To derive the full solution to the heat equation, we note that the initial condition implies that the two variables z and t can be separated. We therefore look for a function of the form $T(z,t) = T_0 \cos\{kz\} f(t)$.

$$\frac{\partial f}{\partial t} = -\kappa k^2 f(t). \tag{4.77}$$

This yields the solution,

$$T(z,t) = T_0 \cos\{kz\} \exp\{-\kappa k^2 t\} = T_0 \cos\left\{\frac{2\pi z}{\lambda}\right\} \exp\left\{-\frac{4\pi^2 \kappa t}{\lambda^2}\right\}. \tag{4.78}$$

This shows that the temperature fluctuation is damped exponentially with time. Damping depends on the wavelength λ (i.e. the length scale) and the relaxation time increases as $\tau_R = \lambda^2/4\pi^2\kappa$. In practice, at time t, fluctuations over wavelengths that are shorter than $\lambda_c \approx 4\pi\sqrt{\kappa t}$ have been smoothed out.

An important consequence of such damping is that the initial conditions are progressively erased and it is therefore impossible to determine from the temperature distribution at a given time what was the initial temperature distribution. The solution of the heat equation backward in time is unstable because short-wavelength components are exponentially amplified.

Time variations

Consider that, at the surface of half-space, temperature varies periodically:

$$T(z=0,t) = T_0 \exp\{i\omega t\} = T_0 \exp\left\{\frac{i2\pi t}{\tau}\right\}, \tag{4.79}$$

where ω and τ are the angular frequency and period, respectively. This introduces a time scale, τ, and hence a length scale, $\sqrt{\kappa\tau}$. We thus expect that the temperature oscillation is undetectable at a distance that scales with $\sqrt{\kappa\tau}$. When using complex exponential rather than trigonometric functions, we retain only the real part of the complex function.

Because of the boundary condition, we seek a solution through the separation of variables: $T(z,t) = \exp(i\omega t) f(z)$. Substituting into the heat equation, we find,

$$f(z) = A\exp\left(\sqrt{\frac{i\omega}{\kappa}}z\right) + B\exp\left(-\sqrt{\frac{i\omega}{\kappa}}z\right), \tag{4.80}$$

with $\sqrt{i} = (1+i)/\sqrt{2}$. Because the solution must remain finite, $A = 0$, and the surface boundary condition gives $B = T_0$.

$$T(z,t) = T_0 \exp\left(-\sqrt{\frac{\omega}{2\kappa}}z\right) \exp\left(i\left(\omega t - \sqrt{\frac{\omega}{2\kappa}}z\right)\right). \tag{4.81}$$

The first term in equation 4.81 thus shows that the amplitude of periodic oscillations is damped exponentially with depth. The "skin depth" $\delta = \sqrt{2\kappa/\omega} = \sqrt{\kappa\tau/\pi}$ is the depth where the amplitude is reduced by $1/e$. The second term represents a wave that propagates downward with frequency-dependent velocity $v = \sqrt{2\omega\kappa} = 2\sqrt{\kappa\pi/\tau}$. The wave exhibits dispersion, as each frequency component travels at a different velocity. This is why a sharp temperature pulse at the surface spreads over an increasingly wide distance as it propagates downward. The wavelength of the temperature wave $\lambda = 2\pi\sqrt{2\kappa/\omega} = 2\sqrt{\pi\kappa\tau}$. At depth $z = \sqrt{\pi\kappa\tau}$, the temperature oscillation has the sign opposite to that of the surface temperature.

This calculation can be used to discuss how the ground responds to changes of surface temperature. For example, Figure 4.3 shows how the effect of the annual temperature cycle is attenuated and its phase lags further behind that of the surface temperature with increasing depth.

We have carried out the calculation for a half-space whereas no real physical system is infinitely large. The calculation, however, specifies the domain of validity: if the true physical system extends over distance L, the above results are valid only if $L \gg \delta$. If this condition is not fulfilled, the thermal response depends on conditions at $z = L$. We provide some relevant solutions in the "advanced" section.

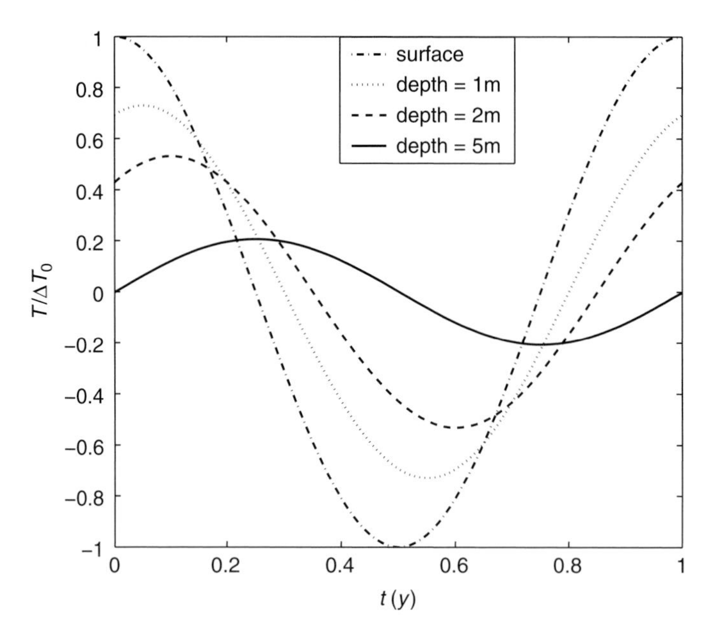

Figure 4.3. Temperature variations at 1, 3 and 5m due to the annual cycle of temperature variations at the surface. The calculations use a standard value $\kappa = 10^{-6}\ \mathrm{m}^2\ \mathrm{s}^{-1}$.

4.4 General solutions to the steady-state heat equation

In the next sections, we shall introduce the standard techniques that are used to obtain solutions to the heat equation. They are based on Fourier series and transforms that are most commonly used to obtain solutions to the steady-state heat equation in Cartesian coordinate systems, and on the Laplace transform, which is most useful in finding transient solutions to initial value problems. Both methods are not exclusive of each other; many initial value problems that can be solved by decomposing the initial temperature distribution in Fourier series and boundary value problems for Laplace's and Poisson's equations in a half-space can also be solved with Laplace transforms. The purpose is to illustrate these methods with examples that are relevant to the geosciences. Many more solutions to the heat conduction equation can be found in Carslaw and Jaeger (1959).

4.4.1 Laterally varying boundary conditions

Here, we do not consider heat sources or sinks and search for solutions of Laplace's equation. In steady state, complex temperature fields can be generated in the interior of a body by heterogeneous boundary conditions. In practice, we consider that the distribution of temperature or heat flux is specified at a boundary. When the boundary is a plane, it is possible to decompose the boundary condition into its Fourier components with the Fourier transform (e.g. Sneddon, 1950):

$$\tilde{f}(k) = \int_{-\infty}^{\infty} f(x) \exp(ikx) dx$$
$$f(x) = \frac{1}{2\pi} \int_{-\infty}^{\infty} \tilde{f}(k) \exp(-ikx) dx, \tag{4.82}$$

where $\tilde{f}(k)$ is the Fourier transform of $f(x)$. The transform variable k is usually referred to as the wave-number. It is related to the wavelength λ by $k = 2\pi/\lambda$. Any function that is sufficiently regular can be decomposed along its individual wave-numbers with the Fourier transform. These equations can be generalized to two or three dimensions, by doing successive Fourier transforms in two or three perpendicular directions. The wave-numbers in each direction are the components of a wave-vector $\vec{k} = (k_x, k_y)$.

Downward continuation of surface temperature variations

We show in Section B.1.4 of Appendix B how to use the theory of analytic functions in the complex plane to solve many 2D problems. If temperature (or heat flux) is known on the line $y = 0$ for instance, the temperature can be calculated for any $y > 0$ (see Appendix A). Here, we use the Fourier transforms, which are easier to calculate numerically than convolution equations. In the following sections, we shall drop the $\tilde{}$ in the notation for the transform when no confusion is possible.

We consider that the surface temperature varies horizontally on the plane $z = 0$:

$$T_0(x,y) = \frac{1}{(2\pi)^2} \int_{-\infty}^{\infty} dk_x \int_{-\infty}^{\infty} dk_y \tilde{T}_0(\vec{k}) \exp(-i(k_x x + k_y y)). \tag{4.83}$$

Writing the solution in the form of an inverse Fourier transform, we obtain the Fourier transform of Laplace's equation,

$$\frac{\partial^2 \tilde{T}}{\partial z^2} - k^2 \tilde{T} = 0, \tag{4.84}$$

with $k = \sqrt{k_x^2 + k_y^2}$. The general solution is $\tilde{T}(\vec{k}, z) = T^{\pm}(\vec{k}) \exp(\pm kz)$. Because the solution must remain finite for $z \to \infty$, the temperature will be written as,

$$\tilde{T}(\vec{k}, z) = \tilde{T}_0(\vec{k}) \exp(-kz). \tag{4.85}$$

This equation shows that surface variations in temperature are attenuated exponentially when they are downward continued. The attenuation is proportional to the wave-number k, i.e. inversely proportional to the wavelength. This is a standard property of potential fields. For the temperature field, this attenuation is the result of lateral diffusion of heat. Such solutions are useful for calculating the shallow temperature field in the Earth's crust when the surface is not homogeneous, such as areas around a shallow lake, for example. They can also be used to calculate disturbances to the surface heat flux due to topography.

In many situations, the surface temperature field is radially symmetric. In cylindrical coordinates r, z, θ, the heat equation is,

$$\frac{1}{r}\frac{\partial}{\partial r}r\frac{\partial T}{\partial r} + \frac{\partial^2 T}{\partial z^2} + \frac{1}{r^2}\frac{\partial^2 T}{\partial \theta^2} = 0. \tag{4.86}$$

If there is cylindrical symmetry $T(r,\theta,z=0) = T_0(r)$, the downward continuation can be done with the use of the Hankel (or Fourier–Bessel) transform,

$$T_0(r) = \int_0^\infty \tilde{T}_0(v) J_0(vr) v dv. \tag{4.87}$$

If there is cylindrical symmetry, the Hankel transform of T is the solution of,

$$\frac{\partial^2 \tilde{T}}{\partial z^2} - v^2 \tilde{T} = 0, \tag{4.88}$$

which is formally identical to equation 4.85. The Hankel and Fourier transforms of the downward continuation operator are identical:

$$\tilde{T}(v,z) = \exp(-vz)\tilde{T}_0(v). \tag{4.89}$$

Note that there is always such a correspondence between the Fourier transform in 2D and the Hankel transform in cylindrical geometry when the boundary conditions are identical.

Note that the conditions that $T(x,z=0) = 0$ and $q(x,z=0) \neq 0$ are inconsistent with Laplace equation for the half-space. They require heat sources in the half-space $z > 0$.

4.4.2 Steady-state heat equation in a layer

For a layer $0 < z < L$, with temperature at the surface $T(x,y,z=0) = 0$, the Fourier transform of the temperature field is,

$$T(\vec{k},z) = A(\vec{k}) \sinh(kz), \tag{4.90}$$

where $k = |\vec{k}|$, and $A(\vec{k})$ depends on the boundary condition at the base of the layer.

Varying heat flux at the base of a layer

If $q(x,y,z=L) = q_L(x,y)$, the Fourier transform of the temperature is obtained from the boundary condition at the base:

$$T(\vec{k},z) = \frac{q_L(\vec{k})\sinh(kz)}{k\lambda\cosh(kL)}, \tag{4.91}$$

and the Fourier transform of the surface heat flux is,

$$q_0(\vec{k}) = \frac{q_L(\vec{k})}{\cosh(kL)}. \tag{4.92}$$

We can reconstruct the surface heat flux by integrating the inverse tramsform.

Varying temperature at the base of a layer

If $T(x,y,z=L) = T_L(x,y)$, the Fourier transform of the temperature is obtained from the boundary condition at $z=L$. It gives,

$$T(\vec{k},z) = \frac{T_L(\vec{k})\sinh(kz)}{\sinh(kL)}, \tag{4.93}$$

and the surface heat flux is,

$$q_0(\vec{k}) = \frac{T_L(k)}{L}\frac{|k|L}{\sinh(kL)}. \tag{4.94}$$

In geological applications, relationships 4.92 and 4.94 are useful, for instance, to analyze the relationship between variations in surface heat flux and the thermal structure at depth. These equations show the effect of horizontal diffusion of heat. As a consequence of this, horizontal variations in heat flux or temperature at the base of the lithosphere can not be detected at the surface unless they involve very long wavelengths, relative to thickness L i.e. $kL < 1$.

4.4.3 Horizontal variations in heat production

In a layer ($0 < z < L$) with heat production varying horizontally, $H(x,y,z) = H(x,y)$, the Fourier transform of the heat equation:

$$\frac{\partial^2 T}{\partial z^2} - k^2 T = -\frac{H(\vec{k})}{\lambda} \tag{4.95}$$

has the solution:

$$T(\vec{k},z) = A(\vec{k})\sinh(kz) + B(\vec{k})\cosh(kz) + \frac{H(\vec{k})}{\lambda k^2}. \tag{4.96}$$

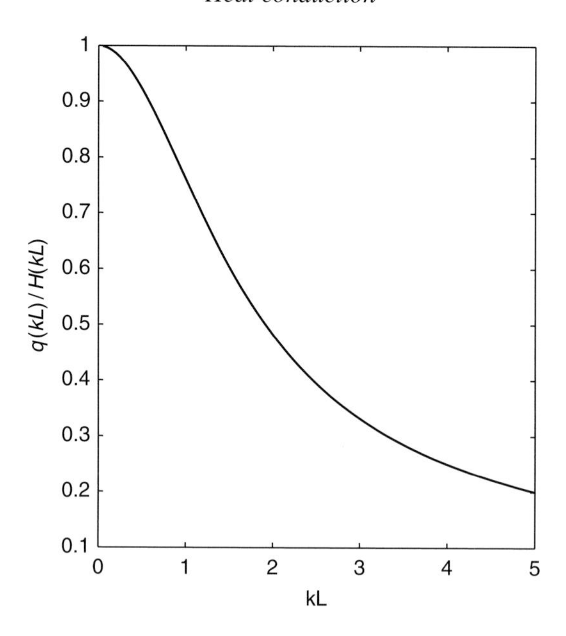

Figure 4.4. Ratio of surface heat flux to the total heat production when heat production varies laterally in a layer of thickness L. The heat production is constant from the surface down to depth L. kL is a dimensioness wave-number.

If temperature $T = 0$ at $z = 0$ and heat flow $q = 0$ at $z = L$, the solution is,

$$T(\vec{k}, z) = \frac{H(\vec{k})}{\lambda k^2}\left(1 - \cosh(kz) + \frac{\sinh(kL)\sinh(kz)}{\cosh(kL)}\right). \tag{4.97}$$

The relationship between surface heat flux and heat production is,

$$q(\vec{k}) = \frac{H(\vec{k}) \times L}{kL}\frac{\sinh(kL)}{\cosh(kL)}. \tag{4.98}$$

Because of lateral diffusion, the local contribution to surface heat flux of a heat producing layer is always less than the vertically integrated heat production, i.e. heat production times crustal thickness (Figure 4.4). Using wavelength as the horizontal distance scale, we see that variations in heat production are attenuated over short distances unless $kL < 1$. Fr $kL > 1$, we get $q(k) \propto H(k)/k$. The effect of heat sources is attenuated by lateral diffusion of heat over a characteristic distance proportional to the thickness of the heat producing layer.

4.4.4 Changes in thermal conductivity: Refraction of heat

As shown in Appendix D, the thermal conductivity of common rocks varies within a range of about one order of magnitude. The crust, in particular, is made of rocks of different origins and compositions, with different values of thermal conductivity. These rocks are juxtaposed in many different geometrical configurations, inducing refraction effects which may account for large temperature variations and affect surface heat flux observations. We shall consider simple examples that demonstrate the amplitude of these effects.

Two semi-infinite layers with different thermal conductivities

We consider a layer of thickness L extending to large horizontal distances in the x direction (Figure 4.5). The layer is made of two halves that join at $x = 0$. Thermal conductivity values are λ_1 and λ_2 in regions $x < 0$ and $x > 0$, respectively. A uniform heat flux Q_B is supplied at the base, at $z = L$, and the surface at $z = 0$ is maintained at zero temperature. The lateral contrast in thermal conductivity deviates the vertical flow of heat near the vertical discontinuity. Dimensional analysis dictates that temperature variations occur over horizontal distance $\approx L$ on either side.

The solution to this problem is first obtained by separating the temperature into two components, one corresponding to the homogeneous case with fixed heat flux at the base, and the other corresponding to a laterally varying temperature field with zero heat flux at the base. The second problem can be solved using the separation of variables: $T(x,z) = X(x)Z(z)$. Substitution into the heat equation, which in this case takes the form of Laplace's equation $\nabla^2 T = 0$, leads to,

$$X \propto \exp(\pm kx)\,;\; Z \propto \exp(\pm ikz). \tag{4.99}$$

Values of k must be chosen to meet the boundary conditions. In this case, with fixed temperature at $z = 0$ and zero heat flux at $z = L$:

$$Z(z) \propto \sin(k_n z) \text{ with } k_n = (2n+1)\pi/2L. \tag{4.100}$$

From above, we deduce that $X(x) \propto \exp k_n x$. Thus,

$$\begin{aligned} T(x,z) &= \frac{Q_B z}{\lambda_1} + \sum_{n=0}^{\infty} A_n \sin(k_n z)\exp(k_n x) \quad x < 0 \\ T(x,z) &= \frac{Q_B z}{\lambda_2} + \sum_{n=0}^{\infty} B_n \sin(k_n z)\exp(-k_n x) \quad x > 0. \end{aligned} \tag{4.101}$$

Using the identity

$$\frac{z}{L} = \frac{8}{\pi^2}\sum_{n=0}^{\infty}\frac{(-)^n \sin(k_n z)}{(2n+1)^2}. \tag{4.102}$$

The constants A_n and B_n are determined from the conditions of continuity of temperature and horizontal heat flow at $x = 0$. This gives,

$$T(x,z) = \frac{Q_B L}{\lambda_1} \frac{8}{\pi^2} \sum_{n=0}^{\infty} \frac{(-)^n \sin(k_n z)}{(2n+1)^2} \left(1 + \frac{\lambda_1 - \lambda_2}{\lambda_1 + \lambda_2} \exp(k_n x)\right) \quad x < 0$$

$$T(x,z) = \frac{Q_B L}{\lambda_2} \frac{8}{\pi^2} \sum_{n=0}^{\infty} \frac{(-)^n \sin(k_n z)}{(2n+1)^2} \left(1 + \frac{\lambda_2 - \lambda_1}{\lambda_2 + \lambda_1} \exp(-k_n x)\right) \quad x > 0. \tag{4.103}$$

Figure 4.5 shows the horizontal variation of surface heat flux. One notes that the heat flow field is perturbed over lateral distance $\approx L$ on both sides of the discontinuity and that the heat flow jump across the discontinuity increases with increasing conductivity contrast. One also notes that the heat flux anomaly is an odd function of x, such that the net integrated heat flux anomaly is zero: heat which gets diverted to the one side of the discontinuity is lost by material on the other side. The vertical temperature profile along the discontinuity is

$$T(0,z) = 2\frac{Q_B z}{\lambda_1 + \lambda_2}. \tag{4.104}$$

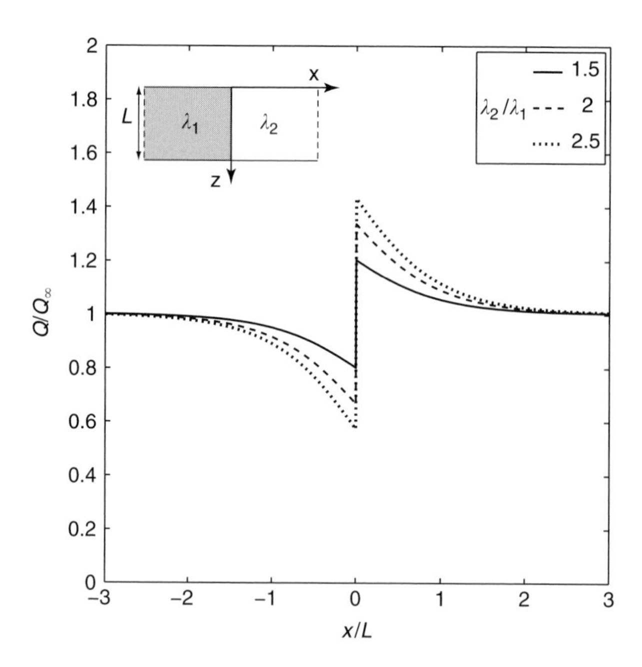

Figure 4.5. Heat flow across a discontinuity in thermal conductivity. The conductivity of a layer of thickness L jumps from λ_1 to λ_2 across the $x = 0$ plane.

Thus, temperatures in the low conductivity region are decreased near the discontinuity whereas those on the high conductivity side are increased.

The region $|x| < a$ with thermal conductivity λ_1

Within a layer $0 < z < L$, is thermal conductivity is λ_1 for the region $|x| < a$ and λ for $|x| > a$. Figure 4.6 shows a similar arrangement for a cylindrically symmetric conductivity structure. The temperature $T = 0$ at $z = 0$ and the heat flux is Q_B for $|x| \to \infty$. A solution to Laplace's equation, symmetric about the origin, is written in terms of the Fourier series expansion:

$$
\begin{aligned}
T(x,z) &= \frac{Q_B z}{\lambda_1} + \sum_{n=0}^{\infty} A_n \sin(k_n z)\cosh(k_n x) \quad |x| < a \\
T(x,z) &= \frac{Q_B z}{\lambda} + \sum_{n=0}^{\infty} B_n \sin(k_n z)\exp(-k_n|x|) \quad |x| > a,
\end{aligned}
\tag{4.105}
$$

with $k_n = (2n+1)\pi/2L$. Introducing the Fourier series 4.102, the constants A_n, B_n are determined from the condition of continuity of temperature and horizontal heat

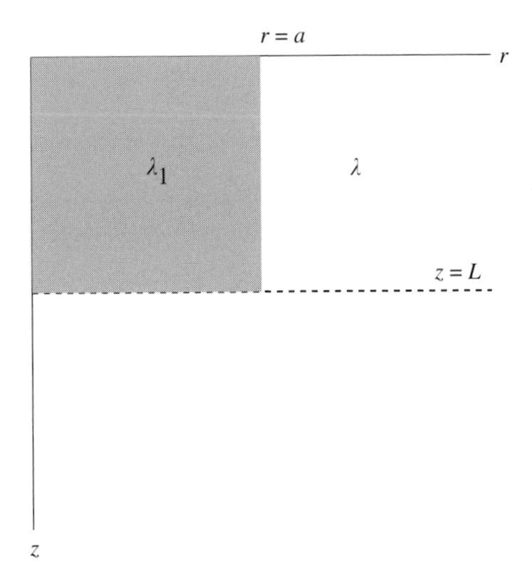

Figure 4.6. Geometry of a finite vertical cylinder embedded in a medium with different conductivity.

flow at $x = a$:

$$T(x,z) = \frac{8Q_BL}{\pi^2\lambda_1}\sum_{n=0}^{\infty}\frac{(-)^n\sin(k_nz)}{(2n+1)^2}\left(1+\frac{(\lambda_1/\lambda-1)\cosh(k_nx)}{\cosh(k_na)+\lambda_1/\lambda\sinh(k_na)}\right) \quad |x|<a$$

$$= \frac{8Q_BL}{\pi^2L\lambda}\sum_{n=0}^{\infty}\frac{(-)^n\sin(k_nz)}{(2n+1)^2}\left(1+\frac{(1-\lambda_1/\lambda)\sinh(k_na)\exp(-k_n(|x|-a))}{\cosh(k_na)+\lambda_1/\lambda\sinh(k_na)}\right)$$

$$|x|>a. \tag{4.106}$$

Vertical cylinder embedded in a medium

We consider a vertical cylinder of length L embedded in a medium of conductivity λ (Figure 4.6). The cylinder has radius a and conductivity λ_1. For radial symmetry, the solution will be obtained in terms of the modified Bessel functions I_0 and K_0 (see equation A.105 for a definition):

$$\begin{aligned} T(r,z) &= \frac{Q_Bz}{\lambda_1} + \sum_{n=0}^{\infty} A_n\sin(k_nz)I_0(k_nr), \quad r<a \\ T(r,z) &= \frac{Q_Bz}{\lambda} + \sum_{n=0}^{\infty} B_n\sin(k_nz)K_0(k_nr), \quad r>a. \end{aligned} \tag{4.107}$$

The constants are determined from the continuity conditions at $r = a$. This gives,

$$\begin{aligned} T(r,z) &= \frac{8Q_BL}{\pi^2\lambda_1}\sum_{n=0}^{\infty}\frac{(-)^n\sin(k_nz)}{(2n+1)^2}\times\ldots \\ &\ldots\left(1+\frac{(\lambda_1/\lambda-1)K_0(k_na)I_1(k_nr)}{\lambda_1/\lambda K_0(k_na)I_1(k_na)+K_1(k_na)I_0(k_na)}\right), \quad r<a \\ &= \frac{8Q_BL}{\pi^2\lambda}\sum_{n=0}^{\infty}\frac{(-)^n\sin(k_nz)}{(2n+1)^2}\times\ldots \\ &\ldots\left(1+\frac{(1-\lambda_1/\lambda)I_1(k_na)K_0(k_nr)}{\lambda_1/\lambda K_0(k_na)I_1(k_na)+K_1(k_na)I_0(k_na)}\right), \quad r>a. \end{aligned} \tag{4.108}$$

The heat flux can be very much enhanced over a narrow conductive cylinder as seen in Figure 4.7. At the discontinuity, the ratio between inner and outer heat flux is equal to the conductivity ratio. The total heat flux is constant as higher heat flux inside is balanced by lower heat flux outside the cylinder.

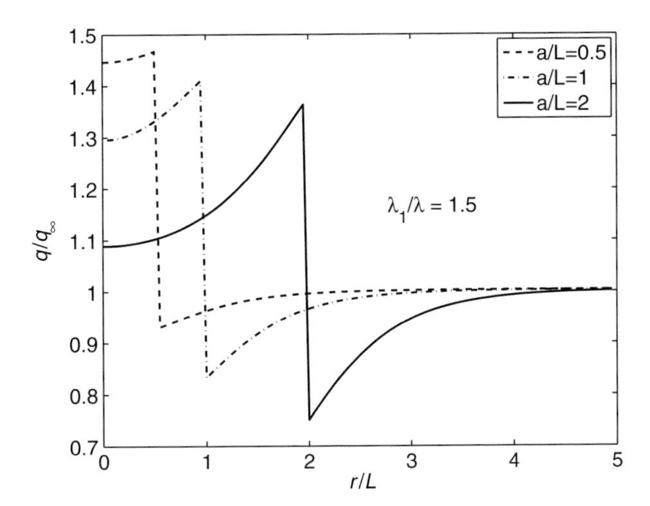

Figure 4.7. Heat flux near a vertical cylinder when the conductivity is 1.5 times higher inside than outside the cylinder. The cylinder extends to finite depth L and its radius is a.

4.4.5 *Non-symmetric temperature in a sphere*

If the heat flux and temperature fields vary on the surface of a sphere, the solutions of Poisson's and Laplace's equations involve spherical harmonics. In theory, the spherical harmonic expansion can be used to continue the temperature field downward within the sphere, but this is of no practical interest because it applies only to a very thin layer near the Earth's surface where heat is transported by conduction. Because the conductive layer thickness is small compared to the Earth's radius, we can use half-space or slab approximations. However, spherical harmonic analysis of the heat flux field is useful to filter out its short-wavelength components and compare its large-scale features with other geophysical fields, such as gravity and the geoid.

4.5 Transient problems

We have already derived with simple physical arguments solutions to several important problems in heat conduction, in particular we have derived the Green's functions for different problems. Many different methods are available to obtain these solutions. However we shall not derive the same results, but show with some examples how to solve initial value problems with integral transform techniques.

Fourier transforms or Fourier series can be used to solve initial value problems by decomposing the initial condition in Fourier components, and superposing the individual solutions. Alternatively, one can use the Laplace transform of the time variable to obtain an ordinary (or partial) differential equation that includes the initial condition. The Laplace transform $F(s)$ of a function $f(t)$ is defined by,

$$F(s) = \int_0^\infty f(t)\exp(-st)dt, \tag{4.109}$$

and the inverse Laplace transform is given by,

$$f(t) = \frac{1}{2\pi i}\int_{\gamma-i\infty}^{\gamma+i\infty} F(s)\exp(st)ds, \tag{4.110}$$

where γ is such that the contour of integration in the complex plane is to the right of all the singularities of $F(s)$. More details are included in Appendix A.

Obviously, the Fourier decomposition and Laplace transform methods must give identical results. The boundary conditions determine the analytical form of the Laplace transform as well as the coefficients of the Fourier series expansion (see Appendix A). The Fourier method is usually simple and straightforward. Calculating the inverse Laplace transform is often tedious when the function is not in tables, but the method is useful for obtaining asymptotic approximations to the solution.

4.5.1 Half-space with time-varying surface boundary condition

Varying surface temperature

For the half-space $z > 0$, with initial condition $T(z, t = 0) = 0$ and boundary condition $T(z = 0, t) = T_0(t)$, a solution is obtained using the Laplace transform. The Laplace transform $T(s, z)$ of the 1D heat equation is,

$$sT(s,z) - T(t=0,z) = \kappa\frac{d^2T}{dz^2}, \tag{4.111}$$

and the boundary conditions are $T(s, z \to \infty) = 0$ and $T(s, z = 0) = T_0(s)$. The general solution is,

$$T(s,z) = A(s)\exp\left(\sqrt{\frac{s}{\kappa}}z\right) + B(s)\exp\left(-\sqrt{\frac{s}{\kappa}}z\right). \tag{4.112}$$

For the solution to remain finite for $z \to \infty$, we must have $A(s) = 0$. The boundary condition at $z = 0$ gives $B(s) = T_0(s)$. Thus,

$$T(s,z) = T_0(s)\exp\left(-\sqrt{\frac{s}{\kappa}}z\right). \tag{4.113}$$

If the surface temperature is a Dirac delta function, its Laplace transform $T_0(s) = 1$ and the inverse transform gives the half-space Green's function that we calculated before (equation 4.59). The advantage of equation 4.113 is that we can use it to obtain the solution for any temperature surface boundary condition without calculating a convolution integral. The task is easy because many Laplace transforms have been compiled in tables.

For a sudden change in surface temperature ΔT at $t = 0$, the Laplace transform is $\Delta T/s$. Introducing this into equation 4.113 gives,

$$T(z,t) = \Delta T \operatorname{erfc}\left\{\frac{z}{2\sqrt{\kappa t}}\right\}, \tag{4.114}$$

where we recognize the solution 4.43.

The solution for temperature linearly increasing with time is likewise obtained by introducing the transform of the boundary condition $\dot{T}_0/s^2$ in equation 4.113.

Heat flux pulse at the surface

For the half-space $z > 0$, initially at temperature $T(z, t = 0) = 0$, the temperature due to an impulse of heat flux at the surface

$$\lambda\frac{\partial T}{\partial z} = q_0 \delta|t|_z. \tag{4.115}$$

The Laplace transform of the surface heat flow is obtained from the transform of the general solution (4.112) with the boundary condition,

$$\lambda\sqrt{\frac{s}{\kappa}}A(s) = q_0\tau, \tag{4.116}$$

which yields the transform of the temperature,

$$T(s,z) = \frac{q_0\tau}{\lambda}\sqrt{\frac{\kappa}{s}}\exp\left(-\sqrt{\frac{s}{\kappa}}z\right). \tag{4.117}$$

The inverse Laplace transform can be found in tables. This gives the solution:

$$T(z,t) = \frac{q_0\tau}{\lambda}\sqrt{\frac{\kappa}{\pi t}}\exp\left(\frac{-z^2}{4\kappa t}\right). \tag{4.118}$$

We recognize the solution 4.56 with $q_0\tau = \Delta E/\delta C_\rho$.

Half-space below a blanketing layer

We shall calculate the effect of rapidly depositing a layer of sediment at the surface of the Earth. We assume that deposition was done so rapidly that temperature did not change, so that it is 0 in the layer $0 < z < a$, and it is $\Gamma(z - a)$ in the region below. A gradient Γ is maintained at infinity. After returning to equilibrium, the temperature will be Γz. We can decompose the temperature into its equilibrium and transient components, and solve for the transient δT. At $t = 0$, the region $0 < z < a$ is at temperature $\delta T = -\Gamma z$, the region $z > a$ is at temperature $\delta T = -\Gamma a$. The solution, obtained by the Laplace transform is,

$$\begin{aligned} \delta T(z,t) &= \Gamma\sqrt{\kappa t}\left(\text{ierfc}\frac{(a-z)}{2\sqrt{\kappa t}} - \text{ierfc}\frac{(a+z)}{2\sqrt{\kappa t}}\right) - \Gamma z, \quad 0 < z < a \\ &= \Gamma\sqrt{\kappa t}\left(\text{ierfc}\frac{(z-a)}{2\sqrt{\kappa t}} - \text{ierfc}\frac{(z+a)}{2\sqrt{\kappa t}}\right) - \Gamma a, \quad z > a, \end{aligned} \tag{4.119}$$

where *i*erfc is the integral of the complementary error function. The transient decays to zero in a characteristic time a^2/k.

4.5.2 Moving half-space with constant temperature on the plane $z = 0$

In order to calculate how erosion or deposition of sediment affects a pre-established steady-state temperature gradient, we assume that the region moves relative to the surface (i.e. the plane $z = 0$) kept at fixed temperature $T = 0$ with a constant velocity v. We neglect heat sources and assume that the initial condition is,

$$T(z, t = 0) = \Gamma z. \tag{4.120}$$

The solution for $t > 0$ is obtained

$$\begin{aligned} T(z,t) =& \Gamma(z - vt) + \frac{\Gamma}{2}(z + vt)\exp\left(\frac{vz}{\kappa}\right)\text{erfc}\left(\frac{z+vt}{2\sqrt{\kappa t}}\right) \\ &+ \frac{\Gamma}{2}(vt - z)\text{erfc}\left(\frac{z - vt}{2\sqrt{\kappa t}}\right). \end{aligned} \tag{4.121}$$

The gradient at the surface $z = 0$ is,

$$\frac{1}{\Gamma}\frac{\partial T}{\partial z} = \left(1 + \frac{v^2 t}{2\kappa}\right)\text{erfc}\left(\frac{v}{2}\sqrt{\frac{t}{\kappa}}\right) - v\sqrt{\frac{t}{\pi\kappa}}\exp\left(-\frac{v^2 t}{4\kappa}\right). \tag{4.122}$$

The characteristic time and length scales are given by κv^{-2} and κv^{-1} respectively. For deposition of sediment $v > 0$ and for erosion, $v < 0$. Note that, for $v > 0$, $\lim_{t\to\infty} \partial T/\partial z = 0$ and for $v < 0$, $\lim_{t\to\infty} \partial T/\partial z = (2 + v^2 t/\kappa) \to \infty$.

The time scales may be very large and the effect of erosion–sedimentation on the gradient is usually small (Figure 4.8). For an erosion rate of 100 m My^{-1}, the characteristic time is 3 Gy; the gradient will increase by less than 10% after 30 My.

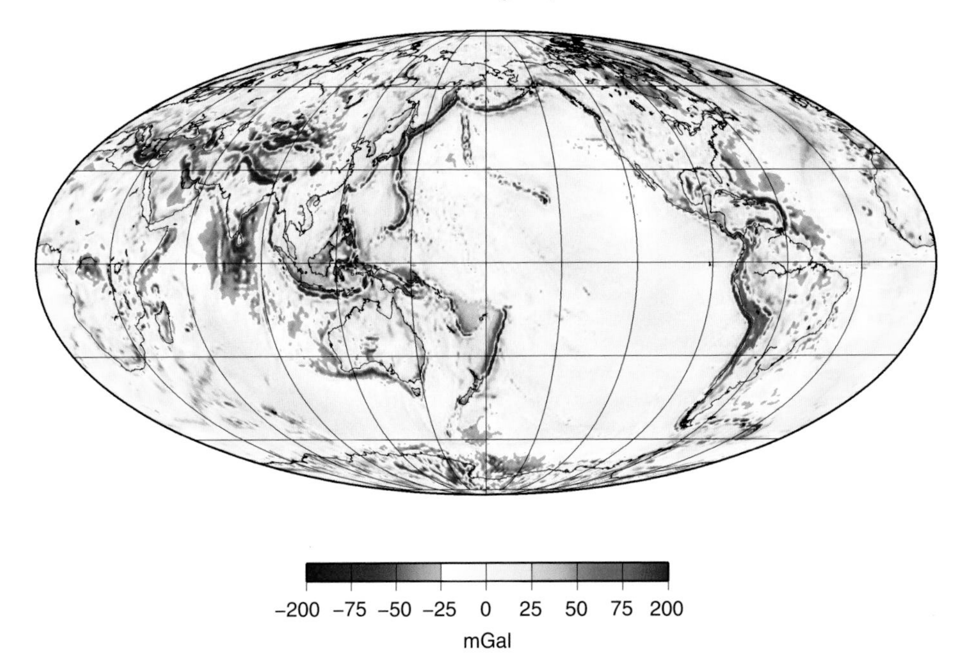

Figure 2.1. Free-air gravity map based on model GGM02 including data from the GRACE satellite mission (Tapley *et al.*, 2005). Note that the major gravity anomalies are in narrow belts.

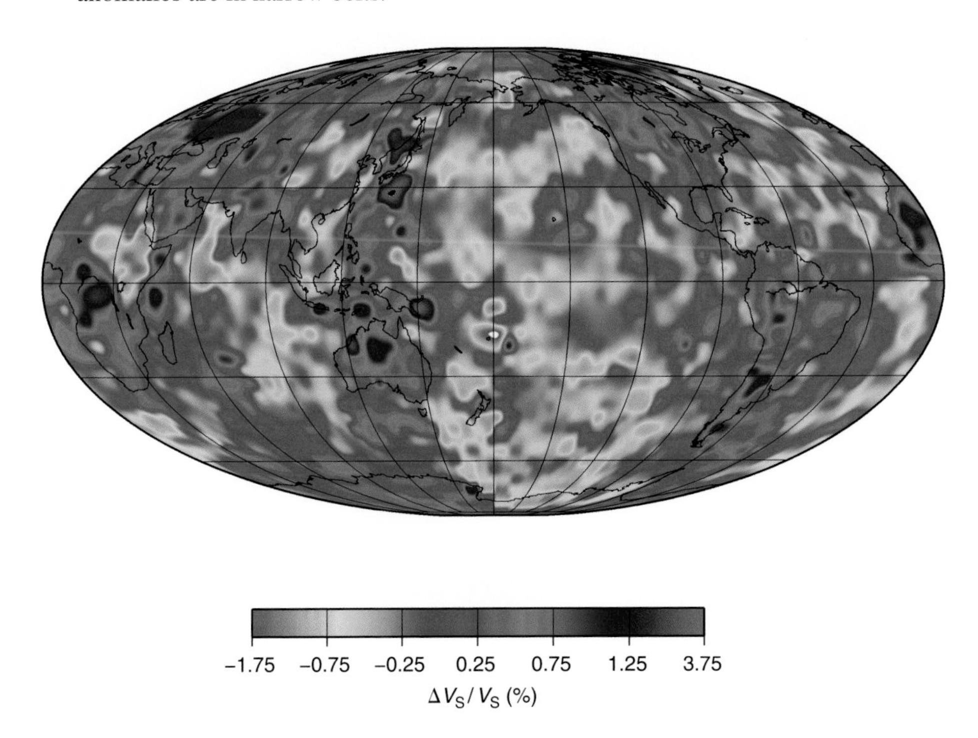

Figure 2.9. Shear wave velocity anomalies at 350 km depth. The anomalies relative to the mean shear wave velocity at that depth are given in %. By convention, the color scale is inverted as negative velocity anomalies are interpreted as hot. The seismic model is that of Grand (1997).

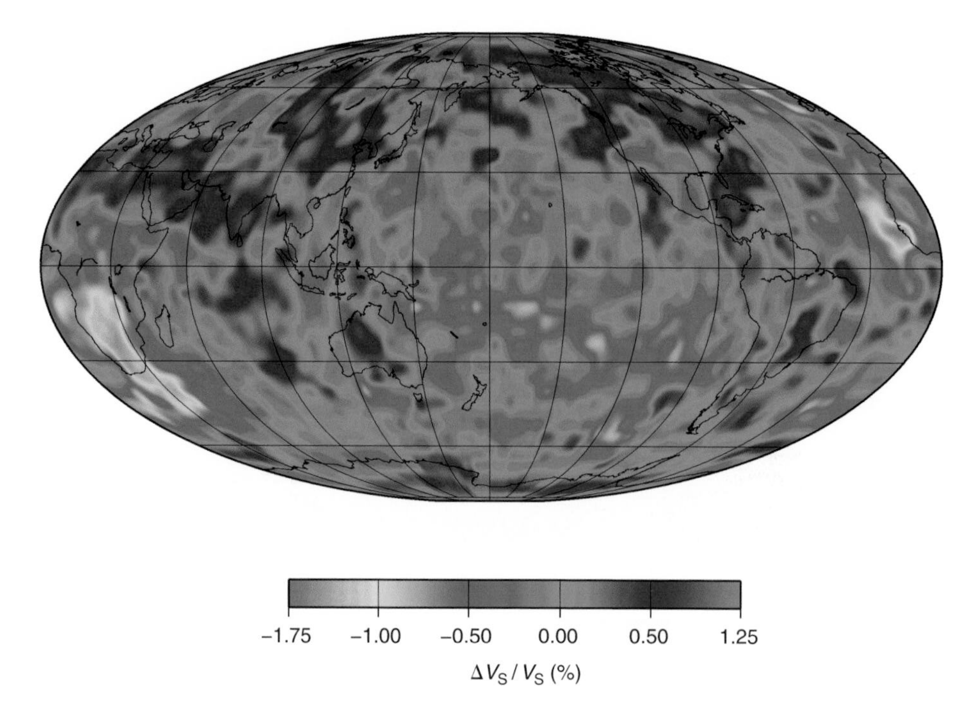

Figure 2.10. Shear wave velocity anomalies at 1810 km depth (Grand, 1997). Note the reduced scale relative to Figure 2.9.

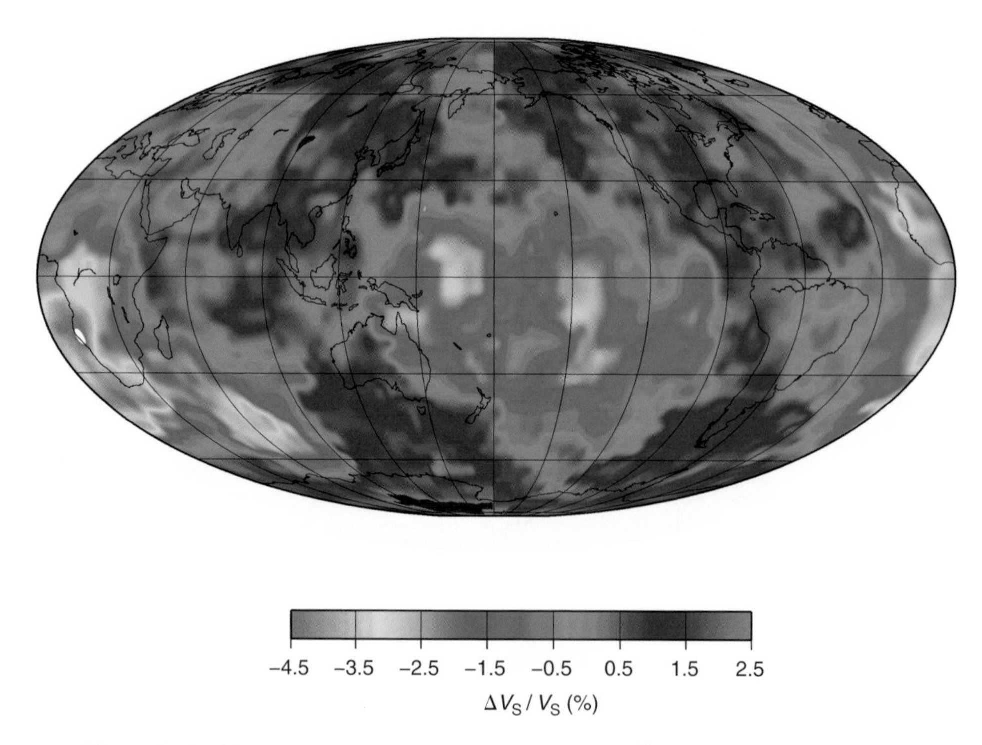

Figure 2.11. Shear wave velocity anomalies in the D″ region at 2800 km depth (Grand, 1997). Anomalies are enhanced relative to the lower mantle.

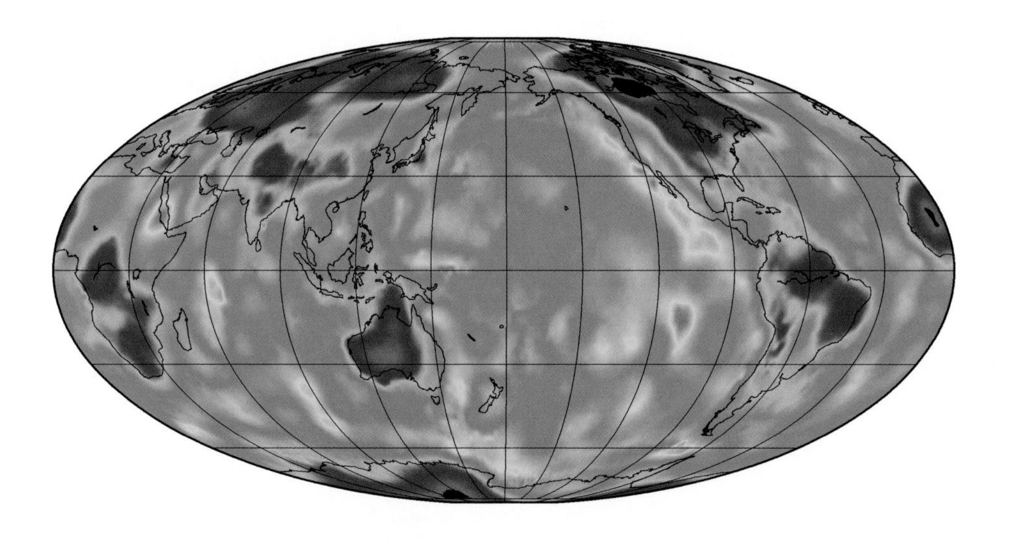

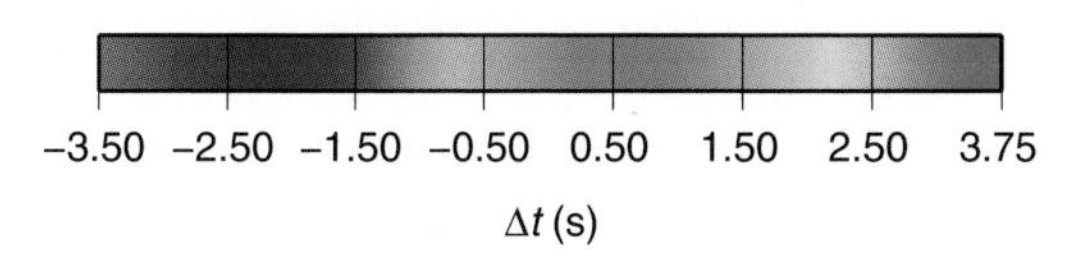

Figure 2.12. Shear wave travel time anomalies across the 60 to 300 km region. The travel times are calculated from the CUB2.0 model of Shapiro and Ritzwoller (2002), and the anomalies are relative to the world average. The value of 60 km for the upper limit is selected to exclude the crust which is not very well resolved by the seismic model.

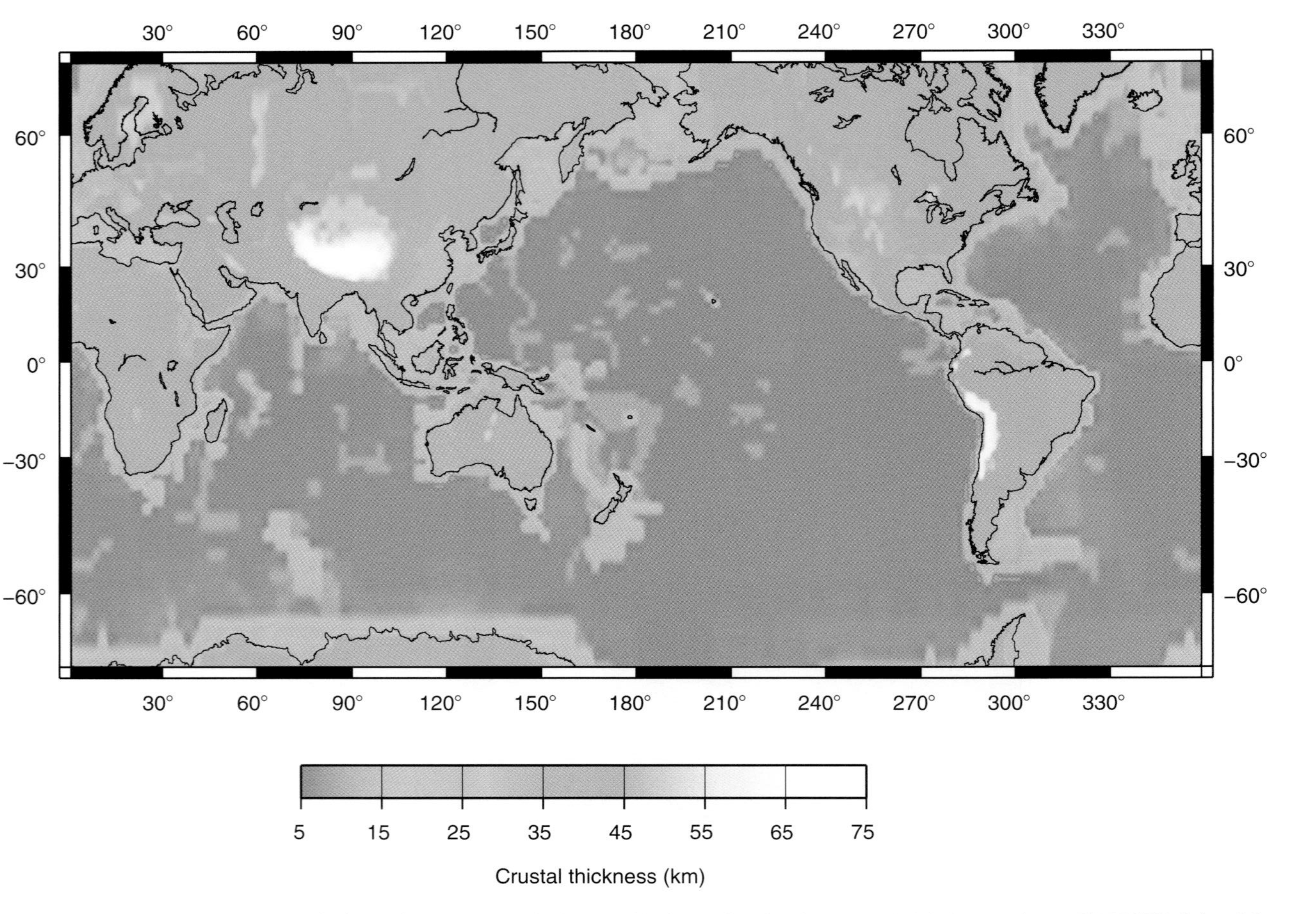

Figure 2.13. Crustal thickness variations from the global compilation of seismic crustal thickness data, CRUST2.0 by Mooney *et al.* (1998).

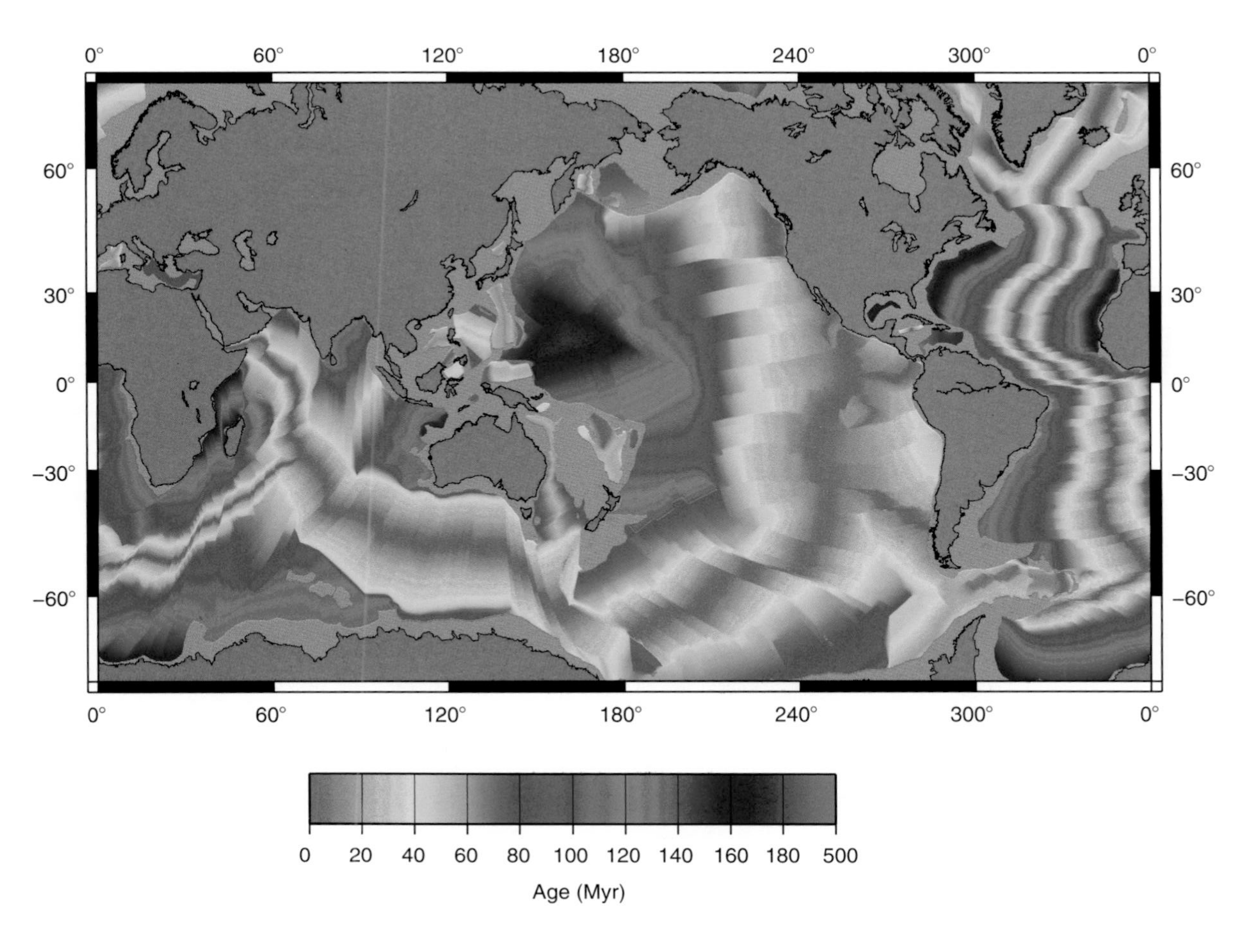

Figure 2.14. Map of the ages of the sea floor from Müller *et al.* (2008)

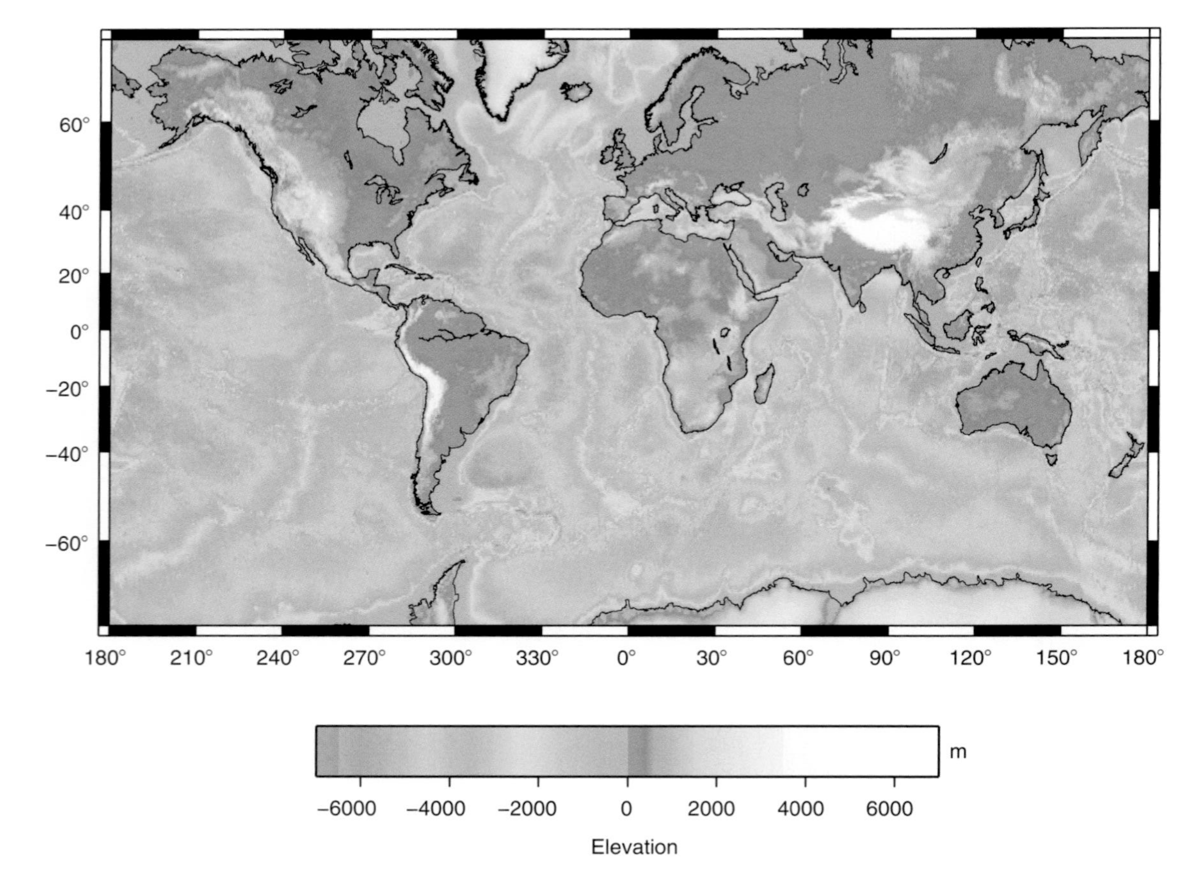

Figure 2.15. Global elevation map – bathymetry of the sea floor. Note the correlation of continental elevation with crustal thickness and sea-floor bathymetry with age.

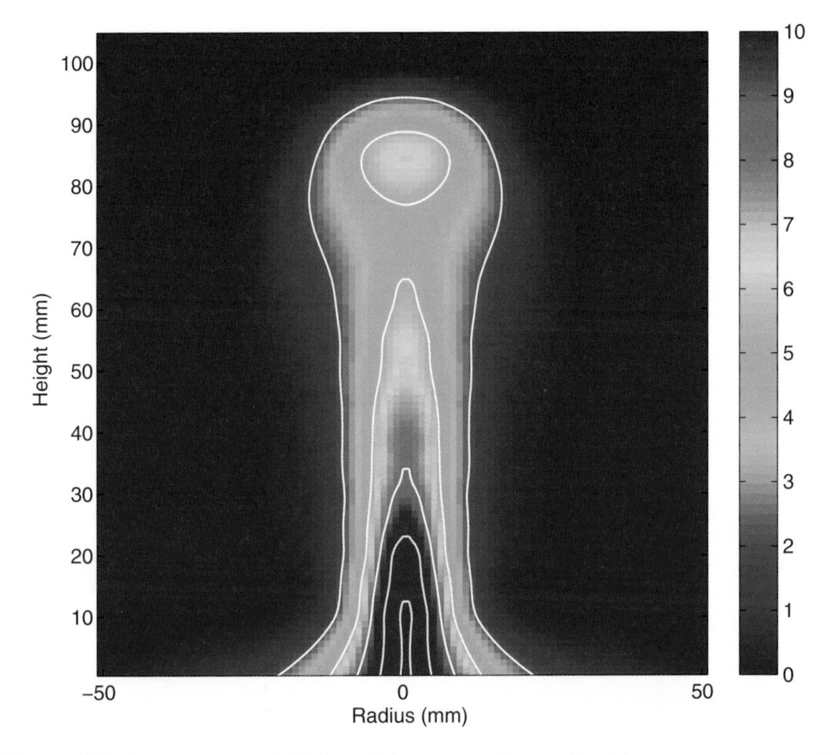

Figure 5.2. Temperature field in a laboratory plume (A. Limare, pers comm.). Experimental techniques are described in Davaille *et al.* (2010). Note the large horizontal temperature gradients and a cap that is wider than the stem.

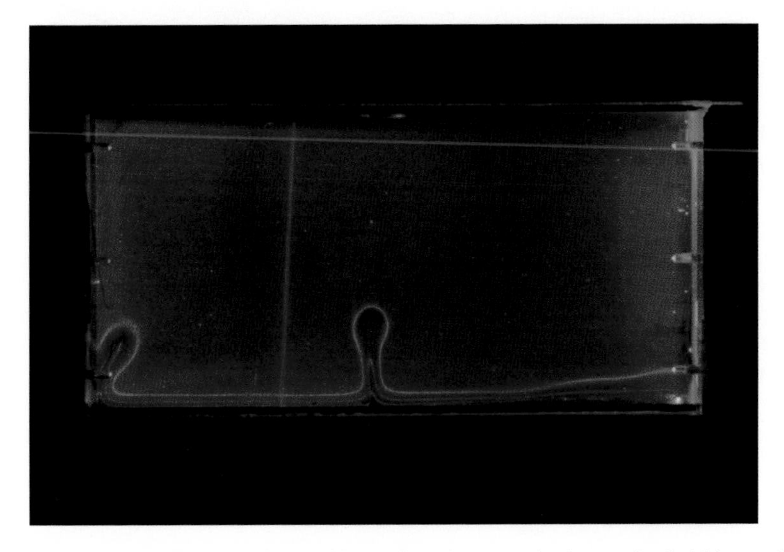

Figure 5.8. Instability of a thermal boundary layer at the base of a fluid layer that is heated from below in the laboratory (from Davaille *et al.*, 2010). The bright contours correspond to isotherms imaged by temperature-sensitive liquid crystals.

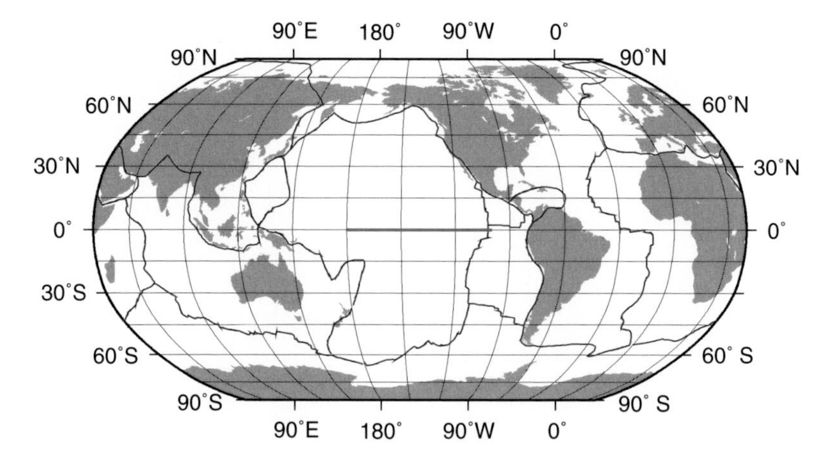

Perturbation (%) relative to AK135, ON, CU_SDT1.0

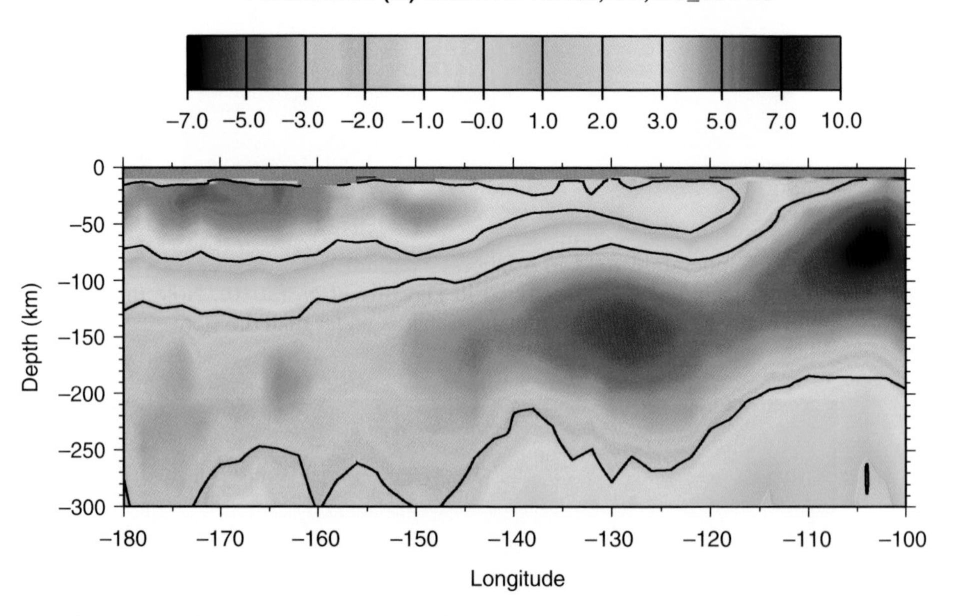

Figure 6.14. Shear wave velocity profile of the Pacific plate along the equator. The velocity increases westward with increasing distance from the East Pacific rise, in good agreement with cooling models for the oceanic lithosphere. The profiles were obtained from CUB2.0, the surface wave tomography model by Shapiro and Ritzwoller (2002).

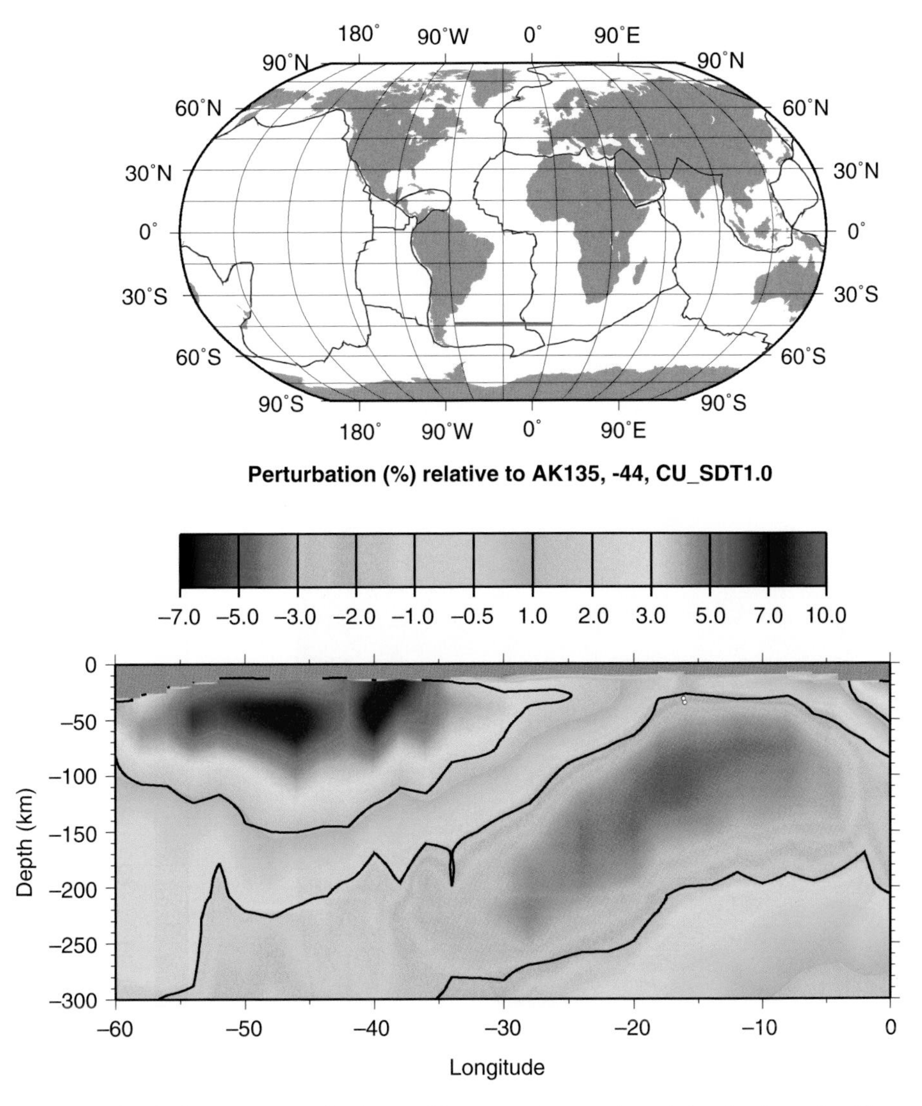

Figure 6.15. Shear wave velocity profile across the southern Atlantic along the 45 S parallel. The velocity increases on both sides away from the Mid-Atlantic ridge. The profiles were obtained from CUB2.0, the surface wave tomography model by Shapiro and Ritzwoller (2002).

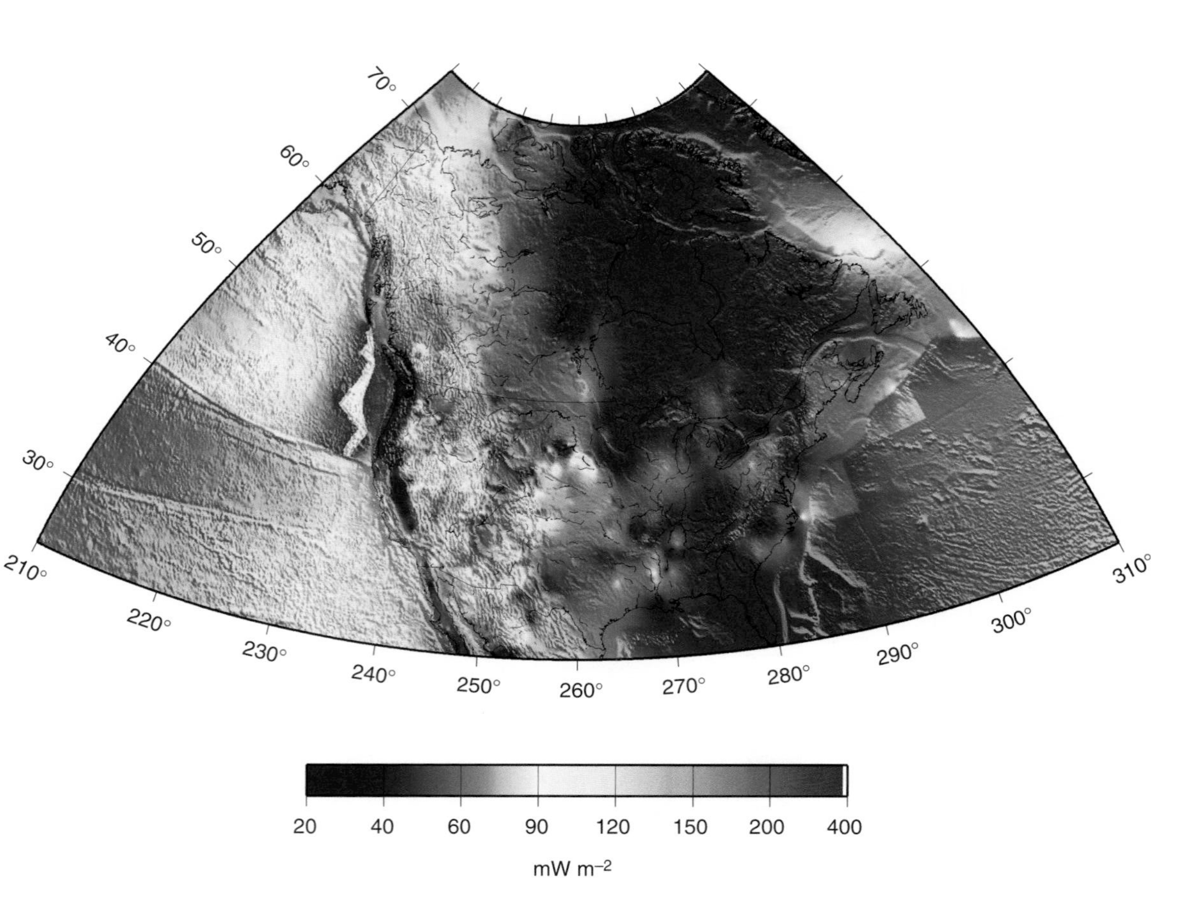

Figure 7.1. Heat flow map of North America.

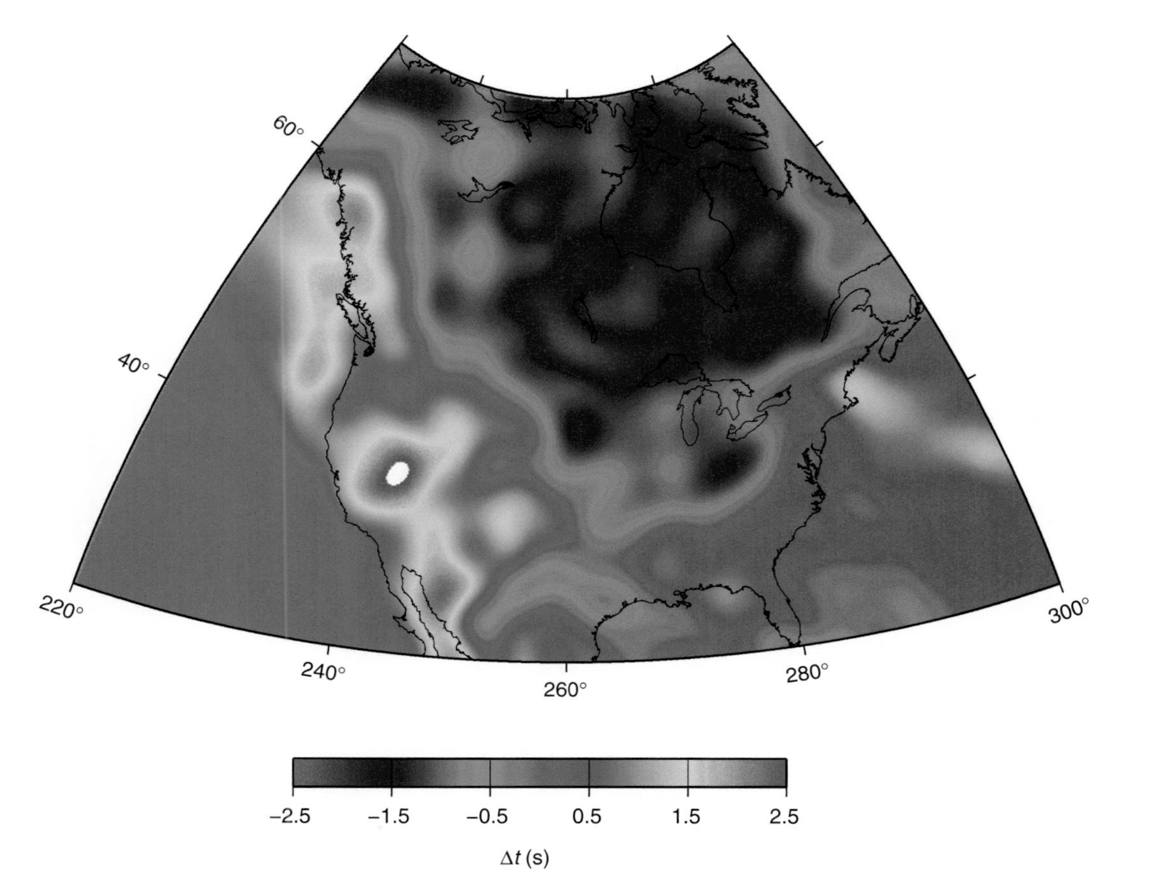

Figure 7.28. Shear wave travel time delays for North America. The delays in the arrival time of shear waves are calculated between 60 and 300 km from the surface wave tomography model of van der Lee and Frederiksen (2005).

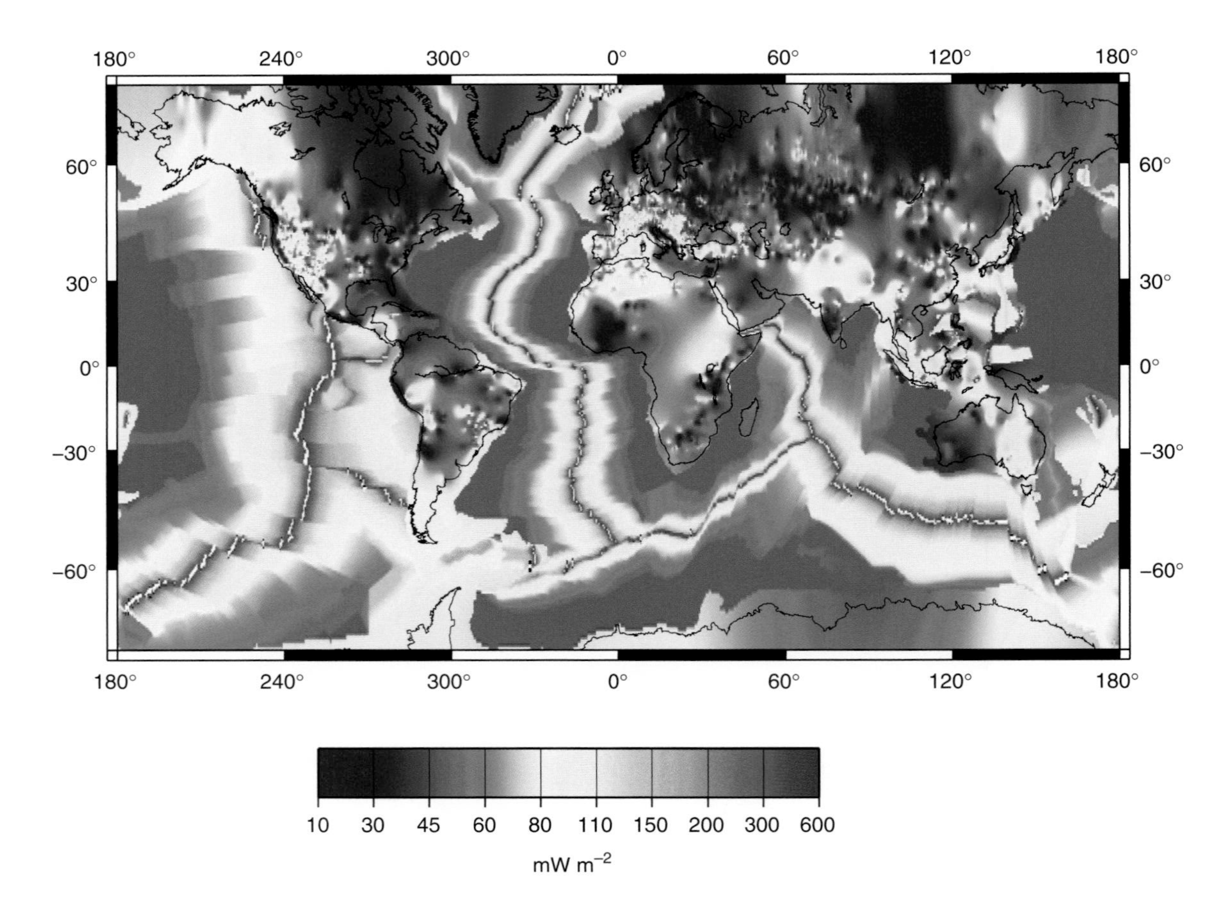

Figure 8.1. Global heat flow map combining land heat flux measurements and the cooling plate model for the oceans, where heat flux is calculated as the maximum of 48 mW m^{-2} and $490/\sqrt{\tau}$ mW m^{-2}, where τ is age in My.

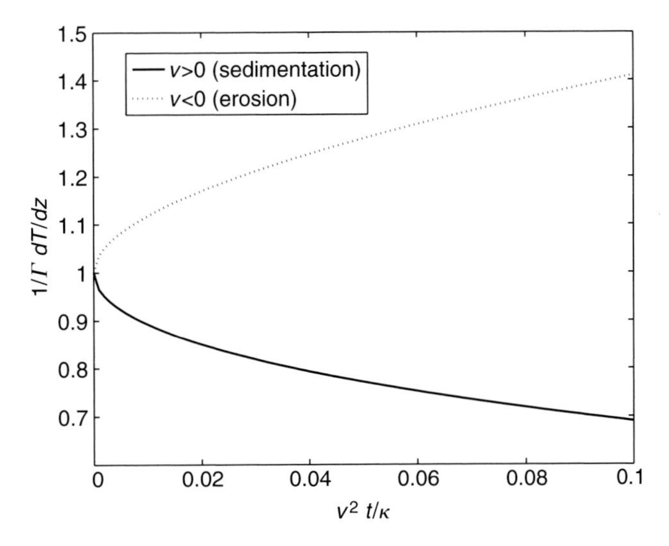

Figure 4.8. Surface temperature gradient below the (isothermal) surface $z = 0$ when the region $z > 0$ moves with velocity v ($v > 0$ sediment deposition, $v < 0$ erosion). For common erosion or sedimentation rates, κ/v^2 is in the range 3–300 Gy.

This holds also for basins where sedimentation rates are $v = 10^{-5} - 10^{-4}$ m y^{-1}, which yields time scales > 3 Gy. The effect is even less for sedimentation in deep ocean basins where sedimentation rates are a few m My^{-1}, and time scales are >100 Gy.

For active mountains, where erosion rates may be as large as 1 km My^{-1}, the gradient is significantly enhanced. One-dimensional solutions are simplification because mountain building is accompanied by overthrusting and crustal thickening.

Solution (4.122) can also be used to estimate the effect of ice accumulation on a glacier. With accumulation rates, $v = 0.1\,\text{m}\,\text{y}^{-1}$, the time scale is small (3,000 y), and the shallow temperature gradient in many glaciers is close to zero (see Figure 12.13 for an example).

4.5.3 Cooling of a layer in infinite or semi-infinite medium

We shall consider the cooling of a hot layer (dike or sill) in the Earth. At this stage we shall only consider the diffusion of the temperature perturbation and neglect the effect of the heat of solidification, which will be considered in Chapter 11. We consider a layer $-a < z < a$, with initial temperature ΔT imbedded in an infinite medium at temperature 0. In this situation, it is convenient to directly apply the

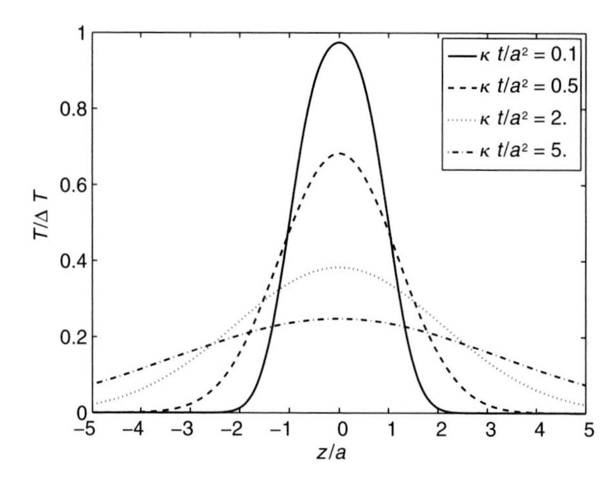

Figure 4.9. Temperature profiles across a layer $-a < z < a$ initially at temperature ΔT. Note that the temperature at the edges is 0.5 ΔT at short times.

convolution theorem with the full space or the half-space Green's functions (4.78). For the half-space, we get,

$$\begin{aligned} T(z,t) &= \frac{\Delta T}{\sqrt{\pi}} \int_{-a}^{+a} \frac{1}{2\sqrt{\kappa t}} \exp(-(z-z')^2/4\kappa t) dz' \\ &= \frac{\Delta T}{2} \left(\operatorname{erf}\left(\frac{z+a}{2\sqrt{\kappa t}} \right) - \operatorname{erf}\left(\frac{z-a}{2\sqrt{\kappa t}} \right) \right). \end{aligned} \tag{4.123}$$

This solution (Figure 4.9) applies, for instance to the cooling of a dike at depth. The time scale is determined only by the thickness of the dike.

The effect of latent heat can be included in the form of additional sensible heat in the layer, i.e. by adding L/C_p to the initial temperature of the layer, where L is the latent heat (on Chapter 11). For most rocks, the ratio L/C_p is on the order of 300 K, and this approach is clearly not valid near the intrusion where it yields too high temperatures. This approximate treatment applies only at a distance > layer thickness a.

The solution is applicable to a sill or a horizontal intrusion only when its depth is much larger than its thickness. For a shallow intrusion, we need to consider the effect of the surface. The solution for a layer $a < z < b$ in the half-space $z > 0$ is obtained by the method of images. The solution is the superposition of the solution for a layer $a < z < b$ with temperature ΔT and a layer $-b < z < -a$ with temperature $-\Delta T$ (Figure 4.10). The cooling is asymmetric, and there are two time scales, one is determined by the layer thickness, the other by the distance to the surface. This solution applies to the cooling of a sill or to model the thermal effect of magma under-plating.

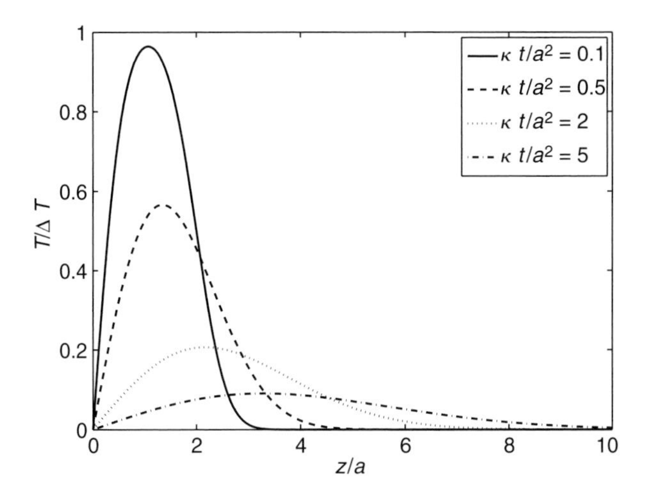

Figure 4.10. Temperature profiles across a horizontal layer cooling in a half-space when the depth to the top is equal to the thickness and the temperature at the surface of the half space is 0.

4.5.4 Cooling of a layer

We examine the cooling of a layer initially at temperature ΔT, with the upper surface kept at temperature $T = 0$ and different boundary conditions at the lower boundary. The heat equation can be solved either by expanding the initial condition in a Fourier series that satisfies the boundary conditions, or by the Laplace transform technique. Such solutions will be used later for calculating the cooling of the oceanic lithosphere.

Constant initial temperature. Surface and base at temperature 0

The initial temperature distribution can be expanded in a Fourier series as,

$$T(z,t=0) = \frac{4\Delta T}{\pi}\sum_{n=0}^{\infty}\frac{\sin((2n+1)\pi z/L)}{(2n+1)}. \tag{4.124}$$

Solving the heat equation for each of the Fourier components gives,

$$T(z,t) = \frac{4\Delta T}{\pi}\sum_{n=0}^{\infty}\frac{\sin((2n+1)\pi z/L)}{(2n+1)}\exp\left(\frac{-(2n+1)^2\pi^2\kappa t}{L^2}\right). \tag{4.125}$$

With the temperature boundary condition at the base, the layer cools symmetrically from the surface and the base.

Constant initial temperature. Surface at temperature 0, no heat flux at the base

In order to match the lower boundary condition, the initial temperature is expanded in a Fourier series as,

$$T(z,t=0) = \frac{4\Delta T}{\pi}\sum_{n=0}^{\infty}\frac{\sin((2n+1)\pi z/2L)}{(2n+1)}. \tag{4.126}$$

Solving the heat equation for each of the Fourier components gives,

$$T(z,t) = \frac{4\Delta T}{\pi}\sum_{n=0}^{\infty}\frac{\sin((2n+1)\pi z/2L)}{(2n+1)}\exp\left(\frac{-(2n+1)^2\pi^2\kappa t}{4L^2}\right). \tag{4.127}$$

With no heat flux at the base, the layer can only cool from its surface. Comparing with the situation when the layer cools symmetrically from surface and base (equation 4.125), we note that the relaxation time constant is four times as long. This comparison illustrates the importance of the boundary conditions in determining the thermal relaxation time. It is important for many geological applications, from the cooling of the sea floor to the subsidence of sedimentary basins.

Layer with time-varying boundary conditions

When the temperature is initially 0, for boundary conditions applied uniformly at the surfaces, temperature depends only on distance z to the surface. The Laplace transform of the temperature in the layer is given by,

$$T(s,z) = A\exp\left(\sqrt{\frac{s}{\kappa}}z\right) + B\exp\left(-\sqrt{\frac{s}{\kappa}}z\right). \tag{4.128}$$

$A(s)$ and $B(s)$ are determined by the boundary conditions.

Constant surface temperature. Varying temperature at the base

If the temperature on the surface $z=L$ varies with time $T(z=0,t)=T_L(t)$ and the temperature on the surface $z=0$ is constant $T(z=0,t)=0$, the Laplace transform of the temperature in the layer is obtained as,

$$T(z,s) = T_0(s)\frac{\sinh\left(\sqrt{\frac{s}{\kappa}}z\right)}{\sinh\left(\sqrt{\frac{s}{\kappa}}L\right)}. \tag{4.129}$$

For a temperature pulse at the base, $T_L(s) = 1$, the inverse transform is,

$$T(z,t) = \frac{2\pi\kappa}{L^2}\sum_{n=0}^{\infty}(-)^n n \sin\left(\frac{n\pi z}{L}\right)\exp\left(\frac{-n^2\pi^2\kappa t}{L^2}\right). \tag{4.130}$$

For a stepwise change in temperature ΔT, we get,

$$T(z,t) = \Delta T\left(\frac{z}{L} + \frac{2}{\pi}\sum_{n=0}^{\infty}\frac{(-)^n}{n}\sin\left(\frac{n\pi z}{L}\right)\exp\left(\frac{-n^2\pi^2\kappa t}{L^2}\right)\right). \tag{4.131}$$

This solution could also have been obtained by integrating between 0 and t the temperature due to a pulse (4.130). Each term in this series shows a relaxation with characteristic time constant depending on the wavelength of the Fourier component. For the leading term of the series, the characteristic relaxation time is $L^2/\pi^2\kappa$.

The transform can be expanded in the following series expansion:

$$T(z,s) = T_0(s)\left(\sum_{n=0}^{\infty}\exp\left(\sqrt{\frac{s}{\kappa}}(z-(2n+1)L)\right)\right.$$
$$\left. - \sum_{n=0}^{\infty}\exp\left(-\sqrt{\frac{s}{\kappa}}(z+(2n+1)L)\right)\right). \tag{4.132}$$

For a stepwise change at $t = 0$ ($T_0(s) = \Delta T/s$), the inverse transform gives,

$$T(z,t) = \Delta T\left(\sum_{n=0}^{\infty}\operatorname{erfc}\left(\frac{(z-(2n+1)L)}{2\sqrt{\kappa t}}\right) - \sum_{n=0}^{\infty}\operatorname{erfc}\left(\frac{(z+(2n+1)L)}{2\sqrt{\kappa t}}\right)\right). \tag{4.133}$$

This gives a different form to the result above. This series can be interpreted in terms of images. The first term is the half-space solution for a source on $z = L$, an image is added to match the boundary condition at $z = 0$, its image must then be added to match the condition at $z = L$, and so on.

Constant temperature at the surface. Varying heat flux at the base

For the slab initially at temperature $T(z,t=0) = 0$, with the temperature gradient $q_L(t) = \lambda\frac{\partial T}{\partial z} = T'_L(t)$ at the base $z = L$, and $T(z=0,t) = 0$, the Laplace transform of the temperature is,

$$T(z,s) = \frac{T'_L(s)\sinh\left(\sqrt{\frac{s}{\kappa}}(z)\right)}{\sqrt{\frac{s}{\kappa}}\cosh\left(\sqrt{\frac{s}{\kappa}}L\right)}. \tag{4.134}$$

For a heat flux pulse on the surface $z = L$, $q_L = \frac{\lambda \Delta T}{L} \delta(t/\tau)$

$$T(z,t) = \frac{2\kappa\tau\Delta T}{L} \sum_{n=0}^{\infty} (-)^n \sin\left(\frac{(2n+1)\pi z}{2L}\right) \exp\left(-\frac{(2n+1)^2\pi^2\kappa t}{4L^2}\right). \quad (4.135)$$

For a stepwise change in heat flux, we obtain the solution by integrating 4.135:

$$\frac{T(z,t)}{\Delta T} = \frac{z}{L} - \frac{8}{\pi^2} \sum_{n=0}^{\infty} \frac{(-)^n}{(2n+1)^2} \sin\left(\frac{(2n+1)\pi z}{2L}\right) \cdots$$
$$\times \exp\left(-\frac{(2n+1)^2\pi^2\kappa t}{4L^2}\right). \quad (4.136)$$

Periodic variations in the boundary condition at the base of the lithosphere

As the lithosphere moves over the convecting mantle, the basal boundary conditions change. These time fluctuations are damped out and are not likely to affect temperature and heat flux near the surface. If the lithosphere moves over hot spots, it experiences variations of temperature and heat flux at its base. If the variation in heat flux at the base of the lithosphere is approximated by a periodic function, $q_L(t) = \Delta Q \cos(\omega t)$, and the surface temperature is 0. Decomposing the solution into a transient and quasi-state components, we obtain the quasi-steady-state variation in surface heat flux,

$$\frac{q_0(t)}{\Delta Q} = \frac{1}{2\left(\cosh^2\sqrt{\frac{\omega\tau}{2}} - \sin^2\sqrt{\frac{\omega\tau}{2}}\right)}$$
$$\times \left(\cos\left(\omega t - \sqrt{\frac{\omega\tau}{2}}\right) \exp\sqrt{\frac{\omega\tau}{2}} + \cos\left(\omega t + \sqrt{\frac{\omega\tau}{2}}\right) \exp -\sqrt{\frac{\omega\tau}{2}}\right) \quad (4.137)$$

where $\tau = L^2/\kappa$ and κ is the thermal diffusivity. The series for the transient is made of exponentially decreasing terms that can be neglected for large t when the effect of the initial condition has decayed. For $\omega\tau \gg 1$, that is, the period of the heat flux variation is much smaller than the heat conduction time for the lithosphere, the surface heat flux variation is $\mathcal{O}(\exp(-\sqrt{\omega\tau/2}))$ and can be neglected.

If a periodic change in temperature $\Delta T \cos \omega t$ is assumed at the lower boundary, the solution of the heat equation gives for the surface heat flux,

$$\frac{q_0(t)}{\lambda \Delta T/L} = \frac{\sqrt{\omega\tau}}{\sinh^2\sqrt{\frac{\omega\tau}{2}} + \sin^2\sqrt{\frac{\omega\tau}{2}}}$$

$$\times \left(\cos\left(\omega t - \sqrt{\frac{\omega\tau}{2}} - \frac{\pi}{4}\right) \exp\sqrt{\frac{\omega\tau}{2}} - \cos\left(\omega t + \sqrt{\frac{\omega\tau}{2}} + \frac{\pi}{4}\right) \exp - \sqrt{\frac{\omega\tau}{2}} \right)$$

$$+ \cdots \tag{4.138}$$

where the ... stand for the initial transient.

For $\omega\tau \gg 1$, that is, the period of the basal heat flux variation is smaller than the heat conduction time for the lithosphere, the surface heat flux is $\mathcal{O}(\exp(-\sqrt{\omega\tau/2}))$ and can be neglected.

For example, if the period of fluctuations at the base of the lithosphere is one half of L^2/κ, the amplitude of basal heat flux variations is reduced to $< 9\%$ when it reaches the surface. For a lithosphere 200 km thick, $L^2/\kappa \approx 1000$ My; we can safely neglect periodic variations at the base.

Layer with horizontal variations in temperature

If the initial condition varies in the horizontal direction for the two-dimensional case, or with distance to a vertical axis of symmetry for the axially symmetric case, the initial condition can be written as a Fourier or a Hankel transform.

The Fourier and the Hankel transforms of the heat equation take the same form:

$$\frac{\partial T}{\partial t} = \kappa \left(\frac{\partial^2 T}{\partial z^2} - k^2 T \right). \tag{4.139}$$

The Laplace transform takes the form,

$$\frac{\partial^2 T}{\partial z^2} - (s + k^2\kappa)T = T_0(k, z). \tag{4.140}$$

For the 2D case, k is the Fourier transform variable; for the axially symmetric case it is the Hankel transform variable.

The Laplace transform of solution is the same as the solution of the one-dimensional problem with $s \to s + \kappa k^2$. When taking the inverse Laplace transform, for each wave-number, the one-dimensional solution (i.e. the solution for wave-number 0) is multiplied by $\exp(-\kappa k^2 t)$. This term accounts for the horizontal or radial diffusion of heat.

4.5.5 Transient heat equation in spherical geometry

In spherical geometry, the heat equation is,

$$\frac{\partial T}{\partial t} = \kappa \left(\frac{1}{r^2} \frac{\partial}{\partial r} r^2 \frac{\partial T}{\partial r} + \frac{1}{r^2 \sin^2 \theta} \frac{\partial^2 T}{\partial \theta^2} + \frac{1}{r^2} \frac{\partial^2 T}{\partial \phi^2} \right). \tag{4.141}$$

Spherical symmetry

If the temperature distribution is spherically symmetric, substituting $u(r,t)/r$ for $T(r,t)$ yields,

$$\kappa \frac{\partial^2 u}{\partial r^2} = \frac{\partial u}{\partial t}, \tag{4.142}$$

which is formally identical to the 1D heat equation. After making the corresponding changes to the boundary and initial conditions, we can thus use the one-dimensional heat equation to obtain solutions for the 3D spherically symmetric heat equation:

$$\frac{\partial T}{\partial t} = \kappa \frac{1}{r^2} \frac{\partial}{\partial r} r^2 \frac{\partial T}{\partial r}. \tag{4.143}$$

The solution in the form $T(t)R(r)$ gives,

$$\frac{1}{T(t)} \frac{\partial T}{\partial t} = \frac{\kappa}{r^2 R(r)} \frac{\partial}{\partial r} r^2 \frac{\partial R(r)}{\partial r}. \tag{4.144}$$

For this equation to hold, each term must be constant. This will hold for:

$$\begin{aligned} R(r) &= \frac{\exp(ikr)}{r} \\ T(t) &= \exp(-\kappa k^2 t). \end{aligned} \tag{4.145}$$

The functions $R(r)$ are used to describe spherical waves; they have the same oscillatory behavior as plane waves and a geometrical dispersion factor $1/r$ which insures the conservation of energy as the wave spreads away from its source. The "wavenumbers" k that are retained in this development depend on the boundary and initial conditions.

Cooling of a sphere with surface kept at fixed temperature

For instance, if the initial temperature is uniformly T_0 for $0 < r < a$ and $T = 0$ at $r = a$, the solution uses the Fourier series expansion,

$$\sum_{n=1}^{\infty} \frac{(-)^{(n-1)}}{n} \sin(nx) = \frac{x}{2}, \quad -\pi < x < \pi, \tag{4.146}$$

or, introducing $x = \pi r/a$,

$$\frac{2aT_0}{\pi r}\sum_{n=1}^{\infty}\frac{(-)^{(n-1)}}{n}\sin(\frac{n\pi r}{a}) = T_0, \quad -a < r < a. \tag{4.147}$$

This gives,

$$T(r,t) = \frac{2T_0 a}{\pi r}\sum_{n=1}^{\infty}\frac{(-1)^{(n-1)}\sin\left(\frac{n\pi r}{a}\right)}{n}\exp(-\frac{\kappa n^2\pi^2 t}{a^2}). \tag{4.148}$$

The heat flux at the surface is given by

$$q(t) = \lambda\frac{\partial T(r=a,t)}{\partial r} = \frac{2\lambda T_0}{a}\sum_{n=1}^{\infty}\exp\left(-\frac{\kappa n^2\pi^2 t}{a^2}\right). \tag{4.149}$$

The series diverges for $t \to 0$, but converges for any $t > 0$; it decreases very rapidly and its value drops < 1 for $\kappa t/a^2 = 0.03$. Using this equation rather than the half-space solution to derive the age of the Earth would not significantly change the value obtained by Kelvin. Indeed, for $t \to 0$, we have,

$$\sum_{n=1}^{\infty}\exp\left(-\frac{\kappa n^2\pi^2 t}{a^2}\right) \approx \frac{a}{\pi\sqrt{\kappa t}}\int_0^{\infty}\exp(-u^2)du = \frac{a}{2\sqrt{\pi\kappa t}}, \tag{4.150}$$

which gives the same heat flux as in the half-space solution.

Cooling of a spherical inclusion within a conductive region

We must solve the heat equation in the region $0 < r < \infty$, with initial conditions $T(r < a, 0) = T_0$ and $T(r > a, 0) = 0$. We search for the solution in the form,

$$\int_0^{\infty} F(k)\frac{\sin(kr)}{r}\exp(-\kappa k^2 t)dk, \tag{4.151}$$

where $F(k)$ is the Fourier sin transform of the initial condition,

$$F(k) = \int_0^a r\sin(kr)dr = \frac{\sin(ka) - ka\cos(ka)}{k^2}. \tag{4.152}$$

The solution is obtained in the form of a Fourier sin transform:

$$\frac{aT_0}{2\pi r}\int_0^{\infty}\frac{\sin(ka) - ka\cos(ka)}{k^2a^2}\exp(-\kappa k^2 t)\sin(kr)dk. \tag{4.153}$$

Approximations for $t \gg a^2/\kappa$ are obtained by expansion in series of ka. As one would expect, the cooling of a spherical intrusion depends on its radius: the evaluation of the sin transform shows that, as it spreads out, the temperature perturbation decays more rapidly for $t > a^2/\kappa$ than for a layer (Figure 4.11).

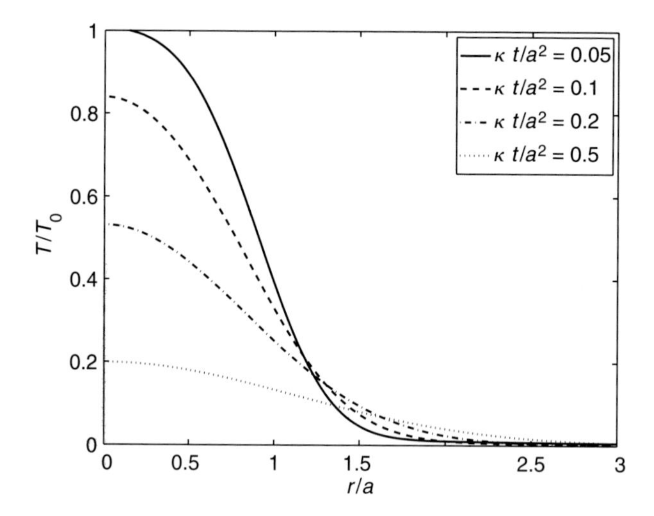

Figure 4.11. Temperature perturbation due to a hot spherical intrusion. As the perturbation spreads out, geometrical dispersion reduces the amplitude of the perturbation relative to that for a layer (Figure 4.9).

Heating of a sphere by heat production

For a sphere initially at temperature 0, with uniform distribution of heat production decreasing exponentially with time $H\exp(-t/\tau_s)$, and surface temperature 0, the temperature distribution is obtained by Laplace transform:

$$T(r,t) = -\frac{\kappa\tau_s H}{\lambda}\exp(-t/\tau_s) \times \left(1 - \frac{a\sin(r/\sqrt{\kappa\tau_s})}{r\sin(a/\sqrt{\kappa\tau_s})}\right)$$
$$-\frac{2a\kappa\tau_s H}{\pi r\lambda}\sum_{n=1}^{\infty}\frac{(-)^n}{n(1-\tau_s n^2\pi^2\kappa/a^2)}\sin\left(\frac{n\pi r}{a}\right)\exp\left(-\frac{n^2\pi^2\kappa t}{a^2}\right). \tag{4.154}$$

The surface heat flux is,

$$q(t) = \frac{\kappa\tau_s H}{a}\left\{\left(1 - \frac{a}{\sqrt{\kappa\tau_s}}\mathrm{cotan}\left(\frac{a}{\sqrt{\kappa\tau_s}}\right)\right) \times \exp(-t/\tau_s)\right.$$
$$\left. +2\sum_{n=1}^{\infty}\frac{1}{1-\tau_s n^2\pi^2\kappa/a^2}\exp\left(-\frac{n^2\pi^2\kappa t}{a^2}\right)\right\}. \tag{4.155}$$

It is easy to verify that $\lim_{t\to 0} q(t) = 0$, as one would expect. For a uniform distribution of heat sources in the sphere, the heat flux due to internal heating is less than quasi-steady state until $t \approx 3\tau_s$. For a conductive sphere of Earth size, we have also seen that the initial temperature condition contributes less than 1 K km^{-1}, or

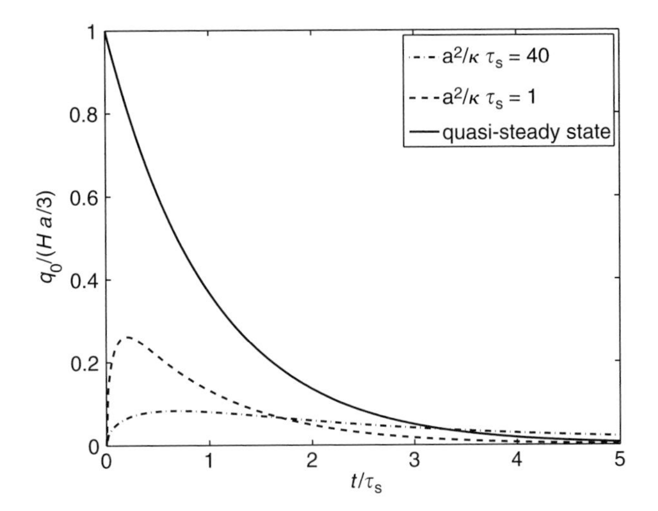

Figure 4.12. Heat flux at the surface of a conductive sphere, when the initial temperature is zero, and heat sources are uniformly distributed and decrease exponentially with time as $\exp(-t/\tau_s)$. Quasi-steady-state heat flux is equal to the total internal heat production at the same time. When the characteristic time for heat conduction $\geq \tau_s$, the heat flux is far from quasi-steady state. For the Earth, $\tau_s = 2.5$ Gy, and the characteristic heat conduction time is $\approx 40\ \tau_s$.

3 mW m^{-2} to the heat flux. For the Earth, the present heat production could still account for the present heat flux, if heat sources are concentrated near the surface.

This example has no relevance to Earth, but it demonstrates the interplay between the decay of heat sources and the inefficiency of conduction to transport heat, in such a way that surface heat flow can never be in equilibrium with heat production at any time.

4.6 Thermal stresses

When an elastic body is heated non-uniformly, stresses are induced by differences in thermal expansion between different parts of the body. This is done adding a term $-\alpha K \Delta T \mathbf{I}$ to the stress tensor $\boldsymbol{\sigma}$ as follows:

$$\boldsymbol{\sigma} = \left(K - \frac{2\mu}{3}\right) \nabla \cdot \mathbf{u}\mathbf{I} + \mu(\nabla\mathbf{u} + \nabla\mathbf{u}^T) - \alpha K \Delta T \mathbf{I}, \tag{4.156}$$

K and μ are the bulk and shear moduli and α is the volumetric thermal expansion coefficient. The equilibrium condition is,

$$\nabla \cdot \boldsymbol{\sigma} = \left(K + \frac{4\mu}{3}\right) \nabla(\nabla \cdot \mathbf{u}) + \mu \nabla \times \nabla \times \mathbf{u} - \alpha K \nabla \Delta T = 0. \tag{4.157}$$

If there is no shear, $\nabla \times \nabla \times \mathbf{u} = 0$ and the equation can be written,

$$\nabla \nabla \cdot \mathbf{u} = \frac{\alpha K}{K + 4\mu/3} \nabla T = \frac{\alpha(1+\sigma)}{3(1-\sigma)} \nabla \Delta T, \tag{4.158}$$

where σ is Poisson's ratio. For a sphere that is not heated uniformly, ΔT is the difference from the average temperature. For a spherically symmetric temperature variation $\Delta T(r)$, the displacement is written as,

$$\frac{d}{dr} \frac{1}{r^2} \frac{d}{dr} r^2 u_r = \frac{\alpha(1+\sigma)}{3(1-\sigma)} \frac{d \Delta T(r)}{dr}, \tag{4.159}$$

which can be solved for different boundary conditions. For instance, if the stress vanishes on the sphere surface $r = a$,

$$u_r = \frac{\alpha(1+\sigma)}{3(1-\sigma)} \left(\frac{1}{r^2} \int_0^r \Delta T(r') r'^2 dr' + \frac{2(1-2\sigma)}{1+\sigma} \frac{r}{a^3} \int_0^a \Delta T(r') r'^2 dr' \right). \tag{4.160}$$

In order to determine the stresses caused by magmatic intrusions, for example, we need to solve the equations above with more complicated boundary conditions. Without going any further into the theory of elasticity, we can see how thermal stresses can be accounted for.

Exercises

4.1 Consider a layer $0 < z < a$, with uniform heat production H and thermal conductivity λ. Find the general solution to 1D steady-state heat equation for that layer. Determine the temperature in the layer when: 1) $T = 0$ for $z = 0$ and $z = a$; 2) $T = 0$ for $z = 0$ and $q = 0$ for $z = a$. Determine the heat flux at the surface $z = 0$ for both conditions.

4.2 Consider an infinite cylinder of radius a with uniform heat production H and thermal conductivity λ. Find the general solution to the heat equation,

$$\frac{1}{r} \frac{\partial}{\partial r} r \frac{\partial T}{\partial r} + \frac{H}{\lambda} = 0.$$

Solve this equation with the conditions that 1) the temperature is finite, and 2) $T = 0$ for $r = a$. Compare the temperature at the center of the cylinder to the temperature at the center of a layer $0 < z < 2a$ when $T = 0$ on $z = 0$ and $z = 2a$.

4.3 Consider a fluid of viscosity μ in a layer $0 < z < a$, when the flow is laminar and there is a horizontal pressure gradient $\partial P/\partial x = P_0'$, the equation for flow 3.54 becomes,

$$\mu \frac{\partial^2 v_x}{\partial z^2} = P_0'.$$

Determine the velocity v_x in the fluid when 1) $v_x = 0$ for $z = 0$ and $z = a$, and 2) $v_x = 0$ for $z = a$ and $\partial v_x/\partial z = 0$ for $z = 0$.

4.4 Consider a layer of thickness L with thermal conductivity λ in steady state with heat sources H uniformly distributed in the region $a < z < b$, and 0 outside. The temperature at the surface $T(z=0)=0$ and the heat flux at the base $q(z=L)=0$. Assuming $a-b=L/4$, determine how the maximum and average temperatures of the layer vary with a $(a < 3L/4)$.

4.5 Assuming that thermal conductivity decreases with temperature as $\lambda(T)=\lambda_0/(1+\alpha T)$ solve the one-dimensional heat equation for $z>0$ with condition $T(z)=0$ and $dT/dz = q_0/\lambda_0$ for $z=0$ with no heat source. If temperature $T_d = q_0 d/\lambda_0$ is obtained for constant conductivity λ_0, calculate a correction for when $\alpha T_d \ll 1$.

4.6 Consider a homogeneous sphere of radius a with uniform heat production H in steady state. Without solving the heat equation, determine how the surface heat flux scales with the radius of the sphere.

4.7 Consider a spherical shell in steady state, $b<r<a$, with uniform heat production H and zero heat flux on the inner surface $(r=b)$. Without solving the heat equation, calculate the heat flux on the outer surface.

4.8 Assuming that thermal conductivity increases with temperature as $\lambda(T)=\lambda_0+\beta T^3$, solve the one-dimensional heat equation with no heat source for $z>0$ with condition $T(z=0)=300$, $dT/dz=q_0/\lambda_0$ at $z=0$.

4.9 Determine the temperature variations in a 2D half space, $z>0$, when the surface temperature is $T(x,z=0)=\Delta T$ for $|z|<a$ and $T(z=0,t)=0$ for $|z|>a$.

4.10 Determine the Fourier transform of the heat flow at the surface of the region $(z>0)$ when the heat production varies as $H(x,y,z)=H_0(x,y)\exp(-z/\delta)$. Determine the relationship between the Fourier spectra of surface heat flux and surface heat production when $|k|\delta \gg 1$ and $|k|\delta \ll 1$.

4.11 Assuming that the glacial–interglacial temperature cycles can be represented by a cosine function with period 100,000 years, determine the depth where the amplitude of the temperature oscillation is reduced to half the surface value ($\kappa = 30\,\mathrm{m^2\,y^{-1}}$).

4.12 During the annual cycle, the underground stores heat during the summer and gives it back during the winter. Assume that this cycle can be described as $T(z=0,t)=\Delta T\cos(2\pi t/\tau)$, with $\tau=1$ year and $\Delta T=15\mathrm{K}$, $\kappa=30\ \mathrm{m^2\ y^{-1}}$. Determine at what depth the temperature is highest when the surface is coldest. What would be the efficiency of an ideal thermodynamic engine extracting power from that temperature difference when the mean surface temperature is 0° Celsius, (273 K)? (The maximum efficiency of a thermodynamic engine is $\Delta W = 1/T_{\min} - 1/T_{\max}$, where T is absolute temperature.) Would such an engine be more efficient in Canada or in India?

4.13 Use the Laplace transform to derive the temperature in the half-space initially at temperature $T=0$ with a constant heat flux condition at the surface,

$$-\lambda\frac{\partial T}{\partial z} = q_0,\ t>0.$$

Find how surface temperature varies with time.

4.14 Use the Laplace transform to obtain a solution for the cooling of a half-space below a blanket of thickness a. i.e. the initial condition is,

$$T(z,t=0)=0,\ 0<z<a$$

$$T(x,t=0)=T_0,\ a<z<\infty,$$

and the surface boundary condition is $T(z=0,t)=0$.

4.15 For a line source of heat of intensity Q in an infinite medium, the temperature at distance r from the line and time t after the start of the source is given by:

$$T(r,t)=\frac{-Q}{4\pi\lambda}Ei\left(\frac{-r^2}{4\kappa\lambda}\right),$$

where the exponential integral is defined as

$$Ei(-\mu)=-\int_{\mu}^{\infty}\frac{\exp(-z)}{z}dz.$$

If the source is active between $0<t<t_d$, find the temperature at r for $t>t_d$. If the temperature at t_d is T_0 for $r=a$, find the temperature perturbation at $t\gg t_d$. (Use the expansion $Ei(z)=\log(z)+0.557+\cdots$ to obtain temperature as a function of t/t_d.) Explain the units for Q in the equation.

4.16 Calculate the series expansion of the temperature gradient at the surface of the moving medium (equation 4.122) for $v^2t/\kappa\ll 1$. Determine the value of the sedimentation/erosion rate for which the correction reaches 10% after 100 My for $\kappa=30\ \mathrm{m^2\ y^{-1}}$.

5

Heat transport by convection

Objectives of this chapter

Heat is transported by convection in magmas, in hydrothermal systems, and in the Earth's mantle. In this section, we investigate thermal convection in the simplest geometrical configurations and boundary conditions, and hence do not deal with specific geological examples. We illustrate important aspects of convective flows and develop the tools that will be used later to analyze a number of natural convective systems. We do not provide a comprehensive account of flow patterns but focus on the characteristics of heat transport. We define dimensionless numbers and use them to characterize the different regimes of convection. We show that, even in highly non-linear convective systems, relationships between physical quantities can be reduced to simple scaling laws. We review experimental data and numerical calculations that support these scaling laws.

5.1 Isolated heat sources: Plumes and thermals

When heat is released from a small area at the base of a fluid layer, such as a heating coil at the bottom of a container, convection develops in an isolated rising element. This is relevant to the atmosphere or the ocean above a lava flow, and to a magma chamber which gets replenished by primitive melt. Over a heated horizontal plate of large width, plumes can be generated by instabilities in a thermal boundary layer. One distinguishes between a plume, which occurs when heat is continuously supplied and which remains connected to the source at all times, and a thermal, which is generated by the release of a finite amount of energy and which detaches from the source (Figure 5.1). In geological reality, convective upwellings are fed from finite sources and are rarely sustained by a constant power input. Thus, one should consider flows that wane as their sources get exhausted, such that an upwelling initially develops as a plume and ends up as a thermal. For clarity purposes, we shall only refer to upwellings due to heating, but the theory and

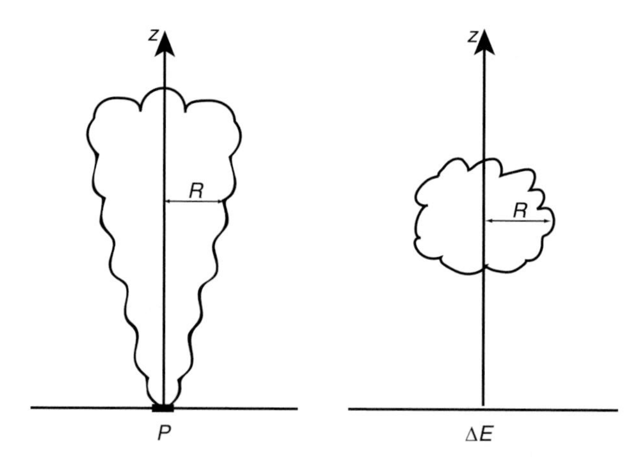

Figure 5.1. Two types of convective flow generated by heating in a small region at the bottom of a fluid layer. Left: a constant power P is generated. A narrow upwelling structure called a plume extends vertically. Right: a fixed amount of energy ΔE is released and a volume of heated fluid called a thermal detaches from the source. The figure is drawn for turbulent flow regimes, such that turbulent eddies at the edges entrain ambient fluid into the convective element.

physical arguments also apply to downwellings generated by cooling at the top of a fluid layer.

5.1.1 Steady plumes

A heat source feeds a constant power P at the base of a fluid volume which is initially at a uniform temperature. Depending on the magnitude of P, flow may be in a laminar or a turbulent regime. The conditions that are required for each regime cannot be specified without knowledge of the plume velocity, which has not yet been determined. We denote by θ the temperature anomaly in the plume and introduce characteristic scales W, R and Θ for the plume velocity, radius and temperature anomaly at height z above the source. We use scaling arguments to derive the relationships between these scales and control variables such as P.

Laminar plumes

A plume is characterized by a large "head" region at the top of a thinner "stem" region (Figure 5.2). Hot plume material loses heat to the surrounding fluid by diffusion implying that the plume grows wider as it rises. In a laminar regime, buoyancy is balanced by viscous forces, so that,

$$W \propto \frac{R^2 \Delta\rho g}{\mu} \propto \frac{R^2 \rho_o \alpha \Theta g}{\mu}. \tag{5.1}$$

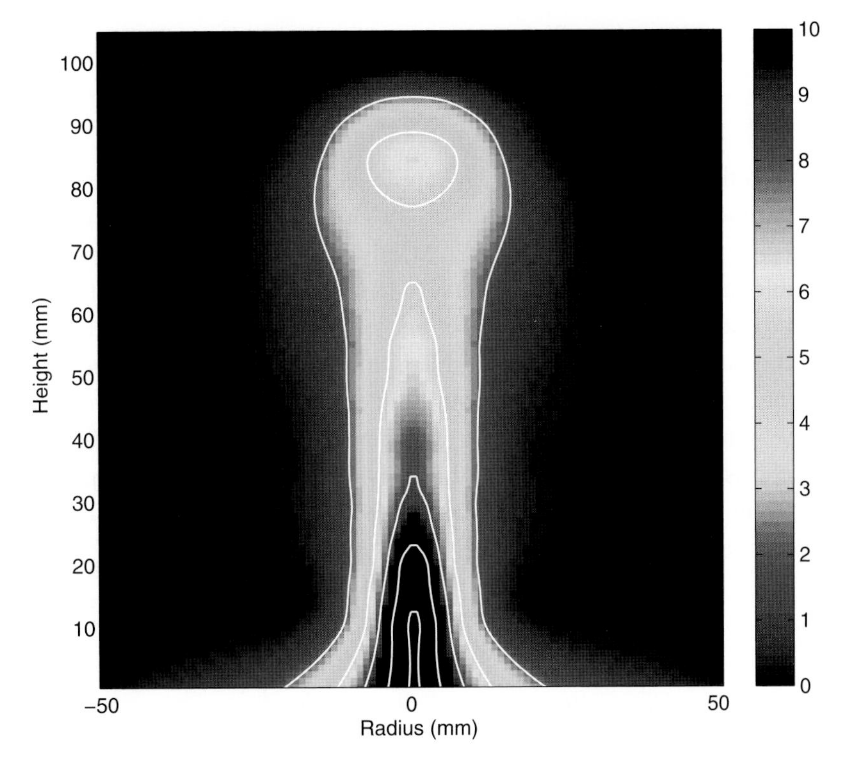

Figure 5.2. Temperature field in a laboratory plume (from A. Limare, pers. comm.). Experimental techniques are described in Davaille *et al.* (2010). Note the large horizontal temperature gradients and a cap that is wider than the stem. See plate section for color version.

We next write that the plume energy flux at any height z is equal to P:

$$P \propto \rho_o C_p W \Theta R^2. \tag{5.2}$$

From these two relationships, we obtain the remarkable result that (Batchelor, 1954),

$$W \propto \sqrt{\frac{\alpha g P}{\mu C_p}}. \tag{5.3}$$

W does not depend on z, which indicates that the plume rises at a constant velocity. This is confirmed by laboratory experiments (Moses *et al.*, 1993; Kaminski and Jaupart, 2003). The third equation comes from the heat transport characteristics: the plume radius increases with height due to diffusion, so that it is $R \approx \sqrt{\kappa t}$ after time t. Writing that $t \propto z/W$, we find that

$$R \propto \sqrt{\kappa \frac{z}{W}}. \tag{5.4}$$

Substituting into the energy flux equation, we finally obtain,

$$\Theta \propto \frac{P}{\lambda z}, \tag{5.5}$$

which shows that the thermal anomaly rapidly decays away from the source.

A full solution of the problem requires numerical calculations. We focus on the "stem" region which can be studied in steady state. The plume is axisymmetric and variables are expressed as a function of radial distance r and height z. Velocity $\mathbf{v}$ has components (u,w) in the (r,z) directions. The governing equations can be simplified greatly by noting that the radial velocity is small and hence the radial component of the momentum equation reduces to $\partial p/\partial r \approx 0$. The pressure field within the plume is therefore close to that in the undisturbed fluid far from the plume, i.e. it is hydrostatic: $p = p_o - \rho_o g z$. The plume develops over large vertical distances, and hence vertical derivatives can be neglected in comparison to radial ones. The governing equations for the steady laminar plume are:

$$\nabla \cdot \mathbf{v} = 0 \tag{5.6}$$

$$u\frac{\partial w}{\partial r} + w\frac{\partial w}{\partial z} = \nu \frac{1}{r}\frac{\partial}{\partial r}\left(r\frac{\partial w}{\partial r}\right) + \alpha\theta g \tag{5.7}$$

$$w\frac{\partial \theta}{\partial z} = \kappa \frac{1}{r}\frac{\partial}{\partial r}\left(r\frac{\partial \theta}{\partial r}\right) \tag{5.8}$$

$$\int_0^\infty \rho_o C_p w\theta 2\pi r dr = P, \tag{5.9}$$

where we have introduced kinematic viscosity, $\nu = \mu/\rho_o$. We note that the energy equation has been reduced to a simple balance between vertical advection and horizontal diffusion. Inserting the scales for the plume radius and velocity into this balance, we deduce that $W/z \sim \kappa R^{-2}$, which is of course what we had already obtained above in equation 5.4.

One may be surprised that inertial terms have been retained in the vertical momentum equation even though we are dealing with a laminar flow regime. The reason is that there are two regions in the flow with different dynamics. Near the axis, heated fluid defines an inner region where flow is driven by buoyancy and where the dominant force balance is between buoyancy and viscous shear. In the outer region, there is no thermal anomaly but the vertical velocity is not negligible: viscous stresses are not zero and must be balanced by inertia. We return to this important point below.

In order to solve these equations, we rely on the scaling laws derived above, which indicate that the plume structure depends on local conditions at height z. The radial profiles of velocity and temperature are identical save for a height-dependent scale factor, a property that is referred to as *self-similarity*. In mathematical terms, we seek solutions of the following form:

$$\theta(r,z) = \frac{P}{2\pi\lambda z} g(\eta), \tag{5.10}$$

where g is a function of similarity variable η;

$$\eta = \frac{r}{R(z)} = \frac{r}{\sqrt{\kappa z/W}}. \tag{5.11}$$

We proceed in the same manner for all variables and, for simplicity, drop the * symbol. In order to satisfy the continuity equation, we introduce the stream function ψ such that,

$$u = -\frac{1}{r}\frac{\partial \psi}{\partial z}, \quad w = \frac{1}{r}\frac{\partial \psi}{\partial r}. \tag{5.12}$$

Substituting for the velocity scale, we find that ψ scales with κz. Thus,

$$\psi = \kappa z f(\eta) \tag{5.13}$$

where f is another dimensionless function to be solved for. The governing equations are now written in compact form as follows:

$$f \frac{d}{d\eta}\left(\frac{d\dot{f}}{\eta}\right) + \mathrm{Pr}\left[f^{(3)} - \frac{d}{d\eta}\left(\frac{\dot{f}}{\eta}\right) + \eta g\right] = 0 \tag{5.14}$$

$$\eta \dot{g} + f g = 0 \tag{5.15}$$

$$\int_0^\infty \dot{f} g \, d\eta = 1. \tag{5.16}$$

These dimensionless equations depend on a single dimensionless number, the Prandtl number Pr. Thus, for example, the axial plume velocity is in fact such that

$$W_o \propto f_W(Pr)\sqrt{\frac{\alpha g P}{\mu C_p}}, \tag{5.17}$$

where $f_W(\mathrm{Pr})$ is some function to be solved for. An asymptotic analysis valid for $\mathrm{Pr} > \approx 10^2$ yields (Worster, 1986):

$$f_W = \sqrt{\log(\epsilon^{-2})/(2\pi)}, \tag{5.18}$$

Table 5.1. *Steady laminar plume characteristics*

Pr	$f_W(Pr)$ †	$g(0)$
1	0.3989	0.6666
$> 10^2$	equation 5.18	1/2

Results from Worster (1986).
† Value of function f_W for the vertical velocity at the axis (equation 5.17).

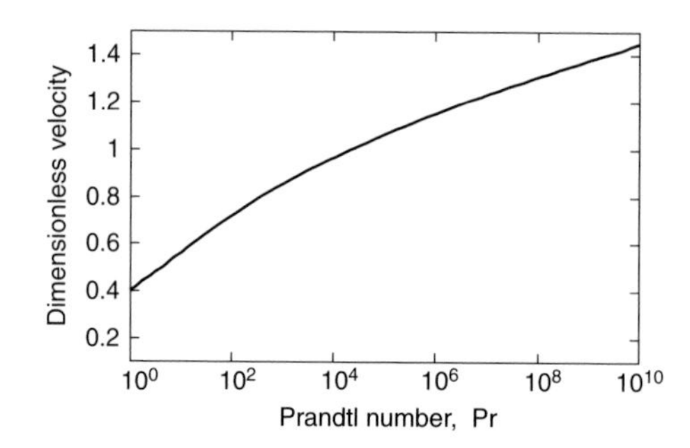

Figure 5.3. Vertical velocity at the axis of a steady laminar plume as a function of Prandtl number (from Kaminski and Jaupart, 2003). Velocity is made dimensionless using the scale in equation 5.3.

where ϵ is the solution of $\epsilon^4 \log \epsilon^{-2} = \mathrm{Pr}^{-1}$. Other quantitative results are listed in Table 5.1 and shown in Figure 5.3. The Pr dependence is weak, but must be accounted for when dealing with fluids that are as diverse as basaltic magmas and the Earth's mantle.

As shown in Figure 5.4, the plume thermal anomaly stretches over radial distance $\delta_\theta \sim R(z)$. Away from the axial region, the dominant dynamical balance is between inertia and viscous forces, such that:

$$w \frac{\partial w}{\partial z} \sim \nu \frac{1}{r} \frac{\partial}{\partial r} \left(r \frac{\partial w}{\partial r} \right). \tag{5.19}$$

We deduce that the vertical velocity drops to zero in a momentum boundary layer of thickness δ_W that scales with $\sqrt{\nu z / W}$. This illustrates the parallel between heat and momentum diffusion that was alluded to in Chapter 3. More specifically, one has,

$$\frac{\delta_W}{\delta_\theta} = \sqrt{\frac{\nu}{\kappa}} = \mathrm{Pr}^{1/2}, \tag{5.20}$$

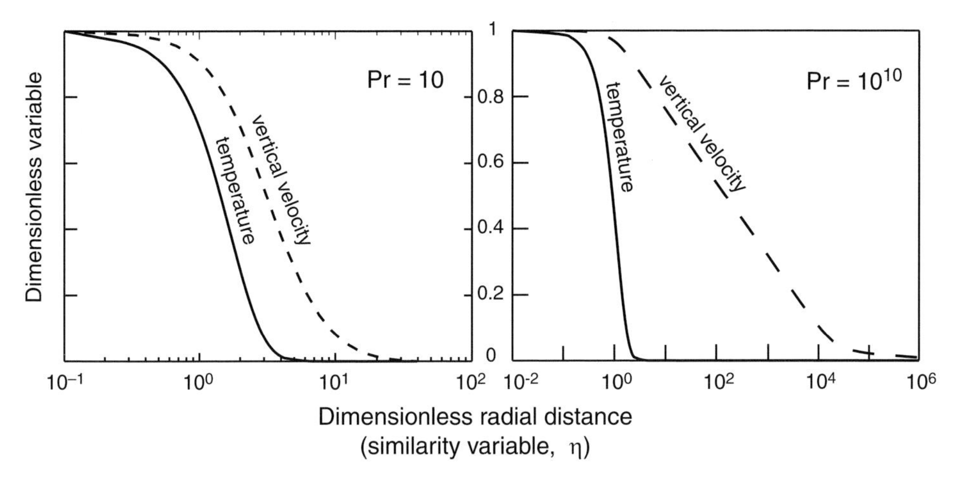

Figure 5.4. Radial profiles of temperature and velocity in steady laminar plumes for two values of the Prandtl number (from Kaminski, 1997). Temperature and velocity have been normalized to their maximum values at the axis. Radial distance is shown as similarity variable η (equation 5.11). In both cases, the thermal anomaly extends to a distance of order one. In contrast, the radial velocity profile stretches over a distance that increases with Pr.

which shows how the Prandtl number affects the flow structure. For large values of Pr, which is relevant for geological fluids, the momentum boundary layer extends to very large distances, so that a plume affects a large volume of fluid. Conversely, of course, a plume is influenced by the presence of other plumes as well as lateral boundaries or irregularities.

The above theory suffers from two shortcomings. It is only valid for a plume that has developed over vertical distances that are much larger than its width and this may not be justified for the mantle, for example. The other shortcoming is that it says nothing about the plume cap above the stem. Laboratory experiments show that the plume cap rises at a velocity that is proportional to that of the stem (Kaminski and Jaupart, 2003).

For the plume to rise in a laminar regime, the Reynolds number must be small. The plume dimensions change with height and one can not define a single value of the Reynolds number that accounts for the flow dynamics at all heights. We define a local Reynolds number at height z:

$$\mathrm{Re}(z) = \frac{\rho W R}{\mu} \propto z^{1/2}. \tag{5.21}$$

This local Reynolds number increases with height and may eventually exceed the threshold value for turbulence.

Turbulent plumes

In a turbulent regime, the flow characteristics do not depend on material molecular properties such as thermal conductivity and viscosity. The flow rapidly adjusts to a self-similar radial structure, which simplifies the writing of the conservation equations. Radial profiles of vertical velocity and temperature can be fit accurately with the same Gaussian function, so that:

$$w(r,z) = W(z)\exp\left(-\frac{r^2}{R^2}\right) \tag{5.22}$$

$$\theta(r,z) = \Theta(z)\exp\left(-\frac{r^2}{R^2}\right), \tag{5.23}$$

where R is a characteristic radius that depends on z. In the Boussinesq approximation, the bulk vertical fluxes of mass, momentum and energy are:

$$\text{Mass} \int_0^\infty \rho w 2\pi r dr = \pi\rho W R^2 \int_0^\infty \exp(-u)du \tag{5.24}$$

$$\text{Momentum} \int_0^\infty \rho w^2 2\pi r dr = \pi\rho W^2 R^2 \int_0^\infty \exp(-2u)du \tag{5.25}$$

$$\text{Energy} \int_0^\infty \rho C_p w\theta 2\pi r dr = \pi\rho C_p W R^2 \Theta \int_0^\infty \exp(-2u)du. \tag{5.26}$$

These relationships involve integrals whose values are independent of height and can be simplified by suitably rescaling W and Θ. This amounts to using average values of velocity and temperature over "boxcar" radial profiles (Turner, 1973).

At the edge of the plume, turbulent eddies entrain surrounding fluid at average volumetric rate ϕ (Figure 5.5), so that mass conservation can be written as,

$$\frac{d}{dz}\left(\rho W \pi R^2\right) = 2\pi R \rho_o \phi. \tag{5.27}$$

Conservation of vertical momentum involves buoyancy only, as shear stresses can be neglected at the edges of the plume:

$$\frac{d}{dz}\left(\rho W^2 \pi R^2\right) = \pi R^2 (\rho_o - \rho) g = \pi R^2 \Delta\rho g, \tag{5.28}$$

where $\Delta\rho = \rho_o - \rho$. Conservation of energy is again written as,

$$P = \pi \rho_o C_p W \Theta R^2. \tag{5.29}$$

For consistency with the other equations, this is rewritten as follows:

$$R^2 W \frac{g\Delta\rho}{\rho_o} = g\alpha R^2 W \Theta = \text{constant} = F = \frac{\alpha g}{\pi \rho_0 C_p} P, \tag{5.30}$$

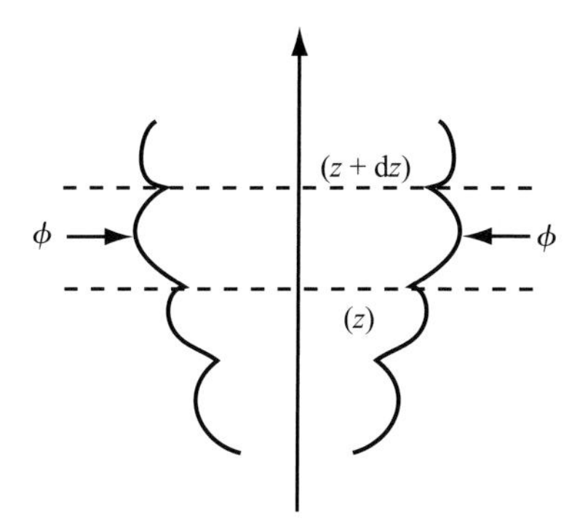

Figure 5.5. Illustration of entrainment into a turbulent plume. Transient eddies that develop at the edge of the plume carry fluid from the exterior into the plume. ϕ denotes the average entrainment rate, which may be thought of as the average horizontal velocity at the edge of the plume.

where quantity F is called the buoyancy flux, with dimensions of L^4T^{-3}. In all these equations, the acceleration of gravity g only appears in combination with the density difference $\Delta\rho$, as expected for buoyancy-driven flow.

One needs a closure equation for ϕ, the entrainment rate of surrounding fluid into the plume. The governing equations involve only one parameter, the buoyancy flux F, so that all variables can be expressed as functions of z and F. ϕ is a velocity and the only combination of z and F that has the dimensions of a velocity is $F^{1/3}z^{-1/3}$. Thus, $\phi \sim F^{1/3}z^{-1/3}$. The same argument holds for the local plume velocity W. Thus, ϕ and W are proportional one to another:

$$\phi = \gamma W, \tag{5.31}$$

where γ is a constant which has been called the *Taylor* entrainment constant in honour of the British physicist G. I. Taylor.

Solutions can be obtained in two different ways. One method is to eliminate variables sequentially in the equations, a cumbersome procedure. The other, far superior, method relies on dimensional analysis. The three unknowns, R, W, Θ depend on only one variable, z, and on the buoyancy flux F. We thus easily deduce the form of the solutions by requiring that the variables have the required dimensions, as was done above for velocity. We obtain:

$$R = C_1 z \tag{5.32}$$

$$w = C_2 F^{1/3} z^{-1/3} \tag{5.33}$$

$$g\frac{\Delta\rho}{\rho_o} = C_3 F^{2/3} z^{-5/3}. \tag{5.34}$$

The turbulent plume decelerates as it rises because of entrainment, in contrast to its laminar counterpart. Constants C_1, C_2, C_3 are easily obtained by substitution into the governing equations and can be expressed as a function of the entrainment constant, γ:

$$C_1 = \frac{6\gamma}{5}, \; C_2 = \frac{5}{6\gamma}\left(\frac{9\gamma}{10}\right)^{1/3}, \; C_3 = \frac{5}{6\gamma}\left(\frac{9\gamma}{10}\right)^{-1/3}. \tag{5.35}$$

The validity of this theory is established by noting that turbulent plumes have the shape of an inverted cone with straight edges, which is indeed such that $R \propto z$, as predicted. Straightforward measurement of the cone aperture, equal to $C_1 = 6\gamma/5$ allows determination of the entrainment constant ($\gamma \approx 0.10$).

As in the laminar case, the theory does not specify the behavior of the plume cap. One may assume, however, that the cap does not grow to dimensions that are large compared to those of the plume. Thus, the ascent rate of the plume which extends to height H is approximately equal to the bulk velocity at $z = H$. Thus,

$$\frac{dH}{dt} \approx W_{z=H} = C_2 F^{1/3} H^{-1/3}. \tag{5.36}$$

Integrating and using initial condition $H(0) = 0$, we find that,

$$H(t) \approx \left(\frac{4C_2}{3}\right)^{3/4} F^{1/4} t^{3/4}, \tag{5.37}$$

which is in good agreement with measurements (Turner, 1973). We note that $dH/dt \propto F^{1/4}$ whereas $W \propto F^{1/3}$, which illustrates the difference between the ascent rate of the bulk plume and the interior velocity.

The local Reynolds number for the turbulent plume at height z is

$$\mathrm{Re} = \frac{\rho W R}{\mu} \propto z^{2/3}, \tag{5.38}$$

which increases with height. Close to the source, therefore, the Reynolds number is small and the flow cannot be turbulent. This is consistent with the laminar plume analysis. In practice, of course, the laminar region may be negligibly small if the power input P (or, equivalently, the buoyancy flux F) is very large.

5.1.2 Thermals

If a finite amount of heat ΔE is released at the base of the fluid, the flow takes the form of an isolated volume of heated fluid that detaches from the boundary (Figure 5.1). Supposing that the fluid is initially at a uniform temperature T_o, ΔE serves to heat a volume V of fluid to temperature $T_o + \Delta T$ such that,

$$\Delta E = \rho_o C_p \Delta T V, \tag{5.39}$$

where V and ΔT are defined at any height above the source. The thermal grows due to diffusion or turbulent entrainment, but the total amount of energy that is transported remains constant. Thus, quantity ΔTV also remains constant. The buoyancy force acting on the thermal is

$$B = g(\rho_o - \rho)V = \rho_o g\alpha \Delta TV = \frac{g\alpha}{C_p}\Delta E, \tag{5.40}$$

and hence also remains constant during ascent. One may introduce a dimensionless number that characterizes the strength of the flow (Griffiths, 1986):

$$\mathrm{Ra}_E = \frac{\rho_o g\alpha \Delta TV}{\kappa\mu} = \frac{g\alpha\Delta E}{C_p\kappa\mu}, \tag{5.41}$$

which is analogous to the Rayleigh number defined for Rayleigh–Benard convection which will be introduced later.

Laminar thermals

For small amounts of energy, the thermal is approximately spherical. At height z, the ascent velocity is W, the thermal diameter is D and the temperature contrast is ΔT. In a laminar regime,

$$W = \frac{1}{12}\frac{\rho_o g\alpha \Delta TD^2}{\mu}. \tag{5.42}$$

In this equation, the proportionality constant is not that for a solid sphere but that for an interior fluid that is less viscous than the exterior one. Using equation 5.41 as well as $V = \pi D^3/6$, this can be recast as,

$$W = \frac{1}{2\pi}\mathrm{Ra}_E\frac{\kappa}{D}. \tag{5.43}$$

As it rises, the thermal heats up exterior fluid that gets swept past it. This fluid becomes buoyant and becomes part of the upwelling. The amount of new fluid that is added to the thermal is determined by heat diffusion. In the reference frame of the rising thermal, exterior fluid takes time $\tau \propto D/W$ to flow past the thermal, so that it becomes heated over thickness $\delta \propto \sqrt{\kappa\tau} \propto \sqrt{\kappa D/W}$. The total volume of heated fluid in time τ is therefore $\Delta V \propto D^2\delta$. Thus, the rate of growth of the thermal volume is (Griffiths, 1986)

$$\frac{dV}{dt} \propto D\delta W \propto \kappa^{1/2}W^{1/2}D^{3/2}, \tag{5.44}$$

which can be recast as an equation for the thermal diameter D using equation 5.43 for W:

$$D\frac{dD}{dt} \propto \kappa\mathrm{Ra}_E^{1/2}. \tag{5.45}$$

This can be integrated for a point source, such that $D(0) = 0$:

$$D(t) = C_E \mathrm{Ra}_E^{1/4} \sqrt{\kappa t}, \tag{5.46}$$

where C_E is a constant to be determined from laboratory experiments or numerical calculations. Substituting for D into equation 5.43 for the velocity, we deduce that,

$$W = \frac{1}{2\pi C_E} \mathrm{Ra}_E^{3/4} \sqrt{\frac{\kappa}{t}}. \tag{5.47}$$

These simple results have been verified by laboratory experiments (Griffiths, 1986). One important consequence is that the Reynolds number for the thermal, which is proportional to WD, remains constant, in contrast to that for a plume. The full expression for the Reynolds number is,

$$\mathrm{Re} = \frac{\rho_o WD}{\mu} = \frac{1}{2\pi} \mathrm{Ra}_E \mathrm{Pr}^{-1}, \tag{5.48}$$

which illustrates the influence of the Prandtl number on the flow regime.

Turbulent thermals

For large energy, the flow is turbulent. Heat conservation is still expressed by equation 5.39 but the conservation equations for mass and vertical momentum take different forms, due to the turbulent regime. The mass conservation equation accounts for turbulent entrainment of the surrounding fluid, as in the case of a plume:

$$\frac{d(\rho V)}{dt} = \gamma \rho_o D^2 W, \tag{5.49}$$

where γ is an entrainment constant which is not equal to that for a plume due to the different flow geometry. The vertical momentum changes due to buoyancy:

$$\frac{d(\rho V W)}{dt} = g(\rho_o - \rho) V. \tag{5.50}$$

As explained above, the buoyancy force remains constant and this equation can be integrated directly:

$$\rho V W = \frac{g\alpha}{C_p} \Delta E t, \tag{5.51}$$

where we have used the fact that $V(0) = 0$ for a point source. The turbulent thermal is not exactly spherical and takes the form of an oblate spheroid such that $V \approx D^3/3$. Using the Boussinesq approximation, the mass conservation equation leads to,

$$\frac{dD}{dt} = \gamma W. \tag{5.52}$$

Substituting this into equation 5.51 and integrating, we get (Scorer, 1957),

$$D = (6\gamma)^{1/4} \left(\frac{g\alpha\Delta E}{\rho_o C_p} \right)^{1/4} t^{1/2} \tag{5.53}$$

$$W = \left(\frac{3}{8\gamma^3} \right)^{1/4} \left(\frac{g\alpha\Delta E}{\rho_o C_p} \right)^{1/4} t^{-1/2}. \tag{5.54}$$

The surprising result is that $D \propto t^{1/2}$ and $W \propto t^{-1/2}$, exactly as in the laminar case (but with different scalings). Thus, the Reynolds number is also constant, implying that the flow is turbulent at all times. We can also calculate the thermal characteristics as a function of height z and find for example that $D \propto z$, as for the turbulent plume.

5.2 Rayleigh–Benard convection

Convection may develop in a fluid layer that is heated from below and cooled from above, such that the top and bottom boundaries are kept at different temperatures. The key difference from the case of an isolated heat source is that it takes a minimum amount of heating for convection to appear. This form of convection has been called *Rayleigh–Benard* convection in honour of Henri Benard, who carried out the first experimental investigation of this phenomenon and Lord Rayleigh, who developed the first theoretical account. In the Earth's mantle, convection is driven in large part by internal heat production, implying different dynamics and a different thermal structure than in the Rayleigh–Benard case. Such characteristics will be discussed in Chapter 9.

5.2.1 Heuristic argument

A simple argument sheds light on the conditions required for the occurrence of convection in a fluid layer. The basic idea is that convection is characterized by temperature fluctuations at the macroscopic scale, in contrast with fluctuations that are always present at the molecular scale.

A fluid layer of thickness h is heated from below, such that its lower and upper boundaries are maintained at temperatures $T_o + \Delta T$ and T_o, respectively (Figure 5.6). Let us consider a sphere of radius a that collects at the upper boundary. It is made of cold fluid and hence sinks into the hotter fluid at a velocity given by,

$$W \propto \frac{g\Delta\rho a^2}{\mu} \propto \frac{\rho_o g\alpha\Delta T a^2}{\mu}. \tag{5.55}$$

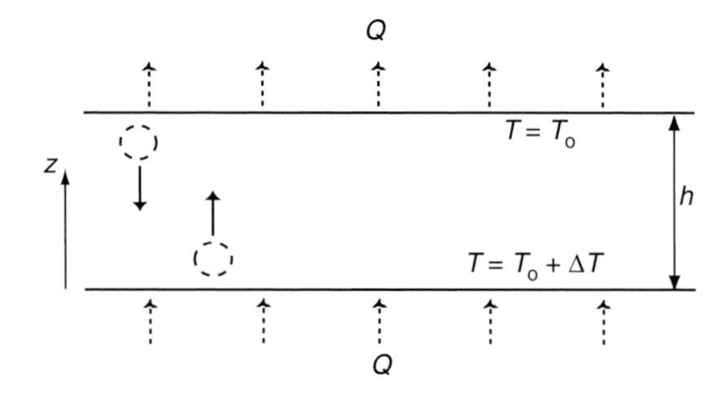

Figure 5.6. Basic set-up for Rayleigh–Benard convection. A fluid layer of thickness h is heated from below and cooled from above with heat flux Q, such that a temperature difference ΔT is maintained across the layer. Two small spheres of radii a are shown on the left, showing how fluid that has been cooled against the upper surface sinks into the interior and how fluid that has been heated against the lower surface rises.

As this sphere goes through the fluid layer, it loses heat to the surroundings by diffusion. Thermal equilibrium is achieved in time $\tau \propto a^2/\kappa$. Over that time, the sphere has traveled a distance d :

$$d \propto W\tau = A\frac{\rho_o g\alpha \Delta T a^4}{\kappa\mu}, \tag{5.56}$$

where A is a proportionality constant. This shows that a very small sphere cannot travel large distances, and hence cannot reach the lower boundary. In this case, the temperature perturbation gets damped efficiently and motion does not affect the whole layer. For convection to occur, motion must develop over thickness h, such that $d \approx h$, and temperature fluctuations must be sustained over the whole layer, such that $a \propto h$. Collecting the various terms, we find that free convection develops in the fluid layer if

$$\text{Ra} = \frac{\rho_o g\alpha \Delta T_c h^3}{\kappa\mu} > R_c \tag{5.57}$$

where ΔT_c is the threshold temperature difference for convection and R_c is some constant. This condition introduces a dimensionless number, Ra, called the Rayleigh number.

The Rayleigh number can be interpreted as the ratio between the characteristic time for transport of heat by convection, h/W, to that by conduction, h^2/κ. Obviously, convection can only develop if the former exceeds the latter. In terms of energy, the Rayleigh number can also be interpreted as the ratio of potential energy made available by generating hot fluid at the base of the layer and dissipation associated with viscous shear and heat diffusion. Convection can only develop if enough potential energy is available to overcome dissipation.

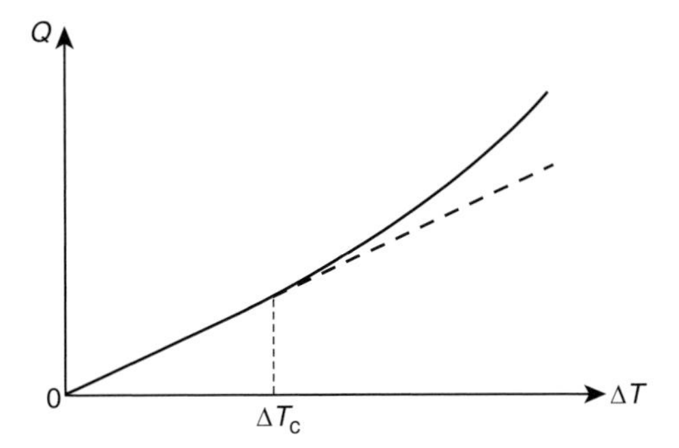

Figure 5.7. Heat flux Q that must be supplied to a fluid layer to sustain temperature difference ΔT across it, as a function of ΔT. At small values of ΔT, Q is proportional to ΔT, which corresponds to heat transport by conduction. Above a threshold temperature difference ΔT_c, the heat flux departs from the conduction trend.

The onset of convection can be detected in various ways, using for example visualization techniques that reveal motions or temperature variations within the fluid. One can also measure the heat flux Q that is required to maintain the temperature difference ΔT (Figure 5.7). For small values of ΔT, Q is proportional to ΔT, showing that heat is transported by conduction. This relationship breaks down above a certain threshold temperature difference, such that Q is larger than the conductive heat flux (Figure 5.7). As ΔT is increased further, Q deviates more and more from the conduction trend, which emphasizes the efficacy of heat transport by convection. Scaling Q with the conductive heat flux across the layer, we define a dimensionless quantity called the *Nusselt* number:

$$\mathrm{Nu} = \frac{Q}{\lambda \dfrac{\Delta T}{h}}, \tag{5.58}$$

which is such that $\mathrm{Nu} > 1$ if $\mathrm{Ra} > R_c$.

5.2.2 Dimensional analysis

In order to derive dimensionless equations for convection, the various variables are scaled as follows:

$$\begin{aligned} &\text{Length } h \\ &\text{Velocity } W \\ &\text{Time } \frac{h}{W} \\ &\text{Pressure } \frac{\mu W}{h} \\ &\text{Temperature } \Delta T \end{aligned} \tag{5.59}$$

Table 5.2. *Physical characteristics of geological convective systems*

System	h	ΔT, K	μ, Pa s	Pr	Ra
Upper mantle	660 km	1300 †	5×10^{20}	10^{23}	10^{6}
Whole mantle	3000 km	3300 †	5×10^{21}	10^{24}	10^{7}
Basaltic lava lake	50 m	50 ‡	10	10^{3}	10^{12}
Basaltic magma reservoir	1 km	50 ‡	10	10^{3}	10^{16}
Dacitic magma reservoir	1 km	50 ‡	10^{6}	10^{8}	10^{11}

Values have been rounded off for clarity.
† True temperature difference deduced from the mantle geotherm of Figure 2.4.
‡ Temperature difference across the actively convecting part of the system.

where velocity scale W has not been specified yet and where we have selected a viscous pressure scale because we will be dealing mostly with low Reynolds number flows. One must deduce the velocity scale from the control variables. One such scale is κ/h, which characterizes heat propagation due to diffusion, but it is obviously not appropriate here because we are interested in convection. We opt for a Stokes' velocity scale,

$$W = \frac{\rho_o g \alpha \Delta T h^2}{\mu}. \tag{5.60}$$

Note that this velocity scale appeared in the heuristic argument developed above. The governing equations are now written in dimensionless form as,

$$\nabla \cdot \mathbf{v} = 0 \tag{5.61}$$

$$\mathrm{RaPr}^{-1}\left[\frac{\partial \mathbf{v}}{\partial t} + \mathbf{v} \cdot \nabla \mathbf{v}\right] = -\nabla P + \nabla^2 \mathbf{v} + T\mathbf{n_z} \tag{5.62}$$

$$\mathrm{Ra}\left[\frac{\partial T}{\partial t} + \mathbf{v} \cdot \nabla T\right] = \nabla^2 T + \frac{Hh^2}{\lambda \Delta T} \tag{5.63}$$

with z positive upwards and n_z the unit vertical vector. $\mathrm{Pr} = \nu/\kappa$ is the Prandtl number and $(Hh^2)/(\lambda\Delta T)$ is a dimensionless rate of internal heat generation. $\mathrm{RaPr}^{-1} = \rho W h/\mu$ is the Reynolds number for the flow and already appeared in our discussion of thermal plumes (equation 5.48). Thus, three dimensionless numbers characterize the flow. In the absence of internal heat generation, one deals with only Pr and Ra. In geological systems, Prandtl numbers vary within a very large range (Table 5.2), and hence we shall pay attention to the influence of Pr. In such systems, Rayleigh numbers are very large and we shall not dwell at length on flow behaviour and pattern at low values of Ra.

5.2.3 The different convective regimes

The critical Rayleigh numbers have been calculated for a large range of boundary conditions in a layer of infinite horizontal extent. In the ideal case of a constant viscosity fluid, the geometry of the convection is well defined at the critical Rayleigh number and corresponds to two-dimensional regular cells with horizontal dimensions that are proportional to the thickness of fluid. With free boundary conditions at both boundaries, the critical Rayleigh number is $27\pi^4/4 \approx 657$, the wavelength is $2 \times \sqrt{2} \times h$ (Chandrasekhar, 1961), so that the width of each convective cell is $\sqrt{2} \times h$. The characteristic length scale of convection is set by the thickness of the fluid layer. In a container of finite dimensions, the width of the fluid layer is a second length scale and the aspect ratio of the container is yet another dimensionless number that controls the dynamics.

The critical Rayleigh number as well as the dimensions of convection cells depend on the boundary conditions. With rigid surfaces, a sharp velocity gradient develops at the boundaries and the critical Rayleigh number is larger than for free boundaries because more energy is required to sustain the enhanced dissipation. In almost all cases, the critical Rayleigh number is larger than 10^2.

As one increases the Rayleigh number, convective motions become increasingly complex due to the increasing energy available and the diminishing stability of simple flow patterns. For Rayleigh numbers larger than $\approx 10^5$, flow is unsteady, and can be described as due to the intermittent breakdown of thin thermal boundary layers at the top and bottom of the fluid (Figure 5.8). This is the case for most geological

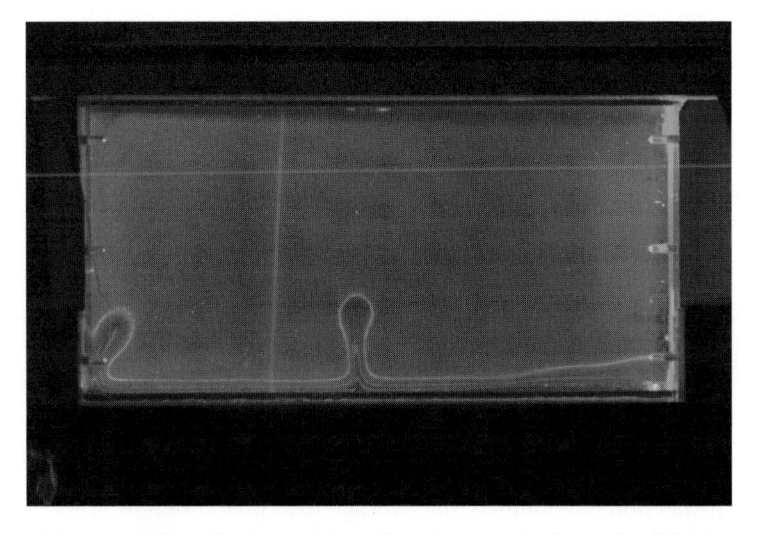

Figure 5.8. Instability of a thermal boundary layer at the base of a fluid layer that is heated from below in the laboratory (from Davaille *et al.*, 2010). The bright contours correspond to isotherms imaged by temperature-sensitive liquid crystals. See plate section for color version.

systems of interest (Table 5.2). In those systems, furthermore, convection is driven at least in part by cooling and is not in steady state. In such conditions, multiple boundary layer instabilities generate complex flow fields that cannot be captured by a single snapshot. One should consider instead a number of flow realizations and adopt a statistical approach with suitably averaged variables.

5.2.4 Convective heat transport

In a convecting fluid, heat transport is effected by upwellings, which carry hot fluid towards the top, and downwellings, which carry cold fluid to the bottom. In order to turn this qualitative statement into an explicit one, we introduce horizontally averaged quantities defined as follows:

$$\overline{f} = \frac{1}{S}\int_S f dS, \tag{5.64}$$

where $\overline{f}$ is an arbitrary variable and S is a horizontal section through the fluid. Temperature is separated into its horizontal average, $\overline{T}$ and fluctuation θ, such that,

$$T = \overline{T}(z,t) + \theta(x,y,z,t). \tag{5.65}$$

The heat equation can now be written as follows:

$$\rho C_p \left[\frac{\partial T}{\partial t} + \frac{\partial (uT)}{\partial x} + \frac{\partial (vT)}{\partial y} + \frac{\partial (wT)}{\partial z} \right] = \lambda \nabla^2 T + H. \tag{5.66}$$

Introducing $\overline{T}$ and θ and averaging over the horizontal, we obtain,

$$\rho C_p \left[\frac{\partial \overline{T}}{\partial t} + \frac{\partial \overline{w\theta}}{\partial z} \right] = \lambda \frac{\partial^2 \overline{T}}{\partial z^2} + H, \tag{5.67}$$

which can be recast as,

$$\rho C_p \frac{\partial \overline{T}}{\partial t} = -\frac{\partial}{\partial z} \left[-\lambda \frac{\partial \overline{T}}{\partial z} + \rho C_p \overline{w\theta} \right] + H. \tag{5.68}$$

This equation can be written in the following form:

$$\rho C_p \frac{\partial \overline{T}}{\partial t} = -\frac{\partial \overline{q}}{\partial z} + H, \tag{5.69}$$

where, by inspection, we identify the average heat flux at height z,

$$\overline{q} = -\lambda \frac{\partial \overline{T}}{\partial z} + \rho C_p \overline{w\theta}. \tag{5.70}$$

The heat flux is broken down into two components, corresponding to conduction and advection by fluid flow. A key point is that heat is transported upwards by both hot upwellings (such that $w > 0$ and $\theta > 0$) and cold downwellings (such that $w < 0$ and $\theta < 0$). In the following, we assume that there are no internal heat sources and hence set $H = 0$.

In steady state, the horizontally averaged heat equation can be integrated over the vertical:

$$\overline{q} = -\lambda \frac{\partial \overline{T}}{\partial z} + \rho C_p \overline{w\theta} = \text{constant} = Q, \quad (5.71)$$

where Q is the heat flux through the layer. At Rayleigh numbers well above the critical value, the vertical profile of the horizontally averaged temperature reveals a simple structure made of a well-mixed interior between thin boundary layers at the top and bottom (Figure 5.9). The boundary layers, with thickness $\delta \ll h$, are characterized by sharp temperature gradients that connect the boundaries to the well-mixed interior at $\overline{T} \approx T_o + \Delta T/2$. In the well-mixed interior, $d\overline{T}/dz \approx 0$, so that $\rho C_p \overline{w\theta} = Q$. In the boundary layers at the top and bottom, the vertical velocity is small and the dominant heat transport mechanism is conduction. At both the upper and lower boundaries, $w = 0$ and hence $-\lambda\,(dT/dz)_{\text{top}} = -\lambda\,(dT/dz)_{\text{bottom}} = Q$. Thus, each heat transport mechanism is active in separate parts of the layer.

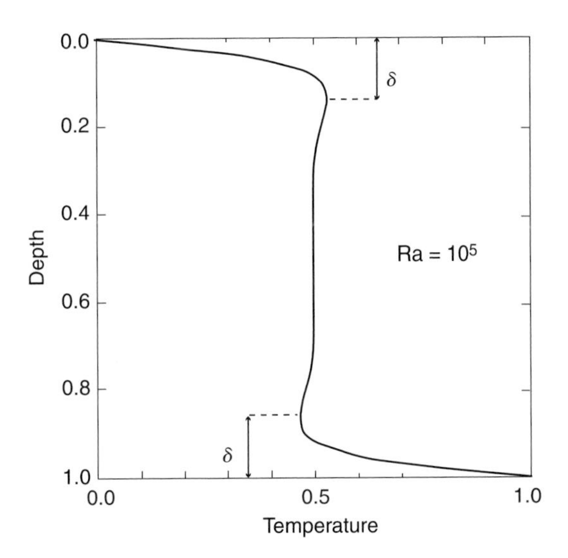

Figure 5.9. Vertical distribution of the horizontally averaged temperature $\overline{T}$ for $\mathrm{Ra} = 10^5$. There is no internal heating and the thermal structure is in steady state. Note the presence of two boundary layers of small thickness δ at the top and bottom, and a well-mixed interior where $\overline{T}$ is approximately constant.

5.2.5 The convective heat flux in a cooling layer

To illustrate further the behavior of the convective heat flux, $\rho_o C_p \overline{w\theta}$, we consider a fluid layer with an adiabatic base that is cooled through its upper boundary. The z coordinate axis is positive upwards, such that $z = h$ at the top and $z = 0$ at the base. The fluid layer is initially at temperature $T_o + \Delta T$ and the upper boundary temperature is set at $T = T_o$ at time $t = 0$. In this case, there is only one thermal boundary layer, at the top, and convection occurs as cold downwellings and a weak return flow. The layer cools down continuously until it is isothermal at temperature T_o. In the well-mixed interior, $\overline{T}$ does not depend on z, so that the heat balance reduces to,

$$\rho C_p \frac{dT_m}{dt} = -\frac{\partial}{\partial z}\left[\rho C_p \overline{w\theta}\right], \tag{5.72}$$

where T_m denotes the temperature of the well-mixed interior. This equation can be integrated between $z = 0$ and z:

$$\rho C_p \overline{w\theta} = -\rho C_p \frac{dT_m}{dt} z, \tag{5.73}$$

where we have used the fact that the base of the layer is insulated, so that there is no temperature gradient, and that the vertical velocity drops to zero at the base, so that $\overline{w\theta}(0) = 0$. Here dT_m/dt is negative and the convective heat flux is therefore positive, i.e. heat is carried upwards. This equation shows that the magnitude of the convective heat flux decreases linearly with depth. In the cold downwellings coming from the top, fluid heats up against the hotter ambient fluid, so that θ decreases with increasing depth. Such behavior has been described above for individual plumes and thermals.

We neglect the thin thermal boundary layer at the top and integrate the bulk heat balance equation (5.69) up to $z = h$, which leads to,

$$\frac{dT_m}{dt} = -\frac{Q}{\rho C_p h}, \tag{5.74}$$

where Q is the heat flux at the top. This specifies the rate of cooling for the fluid. Substituting into equation 5.73, we find that,

$$\rho C_p \overline{w\theta} = Q \frac{z}{h}, \tag{5.75}$$

which is valid everywhere in the fluid, save for the thin thermal boundary layer at the top. As in the case of Rayleigh–Benard convection, heat flux Q has not been specified yet and we address this problem in the next section.

5.2.6 The "4/3" law for the convective heat flux at high Rayleigh number

With no internal heat generation, the convection equations involve only two dimensionless numbers, Ra and Pr. Thus, the dimensionless heat flux Nu (equation 5.58) is a function of Ra and Pr. The Prandtl number depends only on the physical properties of the fluid and hence is set independently of the Rayleigh number, which describes the intensity of heating. Thus, we may treat separately the effects of Ra and Pr on Nu and write,

$$\mathrm{Nu} = C_N(\mathrm{Pr}) f_N(\mathrm{Ra}), \tag{5.76}$$

where C_N and f_N are two independent functions to be solved for. Here, we develop a simple argument that leads to a scaling law for function f_N at large values of Ra.

Near the top and bottom of the fluid layer, heat transport occurs by conduction through thin boundary layers whose thickness δ is such that,

$$Q = \lambda \frac{\Delta T}{2\delta}, \tag{5.77}$$

where we have used the fact that the temperature drop across each boundary layer is $\Delta T/2$ (Figure 5.9). Thus, one has,

$$\mathrm{Nu} = \frac{h}{2\delta}. \tag{5.78}$$

At large Rayleigh numbers, convection proceeds through boundary layer instabilities (Figure 5.8). If $\delta/h \ll 1$, the dynamics of each boundary layer are determined locally, independently of the layer depth, so that heat flux Q and boundary layer thickness δ do not depend on h. This specifies function f_N in (5.76). To demonstrate this, we write,

$$Q = \mathrm{Nu}\lambda \frac{\Delta T}{h} = C_N(\mathrm{Pr})\lambda \frac{\Delta T}{h} f_N\left(\frac{\rho_o g \alpha \Delta T h^3}{\kappa \mu}\right), \tag{5.79}$$

where we see that the layer depth h appears twice. The requirement that Q does not depend on h determines function f_N to an unknown proportionality constant (Townsend, 1959, 1964; Howard, 1966):

$$f_N(\mathrm{Ra}) \propto \mathrm{Ra}^{1/3} \tag{5.80}$$

so that we may write the Nusselt number as follows:

$$\mathrm{Nu} = C_N \mathrm{Ra}^{1/3}. \tag{5.81}$$

This can be turned into an equation for the heat flux Q. In order to emphasize the local character of the scaling law, we should not use ΔT, which is the temperature difference across the whole fluid layer and which includes contributions from both the upper and lower boundary layers. We use instead the temperature drop across a single boundary layer, noted ΔT_δ. In the present case, we know that $\Delta T_\delta = \Delta T/2$ but we shall keep the reference to a local temperature difference for purposes of generality. We find that,

$$Q = C_Q \lambda \left(\frac{g\alpha}{\kappa\nu}\right)^{1/3} \Delta T_\delta^{4/3}, \tag{5.82}$$

where C_Q is another proportionality constant that may depend on Pr. In steady state Rayleigh–Benard convection, $C_Q = 2^{4/3} C_N$.

This relationship is in good agreement with laboratory measurements for Ra $> \approx 5 \times 10^9$ (Figures 5.10, 5.11). Using equations 5.78 and 5.81 together with the experimental value of ≈ 0.065 for C_N (Table 5.3), the requirement that Ra $> 5 \times 10^9$ implies that $\delta/h < 5 \times 10^{-3}$: the thermal boundary layers must be very thin. The purely local behaviour of each boundary layer can be understood if one considers that the exterior flow takes the form of plumes coming from the other boundary. By the time these plumes impact the boundary layer, they are weak and induce negligible perturbations. If one deals with a fluid layer that is cooled from above with an adiabatic base, there is no lower thermal boundary layer and hence the upper boundary layer is only affected by the weak return flow of downgoing plumes. We thus expect that, in this case, the local scaling law (equation 5.82) holds for lower

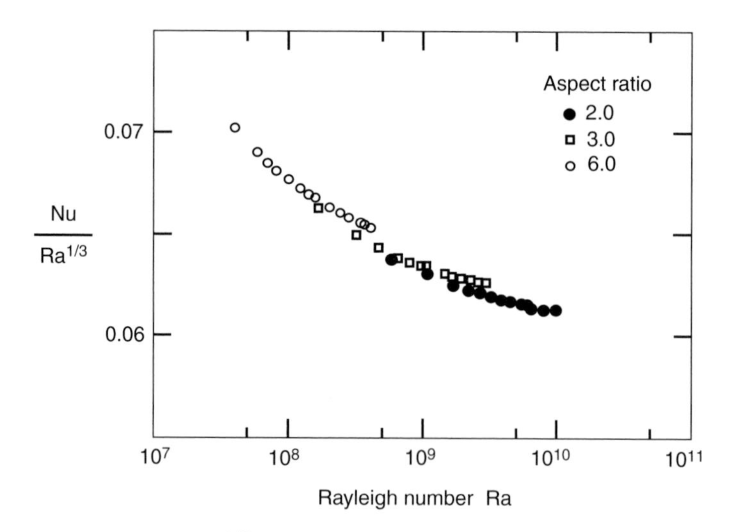

Figure 5.10. Ratio $\mathrm{Nu}/\mathrm{Ra}^{1/3} = C_N$ as a function of Rayleigh number for convection in water, a low Pr fluid, (from Funfschilling *et al.*, 2005). Experiments have been carried out in tanks of different aspect ratios. The local scaling arguments in the thermal boundary layer suggest that this ratio tends to a constant as Ra goes to ∞. This asymptotic regime is achieved for $Ra > 5 \times 10^9$.

Table 5.3. *Data on the convective heat flux in Rayleigh–Benard convection*

Pr	Ra	Bound. cond.	Aspect ratio	C_N $^+$	C_Q $^{++}$	Reference
4–6	5×10^{10}–10^{11}	Rigid	0.98	0.060	0.15	(Funfschilling *et al.*, 2005)
4–6	10^{10}	Rigid	2	0.062	0.16	(Funfschilling *et al.*, 2005)
4–6	3×10^8–4×10^9	Free	1	§	0.16†	(Katsaros *et al.*, 1977)
2750	10^8–10^{13}	Rigid	> 1	0.0659	0.17	(Goldstein *et al.*, 1990)
∞	10^6–10^9 ‡	Free	1.5	0.150¶	0.378¶	(Hansen *et al.*, 1992)

$^+$ Constant in the Nu versus Ra relationship (5.81).
$^{++}$ Constant in the local heat flux scaling law (5.82).
§ Only the boundary layer scaling can be determined in this transient cooling experiment.
† Value re-calculated for a 1/3 scaling exponent (instead of 0.33).
¶ Value re-calculated for a 1/3 scaling exponent at Ra $= 10^9$.
‡ Numerical calculations in 2D.

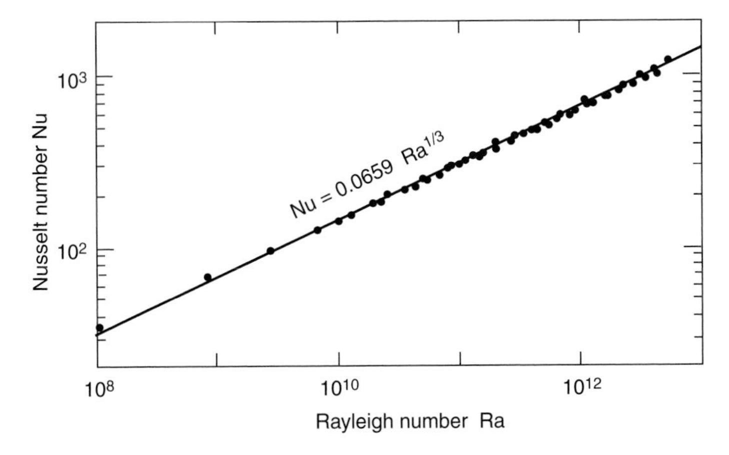

Figure 5.11. Nusselt number as a function of Rayleigh number at Pr $= 2750$ (from Goldstein *et al.*, 1990). The experimental data are consistent with the asymptotic scaling law for Nu with a $1/3$ exponent for Ra $> \approx 5 \times 10^9$.

values of the Rayleigh number. As shown in Figure 5.12, this is indeed so for Ra as low as $\approx 4 \times 10^8$. We shall see later that, in the analogous case of a layer that is heated from within, the local heat flux scaling law (5.82) is valid down to Ra $\approx 5 \times 10^6$.

Table 5.3 lists a few determinations of constants C_N and C_Q. We have selected these data in order to evaluate the influence of the container aspect-ratio, the Prandtl number and the boundary conditions (i.e. free versus rigid surfaces). As shown in Figure 5.10, experiments in water (Pr $\approx$ 4–5) indicate that the heat flux depends weakly on the container aspect ratio. Given the experimental uncertainties and the

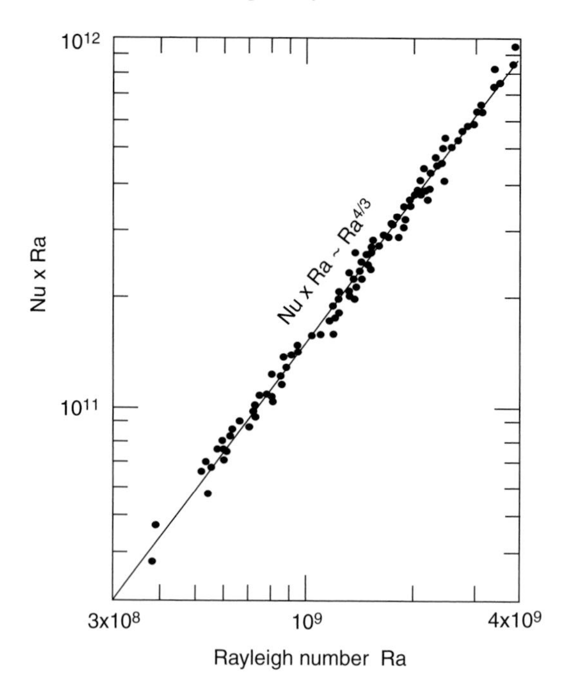

Figure 5.12. Plot of Nu $\times$ Ra as a function of Ra from transient cooling experiments in a large water tank (from Katsaros *et al.*, 1977). In these experiments, water was cooling due to evaporation at the top and there was no basal thermal boundary layer. Ra is calculated with the temperature difference across the upper boundary layer. The asymptotic form of the power-law relationship between Nu and Ra is achieved for Ra $>\approx 3 \times 10^8$.

limited range of Prandtl numbers, the data suggest no dependence on Pr (Table 5.3). The impact of the boundary condition is also difficult to evaluate with the laboratory data. One expects that, all else being equal, a free boundary enhances the heat flux compared to a rigid one. The only laboratory experiment that had a free boundary was made in transient cooling conditions (Table 5.3). This experiment suggests a small increase of C_Q with respect to values obtained in tanks with rigid boundaries and the same aspect ratio, but measurements show rather large scatter (Figure 5.12). There are no experimental data for very large values of the Prandtl number and we resort to numerical calculations. Results for free boundaries in two spatial dimensions indicate that constants C_N and C_Q are significantly larger than those for low Pr fluids in rigid enclosures. We attribute this difference to the change of boundary condition.

At values of Rayleigh number that are less than $\approx 10^9$, the Nu $-$ Ra$^{1/3}$ relationship does not hold. Some authors have sought power laws of the form Nu $\propto$ Ra$^\beta$, where β is some exponent, through best-fit adjustments to data over a finite range of Rayleigh numbers. As shown by the data in Figure 5.10, however, exponents that can be calculated over a small range of Ra gradually increase with Ra until they reach the limit value of $\beta = 1/3$ for Ra $\geq 5 \times 10^9$. Thus, it is not appropriate to use a single value of β for a transient model in which the intensity of convection varies due to cooling or heating. When comparing different determinations of β, one must pay attention to the specific values of Ra that have been investigated in each case.

5.3 Scaling laws for heat flux and velocity in Rayleigh–Benard convection: General theory

Many thermal models rely on the "4/3" scaling law for convective heat flux because it is simple and independent of conditions at remote boundaries. It is based on a qualitative argument, however, and has not been verified systematically. When dealing with systems as diverse as magma reservoirs and the Earth's mantle (Table 5.2), it is desirable to establish robust validity limits. It is not obvious, for example, that a scaling law that has been obtained for water can be applied to a viscous magma because of the large Prandtl number difference (Table 5.2). In addition, it is useful to derive velocity estimates and to determine whether or not convection is in a laminar regime. In a basaltic reservoir with Ra $> 10^{12}$ (Table 5.2), one expects turbulent motions with important consequences for magma mixing. To address these points, one must be able to estimate the Reynolds number from knowledge of Ra and Pr. Furthermore, magma chambers and the Earth's mantle involve different mechanical boundary conditions. The Earth's mantle is bounded by two low viscosity materials, the ocean or the atmosphere at the top, and molten iron at the base, so that its boundaries can deform and move in the horizontal direction. In contrast, magma reservoirs are encased in rigid rocks.

In this section, we review recent theoretical developments from Siggia (1994); Grossmann and Lohse (2000, 2001). One cannot hope to derive simple solutions for the complicated velocity and temperature fields that occur at large Rayleigh numbers. Thus, theory relies on scaling arguments and exact global conservation equations. We refer to Rayleigh–Benard convection with no internal heating and assume steady state. We define a scale for velocity in the fluid interior, U, and use the Reynolds number Re $= Uh/\nu$ as dimensionless velocity. We seek relationships between Re and Nu as a function of the two control variables, Ra and Pr. We are dealing with a heat engine problem involving irreversible processes and investigate the characteristics of dissipation. From these, we define a range of dynamical regimes and extract scaling laws for velocity and heat flux for each of them.

5.3.1 *The dissipation equations*

Kinetic dissipation

Kinetic dissipation is associated with the work of shear forces in the fluid. We subtract from the momentum equation the hydrostatic component of pressure for the horizontally averaged density structure:

$$0 = -\nabla P_h + \rho_o \left[1 - \alpha(\overline{T} - T_o)\right] \mathbf{g}, \tag{5.83}$$

where $\overline{T}$ is the average temperature. Vertical coordinate z is taken to be positive upwards. Taking the scalar product of the momentum equation with velocity, we obtain,

$$\rho_o \mathbf{v} \cdot \frac{\partial \mathbf{v}}{\partial t} + \rho_o \mathbf{v} \cdot (\mathbf{v} \nabla \mathbf{v}) = -\mathbf{v} \cdot \nabla P - \mathbf{v} \cdot (\nabla \cdot \boldsymbol{\sigma}) + \alpha \rho_o w \theta g, \tag{5.84}$$

where pressure $P = p - P_h$, $\boldsymbol{\sigma}$ is the deviatoric stress, w is the vertical component of velocity and $\theta = T - \overline{T}$ as before. In the last term on the right-hand side, we recognize the buoyancy flux that was used to characterize plume dynamics. We recast this equation using $e_c = v^2/2$ the kinetic energy per unit mass:

$$\begin{aligned} \rho_o \frac{\partial e_c}{\partial t} + \nabla \cdot (\rho e_c \mathbf{v}) = & - \nabla \cdot (\mathbf{v} p) + p \nabla \cdot \mathbf{v} \\ & - [\mathbf{v} \cdot (\nabla \cdot \boldsymbol{\sigma}) - \nabla \cdot (\boldsymbol{\sigma} \cdot \mathbf{v})] - \nabla \cdot (\boldsymbol{\sigma} \cdot \mathbf{v}) + \alpha \rho_o w \theta g, \end{aligned} \tag{5.85}$$

where we have used the continuity equation to rearrange the left-hand side. In this equation, we identify the third term on the right-hand side as the irreversible kinetic dissipation, ψ:

$$\psi = \mathbf{v} \cdot (\nabla \cdot \boldsymbol{\sigma}) - \nabla \cdot (\boldsymbol{\sigma} \cdot \mathbf{v}). \tag{5.86}$$

Integrating the kinetic energy equation over the fluid volume and converting volume integrals into surface integrals leads to,

$$\begin{aligned} \int_V \rho_o \frac{\partial e_c}{\partial t} dV + \int_S \rho_o e_c \mathbf{v} \cdot dS = & - \int_S p \mathbf{v} \cdot dS + \int_V p \nabla \cdot \mathbf{v} dV \\ & - \int_V \psi dV - \int_S (\boldsymbol{\sigma} \cdot \mathbf{v}) \cdot dS + \int_V \alpha \rho_o w \theta g dV. \end{aligned} \tag{5.87}$$

Many terms drop out of the equation. The first term on the left-hand side corresponds to time changes of kinetic energy and is equal to zero in steady state. We note in passing that it is the integral that is zero, and not necessarily the local time-derivative of the energy. In turbulent convection, the flow is unsteady, such that the local velocity at any given location fluctuates, but the total amount of kinetic energy can neither grows or decay. Another simplification arises because no fluid is leaving or entering the volume, so that $\mathbf{v} \cdot dS = 0$ on the boundaries. Thus, both the second term on the left-hand side and the first term on the right-hand side are zero. The second term on the right-hand side is zero because of mass conservation, i.e. $\nabla \cdot \mathbf{v} = 0$. A fourth term in this equation is zero:

$$\int_S \mathbf{v} \cdot (\boldsymbol{\sigma} \cdot dS) = 0, \tag{5.88}$$

because $\sigma = 0$ on a free boundary and $v = 0$ on a rigid one. We finally obtain,

$$+\int_V \rho \alpha g w \theta \, dV - \int_V \psi \, dV = 0. \tag{5.89}$$

Equation 5.89 states that viscous dissipation is balanced by the bulk buoyancy flux. We rewrite it using horizontally averaged quantities:

$$+\int_0^h \rho \alpha g \overline{w\theta} \, dz - \int_0^h \overline{\psi} \, dz = 0. \tag{5.90}$$

This equation is very handy because it allows one to include all the dissipative processes that are active simultaneously in the fluid layer. It will be put to extensive use.

From the heat balance equation, we know that,

$$\rho C_p \overline{w\theta} - \lambda \frac{d\overline{T}}{dz} = Q. \tag{5.91}$$

Integrating, we obtain,

$$\int_0^h \rho C_p \overline{w\theta} \, dz = \int_0^h \left(Q + \lambda \frac{d\overline{T}}{dz} \right) dz \tag{5.92}$$

$$= Qh + \lambda [T(h) - T(0)] = Qh - \lambda \Delta T. \tag{5.93}$$

Substituting into (5.90) and introducing dimensionless numbers, the bulk kinetic dissipation per unit area in the layer, ϵ, is calculated:

$$\epsilon = \int_0^h \overline{\psi} \, dz = \mu \frac{\nu^2}{h^3} (\mathrm{Nu} - 1) \mathrm{Ra} \mathrm{Pr}^{-2}. \tag{5.94}$$

Thermal dissipation

A similar equation can be obtained for temperature gradients, involving what is called "thermal dissipation." We begin with the energy equation and multiply it with T:

$$\rho C_p \left[T \frac{\partial T}{\partial t} + \mathbf{v} \cdot (T \nabla T) \right] = \lambda T \nabla^2 T. \tag{5.95}$$

In steady state, this equation can be rearranged:

$$\rho C_p \nabla \cdot (\mathbf{v}\mathbf{T}^2) = 2\lambda \left[\nabla \cdot (T \nabla T) - (\nabla T)^2 \right], \tag{5.96}$$

where we have used the continuity equation in the left-hand side. We integrate this equation over the whole volume and consider each term in turn:

$$\int_V \nabla \cdot (\mathbf{v}T^2) dV = \int_S T^2 \mathbf{v} \cdot dS = 0, \tag{5.97}$$

because no fluid is leaving the volume (i.e. $\mathbf{v} \cdot dS = 0$). For the next term, we begin with the identity:

$$\int_V \nabla \cdot (T \nabla T) dV = \int_S T \nabla T \cdot dS \tag{5.98}$$

$$= \left[T_o \left(\frac{dT}{dz} \right)_o - (T_o + \Delta T) \left(\frac{dT}{dz} \right)_1 \right] S, \tag{5.99}$$

where S is the area in horizontal cross-section. For constant thermal conductivity λ,

$$\left(\frac{dT}{dz} \right)_o = \left(\frac{dT}{dz} \right)_1 = -\frac{Q}{\lambda}, \tag{5.100}$$

and hence,

$$\int_V \nabla \cdot (T \nabla T) dV = \frac{Q \Delta T}{\lambda}. \tag{5.101}$$

We thus obtain an exact integral relation:

$$Q \Delta T = \lambda \int_0^h \overline{(\nabla T)^2} dz, \tag{5.102}$$

which is the exact analog of the mechanical dissipation equation. Thermal dissipation ϵ_θ can now be written as a function of the Nusselt number Nu:

$$\epsilon_\theta = \lambda \int_0^h \overline{(\nabla T)^2} dz = \lambda \frac{\Delta T^2}{h^2} \mathrm{Nu}\, h. \tag{5.103}$$

Temperature gradients are developed in two types of regions with different characteristics: thin boundary layers at the top and bottom, and the fluid interior (Figure 5.9). The thickness of these boundary layers, δ, is such that $\delta \ll h$. Near the boundaries, the vertical velocity drops to zero, implying that heat transport is achieved by conduction. Thus,

$$Q = \lambda \frac{\Delta T}{2\delta}. \tag{5.104}$$

In the following, all arguments will be made at the order of magnitude level, so that factors of order one such as 2 in this equation will be left out. From the definition of the Nusselt number, $\delta \sim h/\text{Nu}$. In each thermal boundary layer, thermal dissipation is :

$$\epsilon_{\theta,B} = \lambda \int_0^{\delta} \overline{(\nabla T)}^2 dz \sim \lambda \frac{\Delta T^2}{\delta^2} \delta \sim \lambda \frac{\Delta T^2}{h^2} \text{Nu} h. \tag{5.105}$$

Recalling the expression for total thermal dissipation (5.103), we find that,

$$\epsilon_{\theta,B} \sim \epsilon_{\theta}, \tag{5.106}$$

which shows that thermal boundary layers always contribute a significant amount of thermal dissipation. We shall see that, in some regimes, an equally important amount of thermal dissipation occurs in the interior.

The kinetic and thermal dissipation equations are valid for all values of Ra and Pr. These equations relate the bulk characteristics of convection, which depend on velocity and temperature fluctuations, to the heat flux into the system, which is driving the flow. They will be used to determine the various regimes that may exist in Rayleigh–Benard convection. Dissipation is achieved in the fluid interior and in boundary layers. In a nutshell, the various regimes correspond to one dominant contribution for each type of dissipation and we expect $2 \times 2 = 4$ regimes. We shall see, however, that one regime is not relevant to geological systems and that some regimes can be subdivided in two different ones depending on the Prandtl number. In the following, we focus on geological cases which involve Prandtl numbers that are larger than one save for the Earth's core, for which $\text{Pr} \ll 1$.

5.3.2 Rigid boundaries

We first tackle convection in magma reservoirs that are encased in rigid walls, on which velocity must vanish. In this case, a momentum boundary layer of thickness δ_u separates the fluid interior from the boundaries (Figure 5.13). At large Prandtl number, as illustrated by the structure of plumes for example (Figure 5.4), the thermal boundary layer is thinner than the momentum one, implying that the horizontal velocity is smaller there than in the bulk fluid. Thus, one must introduce two velocity scales, U_θ for the thermal boundary layer, and U for the bulk interior flow (Figure 5.13).

We begin with kinetic dissipation. As discussed for plumes, the dominant dynamical balance in the fluid interior is,

$$\rho_o \mathbf{v} \cdot \nabla \mathbf{v} = \mu \nabla^2 \mathbf{v}. \tag{5.107}$$

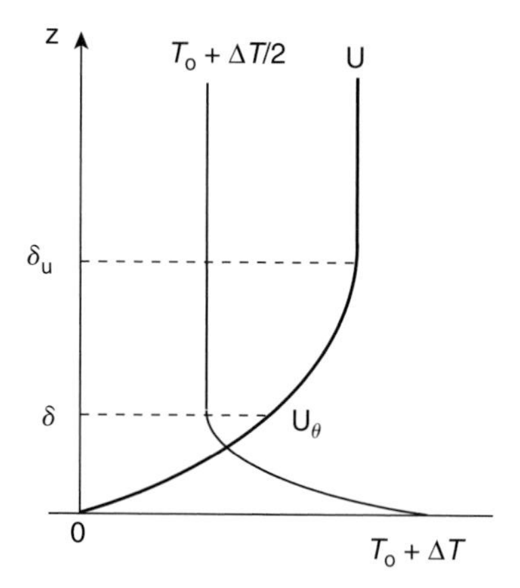

Figure 5.13. Diagram illustrating the thermal and momentum boundary layers at the rigid base of a convecting fluid for Pr > 1. The momentum boundary layer is much thicker than the thermal one, and one must consider different velocity scales for the thermal boundary layer and the fluid interior, denoted by U_θ and U, respectively.

Introducing the interior velocity scale U, we deduce that,

$$\rho_0 \frac{U^2}{h} \sim \mu \frac{U}{h^2}$$
$$\psi \sim \rho_o \frac{U^3}{h}. \tag{5.108}$$

Thus, kinetic dissipation in the interior is,

$$\epsilon_{u,I} \sim \int_0^h \psi \, dz \sim \rho_o U^3 \sim \mu \frac{\nu^2}{h^3} \text{Re}^3. \tag{5.109}$$

In the momentum boundary layer, the vertical velocity gradient U/δ_u is developed over thickness δ_u, by definition (Figure 5.13). Thus, kinetic dissipation is,

$$\psi \sim \mu \frac{U^2}{\delta_u^2}. \tag{5.110}$$

To obtain a scale for the kinetic boundary layer thickness, we use the dominant dynamical balance near the boundary, where flow is horizontal away from the thin plumes that detach from the boundary layer. Thus,

$$\rho_o u \frac{\partial u}{\partial x} \sim \mu \frac{\partial^2 u}{\partial z^2}$$
$$\rho_o \frac{U^2}{h} \sim \mu \frac{U}{\delta_u^2}, \tag{5.111}$$

where we have again supposed that the horizontal flow develops over horizontal distance h. We obtain,

$$\delta_u \sim h\mathrm{Re}^{-1/2}. \tag{5.112}$$

For very small values of Re, this expression is not valid because it predicts that $\delta_u > h$. In this case, one must write $\delta_u \sim h$ because the horizontal flow develops over the whole fluid thickness. Thus, there are two different scalings for δ_u depending on the Reynolds number.

Assuming for the moment that the Reynolds number is large, we use equation 5.112 and find that the boundary layer contribution to kinetic dissipation is,

$$\epsilon_{u,B} \sim \int_0^{\delta_u} \psi \, dz \sim \mu \frac{U^2}{\delta_u^2} \delta_u \sim \mu \frac{\nu^2}{h^3} \mathrm{Re}^{5/2}. \tag{5.113}$$

To evaluate in which part of the fluid kinetic dissipation is largest, we write,

$$\frac{\epsilon_{u,I}}{\epsilon_{u,B}} \sim \mathrm{Re}^{1/2}. \tag{5.114}$$

Thus, at large Reynolds numbers, kinetic dissipation occurs predominantly in the fluid interior, as expected for turbulent flow. For Reynolds numbers of order one, the expression is still valid but the momentum boundary layer may contribute the largest amount of dissipation. We expect that, for given Prandtl number, turbulence sets in above a critical value of the Rayleigh number and that, for given Rayleigh number, this transition occurs when the Prandtl number drops below a threshold value.

Intermediate Prandtl number

We first deal with Prandtl numbers that are not much larger than one, by opposition to the large Prandtl number limit to be studied later.

We begin with thermal dissipation. The thermal boundary layer is thinner than the momentum one, so that the relevant velocity for thermal dissipation, denoted U_θ is less than U (Figure 5.13). A first approximation is that of linear velocity gradient near the rigid boundary. Thus,

$$U_\theta \sim U \frac{\delta}{\delta_u}. \tag{5.115}$$

Substituting for the various variables leads to,

$$U_\theta \sim U \mathrm{Re}^{1/2} \mathrm{Nu}^{-1}. \tag{5.116}$$

In the boundary layer, flow is mostly horizontal, and the main energy balance is between horizontal advection and vertical diffusion:

$$\rho_o C_p u \frac{\partial T}{\partial x} \sim \lambda \frac{\partial^2 T}{\partial z^2}$$
$$\rho_o C_p U_\theta \frac{\Delta T}{h} \sim \lambda \frac{\Delta T}{\delta^2}. \tag{5.117}$$

To estimate the interior contribution to thermal dissipation, we use the fact that, away from the boundaries, temperature gradients are largest over the horizontal, as illustrated by individual plumes. We assume that upwellings are separated by distance $\approx h$, which has been verified many times for both laminar and turbulent regimes. Let δ^* denote the horizontal distance over which temperature gradients develop in the interior, such that $(\nabla T) \sim \Delta T/\delta^*$. The main energy balance is between vertical advection and horizontal diffusion, as discussed for the laminar plumes:

$$\rho C_p w \frac{\partial T}{\partial z} \sim \lambda \frac{\partial^2 T}{\partial x^2}$$
$$\rho C_p U_\theta \frac{\Delta T}{h} \sim \lambda \frac{\Delta T}{\delta^{*2}}, \tag{5.118}$$

where we have used the fact that w scales with U_θ. We note that this balance is different from that of the thermal boundary layer only formally. Thus, thermal dissipation in the interior is,

$$\epsilon_{\theta,I} = \lambda \int_{\delta_\theta}^{h-\delta_\theta} \overline{(\nabla T)}^2 dz \sim \rho C_p U_\theta \frac{\Delta T^2}{h} h, \tag{5.119}$$

where we have used the fact that the boundary layers are thin so that the integral can be estimated over the whole fluid thickness with no significant error. Substituting for the various variables, we find that,

$$\epsilon_{\theta,I} \sim \lambda \frac{\Delta T^2}{h^2} \mathrm{Pr}\,\mathrm{Re}^{3/2} \mathrm{Nu}^{-1}. \tag{5.120}$$

Large Rayleigh numbers As explained above, a significant amount of thermal dissipation always occurs in the thermal boundary layers. At large Rayleigh number, however, we expect thermal dissipation to be important also in the fluid interior, which requires that

$$\lambda \frac{\Delta T^2}{h^2} \mathrm{Nu} h \sim \lambda \frac{\Delta T^2}{h^2} \mathrm{Pr}\,\mathrm{Re}^{3/2} \mathrm{Nu}^{-1}, \tag{5.121}$$

which implies that $\mathrm{Nu} \sim \mathrm{Pr}^{1/2}\,\mathrm{Re}^{3/4}$. Two regimes can be defined depending on the dominant contribution to kinetic dissipation.

In a first regime, noted (IV) following Grossmann and Lohse (2000), most kinetic dissipation occurs in the interior. Thus,

$$\mu \frac{\nu^2}{h^3} \mathrm{NuRaPr}^{-2} \sim \mu \frac{\nu^2}{h^3} \mathrm{Re}^3, \tag{5.122}$$

where we have assumed that $\mathrm{Nu} \gg 1$. This provides a second equation for the two unknowns and, after a few manipulations, leads to,

$$\mathrm{Nu} \sim \mathrm{Ra}^{1/3}, \ \mathrm{Re} \sim \mathrm{Ra}^{4/9}\mathrm{Pr}^{-2/3} \ \text{ for regime IV.} \tag{5.123}$$

We note that the Reynolds number (i.e. the dimensionless velocity) increases with increasing Rayleigh number and decreases with increasing Prandtl number, as expected on intuitive grounds.

The preceding results show that Re decreases as Pr increases. Thus, the flow eventually becomes laminar above a threshold value of Pr, in which case kinetic dissipation is dominated by the boundary layer contribution. In this regime, numbered (III), the kinetic dissipation balance is,

$$\mu \frac{\nu^2}{h^3} \mathrm{NuRaPr}^{-2} \sim \mu \frac{\nu^2}{h^3} \mathrm{Re}^{5/2}. \tag{5.124}$$

Using the thermal dissipation balance (5.121), we find,

$$\mathrm{Nu} \sim \mathrm{Ra}^{3/7}\mathrm{Pr}^{-1/7}, \ \mathrm{Re} \sim \mathrm{Ra}^{4/7}\mathrm{Pr}^{-6/7} \ \text{ for regime III.} \tag{5.125}$$

Small Rayleigh numbers For small values of Ra, the interior contributes a small amount of thermal dissipation and hence does not provide a constraint on the flow dynamics. As shown above, the thermal dissipation equation is automatically satisfied by the boundary layer contribution, so that it brings no information. We turn to the dynamics of the thermal boundary layer, where the dominant heat balance is now,

$$\rho C_p \frac{U_\theta \Delta T}{h} \sim \lambda \frac{\Delta T}{\delta^2}. \tag{5.126}$$

Substituting for the various variables leads to,

$$\mathrm{Nu} \sim \mathrm{Re}^{1/2}\mathrm{Pr}^{1/3}. \tag{5.127}$$

We may again define two regimes, as in the preceding section, by considering the two end-member cases for kinetic dissipation. We may first assume that dissipation is achieved mostly in the interior, but such a regime is unlikely for Pr > 1 because it requires mechanical turbulence with no thermal turbulence in the interior. Turning to the more significant regime where kinetic dissipation is achieved mostly in the momentum boundary layer, we define regime (I) such that,

$$\mathrm{Nu} \sim \mathrm{Ra}^{1/4}\mathrm{Pr}^{-1/12}, \quad \mathrm{Re} \sim \mathrm{Ra}^{1/2}\mathrm{Pr}^{-5/6} \quad \text{for regime I,} \tag{5.128}$$

where we note that the Nusselt number depends very weakly on the Prandtl number. In this regime, Re (i.e. the typical fluid velocity) increases with increasing Ra and decreases with increasing Pr, as expected on intuitive grounds. Thus, for given Prandtl number, mechanical turbulence is achieved above a certain threshold value of Ra. For a specific magma type and hence fixed Pr, turbulent convection can only be achieved if the reservoir is thick enough.

Large Prandtl number

For given Ra, the solutions above are such that Re decreases as Pr increases. Thus, the thickness of the momentum boundary layer, $\delta_u = h\mathrm{Re}^{-1/2}$, also increases. As explained above, below some threshold value of Re and hence above some threshold value of Pr, the kinetic boundary layer extends over the whole fluid layer, so that $\delta_u \sim h$, which modifies the scalings. In this case, turbulence is suppressed and there is in fact no momentum boundary layer separating the interior and a boundary. This situation is achieved for Pr $\gg 1$, which is appropriate for many magmas. We still rely on the fact that the thermal boundary layer is much thinner than the kinetic one, so that,

$$U_\theta \sim U\frac{\delta}{\delta_u} \sim U\frac{\delta}{h} \sim U\mathrm{Nu}^{-1}. \tag{5.129}$$

This new scaling does not affect the calculation of the interior kinetic dissipation, but modifies the expression for the momentum boundary layer contribution:

$$\epsilon_{u,B} \sim \mu\frac{U^2}{\delta_u^2}\delta_u \sim \mu\frac{U^2}{h} \sim \mu\frac{\nu^2}{h^3}\mathrm{Re}^2. \tag{5.130}$$

Thermal dissipation must be calculated with the new velocity scale but we need worry only about the interior contribution:

$$\epsilon_{\theta,I} \sim \lambda\frac{\Delta T^2}{h}\mathrm{PrReNu}^{-1}. \tag{5.131}$$

Following the same logic as before, we should first consider regimes where kinetic dissipation is dominated by the interior component. For large Pr and hence small Re, however, we expect kinetic dissipation to be dominant in the boundary layers. We first treat the case with thermal dissipation in the interior, which corresponds to regime (III_∞). Substituting for the appropriate relationships, we find,

$$\mathrm{Nu} \sim \mathrm{Ra}^{1/3}\ ,\ \mathrm{Re} \sim \mathrm{Ra}^{2/3}\mathrm{Pr}^{-1}\ \text{ for regime } \mathrm{III}_\infty. \tag{5.132}$$

Finally, we consider that thermal dissipation is largest in the thermal boundary layer, corresponding to regime (I_∞). The heat balance in the thermal boundary layer is affected by the change of scaling for δ_u, which modifies the scaling for the horizontal advection velocity there:

$$U_\theta \sim U\delta/\delta_u \sim U\delta/h. \tag{5.133}$$

Substituting for this in the local heat balance in the thermal boundary layer, we obtain,

$$\mathrm{Nu} \sim \mathrm{Re}^{1/3}\mathrm{Pr}^{1/3}. \tag{5.134}$$

Using this and the same kinetic dissipation balance as before, we obtain,

$$\mathrm{Nu} \sim \mathrm{Ra}^{1/5}\ ,\ \mathrm{Re} \sim \mathrm{Ra}^{3/5}\mathrm{Pr}^{-1}\ \text{ for regime } \mathrm{I}_\infty. \tag{5.135}$$

5.3.3 The convective regimes of magma reservoirs

We have not dealt with regime (II) such that kinetic and thermal dissipations occur predominantly in the interior and in the thermal boundary layers respectively, because this requires small values of Pr that are not relevant to magma reservoirs. Using estimates for the various proportionality constants that come into play, one arrives at scaling laws for the variables and at a regime diagram in (Ra, Pr) space (Figure 5.14) (Grossmann and Lohse, 2000, 2001). Table 5.4 lists all the scaling laws with estimates for the proportionality constants, from Grossmann and Lohse (2000, 2001). We can check that the data from Table 5.3 are consistent with this theory. For example, the water experiments fall in regime (IV) and do support the $\mathrm{Nu} \sim \mathrm{Ra}^{1/3}$ relationship predicted by the theory (Table 5.4).

These results can be applied to magma reservoirs. From Table 5.2, one finds that basaltic magma chambers are in regime (IV), characterized by high Reynolds numbers and turbulence, for which the experimental data are very comprehensive. Silicic magma reservoirs are in regime (III_∞). Remarkably, both regimes are such that $\mathrm{Nu} \sim \mathrm{Ra}^{1/3}$ and such that Nu does not depend on Pr. The scalings for the convective velocity, or for Re, differ, due to the different flow regimes.

Table 5.4. *The regimes of Rayleigh–Benard convection for* Pr > 1 *(from Grossmann and Lohse, 2000, 2001)*

Regime	Dominant dissipation §	Nu	Re
I	$(u,B) - (\theta,B)$	$0.31\ \mathrm{Ra}^{1/4}\mathrm{Pr}^{-1/12}$	$0.073\ \mathrm{Ra}^{1/3}\mathrm{Pr}^{-5/6}$
I_∞ ($\mathrm{Pr} \gg 1$)	$(u,B) - (\theta,B)$	$0.35\ \mathrm{Ra}^{1/5}$	$0.054\ \mathrm{Ra}^{3/5}\mathrm{Pr}^{-1}$
III	$(u,B) - (\theta,I)$	$0.018\ \mathrm{Ra}^{3/7}\mathrm{Pr}^{-1/7}$	$0.023\ \mathrm{Ra}^{4/7}\mathrm{Pr}^{-6/7}$
III_∞ ($\mathrm{Pr} \gg 1$)	$(u,B) - (\theta,I)$	$0.027\ \mathrm{Ra}^{1/3}$	$0.015\ \mathrm{Ra}^{2/3}\mathrm{Pr}^{-1}$
IV	$(u,I) - (\theta,I)$	$0.060\ \mathrm{Ra}^{1/3}$	$0.088\ \mathrm{Ra}^{4/9}\mathrm{Pr}^{-2/3}$

§Dominant contributions to kinetic and thermal dissipation (see text). u and θ stand for the kinetic and thermal dissipation, respectively, and symbols I and B indicate interior and boundary-layer contributions, respectively.

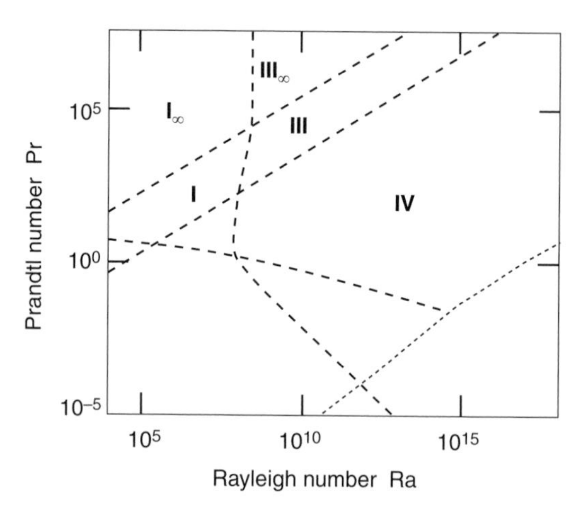

Figure 5.14. The various convective regimes for a fluid layer between rigid boundaries, as a function of Prandtl number and Rayleigh number (from Grossmann and Lohse, 2000, 2001). The numbers in capitals refer to the various regimes in Table 5.4. As indicated by Table 5.2, basaltic and silicic magma reservoirs are predicted to be in regimes IV and III_∞, respectively.

5.3.4 Convection with free boundaries

Here, we will focus on conditions appropriate for the Earth's mantle, such that the Reynolds number is small and flow is in a laminar regime. With free boundaries, furthermore, the velocity field is developed over the whole layer thickness and there are no momentum boundary layers separating the interior from the boundaries (Figure 5.15). Thus, it is pointless to consider two separate contributions to kinetic

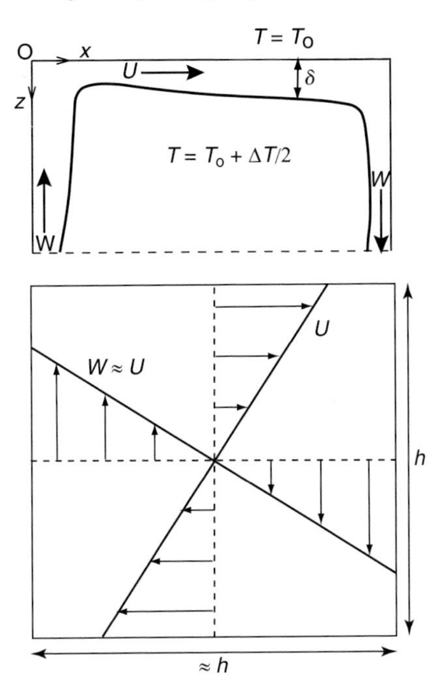

Figure 5.15. Schematic structure of a convection cell with free boundaries at large Prandtl number. Top: near the upper boundary, a thermal boundary layer grows in the horizontal part of the flow that connects upwellings and downwellings. Bottom: simplified representation of the velocity field emphasizing the absence of momentum boundary layers with sharp velocity gradients.

dissipation. Velocity gradients can be scaled with U/h, and hence,

$$\int_0^h \psi \, dz \sim \mu \frac{U^2}{h^2} h \sim \mu \frac{\nu^2}{h^3} \mathrm{Re}^2. \tag{5.136}$$

Using the kinetic dissipation equation (5.94), we obtain,

$$\mu \frac{\nu^2}{h^3} (\mathrm{Nu} - 1) \mathrm{Ra} \mathrm{Pr}^{-2} \sim \mu \frac{\nu^2}{h^3} \mathrm{Re}^2, \tag{5.137}$$

which reduces to $\mathrm{Nu}\mathrm{Ra}\mathrm{Pr}^{-2} \sim \mathrm{Re}^2$ if we assume that $\mathrm{Nu} \gg 1$, as appropriate for large Ra.

As regards thermal dissipation, we expect that the dominant contribution comes from the boundary layers. For free boundaries, the velocity scales with U in both the boundary layers and the fluid interior. This is obvious in the cellular flow structure where streamlines rotate around the edges of the cell (Figure 5.15). The main heat balance in the fluid interior is between vertical advection and horizontal diffusion, as appropriate for laminar plumes. Thus, assuming as before that upwellings and downwellings are separated by a distance that scales with the layer thickness h (which is obviously true for the cellular regime), we write again that temperature gradients in the fluid interior extend over a typical length scale δ^* such that,

$$\rho_o C_p U \frac{\Delta T}{h} \sim \lambda \frac{\Delta T}{\delta^{*2}}. \tag{5.138}$$

If we remember that, in the thermal boundary layers, the main heat balance is between horizontal advection over length h and vertical diffusion over thickness δ, we see that $\delta^* \sim \delta$. Thus, thermal dissipation is evenly distributed between the boundary layers and the interior. This can be understood easily in the case of a convection cell, such that upwellings and downwellings connect horizontal flows with similar dimensions and temperature contrasts. Using the boundary layer scalings, i.e. $\mathrm{Nu} \sim h/\delta$ and $\delta \sim \sqrt{\kappa h/U}$, we find that,

$$\mathrm{Re}^{1/2}\mathrm{Pr}^{1/2} \sim \mathrm{Nu}. \tag{5.139}$$

Using the kinetic dissipation equation derived above (equation 5.137), we find,

$$\mathrm{Nu} \sim \mathrm{Ra}^{1/3} \ , \ \mathrm{Re} \sim \mathrm{Ra}^{2/3}\mathrm{Pr}^{-1}, \tag{5.140}$$

or, equivalently,

$$\delta \sim h\mathrm{Ra}^{-1/3} \ , \ U \sim \frac{\kappa}{h}\mathrm{Ra}^{2/3}. \tag{5.141}$$

A key result is again that the Nusselt number does not depend on the Prandtl number in this regime.

5.3.5 *Summary*

We have shown that it is appropriate to use the $\mathrm{Nu} \sim \mathrm{Ra}^{1/3}$ scaling law for both magma reservoirs and the Earth's mantle, despite the large differences between their Prandtl and Rayleigh numbers. As shown in Table 5.3, however, the proportionality constant is significantly larger for free boundaries than for rigid ones. One other important result is that basaltic magma reservoirs convect in a turbulent regime.

5.4 Convection in porous media

In the Earth's crust, heat can be transported by the movement of fluids in permeable rocks. This affects estimates of the energy budget near the Earth's surface because standard heat flow measurements assume that all the heat is transported by conduction. The movement is maintained by fluid pressure differences either due to density (i.e. temperature) differences in the fluid or due to external factors such as topography and changes in the level of the water table. Hydrothermal circulation requires the temperature gradient and the permeability to be sufficiently high. The permeability of rocks can vary by several orders of magnitude (Table D.9) and it decreases with lithostatic pressure. Hydrothermal circulation has been found to be of paramount importance on the oceanic sea floor (see Chapter 6). Near mid-oceanic ridges, where the temperature gradient is very high, the permeability of the uppermost crust must be sufficient to keep hydrothermal circulation active; below this shallow layer, the permeability is too low to permit hydrothermal circulation and the lithosphere is cooling by conduction only. Therefore, the total heat loss of the lithosphere is well accounted for by conductive cooling models, even though the heat flow near the surface is affected by hydrothermal circulation. The presence of hydrothermal circulation has reconciled the cooling models for the oceanic lithosphere with heat flow observations and it accounts for ≈ 10 TW, i.e. more than 20% of the energy budget of the Earth. In the continents, hydrothermal circulation occurs in regions when the temperature gradient and/or the permeability of the rocks are sufficiently high. Some sedimentary rocks have low permeability and hydrothermal circulation affects many sedimentary basins. The temperature gradient and heat flux are high in zones of active extension, where hydrothermal circulation affects the sediment filled grabens. Very high temperature gradients are also found in regions affected by recent magmatic activity in the continental crust, where the cooling of intrusions drives active hydrothermal systems. Although the permeability of fresh granites is very low, there is good evidence that hydrothermal systems have been active near many granitic intrusions. Heat supplied by the radioactivity of very enriched intrusions may also be sufficient to sustain transport of heat by fluid motion. By contrast with molten cooling intrusions, such systems are less energetic but can last for a very long time.

There are two end models for hydrothermal systems: they can be approximated by pipe flow where fluids move along open fractures and are heated, or as diffuse flow in a porous medium. There are several important differences between convection in a fluid and convection in a porous medium. The physical properties of both the fluid and the matrix must be accounted for. The fluid and the matrix may not be in equilibrium and the energy exchanges between the fluid and the matrix must also be included. Finally, different boundary conditions may be needed to represent real physical systems. For example, in hydrothermal convection near mid-oceanic ridges, the upper boundary may or may not be permeable and the boundary conditions must include the interactions between the hydrothermal fluids and the ocean.

5.4.1 Fluid motion in a porous medium. Darcy's law

The velocity $\vec{v}$ of a fluid in a porous medium is determined by Darcy's law:

$$\vec{\mathbf{v}} = -\frac{k}{\mu}\nabla P = -\frac{kg\rho}{\mu}\nabla h, \tag{5.142}$$

where P is pressure, k is the permeability of the rock matrix (m^2) and μ is the fluid viscosity. The fluid pressure is related to h, the static "hydraulic head" which relates the energy density to the height of an equivalent column of fluid $g\rho_f h = P$. Local changes in the temperature gradient induce vertical fluid motion. The vertical velocity of the fluid due to local pressure variation is obtained as,

$$v_z = -\frac{k}{\mu}\left(\frac{\partial P}{\partial z} - \rho_f g\right). \tag{5.143}$$

In the case of a constant vertical temperature gradient Γ, the fluid density varies with depth:

$$\rho_f = \rho_0(1 - \alpha T) = \rho_0(1 - \alpha \Gamma z), \tag{5.144}$$

the vertical velocity is

$$v_z = -\frac{kg\rho_0}{\mu}\alpha\Gamma z. \tag{5.145}$$

Mass conservation implies a vertically uniform horizontal velocity.

5.4.2 Thermal convection in porous media

Energy conservation

When the fluid and the rock matrix are not in thermal equilibrium, the energy conservation equations for the fluid and the matrix must include the exchange of heat between the fluid and the matrix. If ϕ is the porosity of the rock, and the pores are entirely filled with fluid, a coupled system of equations is needed to determine temperature in the fluid and the solid:

$$\phi\rho_f C_f\left(\frac{\partial T_f}{\partial t} + \vec{\mathbf{v}}_{\mathbf{f}} \cdot \nabla T_f\right) = \phi\lambda_f \nabla^2 T_f + H(T_m - T_f) \tag{5.146}$$

$$(1-\phi)\rho_m C_m \frac{\partial T_m}{\partial t} = (1-\phi)\lambda_m \nabla^2 T_m + H(T_f - T_m), \tag{5.147}$$

where the subscripts f and m refer to fluid and the (void-free) matrix respectively. H is the volumetric heat transfer coefficient between the fluid and the matrix (i.e. H has units of W m^{-3} K^{-1}).

When the two phases are at the same temperature, we must use effective values for thermal conductivity and heat capacity $\lambda_e = \phi\lambda_f + (1-\phi)\lambda_m$ and $\rho_e C_e = \phi\rho_f C_f + (1-\phi)\rho_m C_m$.

In steady state, when $T_m = T_f = T$, we obtain a single equation:

$$\phi\rho_f C_f \vec{\mathbf{v}}_{\mathbf{f}} \cdot \nabla T = \lambda_e \nabla^2 T, \tag{5.148}$$

with $\lambda_e = \phi\lambda_f + (1-\phi)\lambda_m$.

A fluid flowing at constant vertical velocity v in a porous matrix will change the vertical temperature profile. In steady state, the heat equation gives,

$$\phi v \rho_f C_f \frac{\partial T}{\partial z} = \lambda_e \frac{\partial^2 T}{\partial z^2}, \tag{5.149}$$

with ρ_f C_f density and thermal capacity of fluid and λ_e effective thermal conductivity of the matrix and fluid. If $T = 0$ and $dT/dz = \Gamma_0$ for $z = 0$ then the temperature profile is given by,

$$T(z) = \frac{\Gamma_0 \lambda_e}{\phi v \rho_f C_f}\left(\exp\left(\frac{\phi v \rho_f C_f z}{\lambda_e}\right) - 1\right), \tag{5.150}$$

where v is positive downward. If the flow is upward $v = -|v|$.

The temperature at the base of the layer T_L is found as,

$$T_L = \frac{\Gamma_0 L}{\mathrm{Pe}}(\exp(\mathrm{Pe}) - 1), \tag{5.151}$$

where Pe is the Péclet number,

$$\mathrm{Pe} = \frac{\phi\rho_f C_f v L}{\lambda_e}. \tag{5.152}$$

The conductive heat flux across the layer can be determined from,

$$q_c = \frac{\lambda_e T_L}{L} = \frac{\lambda_e \Gamma_0}{\mathrm{Pe}}(\exp(\mathrm{Pe}) - 1). \tag{5.153}$$

Note that for $v > 0$, i.e. the fluid moves downwards, $\mathrm{Pe} > 0$ and $T_L/\Gamma_0 L > 1$, for $v < 0$, we have $\mathrm{Pe} = -|\mathrm{Pe}| < 0$ and $T_L/\Gamma_0 L < 1$. Note that $\lambda_e/(\rho_f C_f)$ has dimensions of thermal diffusivity.

Fluid convection in a porous medium. Rayleigh number.

The conditions for the onset of convection for a fluid in a porous medium can be studied with the same marginal stability analysis used to study thermal convection in a fluid layer, but must account for the physical properties of both the fluid and the matrix.

The Rayleigh number for convection in a porous medium can be defined by writing the dimensionless form of the governing equations, or by comparing the characteristic time for conductive and convective heat transport. A temperature difference ΔT is applied across a layer of thickness L with a solid matrix of permeability k containing a fluid of viscosity μ. The characteristic time for heat conduction is L^2/κ_m where κ_m is the thermal diffusivity of the matrix. Transport by the motion of the fluid in the matrix has a time scale L/v_f. And the velocity is $v_f = k\nabla P/\mu$, with $\Delta P \propto g\rho_f \alpha \Delta T$. This gives the velocity scale $v_f = kg\rho_f \alpha \Delta T/\mu$. The Rayleigh number is the ratio of the characteristic time for convective to conductive transport, thus:

$$\mathrm{Ra} = \frac{kg\rho_f \alpha \Delta T L}{\kappa_m \mu}. \tag{5.154}$$

Horton and Rogers (1945) and Lapwood (1948) have analyzed the conditions for the onset of convection in a porous medium heated from below. The heat equation is,

$$\rho_e C_e \frac{\partial T}{\partial t} + \phi \rho_f C_f (\vec{\mathbf{v}} \cdot \nabla T) = \lambda_e \nabla^2 T, \tag{5.155}$$

where indices m and f refer to matrix and fluid, with also,

$$\nabla \cdot \vec{\mathbf{v}} = 0 \tag{5.156}$$

$$\vec{\mathbf{v}} = -\frac{k}{\mu}(\nabla P + g\rho_f \alpha (T - T_0)). \tag{5.157}$$

The scaling factor for the velocity is thus,

$$\frac{kg\rho_f \alpha \Delta T}{\mu}. \tag{5.158}$$

The heat equation can thus be written in dimensionless variables:

$$\mathrm{Ra}\left(\frac{\rho_e C_e}{\rho_f C_f}\frac{\partial T'}{\partial t'} + \phi \vec{\mathbf{v}}' \cdot \nabla' T'\right) = \nabla'^2 T', \tag{5.159}$$

where the Rayleigh number is,

$$\mathrm{Ra} = \frac{\alpha g \rho_f k L \Delta T}{\mu \lambda_e/(\rho_f C_f)}. \tag{5.160}$$

Note that $\lambda_e/(\rho_f C_f)$ has the dimension of diffusivity but it includes thermal properties of the fluid and the rock matrix. In this form, the Rayleigh number is a

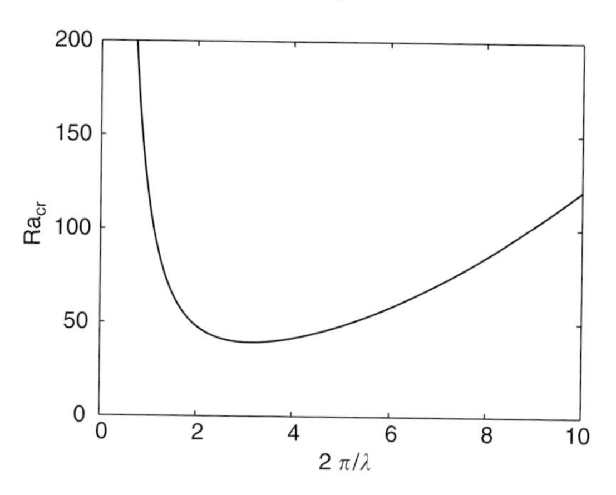

Figure 5.16. Critical Rayleigh number versus wavelength for convection in a porous medium.

measure of the efficacy of heat transport by the convecting fluid relative to heat transport by conduction in the matrix and fluid. The critical Rayleigh number where a perturbation of wavelength λ in the temperature field grows is found by linear stability analysis (Lapwood, 1948). Its form depends on the boundary conditions. For fixed temperature at the upper and lower boundaries, and for impervious upper and lower boundaries, it is found that the critical value for growth of an instability of wavelength λ is,

$$\mathrm{Ra}_{\mathrm{cr}} = \frac{((2\pi L/\lambda)^2 + \pi^2)^2}{(2\pi L/\lambda)^2}. \tag{5.161}$$

The minimum value of the critical Rayleigh number is found for $2\pi L/\lambda = \pi$ where $\mathrm{Ra}_{\mathrm{cr}} = 4\pi^2$ (Figure 5.16). Experiments by Elder (1967) have confirmed that, when the upper and lower boundaries are kept at fixed temperature, convection in a porous medium starts for $\mathrm{Ra} \approx 40$. If only the lower boundary is impervious and there is flow across the upper boundary with water above it (open top system), the critical Rayleigh number is 27.1. Other boundary conditions yield different values for the critical Rayleigh number (Table 5.5).

In a sea-floor hydrothermal system, with a temperature difference of 500 K across the layer and standard values for the physical parameters, we have $\mathrm{Ra} \approx 10^{12}\ kL$. For $L > 1000$ m, the critical Rayleigh number is exceeded for $k > 10^{-13}$ m^2. Rayleigh numbers are smaller than in the case of convection in the mantle or magma chambers.

Heat transport

Heat transport is best characterized by the Nusselt number which measures the ratio of the total to the conductive heat flux. A high Nusselt number can be related to the

Table 5.5. *Critical Rayleigh number for convection of a fluid in a porous medium with different boundary conditions*

Lower		Upper		Ra_c	$2\pi L/\lambda$
Fixed T	Impervious	Fixed T	Impervious	$4\pi^2$	π
Fixed T	Impervious	Fixed Q	Impervious	27.10	2.33
Fixed Q	Impervious	Fixed Q	Impervious	12	–
Fixed T	Impervious	Fixed T	Fixed P	27.10	2.33
Fixed Q	Impervious	Fixed T	Fixed P	17.65	1.75
Fixed T	Impervious	Fixed Q	Fixed P	π^2	$\pi/2$

formation of boundary layers. For the standard vertical temperature profile across the convecting system, with two conductive boundary layers, $\text{Nu} = L/2\delta$.

At Rayleigh numbers much higher than critical, hydrothermal convection can only be studied with numerical or physical experiments (Elder, 1967; Ribando *et al.*, 1976; Lister, 1990; Cherkaoui and Wilcock, 1999, 2001). These experiments have shown that, at low Rayleigh number, the Nusselt number is proportional to the Rayleigh number. At larger values of Ra, the Nusselt number becomes proportional to $\text{Ra}^{1/3}$, as for Rayleigh–Benard convection (Elder, 1967; Lister, 1990). Numerical models of convection in closed-top systems show several transitions controlled by the value of Ra. With increasing Ra, the aspect ratio of the convection cell decreases, and convection becomes time periodic with the Nu oscillating at fixed frequency ($f \propto \text{Ra}^p$); at higher Ra, two frequencies appear before convection becomes chaotic (Kimura *et al.*, 1986). Each regime is characterized by a different exponent for the scaling law of Nu to Ra. The same transitions appear in numerical and analog models of hydrothermal convection in open-top systems (Cherkaoui and Wilcock, 1999, 2001).

Crude estimates of the Nusselt number in hydrothermal systems are in the range 500–50,000. Two observations on sea-floor hydrothermal systems are available to test the model and adjust its parameters: the ratio of maximum to minimum heat flux and the wavelength of the heat flux variations $\approx 7-8$ km. Numerical experiments show that $q_{\max}/q_{\min} \propto \text{Ra}^p$ with $p \approx 2.5$ with permeable top and $p \approx 0.85$ for impermeable top (Ribando *et al.*, 1976).

Numerical methods allow models with variations of permeability with depth $k = k_0 \exp(-z/d)$. One of the problems limiting numerical models is that turbulence may develop in the pores at high Rayleigh number when the Reynolds number $\text{Re} = \rho v \delta/\mu \approx 1$ (with δ the size of the pores).

Sea-floor hydrothermal systems

Near mid-oceanic ridges where new sea floor is emplaced, there is a large temperature difference between the ascending magma and the ocean water. Hydrothermal circulation takes place if the permeability of the oceanic crust is sufficiently large ($k > 10^{-14}$ m^2). As will be seen later, hydrothermal circulation accounts for a large fraction of the energy budget in young sea floor.

Cooling of intrusions by hydrothermal circulation

Intrusions are emplaced in the shallow crust at temperatures several hundred degrees above equilibrium. Hot intrusions could sustain a hydrothermal system before they cool off provided that the effective permeability of the matrix is sufficient. Energy brought by an intrusion represents ≈ 0.5 MJ kg^{-1}; for a 10 km thick intrusion this represents $\approx 1.5 \times 10^{13}$ J m^2.

Hydrothermal convection in radioactive intrusions

Granitic intrusions are normally enriched in radioactive elements; the average heat production in granites is 3 μW m^{-3} (see Table E.2), but values of heat generation as high as 15 μW m^{-3} have been reported. Thick intrusions thus maintain large horizontal temperature variations and could sustain hydrothermal circulation long after reaching a steady-state thermal regime. The amplitude of lateral temperature variations $\Delta T = HL^2/2\lambda_e$ can be used to estimate the fluid pressure gradient $\propto g\rho_f\alpha HL^2/2\lambda_e d$, with d the width of the intrusion. Fluid velocity scales to,

$$v = \frac{k}{\mu} g\rho_f\alpha \frac{HL^2}{2\lambda_e}. \tag{5.162}$$

With temperature differences 150 K, fluid velocities $\approx 10^6 \times k$ can be sustained. Note that the permeability of fresh granite (10^{-17}m^2, Table D.9) only permits very slow flows (mm yr^{-1}) and the temperature in the pluton remains close to conductive equilibrium if the Peclet number $\mathrm{Pe} = \phi\rho_f C_f vL/\lambda_e \ll 1$.

5.4.3 Pipe flow

Bodvarsson and Lowell (1972) proposed a schematic two-dimensional model consisting of a loop of fluid descending along a vertical fracture, moving horizontally and heating along a plane horizontal fracture, and returning along a vertical fracture. Fluid movement is maintained by the pressure difference between the two vertical fractures, $\Delta P = \alpha\rho_f g\Delta Th$, where h is the depth. The pressure gradient maintains fluid motion:

$$v = \frac{1}{2\mu}(z^2 - \delta^2/4)\frac{\Delta P}{L}. \tag{5.163}$$

where ΔP is the pressure drop over the entire loop and δ is the thickness of the horizontal layer of fluid. The fluid flux,

$$q = \frac{\delta^3}{12\mu}\frac{\Delta P}{L}. \tag{5.164}$$

In order to estimate ΔT, assume that it is proportional to the conductive heat flux Q_c times the length of the horizontal fracture:

$$\Delta T \propto \frac{Q_c L}{q \rho_f C_f}, \tag{5.165}$$

which gives,

$$\Delta T \propto \left(\frac{12\mu Q_c}{\alpha g h C_f}\right)^{1/2} \frac{L}{\rho_f \delta^{3/2}}. \tag{5.166}$$

5.4.4 Topography-driven convection

Hydraulic head h (or potential) for steady-state fluid flow satisfies Laplace's equation. The solution for a layer $0 < z < L$ with boundary condition $h(x, z = 0) = h_0(x)$ and no flow across the lower boundary (i.e. $\frac{\partial h}{\partial z} = 0$) is obtained as a Fourier series or transform:

$$h(x,z) = \int_{-\infty}^{\infty} h_0(k)\cosh(kz/L)\exp(-ikx)dk, \tag{5.167}$$

and the fluid velocity is obtained from Darcy's law (5.142). For a single cell of width a and thickness L, such that there is no flow across the lateral boundaries $x = 0$ and $x = a$ and bottom boundary $z = L$, and with the pressure on the upper boundary $P_0(x) = P_0 \cos(\pi x/a)$, the stream function is given by,

$$\Phi(x,z) = P_0 \cosh\left(\frac{\pi(z-L)}{a}\right)\cos\left(\frac{\pi x}{a}\right)/\cosh\left(\frac{\pi L}{a}\right). \tag{5.168}$$

From this the flow field can be determined:

$$\begin{aligned} v_x &= \frac{k\pi P_0}{\mu a}\cosh\frac{\pi(z-L)}{a}\cos\frac{\pi x}{a}/\cosh\left(\frac{\pi L}{a}\right) \\ v_z &= -\frac{k\pi P_0}{\mu a}\sinh\frac{\pi(z-L)}{a}\sin\frac{\pi x}{a}/\cosh\left(\frac{\pi L}{a}\right). \end{aligned} \tag{5.169}$$

Solution for the temperature field has been proposed by Domenico and Palciauskas (1973).

Exercises

5.1 In steady state, the heat equation for a fluid moving vertically in a porous medium can be written as,

$$\frac{\partial}{\partial z}(\Phi \rho_f C_f v_z T + q) = 0. \tag{5.170}$$

Check that this equation is indeed verified. What is the meaning of the constant $\Phi \rho_f C_f v_z T + q$? Discuss how the conductive and advective heat fluxes vary with depth when $v_z > 0$ and $v_z < 0$.

5.2 Determine the steady-state temperature profile in a fluid moving vertically in the layer $0 < z < L$ with the temperature and its gradient Γ_0 fixed at the surface. Discuss solution for $v \gtrless 0$.

5.3 Determine the steady-state temperature in a fluid moving vertically in the layer $0 < z < L$, with both boundaries kept at fixed temperature. Determine how the conductive and convective heat fluxes vary with depth.

6

Thermal structure of the oceanic lithosphere

Objectives of this chapter

Because of their simple structure, the oceanic plates offer an ideal situation for the application of simple thermal models to geodynamics. In this chapter, we present the cooling model of the sea floor. We shall analyze how the model depends on boundary conditions and discuss how well the model's predictions fit the data. We shall also show how the discrepancies between the model's predictions and the data lead to the inference of hydrothermal circulation in the young oceanic crust, and small-scale convection in the mantle beneath old ocean basins. We shall also use the cooling model to analyze the effect of mantle hot spots on the oceanic plates. Finally, we shall show how the cooling model for the oceanic lithosphere explains many other geophysical observations.

6.1 Continental and oceanic heat flow

6.1.1 Introduction

With the sea-floor spreading hypothesis and the advent of plate tectonics, it became clear that oceanic and continental heat flow are fundamentally different. While crustal radiogenic heat production is the largest component of the continental heat flux, oceanic heat flux is due to the transport of heat to the surface of the Earth by mantle convection (Turcotte and Oxburgh, 1967; McKenzie, 1967; Sclater and Francheteau, 1970). The oceanic lithosphere is in a transient thermal state in contrast with continents which are mostly in, or close to, thermal steady-state. Oceanic heat flux is described by a decreasing function of age which parallels that of elevation (or bathymetry). The continental lithosphere has experienced a longer evolution and is characterized by a complicated structure and composition, and there is no relationship with age, except on very young crust that has not reached steady state. Continental elevation is controlled mainly by variations of crustal thickness and composition, and depends weakly on thermal structure.

6.1.2 Lithosphere and thermal boundary layer structure

Heat is brought to the base of the lithosphere by convection in the mantle beneath both continents and oceans. The vertical temperature profile must be divided into two parts: an upper part where heat is transported by conduction and a lower convective boundary layer. In steady state and in the absence of heat-producing elements, heat flux is constant in the conductive upper part, implying a constant temperature gradient for constant thermal conductivity. In contrast, the temperature gradient is not constant in the convective boundary layer and progressively tends to a small value in the mantle beneath. For definition of the thermal lithosphere, we must consider three different depths (Figure 6.1). The shallowest boundary, h_1, corresponds to the lower boundary of the conductive upper part and of what we shall call the thermal lithosphere. The deepest boundary, h_3, corresponds to the lower limit of the thermal boundary layer. This boundary may also be regarded as the transition between the lithospheric regime and the fully convective mantle regime, such that the mantle thermal structure below is not related to that of the lithosphere. An

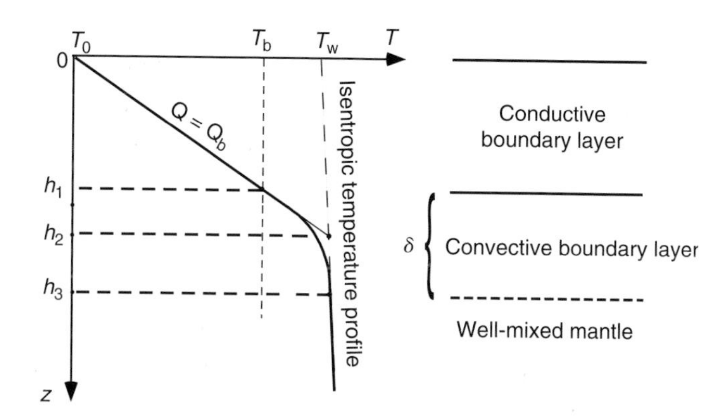

Figure 6.1. Schematic structure of the thermal boundary layer at the top of the Earth's convecting mantle. The boundary layer must be split into two parts. In the effectively rigid upper part of thickness h_1, heat is transported by conduction. In the unstable lower part of thickness $\delta = h_3 - h_1$, heat is brought to the base of the upper part through convection. The vertical profile obtained by downward extrapolation of shallow temperature data intersects the well-mixed isentropic profile at yet another depth, h_2. The temperature at the base of the conductive upper part is T_b, which is significantly smaller than the well-mixed potential temperature T_w. The temperature difference across the unstable boundary layer of thickness δ is $T_w - T_b$.

intermediate depth, h_2, is obtained by downward extrapolation of the conductive geotherm to the adiabatic temperature profile for the convecting mantle. With no knowledge of boundary layer characteristics, one can only determine h_2 and h_3, the former from heat flow data and the latter from seismic velocity anomalies. Such determinations are associated with two major caveats. One is that they cannot be equal to one another, which does not permit cross-checks. The other caveat is that they say nothing about h_1. Yet, it is h_1 which defines the mechanically coherent unit (the "plate") which moves at the Earth's surface and sets the thermal relaxation time. Uncertainty about this thickness has severe consequences because the diffusive relaxation time is $\propto h^2$. A final remark is that h_1 characterizes the upper boundary condition at the top of the convecting mantle.

6.1.3 Basal boundary conditions

In steady state, the heat flux at the Earth's surface is equal to

$$Q_0 = Q_{\text{crust}} + Q_{\text{lith}} + Q_b, \tag{6.1}$$

where Q_{crust} and Q_{lith} stand for the contributions of heat sources in the crust and in the lithospheric mantle and Q_b is the heat flux at the base of the lithosphere. In the oceans, one may ignore the first two with negligible error and surface heat flux is a direct measure of the basal heat flux.

With reference to Figure 6.1, one must introduce three temperatures, T_0 at the upper boundary, which, for all practical purposes, may be taken as fixed and equal to 0°C, T_b at the base of the lithosphere and T_w in the well-mixed convective interior. In steady state and in the absence of heat sources, the heat flux,

$$Q_0 = Q_b = \lambda \frac{T_b - T_0}{h_1}, \tag{6.2}$$

with λ = thermal conductivity. A closure equation relates this flux to the temperature difference across the convective boundary layer, $(T_w - T_b)$ (Figure 6.1). This requires solving for the fully coupled heat transfer problem, with convection models involving a variety of scales. Numerical models of this kind remain tentative because of the uncertainties in the set-up (initial conditions, rheological properties, accounting for continents and oceans, etc.). Some insight may be gained from simple models that are applied locally to a subset of observations. A common procedure is to introduce a heat transfer coefficient B:

$$Q_b = B\,(T_w - T_b)\,. \tag{6.3}$$

Two limiting cases have been considered. For perfectly efficient heat transfer, $B \to \infty$, implying that $T_b \to T_w$. This is the fixed temperature boundary condition.

Another limit case is when the convective mantle can only maintain a fixed heat flux. In this case, Q_b is set to a constant. Both boundary conditions allow straightforward solutions to the heat equation if the lithosphere thickness is fixed. This is not the case in the oceans where the lithosphere thickens as it cools down.

6.2 Cooling models for oceanic heat flux and bathymetry

6.2.1 The oceanic heat flux and bathymetry

A direct consequence of the sea-floor spreading hypothesis is that the sea floor is cooling as it moves away from the mid-oceanic ridges (see Chapter 2). The effect of this cooling on the heat flux and bathymetry of the oceans can be calculated and compared with the observations, providing a test of the sea-floor spreading hypothesis. The cooling model predicts how heat flux decreases and how the depth of the sea floor increases with the age of the sea floor. Different boundary conditions have been used and their effects have been thoroughly tested against heat flux and bathymetry data. For young sea floor, the fit of all models to the heat flux data has remained very crude because these data are noisy and affected by hydrothermal circulation; the cooling models also predict the bathymetry for which an excellent fit has been obtained.

6.2.2 Cooling half-space model

Oceanic ridges are associated with mantle upwellings feeding plate-scale horizontal flow. Such flow occurs in a variety of settings and has been studied extensively. Flow is dominantly horizontal with negligible variations of horizontal velocity with depth. The large wavelengths of heat flux variations imply that heat transfer is dominantly vertical. In a 2D rectangular coordinate system, with x the distance from the ridge and z the depth from the sea floor, the temperature obeys the following equation:

$$\rho C_p \left(\frac{\partial T}{\partial t} + u \frac{\partial T}{\partial x} \right) = \frac{\partial}{\partial z} \left(\lambda \frac{\partial T}{\partial z} \right), \tag{6.4}$$

where u is the horizontal velocity, ρ is the density of the lithosphere, C_p is the heat capacity and λ is the thermal conductivity. We have neglected viscous heat dissipation and radiogenic heat production, which is very small in the oceanic crust and mantle (Figure 6.2). Over the time scale of an oceanic plate, steady state can be assumed (i.e. the temperature remains constant at a fixed distance from the ridge). In steady state with constant thermal conductivity, the heat equation (6.4)

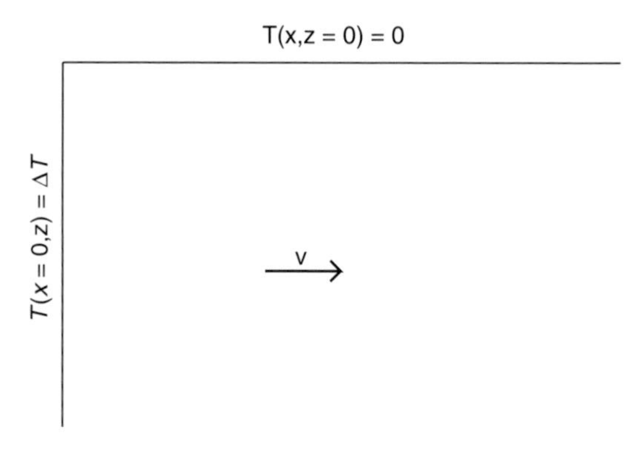

Figure 6.2. The boundary conditions for the half-space model of the cooling oceanic lithosphere. Magma is intruded on the left side ($x = 0$) at temperature $T = \Delta T$. The surface of the region $z > 0$ is kept at temperature 0. The region moves with horizontal velocity v.

reduces to,

$$\rho C_p u \frac{\partial T}{\partial x} = \lambda \frac{\partial^2 T}{\partial z^2}. \tag{6.5}$$

For a constant spreading rate, the age τ is

$$\tau = x/u, \tag{6.6}$$

which leads to,

$$\frac{\partial T}{\partial \tau} = \kappa \frac{\partial^2 T}{\partial z^2}, \tag{6.7}$$

where κ is thermal diffusivity. This is the one-dimensional heat diffusion equation, whose solution requires a set of initial and boundary conditions. The upper boundary condition and the initial condition can be specified with little error. The high efficiency of heat transport in the water column ensures that the sea-floor surface is kept at a fixed temperature T_0 of about 4°C, i.e. $\approx$ 0°C. The initial condition is set by a model for the ascent of hot material (Roberts, 1979). Over the horizontal scale of a mantle upwelling, one may neglect dissipation. Neglecting further lateral heat transfer, one can use solutions for isentropic pressure release (McKenzie and Bickle, 1988). For such an initial geotherm, temperature decreases with increasing height above the melting point. The total temperature drop depends on the starting mantle temperature, which sets the depth at which melting starts, and estimates for the heat of fusion, but never exceeds 200 K. This is much smaller than the difference between the surface and the starting temperature T_M and it may be neglected in a first approximation.

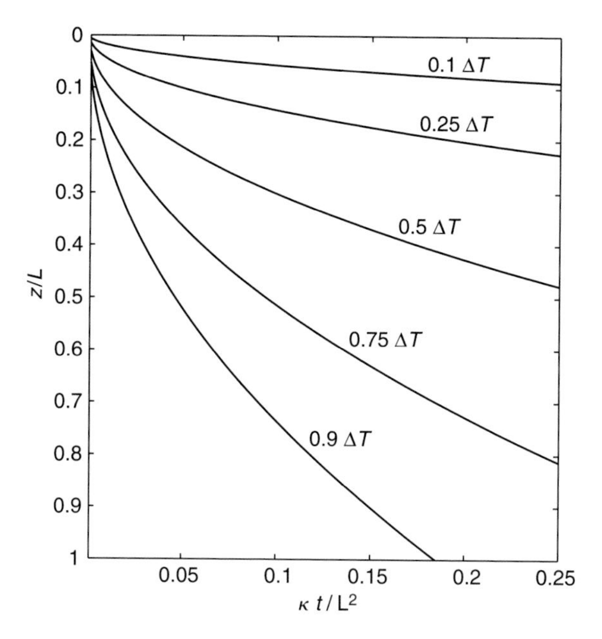

Figure 6.3. Depth of the isotherms as a function of sea-floor age t for the half-space cooling model. The depth scale L is arbitrary. For $L = 100$ km, $\tau = 0.25\ L^2/\kappa = 80$ My.

For uniform initial temperature $T(z, t = 0) = \Delta T$, the solution is identical to that of Kelvin for the cooling half-space (see equation 4.43):

$$T(z,\tau) = \Delta T \text{erf}\left(\frac{z}{2\sqrt{\kappa\tau}}\right) = \frac{2\Delta T}{\sqrt{\pi}} \int_0^{z/2\sqrt{\kappa\tau}} \exp(-\eta^2) d\eta. \tag{6.8}$$

In the half-space model, the depth of isotherms increases with age, or distance to the spreading center (Figure 6.3). It yields a heat flux proportional to $1/\sqrt{\tau}$:

$$q(\tau) = \lambda \frac{\Delta T}{\sqrt{\pi\kappa\tau}} = C_Q \tau^{-1/2}, \tag{6.9}$$

where $C_Q = \lambda \Delta T / \sqrt{\pi\kappa}$ is a constant. The $\sqrt{\tau}$ age dependence also holds when physical properties are temperature dependent (Carslaw and Jaeger, 1959), but C_Q would be different. With reference to Figure 6.1, lithospheric thickness increases with time and is given by,

$$h_2 = \sqrt{\pi\kappa\tau}. \tag{6.10}$$

Bathymetry

A second prediction of the cooling model concerns bathymetry. As the lithosphere cools down, it becomes denser. An isostatic balance condition leads to a simple equation for subsidence with respect to the ridge axis (Sclater and Francheteau, 1970):

$$\Delta h(\tau) = h(\tau) - h(0) = \frac{1}{\rho_m - \rho_w} \int_0^d (\rho[T(z,\tau)] - \rho[T(z,0)])\, dz, \tag{6.11}$$

where $h(\tau)$ and $\Delta h(\tau)$ are the depth of the ocean floor and subsidence at age τ and where ρ_m and ρ_w denote the densities of mantle rocks at temperature ΔT and water, respectively. In this equation, d is some reference depth in the mantle below the thermal boundary layer. This equation neglects the vertical normal stress at depth d, which may be significant only above the mantle upwelling structure, i.e. near the ridge axis. Assuming for simplicity that the coefficient of thermal expansion α is constant, the equation of state for near-surface conditions is,

$$\rho(T) = \rho_m[1 - \alpha(T - \Delta T)]. \tag{6.12}$$

From the isostatic balance condition (6.11), we obtain,

$$\begin{aligned} \frac{dh}{d\tau} &= \frac{-\alpha\rho_m}{\rho_m - \rho_w} \frac{d}{d\tau}\left[\int_0^d T(z,\tau)dz\right] \\ &= \frac{-\alpha}{C_p(\rho_m - \rho_w)} \frac{d}{d\tau}\left[\int_0^d \rho_m C_p T(z,\tau)dz\right], \end{aligned} \tag{6.13}$$

where we have also assumed that C_p is constant. Heat balance over a vertical column of mantle between $z = 0$ and $z = d$ implies that

$$\frac{dh}{d\tau} = \frac{\alpha}{C_p(\rho_m - \rho_w)}[q(0,\tau) - q(d,\tau)], \tag{6.14}$$

which states that thermal contraction reflects the net heat loss between the surface and depth d. Neglecting the heat flux at depth d leads to overestimating the subsidence rate. For the half-space model, however, $d \to \infty$, the equation becomes,

$$\frac{dh}{d\tau} = \frac{\alpha}{C_p(\rho_m - \rho_w)} q(0,\tau). \tag{6.15}$$

Because $q(0,\tau)$ depends on ΔT, the subsidence rate also depends on the initial temperature at the ridge axis. Thus,

$$h(\tau) = h_0 + C_h\sqrt{\tau}, \tag{6.16}$$

with h_0 the depth of the mid-oceanic ridges, and

$$C_h = \frac{2\alpha\rho_m \Delta T}{(\rho_m - \rho_w)}\sqrt{\frac{\kappa}{\pi}} = \frac{2\alpha}{C_p(\rho_m - \rho_w)} \times C_Q. \tag{6.17}$$

Using standard values of these parameters for the mantle ($\alpha = 3.1 \times 10^{-5}$ K^{-1}, $C_p = 1,170$ J kg^{-1} K^{-1}, $\rho_m = 3,330$ kg m^{-3} and $\rho_w = 1,000$ kg m^{-3}), we obtain $C_h/C_Q = 704$ m^3 W^{-1} My^{-1}.

Geoid

A third prediction of the model concerns the geoid anomalies which decrease linearly with age. As before, the observed C_Q value can be used together with established values of physical properties to calculate these anomalies.

The geoid height H is proportional to the gravity potential; in the first approximation, it is the moment of the vertical density distribution:

$$\Delta H = \frac{-2\pi G}{g} \int_0^\infty z \Delta\rho(\tau, z) dz, \tag{6.18}$$

with g the acceleration of gravity, G the gravitational constant, and $z = 0$ refers to sea level. If we use the mid-oceanic ridges as reference:

$$\Delta H = \frac{-2\pi G}{g} \left(\frac{(\rho_m - \rho_w)h^2}{2} - \alpha\rho_m \int_0^\infty (h + z)(\Delta T - T(z))dz \right), \tag{6.19}$$

we obtain,

$$\Delta H = \frac{-2\pi G \rho_m \alpha \Delta T \kappa \tau}{g} \left(1 + \frac{2\rho_m \alpha \Delta T}{\pi(\rho_m - \rho_w)} \right) = C_H \tau. \tag{6.20}$$

With standard values for the physical properties of the mantle, and $\Delta T = 1300$ K, we find $C_H \approx 0.1$ m My^{-1}, which is close to the observed values.

6.2.3 Heat flux and bathymetry data

Heat flux data

In order to compare the cooling model with heat flux data, it is useful to group the data within age intervals and plot the averages as a function of age (Figure 6.4). The main conclusion that can be drawn from the statistics displayed on this figure is that the global oceanic heat flux data set does not allow precise studies of the oceanic lithosphere.

Heat flux data on the flanks of the ridges are very scattered, with very high and very low values. Furthermore, for young sea floor, the mean heat flux when

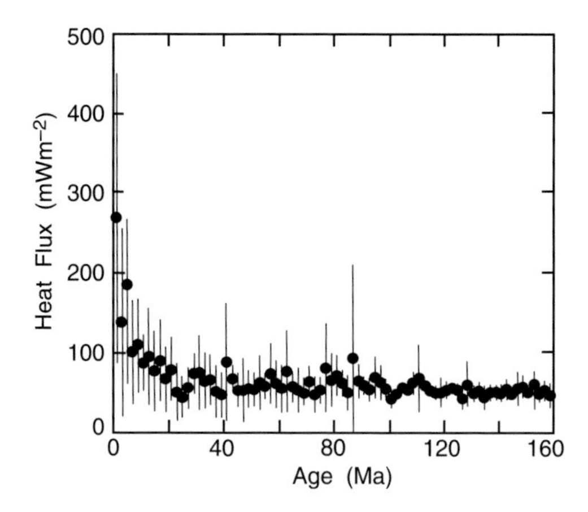

Figure 6.4. Oceanic heat flux data binned in 2 My intervals as a function of age, from a compilation by Stein and Stein (1992). Vertical bars show the standard deviations in each bin that are very large at young ages.

binned by age intervals is much lower than predicted by all the cooling models. The discovery of hydrothermal vents on the sea floor confirmed the suspicion that near the mid-oceanic ridges, hydrothermal circulation accounts for some of the heat transported (see Section 5.4). Hydrothermal circulation develops in fractures and pores, which get closed by the confining pressure deeper than 10 km. Transport of heat by hydrothermal circulation plays a crucial role at shallow depth where heat flux measurements are made. At the lithospheric scale, conduction remains the heat transport mechanism below the superficial zone where hydrothermal circulation is active.

Young sea floor

In order to compare heat flux data with model predictions, it is necessary to select measurements where the conductive heat flux represents the total flux, for instance areas that are well sealed by sediments.

For very young sea floor, a very detailed heat flux survey was conducted near the Juan de Fuca ridge by Davis *et al.* (1999) with three specific goals: evaluate the intensity and characteristics of hydrothermal circulation, assess local thermal perturbations due to basement irregular topography and test cooling models for the lithosphere. A profile over a well-sedimented region with basement age between 1 and 3 My conformed to the expected $\tau^{-1/2}$ relationship (Figure 6.5). The heat flux values can be fitted by a $\tau^{-1/2}$ relationship with the constant C_Q (in equation 6.9) between 470 and 510 mW m^{-2} My$^{1/2}$. Using standard values of thermal conductivity and thermal diffusivity, we obtain an estimate for the mantle

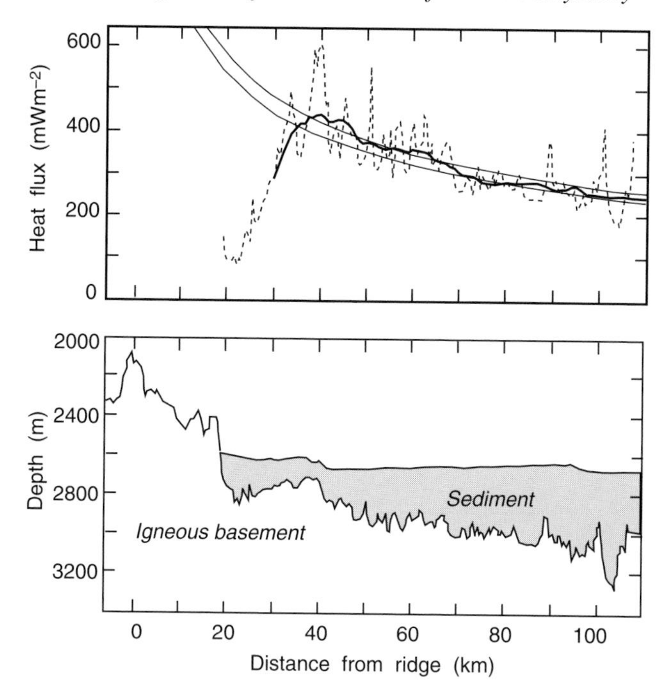

Figure 6.5. High resolution heat flux profile near the Juan de Fuca ridge from Davis *et al.* (1999). The full spreading rate is 5–6 cm y^{-1}. Two parallel lines stand for two predictions of the half-space cooling model with constant C_Q in the $\tau^{-1/2}$ heat flux–age relationship equal to 470 and 510 (with heat flux in mW m^{-2} and age in My).

temperature $T_M = 1350\,°C$, which is very close to values deduced from the chemical composition of mid-ocean ridge basalt.

This value can be corroborated using heat flux data over a wider age range. To this aim, we use the "reliable" heat flux data from selected sites where thick sedimentary cover is hydraulically resistive and seals off hydrothermal circulation which may still be effective in the igneous basement (Sclater *et al.*, 1976). Thus, there are no localized discharge zones and the average heat flux is equal to the rate at which the basement loses energy. Adding the constraint that, for the half-space model, heat flux tends to zero as age tends to ∞ tightens the estimate (Harris and Chapman, 2004). Figure 6.6 shows that values for C_Q are between 475 and 500 mW m^{-2} $My^{1/2}$, in remarkable agreement with the value deduced from the local Juan de Fuca survey. Combining these two independent determinations, we conclude that $C_Q = 490 \pm 20$ mW m^{-2} $My^{1/2}$. This value will be used later to calculate the total heat loss through the sea floor.

When all ages are considered, the average heat flux fits the half-space cooling model for ages <80 My; but observed values are systematically higher for ages >120 My (Figure 6.7).

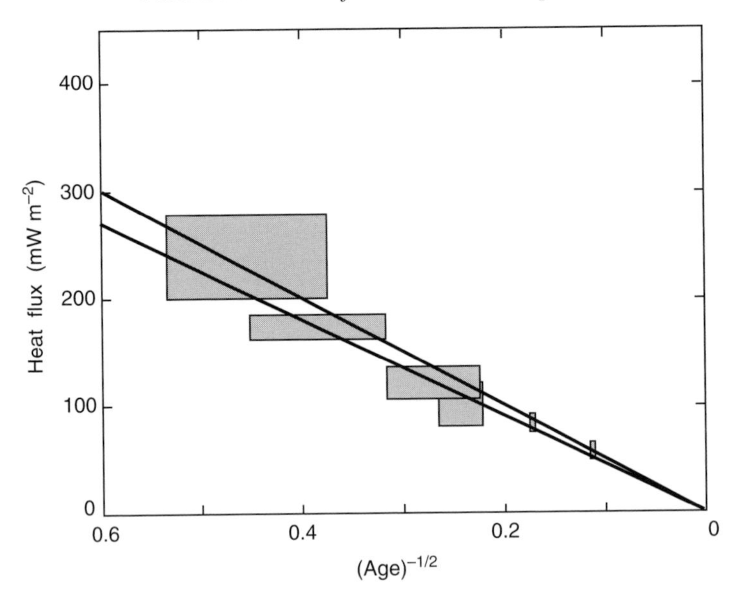

Figure 6.6. Averaged heat flux over well-sedimented areas excluding ocean floor older than 80 My (Sclater *et al.*, 1976). Solid lines correspond to values of 475 and 500 $\mathrm{mW\,m^{-2}My^{-1/2}}$ for constant C_Q in equation 6.9. Both lines are forced through zero heat flux at ∞ age.

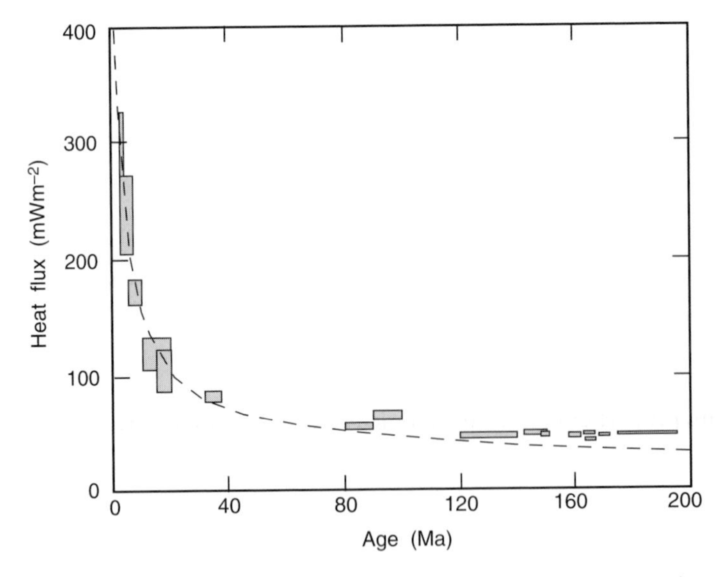

Figure 6.7. Reliable oceanic heat flux data as a function of age, from Lister *et al.* (1990). The boxes represent the average heat flux in well-sedimented areas of the sea floor. The height of each box is the 90% confidence interval for the mean heat flux, and the width represents the age range, or uncertainty. The dashed line is the best-fit half-space cooling model. Note that heat flux data through old sea floor are systematically higher than the model predictions.

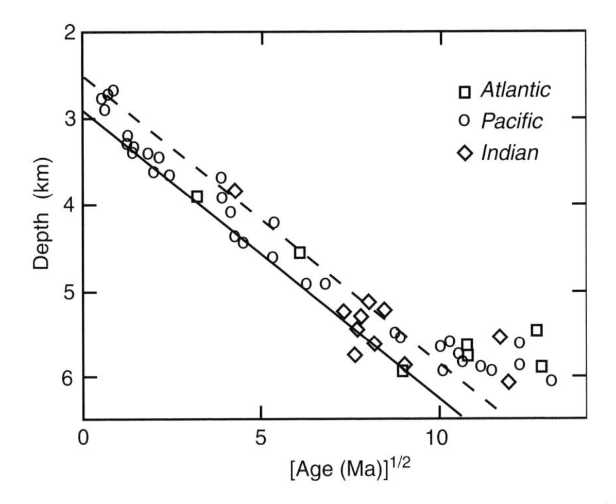

Figure 6.8. Depth to the sea-floor basement as a function of the square root of age, from Carlson and Johnson (1994). These data correspond to DSDP holes and are not affected by uncertainties on sediment thickness. The dashed lines represent the two extreme linear relationships that are consistent with data.

Bathymetry records the total cooling of the oceanic lithosphere since its formation at the mid-oceanic ridges. Bathymetry data are far less noisy than heat flux data and fit extremely well the predictions of the half-space cooling model for oceanic lithosphere younger than $\approx$100 My. The constant C_h in equation 6.21 can be determined by fitting the observed depth to the sea floor basement to $\sqrt{\tau}$. The constant C_Q is simply proportional to C_h, as discussed above.

Deep sea drill-holes have been used to get the most precise estimates of basement depth and sediment thickness (Figure 6.8). After a correction for isostatic adjustments to sediment loading, the basement depth can be fitted to sea-floor age. For ages less than 80 My, the best fit depth versus age relationship is,

$$h(\tau) = (2600 \pm 20) + (345 \pm 3)\tau^{1/2}, \text{ with h in meters and } \tau \text{ in My.} \qquad (6.21)$$

From the value of C_h and using the value $C_h/C_Q = 704\ \mathrm{m^{-3}mW^{-1}My^{-1}}$ calculated for standard properties of oceanic mantle rocks, the predicted heat flux over the same age range is,

$$q(\tau) = (480 \pm 4)\tau^{-1/2}, \text{ with } q \text{ in mW m}^{-2} \text{ and } \tau \text{ in My.} \qquad (6.22)$$

This value derived from bathymetry is thus consistent with those derived from heat flux data in well-sedimented areas. (Note that the uncertainty on these estimates is much larger than the standard error, given above.) The cooling half-space model therefore accounts for all the observations on young sea floor. The value $C_Q = 480$ mW m^{-2} My$^{1/2}$ in the half-space model implies a mantle temperature of 1370 °C.

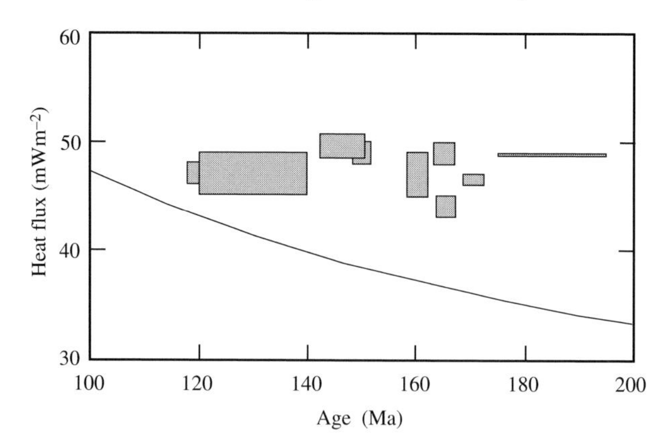

Figure 6.9. Heat flux data and prediction of the half-space cooling model for ages older than 100 My. From Lister *et al.* (1990)

Old ocean basins. Flattening of the heat flux versus age curve

The fit of both bathymetry and heat flux data to half-space cooling model predictions can only be observed for sea-floor ages less than 80 My. For ages greater than 100 My, the heat flux remains approximately constant at about 48 mW m^{-2} (Figure 6.9). For old sea-floor ages, depth to the ocean floor also departs from theoretical predictions and tends to a constant value (Figure 6.8). The interpretation of bathymetry data is ambiguous for several reasons. Depth values exhibit some scatter due to inaccurate estimates of sediment thickness and inherent basement roughness (Johnson and Carlson, 1992). Another issue is the influence of sea-mounts and large hot spot volcanic edifices, which obscure the behavior of the "normal" lithosphere. Depth to the sea floor is also sensitive to the thermal structure of the whole upper mantle and to stresses at the base of the lithosphere, which depend on the dynamics of plate-scale convection (Davies, 1988). It turns out, therefore, that the flattening of the bathymetry does not allow straightforward conclusions.

Heat flux is sensitive neither to deep mantle temperature anomalies nor to convective stresses. It records shallow thermal processes within and just below the lithosphere. Detailed and accurate surveys were conducted through old sea floor with the objective of detecting systematic variations of the heat flux. Figure 6.9 shows reliable heat flux data including designated measurements to determine whether heat flux is indeed constant for old ocean basins. It is clear that heat flux departs from the $1/\sqrt{\tau}$ behavior and exhibits no detectable variation at ages larger than about 120 My. This indicates that heat is supplied to the lithosphere from below.

6.2.4 Plate models for the oceanic lithosphere

For ages >80–100 My, the depth to the sea floor no longer increases as predicted by the half-space cooling model and heat flux levels off at $\approx$48 mW m^{-2}. This

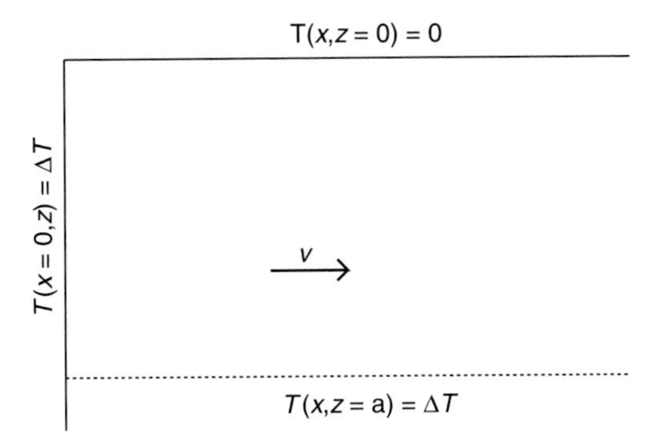

Figure 6.10. Boundary conditions for the cooling plate model of the oceanic lithosphere. Magma is intruded on the left at temperature ΔT, the plate moves with horizontal velocity v, its surface is kept at $T = 0$ and its base $z = a$ is maintained at constant temperature ΔT.

flattening of the bathymetry and heat flux indicates that heat is brought into the lithosphere from below at the same rate as it is lost at the surface. This has led to the "plate" model for which a boundary condition is specified at some fixed depth (the base of the plate). The relationships between the thermal boundary layer structure and the plate model characteristics are shown in Figure 6.10.

Plate with fixed temperature at the base

The original plate model of McKenzie (1967) is such that the plate is initially at a fixed temperature ΔT_T, the surface is maintained at $T = 0$ and the base of the plate at depth a_T is maintained at ΔT_T. With reference to Figure 6.1, one has $a_T = h_2$, showing that the fixed temperature model does not specify the thickness of the unstable boundary layer at the base of the lithosphere.

Assuming for simplicity that physical properties are constant, the temperature within the plate ($0 < z < a_r$) obeys the following equation:

$$T(z,t) = \Delta T_T \left(\frac{z}{a_T} + \frac{2}{\pi} \sum_{n=1}^{\infty} \frac{1}{n} \sin\left(\frac{n\pi z}{a_T}\right) \exp\left(\frac{-n^2\pi^2\kappa t}{a_T^2}\right)\right), \tag{6.23}$$

which defines a characteristic time:

$$\tau_T = \frac{a_T^2}{\kappa}. \tag{6.24}$$

For $t \gg \tau_T$, the series can be approximated by its leading term, as follows:

$$T(z,t) = \Delta T_T \left(\frac{z}{a_T} + \frac{2}{\pi} \sin\left(\frac{\pi z}{a_T}\right) \exp\left(-\frac{\pi^2\kappa t}{a_T^2}\right)\right), \tag{6.25}$$

which shows straightforward relaxation behavior.

The surface heat flux is given by,

$$q_0(t) = \frac{\lambda \Delta T_T}{a_T}\left(1 + 2\sum_{n=1}^{\infty} \exp\left(\frac{-n^2\pi^2\kappa t}{a_T^2}\right)\right), \tag{6.26}$$

which diverges at $t \to 0$. For $\tau \ll a_T^2/\kappa$, the heat flux

$$\begin{aligned} q_0(t) &= \frac{\lambda \Delta T_T}{a_T} \sum_{n=-\infty}^{\infty} \exp\left(\frac{-n^2\pi^2\kappa t}{a_T^2}\right) \\ &\approx \frac{\lambda \Delta T_T}{a_T} \int_{-\infty}^{\infty} \exp\left(\frac{-u^2\pi^2\kappa t}{a_T^2}\right) du \approx \frac{\lambda \Delta T_T}{\sqrt{\pi\kappa t}}, \end{aligned} \tag{6.27}$$

which is identical to the half-space solution. The subsidence due to thermal contraction of the plate can not be estimated directly from the integrated surface heat loss because the heat flux varies at the base of the plate. The subsidence is obtained as,

$$h(t) = \alpha \int_0^{a_T} \{T(z,t) - T(z,0)\} dz, \tag{6.28}$$

this gives,

$$h(t) = \frac{\alpha \Delta T_T a_T}{2}\left(1 - \frac{8}{\pi^2}\sum_{n=1}^{\infty} \frac{1}{(2n-1)^2} \exp\left(\frac{-(2n-1)^2\pi^2\kappa t}{a_T^2}\right)\right). \tag{6.29}$$

Plate with fixed heat flux at the base

An alternative plate model corresponds to the fixed flux boundary condition. In this case, the plate is initially at temperature ΔT_Q and fixed flux $\lambda \Delta T_Q/a_Q$ is maintained at the base a_Q (which corresponds to h_1 in Figure 6.1).

With fixed heat flux at the base, $Q(a_Q,t) = \lambda \Delta T_Q/a_Q$, the temperature of the plate is given by,

$$\begin{aligned} T(z,t) = \frac{\Delta T_Q z}{a_Q} &+ \frac{4\Delta T_Q}{\pi} \sum_{n=0}^{\infty} \left(\frac{1}{(2n+1)} - \frac{2}{\pi}\frac{(-1)^n}{(2n+1)^2}\right) \cdots \\ &\times \sin\left(\frac{(2n+1)\pi z}{2a_Q}\right) \exp\left(\frac{-(2n+1)^2\pi^2\kappa t}{4a_Q^2}\right). \end{aligned} \tag{6.30}$$

Because the temperature below the base of the plate varies with time, the subsidence due to thermal contraction can not be directly estimated from the temperature in the plate.

The surface heat flux is given by,

$$q(t) = \frac{\lambda \Delta T_Q}{a_Q} \left(1 + 2 \sum_{n=0}^{\infty} \left(1 + \frac{(-)^n 2}{(2n+1)\pi} \right) \exp \left(\frac{-(2n+1)^2 \pi^2 \kappa \tau}{4 a_Q^2} \right) \right), \tag{6.31}$$

which tends to ∞ for $t \to 0$. It can be shown that for $t \ll a_Q^2/\kappa$,

$$q(t) = \frac{\lambda \Delta T_Q}{\sqrt{\pi \kappa t}}. \tag{6.32}$$

The surface heat flux for the half-space and the plate models with different boundary conditions are compared on Figure 6.11, with the same time scale τ for the half-space and the two plate models' boundary condition and for the heat flux boundary condition.

The temperature difference ΔT_Q corresponds to steady state in the plate with basal heat flux Q_b:

$$\Delta T_Q = \frac{Q_b a_Q}{\lambda}, \tag{6.33}$$

and the characteristic relaxation time is

$$\tau_Q = 4 \frac{a_Q^2}{\kappa}. \tag{6.34}$$

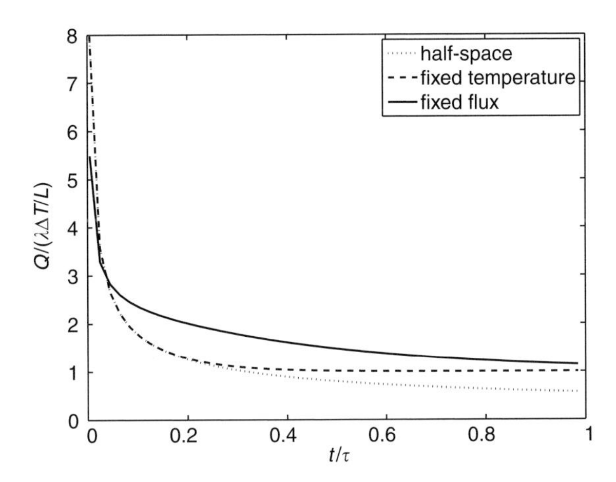

Figure 6.11. Surface heat flux calculated for the half-space and plate cooling models with fixed temperature and fixed flux at the base. For comparison, the same plate thickness is used for the flux and temperature boundary conditions (i.e. $a_Q = a_T$). This leads to much longer re-equilibration time for the fixed flux condition. If $a_Q = a_T/2$, the two plate models predict the same values for $t/\tau > 0.25$ but the fixed flux boundary condition implies lower surface heat flux than for the half-space and fixed temperature for $t/\tau \ll 1$.

The two plate models exhibit the same type of thermal relaxation, but the value of the characteristic relaxation time depends on the boundary condition. The two relaxation times must have the same value of about 80 My, the age at which the subsidence departs from boundary layer cooling subsidence. This implies a different depth to the lower boundary and leads to the relationship,

$$a_Q = \frac{a_T}{2} \approx 50 \text{ km}. \tag{6.35}$$

With reference to Figure 6.1, one has $a_T = h_2$ and $a_Q = h_1$. Thus, the solutions for the two plate models are consistent with the fundamental inequality $h_2 > h_1$. The estimate for h_1 may be compared to the thickness of depleted mantle, which depends on the well-mixed mantle potential temperature (T_w). For $T_w \approx 1300\,°\mathrm{C}$, this is $\approx$60 km. Accounting for the various uncertainties involved, the two estimates may be considered in satisfactory agreement. Another point concerns the temperature at the base of the plate. For the heat flux to tend to the same value of 48 mW m^{-2}, the temperature gradient must be the same for the two boundary conditions, which implies that the temperature, $\Delta T_Q = \Delta T_T/2$, is much too low a value.

Surface heat flux and depth to the sea floor are readily calculated from the fixed temperature condition and are in good overall agreement with the observations (Parsons and Sclater, 1977). There are small systematic differences between model predictions and observations, however, indicating that the model is only an approximation (Johnson and Carlson, 1992). The model's parameters are determined by a best fit to the bathymetry and/or heat flux data. The analysis provides estimates of $\approx$100 km and 1300 °C for a_T and ΔT_T respectively, but these values are sensitive to the data set as well to the model specifications (such as temperature-dependent physical properties). The widely used model of Stein and Stein (1992) requires $\Delta T = 1450\,°\mathrm{C}$, which is inconsistent with petrological estimates of mantle temperature (Table 6.1).

From a physical standpoint, the fixed temperature and fixed heat flux models must both be considered as crude simplifications. The former requires a specific time variation of heat flux at the base of the lithosphere, which may not be consistent with true mantle dynamics. The latter model requires that heat is brought into the lithosphere at small ages. From a practical viewpoint, the half-space cooling model provides an excellent fit to the heat flux and bathymetry data for ages $<$80 My. Over the same age range, the plate model does not provide as good a fit to the bathymetry data. The half-space model does not apply for ages $>$100 My, where heat flux and bathymetry are constant and no longer need to be calculated from a model. More complex models have a fixed heat flux brought to the base of the growing thermal boundary layer.

Table 6.1. *Different estimates of the parameters of the oceanic cooling model*

ΔT (°C)	a (km)	Method	Reference
1370	–	bathymetry (age <80 My)	
		Half-space model	Johnson and Carlson (1992)
1333	125	Constant properties – fixed T	Parsons and Sclater (1977)
1450	95	Constant properties – fixed T	Stein and Stein (1992)
1350	118	T-dependent properties	
		fixed Q at variable depth†	Doin and Fleitout (1996)
1315	106	T-dependent properties – fixed T	McKenzie *et al.* (2005)

† In this model, heat flux is fixed at the base of the growing thermal boundary layer.

Thickening of the lithosphere with time

Oldenburg (1975) considered that the base of the plate is a phase change boundary between a solid slab and a partially molten isothermal half-space. This involves the solution of a Stefan problem, which will be treated in Chapter 11. The solution predicts that the lithosphere will thicken as $t^{1/2}$. The simplest model follows from the half-space cooling model, with the surface $z=0$ at temperature 0, and the region $z>0$, initially at $T=T_M$, the "freezing" temperature of the lithosphere. Following the same approach as in Chapter 11, we obtain the temperature field as,

$$\begin{aligned} T(z,t) &= A \times \operatorname{erf}\left(\frac{z}{2\sqrt{\kappa t}}\right), \quad 0<z<z_m(t) \\ &= T_M, \quad z_m(t)<z. \end{aligned} \tag{6.36}$$

For the boundary condition at $z_m(t)$ to be verified for all t, we must have $z_m(t)=\eta\sqrt{t}$. The temperature condition at $z=z_m(t)$ gives $A=T_M/\operatorname{erf}(\eta(2\sqrt{\kappa}))$, where the constant η is obtained from the Stefan boundary condition:

$$\frac{\eta}{2\sqrt{\kappa}} \operatorname{erf}\left(\frac{\eta}{2\sqrt{\kappa}}\right) \exp\left(\frac{\eta^2}{4\kappa}\right) = \frac{\lambda T_M}{\kappa\sqrt{\pi} L\rho} = \frac{C_p T_M}{\sqrt{\pi} L}. \tag{6.37}$$

Numerical values for $\eta/(2\sqrt{\kappa})$ depend on the latent heat and the degree of partial melting, for 100% melting, $C_p T_M/(\sqrt{\pi}L) \approx 1.85$, and $\eta/2\sqrt{\kappa}$=0.9, for 5%, we have $\eta/(2\sqrt{\kappa})$=1.75. The surface heat flux is given by,

$$q_0(t) = \frac{\lambda T_M}{\sqrt{\pi\kappa t}} \frac{1}{\operatorname{erf}\left(\dfrac{\eta}{2\sqrt{\kappa}}\right)}. \tag{6.38}$$

It is higher than for the cooling half-space model by a factor 1.25 for 100% melting but only 1.02 for more realistic values of partial melting (5%). The time necessary for the lithosphere to reach a thickness of 100 km can be estimated as,

$$t_{100} \approx 320 \frac{4\kappa}{\eta^2} \tag{6.39}$$

in My, for $\kappa = 31.5\ \mathrm{km^2\ My^{-1}}$, i.e. it would be more than 350 My for 100% melting and ≈100 My for 5%. For the amount of fractional melting inferred from the thickness of the crust, these effects are too small to be observed.

6.2.5 Large-scale variations of mantle temperature

The chemical composition of mid-ocean ridge basalts varies within a restricted range, but these variations are highly significant because they require the mantle temperature to be not uniform along a ridge axis (Klein and Langmuir, 1987). Detailed petrological studies yield a range of 1300–1450 °C for the source temperature (Kinzler and Grove, 1992). This range does not correspond to experimental errors, but to compositional variations amongst mid-ocean basalts which are due to variations in source temperature. Depending on which scale these variations occur, their causes and consequences differ strongly. Small-scale heterogeneities would be of local significance only and would have no effect on geophysical observables. Large-scale heterogeneities would indicate the existence of large mantle domains and would imply variations in plate temperatures and physical properties.

Heat flux data allow accurate determination of the heat flux cooling constant, C_Q, between 470 and 510 $\mathrm{mWm^{-2}My^{-1/2}}$, corresponding to an uncertainty of only $\approx \pm 4\%$. In terms of mantle temperature, assuming no uncertainty on the thermal properties entering the expression for C_Q, this corresponds to an uncertainty for the initial temperature of $\approx \pm 60$ °C. This is compatible with the petrological estimates but does not allow an independent constraint. Bathymetry data exhibit large-scale trends with along-strike variations of ridge topography as well as of subsidence rate (Marty and Cazenave, 1989). With a modified plate model, one may estimate that the mantle temperature varies by about ±100 °C (Lago *et al.*, 1990). Such variations occur on a large scale and may be traced from the ridge to old sea floor (Humler *et al.*, 1999).

6.2.6 Hydrothermal circulation

There are many direct observations of hydrothermal vents on young oceanic sea floor. In addition, to this focused fluid flow, there is also diffuse flow through the permeable rock matrix. Hydrothermal circulation and heat transport include both diffuse flow and focused vents. Many experiments have been conducted on different spreading centers that demonstrate hydrothermal activity. But not all hydrothermal vents are visible and calculating the diffuse hydrothermal heat transport has not proven feasible. Determining directly from a very dense network of heat flux measurements the total heat loss through hydrothermal circulation has thus proven a quasi-impossible task. An additional problem is that, although active hydrothermalism involving very hot fluids seems to be found exclusively on very young sea floor (i.e. <10 My), transport of heat by fluids in the rock matrix has been shown to operate on sea floor as old as 80 My, hence the importance of determining the heat flux–age relationship in areas where the crust has been sealed by thick sediments.

The heat loss by hydrothermal circulation can be obtained as the difference between the predicted and the average measured heat flux within each age bin. The difference is very large at young ages. Integrating the heat flux in equation 6.9 for ages <2 My gives an average of $\sqrt{2} \times C_Q$ for that age bin, i.e. ≈ 680 mW m^{-2}, for $C_Q = 480$, compared with an average of 136 mW m^{-2} for all the measured values in that age bin. Integrating the difference between predicted and observed heat flux over the entire sea floor yields the total heat loss by hydrothermal circulation: 11TW.

6.3 Hot spots and thermal rejuvenation of the oceanic plates

Swells are isolated elevation anomalies, a few hundred km wide, on the sea floor. Swells are usually associated with persistent volcanic activity near their center. Wilson (1963) suggested that long chains of volcanoes on the sea floor are the trace of the motion of the oceanic plates as they move along a "hot spot" in the mantle. This has been supported by geochronology studies that show that the ages of volcanoes increase away from the present active center. The long wavelength swells are due to heating and thinning of the oceanic lithosphere as it moves over the hot spot. Thinning of the lithosphere must be fast relative to plate motion. Conductive heating of the lithosphere must be excluded because it is too slow. Rapid thinning thus requires some form of advection, and several mechanisms have been proposed. The amplitude of the elevation anomaly and the lack of a large heat flux anomaly over the swells imply that only the lowermost part of the lithosphere is involved.

6.3.1 Thermal model: reheating of a half-space

If the direction of plate motion relative to the hot spot and the age trend are parallel, the amplitude of the thermal anomaly in the lithosphere depends only on the age of the plate. Where τ is the plate age over the hot spot and the hot mantle material has reached a depth z_c, the temperature at the hot spot location has been changed to,

$$T(z,\tau) = T_M \operatorname{erf}\left(\frac{z}{2\sqrt{\kappa\tau}}\right),\ z < z_c$$
$$T(z,\tau) = T_M \quad z > z_c. \tag{6.40}$$

The elevation anomaly due to the temperature below z_c being higher than normal for age τ is,

$$\begin{aligned}\delta h &= \frac{\alpha T_M \rho_m}{\rho_m - \rho_w}\int_{z_c}^{\infty} \operatorname{erfc}\frac{z}{2\sqrt{\kappa\tau}}dz \\ &= \frac{\alpha T_M \rho_m}{\rho_m - \rho_w}\left(2\sqrt{\frac{\kappa\tau}{\pi}}\exp\left(\frac{-z_c^2}{4\kappa\tau}\right) - z_c \operatorname{erfc}\left(\frac{z_c}{2\sqrt{\kappa\tau}}\right)\right).\end{aligned} \tag{6.41}$$

After emplacement of the hot material below z_c, the temperature field can be treated as the superposition of the cooling half-space and a thermal perturbation starting at $t = \tau$:

$$\delta T(z,\tau) = 0 \quad z < z_c$$
$$\delta T(z,\tau) = T_M \operatorname{erfc}\left(\frac{z}{2\sqrt{\kappa\tau}}\right),\ z > z_c. \tag{6.42}$$

The decay of that perturbation can be calculated with the half-space Green's function (equation 4.20). This gives,

$$\begin{aligned}\delta T(z,t') = T_M &\int_{z_c}^{\infty} \operatorname{erfc}\left(\frac{z'}{2\sqrt{\kappa\tau}}\right) \times \frac{1}{2\sqrt{\pi\kappa t'}} \\ &\times \left(\exp\left(\frac{-(z-z')^2}{4\kappa t'}\right) - \exp\left(\frac{-(z+z')^2}{4\kappa t'}\right)\right) dz',\end{aligned} \tag{6.43}$$

where $t' = t - \tau$. This yields the surface heat flux perturbation as,

$$\begin{aligned}\delta q(t') &= \frac{2\lambda T_M}{\sqrt{\pi\kappa t'}}\int_{z_c/2\sqrt{\kappa t'}}^{\infty} \operatorname{erfc}\left(u\sqrt{\frac{t'}{\tau}}\right)\exp(-u^2)u\,du \\ &= \frac{\lambda T_M}{\sqrt{\pi\kappa t'}}\left(\operatorname{erfc}\frac{z_c}{2\sqrt{\kappa\tau}}\exp\left(\frac{-z_c^2}{4\kappa t'}\right) - \sqrt{\frac{\tau}{t'+\tau}}\operatorname{erfc}\left(\frac{z_c}{2}\sqrt{\frac{t'+\tau}{\kappa\tau t'}}\right)\right).\end{aligned} \tag{6.44}$$

6.3.2 Thermal model of hot spot: reheating of a plate

For a plate with fixed temperature at the base, the temperature at time τ of hot spot activity has been changed to,

$$T(z,\tau) = T_M\left(\frac{z}{a} + \frac{2}{\pi}\sum_{n=1}^{\infty}\frac{1}{n}\sin\left(\frac{n\pi z}{a}\right)\exp\left(\frac{-n^2\pi^2\kappa\tau}{a^2}\right)\right),\ z < z_c \tag{6.45}$$

$$T(z,\tau) = T_M,\ z_c < z < a,$$

where a is the plate thickness and z_c is the depth to isotherm T_M. The elevation anomaly is,

$$\delta h(\tau) = \frac{\alpha T_M a \rho_m}{2(\rho_m - \rho_w)}\left((1-l_t)^2 - \frac{4}{\pi^2}\sum_{n=1}^{\infty}\frac{1}{n^2}(\cos(n\pi l_t) - (-)^n)\exp\left(\frac{-n^2\pi^2\kappa\tau}{a^2}\right)\right), \tag{6.46}$$

with $l_t = z_c/a$ the dimensionless lithospheric thickness after re-heating. The solution for the temperature perturbation that fits the plate boundary conditions can be written as,

$$\delta T(z,t') = T_M\sum_{n=1}^{\infty}A_n\sin\left(\frac{n\pi z}{a}\right)\exp\left(\frac{-n^2\pi^2\kappa t'}{a^2}\right), \tag{6.47}$$

where $t' = t - \tau$ and the Fourier coefficients A_n are determined to fit the initial condition (i.e. the perturbation at time $t = \tau$). Limiting the series 6.45 to its leading terms gives,

$$\begin{aligned}
A_n &= \frac{1}{2a}\int_{z_c}^{a}\left(\left(1-\frac{z}{a}\right) - \frac{2}{\pi}\sin\left(\frac{\pi z}{a}\right)\exp\left(\frac{-\pi^2\kappa\tau}{a^2}\right)\right)\sin\left(\frac{n\pi z}{a}\right)dz\\
A_1 &= (1-l_t)\cos(\pi l_t) + \frac{\sin(\pi l_t)}{\pi} + \exp\left(\frac{-\pi^2\kappa\tau}{a^2}\right)\left((l_t - 1) - \frac{\sin(2\pi l_t)}{2\pi}\right)\\
A_n &= (1-l_t)\frac{\cos(n\pi l_t)}{n} + \frac{\sin(n\pi l_t)}{\pi}\\
&\quad + \frac{1}{\pi}\exp\left(\frac{-\pi^2\kappa\tau}{a^2}\right)\left(\frac{\sin((n-1)\pi l_t)}{n-1} - \frac{\sin((n+1)\pi l_t)}{n+1}\right).
\end{aligned} \tag{6.48}$$

The perturbation to the surface heat flux is given by,

$$\delta q(t) = \frac{\lambda T_M}{a}\left(2\sum_{n=1}^{\infty}nA_n\exp\left(\frac{-n^2\pi^2\kappa(t-\tau)}{a^2}\right)\right). \tag{6.49}$$

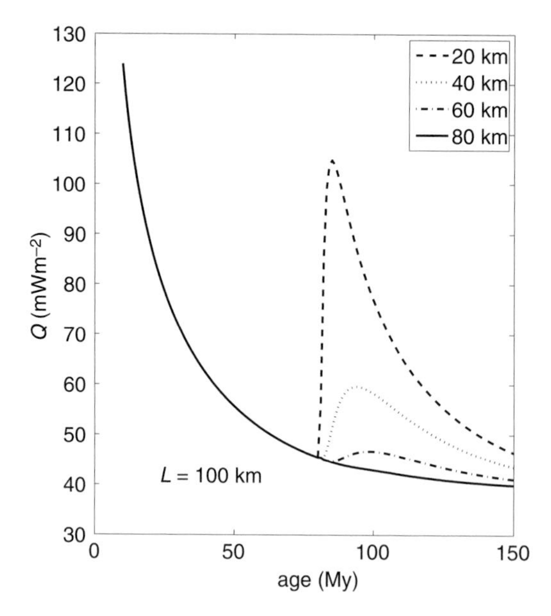

Figure 6.12. Heat flux variation with age following the replacement of the lowermost lithosphere by hot material at 80 My calculated for different depths of the intrusion. Standard values are used for all the parameters.

The maximum of this heat flux anomaly is retarded by several My relative to the time of lithospheric thinning, and it is also much reduced in amplitude (Figure 6.12). Depending on the extent of lithospheric thinning, the amplitude of the anomaly may be much smaller than the background heat flux.

The anomalous sea-floor topography is,

$$\delta h(t) = \frac{\alpha T_M a \rho_m}{\rho_m - \rho_w} \sum_{n=1}^{\infty} \frac{2A_{2n-1}}{(2n-1)\pi} \exp\left(\frac{-(2n-1)^2\pi^2\kappa(t-\tau)}{a^2}\right). \tag{6.50}$$

Contrary to the surface heat flux, the bathymetry changes instantly at the time of lithospheric thinning. The amplitude of the swell also depends on age and amount of lithospheric thinning, but it is quite distinctive on the sea floor (Figure 6.13).

6.3.3 Heat flux of hot spots

Unless the oceanic lithosphere is thinned by about 50%, the heat flux anomaly over a hot spot is relatively small (Figure 6.12). It is not surprising that the search for heat flux anomalies over hot spot swells has led to mixed results. Courtney and White (1986) found a weak anomaly over Cape Verde. Bonneville *et al.* (1997) had to resort to closely spaced measurements to detect a small anomaly of about 8 mW m^{-2}over

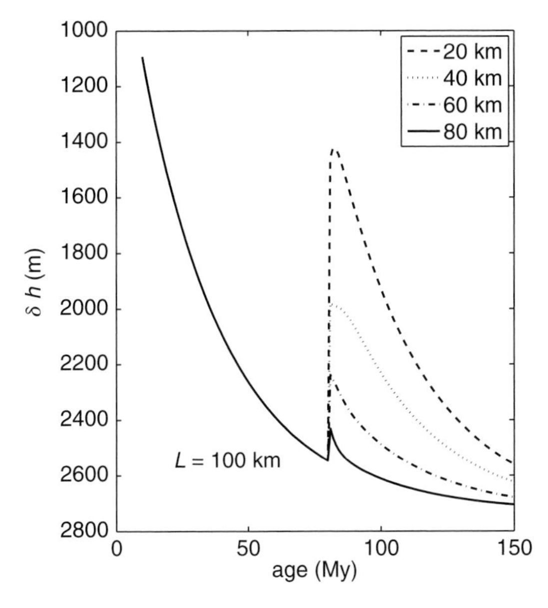

Figure 6.13. Variation of the bathymetry following replacement of the lowermost lithosphere by hot material at 80 My calculated for different depths of the intrusion. Standard values are used for all parameters.

the Reunion hot spot track. A global analysis shows that swells are associated with above normal heat flux values (Stein, 1995). Without lithosphere thinning, it would take about 100 My for a basal thermal perturbation to be detectable in surface heat flux data. No heat flux anomaly is above the error level along the Hawaiian hot spot track, probably due to hydrothermal convection within the sedimentary moat surrounding the island (Von Herzen *et al.*, 1989; Harris *et al.*, 2000). A recent seismic study by Li *et al.* (2004) indicates thinning by ≈ 50 km of the lithosphere beneath the island of Kauai. From Figure 6.12, we see that the maximum heat flux anomaly is <10 mW m^{-2} and lags several My behind the hot spot activity.

It is not possible to directly estimate from surface heat flux measurements the amount of heat introduced into the lithosphere by the hot spot. The swell of the sea floor is the surface expression of the thinning of the lithosphere, or the heat added by the replacement of the lower part of the lithosphere by hot material. The heat flux of the hot spot into the lithosphere can thus be estimated from the shape of the swell and the plate velocity. A swell height of δh relative to sea floor of the same age must be supported be thermal expansion of the lithospheric column $\delta h = \alpha h {<}\Delta T{>}$, where ${<}\Delta T{>}$ is averaged over the entire lithospheric column. This corresponds to an input of heat $\rho_m C_p h {<}\Delta T{>}$:

$$\Delta Q = \frac{\rho_m C_p \delta h}{\alpha}. \qquad (6.51)$$

Assuming a plate velocity v and that the the swell height varies laterally as $\delta h = h_0 \exp(-x^2/w^2)$, with x in the direction perpendicular to plate motion, we get the total energy flux into the plate as,

$$Q = \frac{\sqrt{\pi} h_0 w v \rho_m C_p}{\alpha}. \tag{6.52}$$

It is in the order of 50–100 GW, for typical plate velocities and swell geometries.

The data indicate large modifications of lithospheric thickness and thermal structure above mantle plumes, as expected on physical grounds (Moore *et al.*, 1999; Jurine *et al.*, 2005). Such modifications, however, depend on plume strength and lithosphere thickness and must be evaluated on a case-by-case basis.

6.4 Other effects of oceanic plate cooling

Although bathymetry has provided the strongest evidence for the sea-floor cooling models, many other geophysical data are now available to study in detail the thermal structure of oceanic mantle. Because the composition of oceanic plates is very uniform, temperature is the main control parameter on their seismic velocities and their rheological properties. Results from seismic tomography in the oceans can be compared with the cooling model. Likewise observations on the maximum depth of earthquakes, or the effective elastic thickness of oceanic plates can be compared with the predictions of the cooling models.

6.4.1 Seismic tomography of the oceanic plates

Seismic wave velocities are strongly dependent on temperature. Several global models of surface wave seismic tomography are now available and show consistent results (Shapiro and Ritzwoller, 2002; Debayle *et al.*, 2005). With their increased resolution, seismic tomography models of the upper mantle show an increase in velocity with distance to the spreading centers. The increase in velocity fits well cooling models for the lithosphere beneath all the oceans (Figures 6.14 and 6.15). One could trace the base of the cooling plate by fitting isotherms to the velocity contours.

These shear wave velocity models of the oceanic upper mantle from surface wave tomography provide clear images of the cooling oceanic lithosphere and can be used to study thermal processes within and beneath the lithosphere. Departures from the cooling model have been observed over some hot spots in the North Atlantic (Pilidou *et al.*, 2005).

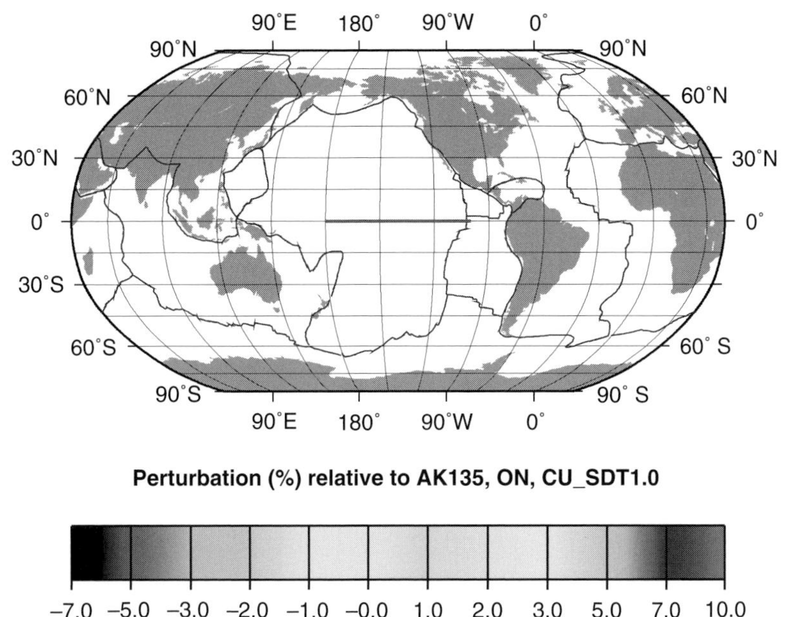

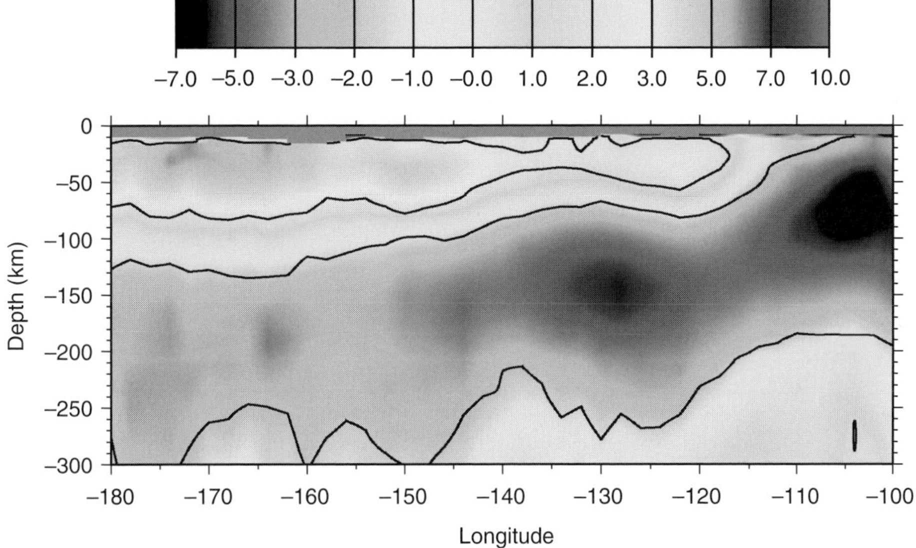

Figure 6.14. Shear wave velocity profile of the Pacific plate along the equator. The velocity increases westward with increasing distance from the East Pacific rise, in good agreement with cooling models for the oceanic lithosphere. The profiles were obtained from CUB2.0, the surface wave tomography model by Shapiro and Ritzwoller (2002). See plate section for color version.

6.4.2 Rheological profiles

Because changes in physical properties are predominantly controlled by temperature, it is thus possible to predict how the rheology and the strength of the oceanic lithosphere will vary with age. At low temperature, permanent deformations in

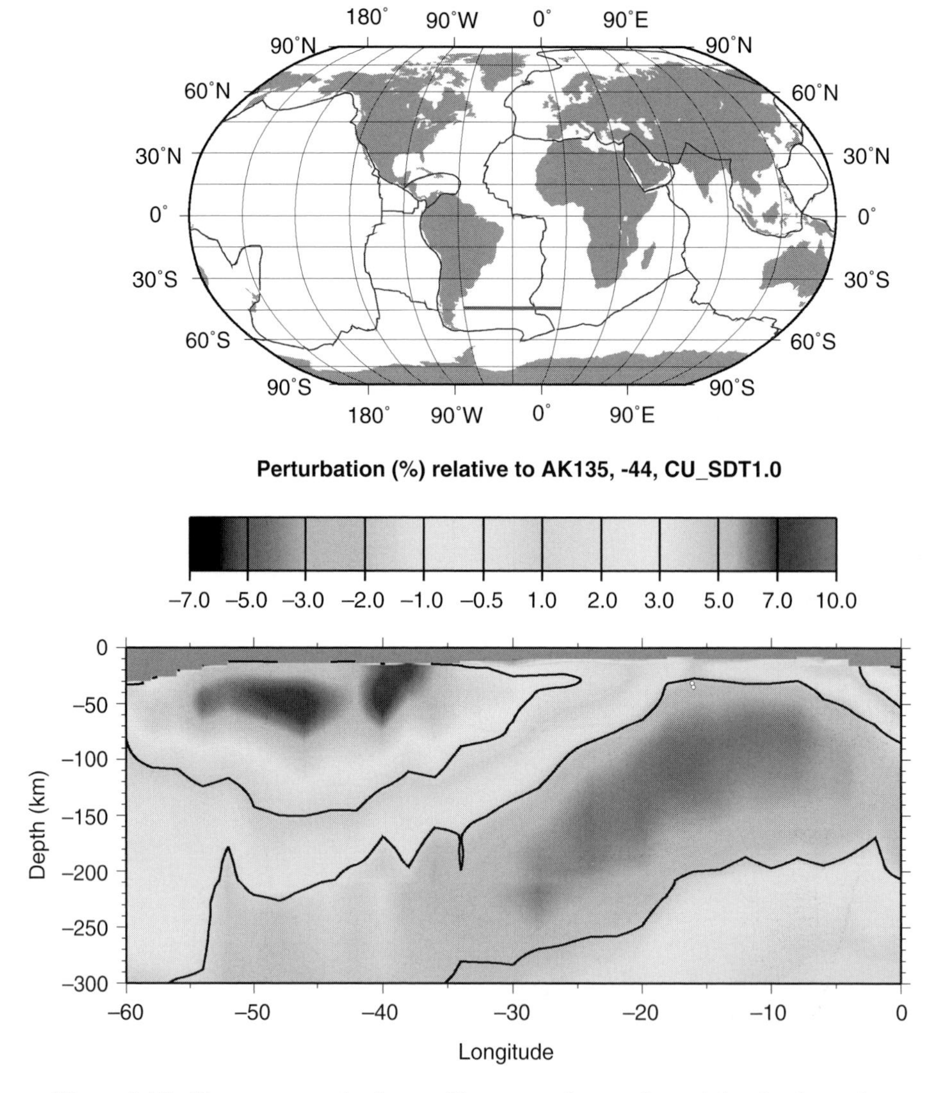

Figure 6.15. Shear wave velocity profile across the southern Atlantic along the 45 S parallel. The velocity increases on both sides away from the Mid-Atlantic ridge. The profiles were obtained from CUB2.0, the surface wave tomography model by Shapiro and Ritzwoller (2002). See plate section for color version.

rocks occur by brittle failure. Experiments by Byerlee (1978) have shown that the shear stress σ_t required to overcome friction along a fracture is independent of rock type and depends only on the effective stress normal to the plane of fracture σ_n':

$$\begin{aligned}\sigma_t &= 0.85\sigma_n', \ \sigma_n' < 200 \text{ MPa} \\ \sigma_t &= 50 + 0.6\sigma_n', \ \sigma_n' < 200 \text{ MPa},\end{aligned} \tag{6.53}$$

where the effective normal stress is the normal stress less the pore fluid pressure, often assumed to be hydrostatic $\sigma_n' = \sigma_n - P_f$. The difference between the maximum and minimum principal stresses depends on their orientations. If the maximum (most compressive) principal stress σ_1 is vertical and the minimum σ_3 is horizontal, we must have,

$$\sigma_3 - \sigma_1 = 0.8(P - P_f), \tag{6.54}$$

or alternatively, if the most compressive principal stress σ_1 is horizontal and σ_3 vertical, brittle failure occurs for

$$\sigma_1 - \sigma_3 = 4(P - P_f). \tag{6.55}$$

At high temperature, most rocks become ductile and can flow. At relatively low stress, the deformation rate $\dot{\epsilon}$ is related to the deviatoric stress σ by a power law (Kohlstedt and Goetze, 1974):

$$\dot{\epsilon} = A\sigma^n \exp(-Q_P/RT), \tag{6.56}$$

where A and n are constants and $n = 1-4$, Q_P is the activation energy, R is the gas constant and T is temperature.

At shallow depths where the frictional strength is less than the ductile strength, rocks fail by brittle failure. At greater depths (with higher temperatures), the ductile strength becomes lower and ductile deformation can be sustained with a stress less than that required for brittle failure.

6.4.3 Seismicity

The maximum depth of earthquakes in the oceanic lithosphere is thought to correspond to a change of rheology associated with the temperature controlled brittle–ductile transition.

Observations show that the maximum depth of oceanic earthquakes increases with the age of the sea floor. The maximum depth of earthquakes depends on their focal mechanisms but for each type of faulting, it is bounded by the $\approx 750\,°C$ isotherm predicted by the cooling models (Figure 6.16, Wiens and Stein (1983, 1984)).

6.4.4 Effective elastic thickness

The strength of the cooling plate increases as it moves away from the ridge. One measure of the integrated strength of the lithosphere is its effective elastic thickness. The effective elastic thickness measures the ability of the plate to support

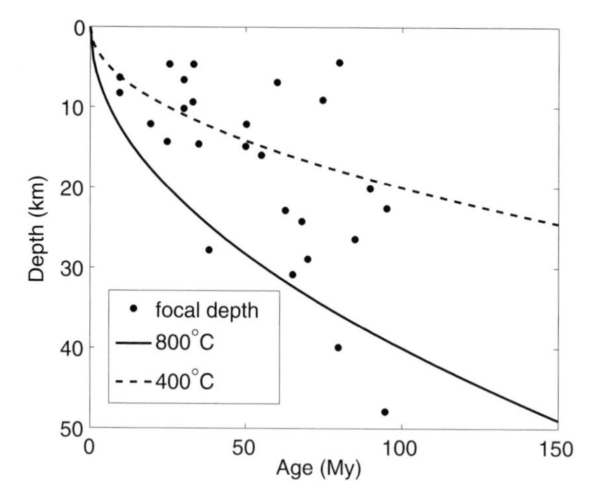

Figure 6.16. Depth of earthquakes in the oceans versus age of the sea floor. The two isotherms are calculated for a half-space cooling model. Data from Wiens and Stein (1983).

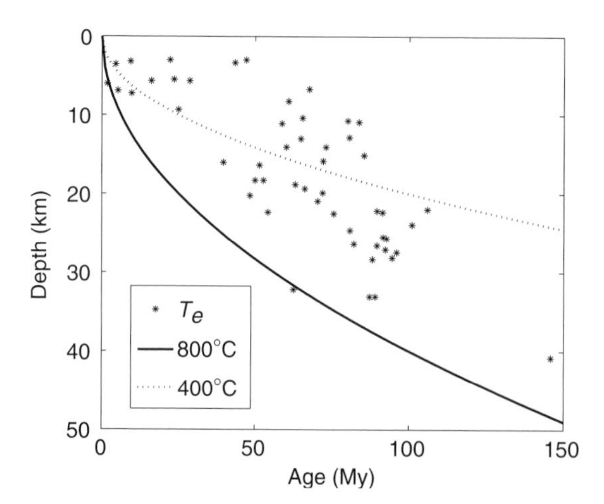

Figure 6.17. Effective elastic thickness values reported as a function of sea-floor age, compared with the isotherms for a half-space cooling model. Data from Watts (2001).

loads on its surface. The effective elastic thickness can be determined from the relationship between the topography/bathymetry and the Bouguer gravity anomaly spectra (Watts, 2001, 2007). In the oceans, the effective elastic thickness is often measured by the flexural response to sea-mounts that are emplaced on the sea floor. The elastic thickness depends on the age of the sea floor at the time the sea-mounts were emplaced. Effective elastic thickness increases with sea floor age and falls between the 450 and 600 °C isotherms (Figure 6.17 and Watts and Daly (1981)).

Exercises

6.1 Find the Laplace transforms of the temperature and surface heat flow and bathymetry for the cooling plate models. Calculate the surface heat flux and subsidence for $t \to 0$ for the different plate models.

6.2 Estimate the latent heat contribution in the energy budget of a thickening plate (consider 5% partial melting).

6.3 Calculate the heat flux at depth d for the half-space cooling model. If $d = 100$ km, find the age τ, when $q(d, \tau) < 0.01 \times q(0, \tau)$. ($k = 30\ \mathrm{km^2 My^{-1}}$.)

6.4 Write the isostatic balance equation for the cooling plate (with fixed temperature at the base); derive an equation for bathymetry as a function of age. Compare the solutions for different values of the plate thickness and the half-space solution.

6.5 Determine the amplitude of the geoid anomaly for the plate cooling model (with fixed temperature at the base), i.e. determine the difference in geoid height between the mid-oceanic ridges and old ocean basins. How does the anomaly depend on the thickness of the plate?

6.6 Find the relationship between the total heat flux brought by a hot spot, the height of the swell and the plate velocity.

7

Thermal structure of the continental lithosphere

Objectives of this chapter

In this chapter, we shall pursue several objectives. The first one is to discuss how to determine the distribution of heat-producing elements (HPEs: K, Th, U) in the continental crust and the steady-state temperatures in the lithosphere. A second objective is to discuss whether and how varying thermal regimes might have affected geological processes with time, and the conditions that allowed the formation and stabilization of cratonic roots. A third objective is to examine the relationships between crustal and lithospheric deformations and the thermal regime. Finally we shall present some direct applications of thermal models to geological problems: interpretation of metamorphic conditions in terms of tectonics, or evolution of sedimentary basins, etc.

7.1 Continental heat flux

The continents differ from the oceans because their thick crust is enriched in radioelements, because they are affected by deformations that modify the temperature field, and because their central cores (the cratons) are underlain by thick and cold roots. These major differences between continental and oceanic heat flow are illustrated by the heat flux map of North America (Figure 7.1). In the oceans, there is a contrast between the eastern margin with old sea floor and low heat flux, and the western margin where the sea floor is young and heat flux is high. In the continent, the heat flux is higher in the active western part than in the eastern part that has been stable for at least 200 My. On the continental scale, there is a trend of decreasing heat flux towards the center of the Canadian Shield, near Hudson Bay. Much of our discussion will be focused on North America because many data are available, and also because it contains both stable and very active geological provinces, and includes both the steady-state and transient regimes. But the approaches that we shall outline are applicable to continental lithospheres throughout the

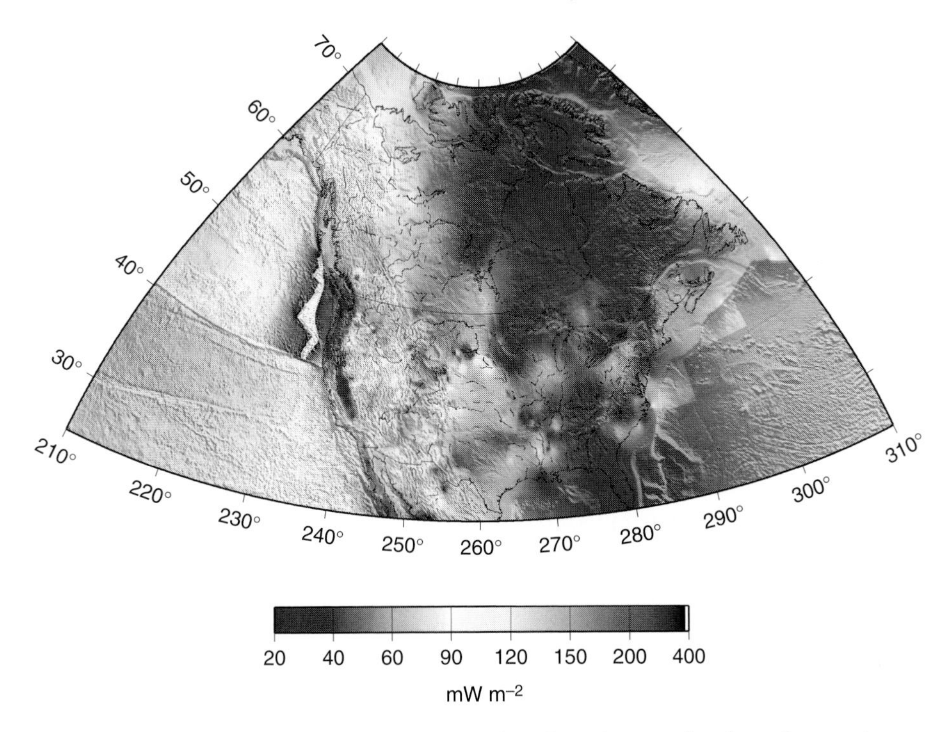

Figure 7.1. Heat flow map of North America. See plate section for color version.

world and we shall bring examples from different continents to illustrate this point.

We shall first focus on interpreting the heat flux in the stable cratons that are close to steady state. In the oceans, the age of the sea floor is the main control variable on the heat flow and lithospheric thickness. The continental lithosphere has experienced a longer evolution and is characterized by a complicated structure and composition. The continental crust is enriched in radioactive elements which contribute the largest component to the surface heat flux. In stable continental regions, the variations in surface heat flux are mostly due to large changes in crustal radioactivity. A relationship with age is difficult to establish, because crustal age is an ambiguous variable. The continental crust has experienced reactivation (thrusting or rifting events, episodes of uplift or subsidence) and the "age" of the rocks exposed at the surface is not always identical to the "age" of the rocks at depth. Variations of heat flux with age are obscured by large changes of radiogenic heat production in the crust. In contrast to the bathymetry of the sea floor, continental elevation is controlled mainly by variations of crustal thickness and density: crustal thickening by 5 km results in $\approx$500 m higher elevation.

The main objectives of this chapter are to interpret continental heat flux data and to discuss the relationships between tectonic and thermal regimes. How the continental lithosphere stabilizes and remains stable for billions of years depends on its thermal structure, which in turns depend on surface heat flux and on the vertical distribution of the radioactive elements. Tectonic deformation of the continental lithosphere occurs at plate boundaries but, sometimes also in intra-plate settings. In compressional orogens, the deformations induce transient perturbations of the thermal field. In regions of extension, perturbations of the temperature field may be the primary cause of stresses and deformations in the continental lithosphere. In both situations, how the lithosphere deforms depends on its mechanical strength and thus on the temperature field.

For the stable continents, we shall see that lithospheric thickness and strength are determined as much by surface heat flux as by the vertical distribution of heat-producing elements. In addition to systematic measurements of heat flux and heat production, determinations of (P,T) conditions in the lithospheric mantle through xenolith studies have contributed to our understanding of the thermal structure of the lithosphere. In active regions, the surface heat flux contains a large transient component due to the deformation of the steady-state temperature field. The ultimate cause for the development of zones of extension and intra-continental rifts is still debated. Thermal models can be used to test the proposed mechanisms of evolution in these zones of extension. The decay of thermal transients and the return to equilibrium of the continental lithosphere is accompanied by subsidence of sedimentary basins. The sedimentary record provides further constraints on the parameters of thermal models. Shear wave velocity is very sensitive to temperature and vertical profiles of shear wave velocity from seismic tomography studies have also been used to determine mantle temperatures.

7.2 Continental lithosphere in steady state

In steady state, the heat flux at the Earth's surface is equal to

$$Q_0 = Q_{\text{crust}} + Q_{\text{lith}} + Q_b, \tag{7.1}$$

where Q_{crust} and Q_{lith} stand for the contributions of heat sources in the crust and in the lithospheric mantle and Q_b is the heat flux at the base of the lithosphere. Determining the basal heat flux involves estimating the potentially large crustal contribution, Q_{crust}. Steady state cannot be taken for granted because it depends on the lithosphere thickness, which is part of the solution.

7.2.1 Vertical temperature distribution

The continental lithosphere is much thicker than its oceanic counterpart. The choice of a boundary condition at the base of the lithosphere is important when considering the transient regime and the return to equilibrium. For a 200 km thick lithosphere with constant temperature at its base, the thermal relaxation time is $\approx$300 My. As shown in Chapter 6, it is four times as long for a constant heat flux boundary condition. Gass *et al.* (1978) and Jaupart *et al.* (1998) have also pointed out that, for lateral changes in basal heat flux to reach the surface, they must remain immobile relative to the lithosphere for more than 1 Gy. Basal boundary conditions varying on a time scale $<$500 My have little or no effect on surface heat flux. A thick lithosphere introduces two additional complexities. One is that even very small concentrations of radioelements in the lithospheric mantle may account for a non-negligible fraction of the total heat flow. Another is that the surface heat flux cannot be in equilibrium with the present heat production in the lithospheric mantle because the thermal relaxation time is on the same order as the half-life of radioelements (Michaut and Jaupart, 2004).

In the crust, the progressive rundown of radioactivity is slow compared to thermal equilibration time and steady state may be assumed in many situations. Because heat production varies at all scales, the approach must depend on the scale of the study. Deep horizontal variations of heat production (and also of basal heat flux) are smoothed out by diffusion. Conventional wisdom from potential theory is that short-wavelength variations are associated with shallow sources whereas long-wavelength ones might be due to deep sources. For heat flow, the former is certainly true but not the latter: crustal sources reflect the surface geology which follows a long-wavelength pattern. Starting from the surface, downward continuation is unstable for small wavelengths (see equation 4.92 in Chapter 4). In practice, one must use an averaging window 100-km wide for crustal temperatures and 500-km wide for the deep lithosphere.

With a reliable model for vertical variation of the horizontally averaged heat production, $A(z)$, we can integrate the heat conservation equation (4.14) from Moho to the surface and obtain the surface heat flux Q_0 as,

$$Q_0 = Q_m + \int_0^{z_m} A(z')dz', \tag{7.2}$$

where Q_m is Moho heat flux and z_m is depth to Moho. Note that one may not assume that Q_b, the heat flux at the base of the lithosphere, is equal to Q_m, because of long thermal transients and heat production in the lithospheric mantle. These issues will be discussed later. In steady state, vertical temperature profiles are obtained by integration of equation 4.13:

$$\lambda(T)\frac{dT}{dz} = Q_0 - \int_0^{z} A(z')dz', \tag{7.3}$$

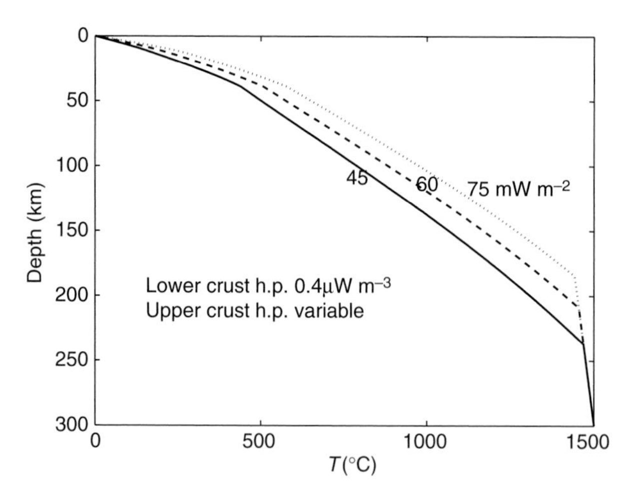

Figure 7.2. Three continental geotherms calculated for surface heat flux 45, 60 and 75 mW m^{-2}. Calculations are made with temperature- and pressure-dependent thermal conductivity (see Appendix D) and for the same Moho heat flux of 15 mW m^{-2}. Crustal heat production is assumed to be distributed in two layers of equal thickness with a fixed value of 0.4 μW m^{-3} in the lower crust. Heat production in the mantle is assumed to be negligible. With such models, changing the surface heat flux by ± 15 mW m^{-2} leads to changes of ≈ ± 100 K for temperatures at the Moho and at 200 km depth. The profiles are truncated beneath the depth where they intersect the 1350 °C isentrope.

where $\lambda(T)$ is the temperature-dependent thermal conductivity. In practice, the function $A(z)$ is not perfectly known but one may obtain constraints on the Moho heat flux, as will be explained below. As a first approximation, one may neglect heat production in the lithospheric mantle. Specifying the values of heat flux at the surface and at the Moho then sets the total amount of crustal heat production, which leaves only one unknown: the vertical variation of heat production in the crust.

Figures 7.2–7.4 illustrate the effects of changing the three main variables: the surface heat flux, the Moho heat flux and the vertical distribution of crustal heat production. These calculations account for temperature- and pressure-dependent conductivity and the mantle temperatures depend on the assumed conductivity function (see Appendix D.1 for the conductivity variations with temperature and pressure). For a stratified crust with an enriched upper layer and a fixed Moho heat flux, increasing the surface heat flux from 45 mW m^{-2} to 75 mW m^{-2} leads to temperature differences in the mantle (≈ 200K). Temperatures are more sensitive to changes of the vertical distribution of heat-producing elements (Figure 7.3). Mantle temperatures are also very sensitive to small variations in Moho heat flux (Figure 7.3). Changing Moho heat flux from 12 to 18 mW m^{-2}, i.e. within the range allowed by the data (see below), leads to ≈ 250 K differences in temperature

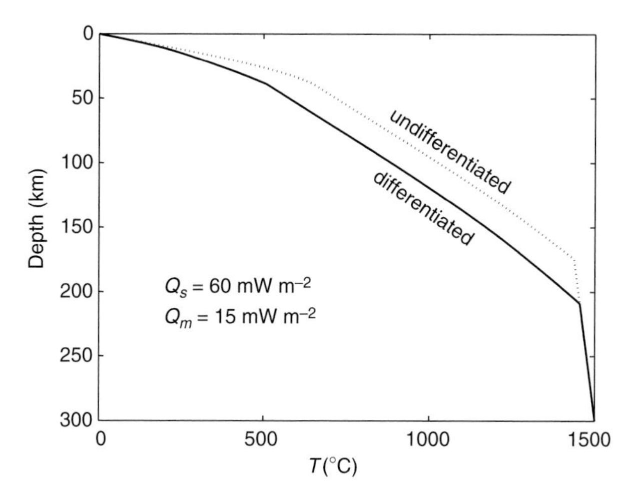

Figure 7.3. Two continental geotherms for the same heat flux value of 90 mW m^{-2}and two different vertical distributions of crustal heat production. Calculations are made with temperature-dependent conductivity (see text) and for the same Moho heat flux value of 15 mW m^{-2}. One curve corresponds to a stratified continental crust, as in Figure 7.2, and the other to a homogeneous crust. Changing the vertical distribution of heat production leads to changes of $\approx$ 220K and 140K for temperatures at the Moho and at 200 km depth.

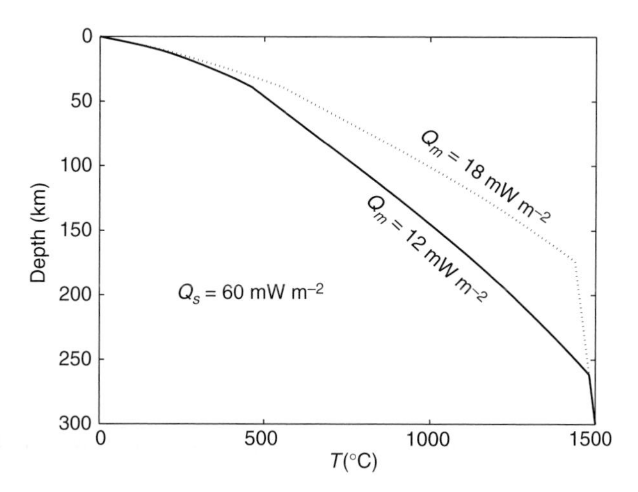

Figure 7.4. Two continental geotherms for the same heat flux value of 60 mW m^{-2} and two different values of the Moho heat flux (12 and 18 mW m^{-2}). Calculations are made with temperature-dependent conductivity (see text) and for the same stratified crustal model as in Figure 7.2. Changing the Moho heat flux leads to changes of $\approx$ 100 K and 260 K for temperatures at the Moho and at 150 km depth.

in the lowermost lithosphere. The numbers depend on the conductivity function but the trends remain the same, regardless of the assumed variations of conductivity.

7.2.2 Crustal heat production

In principle, the distribution of radiogenic heat production could be obtained by sampling all representative rocks in a geological province. This is rarely feasible in practice because samples from the lower crust are seldom available. The vagaries of geochemical studies are seldom consistent with those of heat flux studies, implying that one must deal with poorly matched data sets. For these reasons, some authors have searched for shortcuts and have tried to derive the crustal heat production directly from the heat flux data.

Early work on continental heat flux revealed, within certain provinces, a linear relationship between the local values of heat flux Q and heat production A_0:

$$Q = Q_r + A_0 D. \tag{7.4}$$

The slope D, which has the dimension of length and is usually ≈ 8 km, is related to the thickness of a shallow heat-producing layer. Q_r is called the reduced heat flux. The region where the relationship holds defines a heat flux province characterized by D and Q_r. This relationship was first developed from heat flux measurements on the very radioactive granitic plutons in New England (Figure 7.5). Within the relatively large errors in both heat flux and heat production values, several other provinces were defined, including the Sierra Nevada and the Basin and Range, which are not in steady state. Among the many heat source distributions that fit this relationship, the exponentially decreasing one, $A(z) = A_0 \exp(-z/D)$, was favored because it is independent of the erosion level (Lachenbruch, 1970). With this model, the total heat production in the crust is $\approx A_0 \times D$ and $Q_m \approx Q_r$ if D is less than 10 km. This (and other) simple models based on the linear relationship imply that the vertical distribution of heat production can be described by one universal function with well-defined parameters within each province. If it were true, it would be possible to infer the vertical distribution of heat production as well as the mantle heat flux directly from surface heat flux. New England, where the relationship was first established, is exceptional because the crust has been intruded by highly radioactive plutons (with heat production as high as 10 μW m^{-3}) whose contribution overwhelms the crustal contribution to the heat flux signal. Over other rock types, heat flux and surface heat production are only weakly correlated. The example of the 1.8 Gy Trans-Hudson orogen in Canada (Figure 7.6) illustrates that there is no clear heat flow–heat production relationship in Precambrian shields. This was confirmed by data from several regions of the world, such as the Indian, Ukrainian, South-African, or Baltic shields (Roy and Rao, 2000; Mareschal *et al.*, 1999; Jones, 1987, 1988).

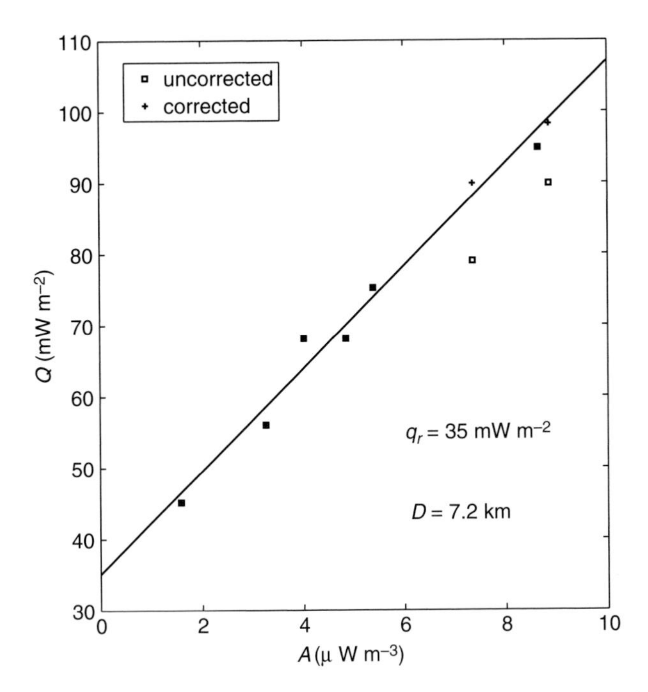

Figure 7.5. Heat flow heat production relationship for New England from Birch *et al.* (1968). Two heat flux values that did not fit were corrected for refraction effects.

From a more theoretical perspective, for the rather small wavelengths involved, surface heat flux is only sensitive to shallow heat production contrasts. The linear relationship is an artefact because horizontal heat transport smoothes out deep differences in heat production rates (see equation 4.98 and Figure 4.4 in Chapter 4). Values of D are related to the horizontal correlation distance of heat production (Jaupart, 1983a; Vasseur and Singh, 1986; Nielsen, 1987). Consequently, the reduced heat flux is heat flux at some intermediate crustal depth. There is no universal model of formation and evolution for the continental crust, and there can be no "universal" law that relates crustal heat production to surface heat flux, or that relates heat flux and heat production to crustal age. One cannot get around the problem of estimating heat production in the mid and lower crust.

Sampling of different structural levels in the crust, studies of lower crustal xenoliths, and studies in very deep boreholes have given us much needed estimates of the vertical distribution of heat generation. Samples from the lower crust show that heat production is not negligible. Xenoliths lead to a global average heat production of 0.28 $\mu W\ m^{-3}$ (Rudnick and Fountain, 1995) while exposed rocks from lower crustal levels yield 0.4–0.5 $\mu W\ m^{-3}$ (Figure 7.7). These values are too high to be consistent with an exponential decrease of heat sources.

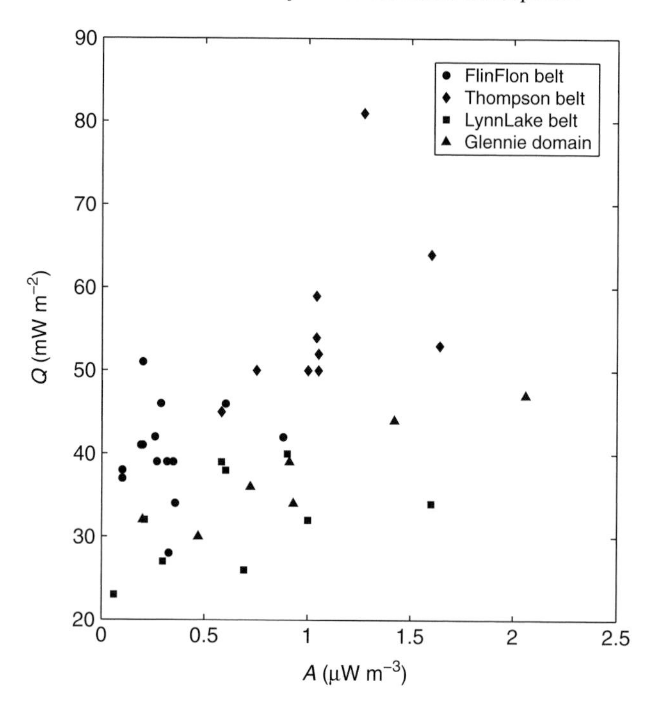

Figure 7.6. Heat flux–heat production relationship for the 1.8 Gy old Trans-Hudson orogen in the Canadian Shield. Data from Mareschal *et al.* (2004). A linear heat flow heat production relationship cannot be defined for the entire orogen, nor for the individual belts.

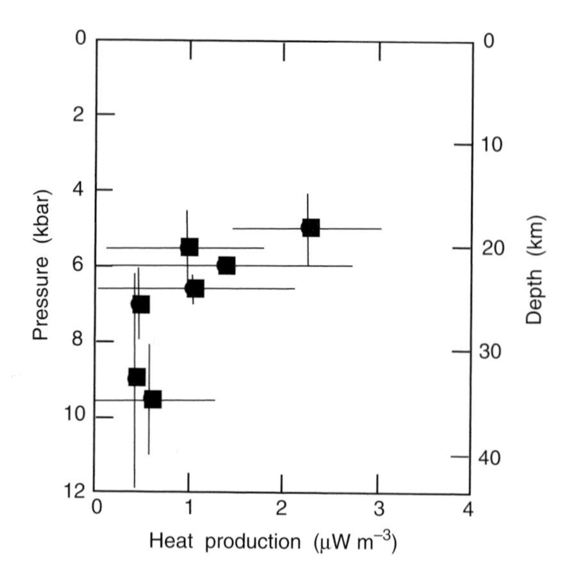

Figure 7.7. Heat production as a function of peak metamorphic pressure in samples from exposed crustal sections. Data from Joeleht and Kukkonen (1998).

Studies of exposed crustal sections suggest a general trend of decreasing heat production with depth, but this trend is not a monotonic function (Ashwal *et al.*, 1987; Fountain *et al.*, 1987; Ketcham, 1996). In the Sierra Nevada, heat production first increases, then decreases and remains constant in the lower crust beneath 15 km (Brady *et al.*, 2006). Measurements in very deep boreholes have shown that the concentration of heat sources does not systematically decrease with depth. At Kola, in the Baltic Shield, the Proterozoic supracrustal rocks (above 4 km) have much lower heat production (0.4 μW m^{-3}) than the Archean basement (1.47 μW m^{-3}) (Kremenentsky *et al.*, 1989). At KTB, in Bavaria, heat production decreases with depth at shallow levels, reaches a minimum between 3 and 8 km and increases again in the deepest part of the borehole (see Figure E.4 in Appendix E). In the 5100 m Chinese Continental Scientific Drilling (CCSD) main hole, heat production is highest in the deepest part of the hole between 3000 and 5000 m (He *et al.*, 2008). For all these examples, the distribution has been modified by the tectonics of the region.

As will be shown, one approach to derive crustal heat production is to obtain reliable average values for both the surface and Moho heat fluxes. Their difference yields the total amount of heat produced in the crust. Comparing this with the average heat production measured at the surface, one can obtain a measure of its vertical variations. For that reason, Perry *et al.* (2006) introduced a differentiation index:

$$D_I = \frac{A_0}{< A >}, \tag{7.5}$$

where A_0 is the surface heat production and $< A >$ is the mean crustal heat production. If Moho heat flux is Q_m and crustal thickness H,

$$D_I = \frac{A_0 H}{Q_0 - Q_m}. \tag{7.6}$$

How one estimates the Moho heat flux is discussed in the next section. Crustal stabilization requires vertical differentiation of the radioelements. This operates as a self-regulating system: high heat production with uniform vertical distribution gives rise to elevated temperatures favoring melting in the lower crust and differentiation (Sandiford and McLaren, 2002). One thus expects the differentiation index to be high when average crustal heat production and surface heat flux are high. For example, $D_I \approx 3$ and 1 for the Phanerozoic Appalachians and Grenville provinces, North America, respectively (Figure 7.8). It is expected that crustal differentiation should lead to $D_I > 1$. This is not always the case, because the rocks exposed at the surface can be brought up by other processes than magmatic differentiation. For instance, in the Flin Flon volcanic belt of the Proterozoic Trans-Hudson orogen, North America, $D_I \approx 0.4$. A similar

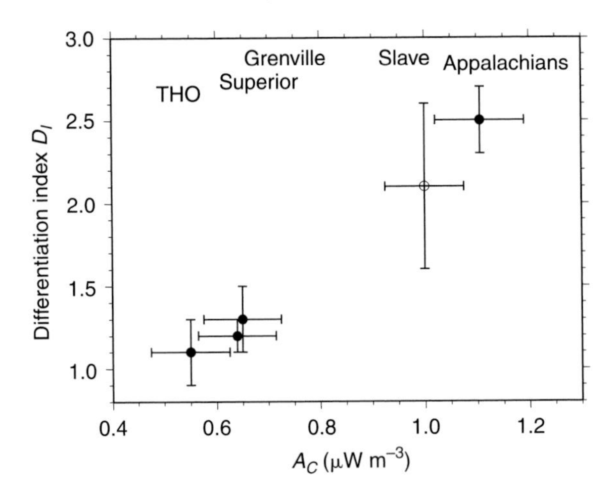

Figure 7.8. Crustal differentiation index D_I as a function of average crustal heat production in five large geological provinces of North America, from Perry *et al.* (2006).

value is obtained at the Kola super-deep hole. In both cases, the Proterozoic supracrustal rocks were tectonically transported over a more radiogenic Archean basement.

7.2.3 Estimating mantle heat flux

When the lithosphere is thick, variations in basal heat flux are unlikely to be reflected in the surface heat flux. Indeed, except for very long wavelengths, any basal heat flux variation of amplitude ΔQ_b is attenuated when upward continued to the surface (equation 4.92):

$$\Delta Q_0 = \frac{\Delta Q_b}{\cosh(2\pi L/\lambda)}, \tag{7.7}$$

where L is the lithosphere thickness and λ the wavelength of the variation. As mentioned above, ΔQ_b must be understood as a time average over >500 My. Likewise, for wavelengths less than 500 km, horizontal temperature variations in the lithosphere induced by changes in the basal heat flux would be larger than allowed for by other geophysical data. Mareschal and Jaupart (2004) concluded that variations in the basal heat flux accounts for less than ± 3 mW m^{-2} of the surface heat flux variations, i.e. they are comparable to the uncertainty on the heat flux determination.

As seen above, a local relationship between heat flux and surface heat production can seldom be established, because surface heat production varies on a very

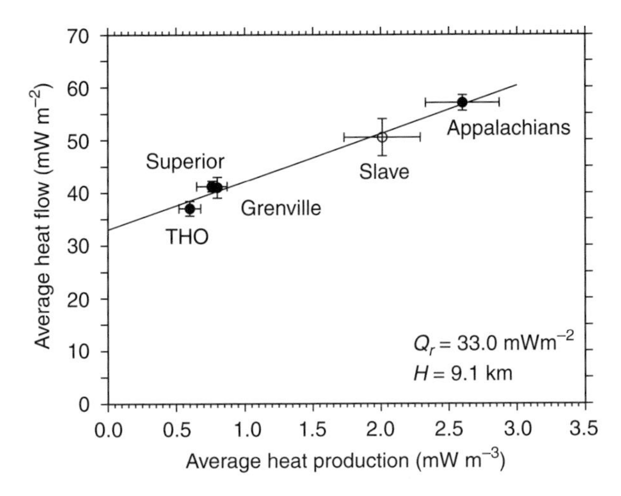

Figure 7.9. Relationship between the average values of heat flux and surface heat production for five large geological provinces of North America, from Perry *et al.* (2006).

short scale. On a wider scale, there is a relationship between heat flux and heat production when they are averaged on a province-wide scale (Figure 7.9). Such a relationship implies that heat flux is about the same at a depth given by the slope of this relationship, about 9 km. This leaves little room for variations of the mantle heat flux. The variations in surface heat flux between geological provinces occur on a very short distance and this requires that the differences originate in the crust. For example, the transition between low heat flux values in the Grenville to higher values in the Appalachians takes place over a distance less than 50 km (Mareschal *et al.*, 2000).

Crustal heat production and Moho heat flux

Various estimates of Moho heat flux have been obtained in stable cratons from several regions of the world by different methods that we shall explain (Table 7.1).

Regions of low surface heat flux provide a strong constraint on the Moho heat flux. In several parts of the Canadian Shield, heat flux values as low as 22 mW m^{-2} have been measured (Jaupart and Mareschal, 1999; Mareschal *et al.*, 2000). Similar values have also been reported for the Siberian Shield (Duchkov, 1991), the Norwegian Shield (Swanberg *et al.*, 1974) and western Australia (Cull, 1991). These regions provide an upper bound to the mantle heat flux. One may refine this estimate further by subtracting the lowest possible crustal heat production. For no crustal material are heat production estimates lower than 0.1 μW m^{-3} (Pinet and Jaupart, 1987; Joeleht and Kukkonen, 1998; Rudnick and Fountain, 1995). Over the average thickness of $\approx$ 40 km, the contribution of the crust must be at least

Table 7.1. *Various estimates of the heat flux at Moho in stable continental regions*

Region	Moho heat flux ($mW\ m^{-2}$)	References
Norwegian Shield	11 †	Pinet and Jaupart, 1987
Vredefort (South Africa)	18 †	Nicolaysen *et al.*, 1981
Kapuskasing (Canadian Shield)	11–13 †	Ashwal *et al.*, 1987; Pinet *et al.*, 1991
Grenville (Canadian Shield)	13 †	Pinet *et al.*, 1991
Abitibi (Canadian Shield)	10–14 †	Guillou *et al.*, 1994
Siberian craton	10–12 †	Duchkov, 1991
Dharwar craton (India)	12–19 †	Roy and Rao, 2003
Trans-Hudson orogen (Canadian Shield)	11–16 †*	Rolandone *et al.*, 2002
Slave province (Canada)	12–24 ‡	Russell *et al.*, 2001
Baltic Shield	7–15 ‡	Kukkonen and Peltonen, 1999
Kalahari craton (South Africa)	17–25 ‡	Rudnick and Nyblade, 1999

† Estimated from surface heat flux and crustal heat production.
∗ Estimated from condition of no melting in the lower crust at the time of stabilization.
‡ Estimated from geothermobarometry on mantle xenoliths.

4 $mW\ m^{-2}$, and the mantle heat flux must be < 18 $mW\ m^{-2}$. In Norway, Swanberg *et al.* (1974) obtained a heat flux value of 21 $mW\ m^{-2}$ over an anorthosite body; after estimating the crustal heat production, they concluded that mantle heat flux is about 11 $mW\ m^{-2}$. The same value of mantle heat flux was obtained from the analysis of all the heat flux and radiogenic heat production data in the Norwegian Shield (Pinet and Jaupart, 1987). Different arguments from these have led to the same range of values in other Precambrian areas (Table 7.1). In the Baltic and in the Siberian shields, the lowest regional average heat flux values are 15 and 18 $mW\ m^{-2}$ respectively. Such measurements provide upper limits on mantle heat flux that are even lower than in Canada (Table 7.2).

In several regions of the world, a large fraction of the crustal column has been exposed by tectonic processes. Sampling of such exposed cross-sections allows determination of the vertical distribution of radiogenic elements. If heat flux and seismic data are also available, it is possible to determine the total crustal heat production. For the Kapuskasing structure in the Canadian Shield where the crustal contribution could be determined, the Moho heat flux was calculated to be 13 $mW\ m^{-2}$. Another good example is the Closepet granite, in southern India, where different crustal levels are exposed and mantle heat flux was estimated to be in the 12–14 $mW\ m^{-2}$ range (Roy *et al.*, 2008). The average crustal heat production can also be estimated in provinces where all crustal levels can be found at the surface. In these provinces, systematic sampling will yield an estimate of the

Table 7.2. *Regional variations of the heat flux in different cratons. Minimum and maximum values obtained by averaging over 200 km × 200 km windows*

	Heat flux ($mW\ m^{-2}$)	
	minimum	maximun
Superior province	22	48
Trans-Hudson orogen	22	50
Australia	34	54
Baltic Shield	15	39
Siberian Shield	18	46

average bulk crustal heat production. In the Grenville province of the Canadian Shield, the average crustal heat production was determined to be 0.65 $\mu W\ m^{-3}$ for an average surface heat flux of 41 $mW\ m^{-2}$. This yields a Moho heat flux of 15 $mW\ m^{-2}$ (Pinet *et al.*, 1991). Similar results have been reported for other shields in the world, including South Africa (Nicolaysen *et al.*, 1981) and India (Roy and Rao, 2000), and are listed in Table 7.1. Other methods have combined heat flux with other geophysical data, mainly long-wavelength Bouguer gravity, and seismic data on crustal thickness, to estimate changes in crustal composition. A search for all models consistent with the data, including bounds on heat production rates for the various rock types involved, leads to a range of 7–15 $mW\ m^{-2}$ for the mantle heat flux (Guillou *et al.*, 1994).

The above estimates were derived using local geophysical and heat production data in several provinces and rely on knowledge of crustal structure. Independent determinations of the mantle heat flux may be obtained by considering the lithospheric thickness determined by seismic and xenolith studies. Pressure and temperature estimates from mantle xenoliths may be combined to determine a best-fit geotherm consistent with heat transport by conduction. Mantle heat flux estimates obtained in this manner depend on the value assumed for thermal conductivity. Such estimates are consistent with those deduced from crustal models (Table 7.1). These estimates come from Archean and Proterozoic cratons where heat flux values are generally low. Heat flux values tend to be higher in younger stable continental regions. For example, heat flux is higher (57 $mW\ m^{-2}$) in the Appalachians than in the Canadian Shield. The crust of the Appalachians contains many young granite intrusions with very high heat production ($>3\ \mu W\ m^{-3}$). The elevated heat flux can be accounted for by the contribution of these granites and does not require mantle heat flux to be higher than in the

Shield (Pinet *et al.*, 1991; Mareschal *et al.*, 2000). Throughout stable North America, including the Appalachians, variations of the mantle heat flux may not be exactly zero but must be less than departures from the best-fitting relationship (Figure 7.9), or about ± 2 mW m^{-2}. This estimate is close to the intrinsic uncertainty of heat flux measurements.

Allowing for the uncertainties and requiring consistency with low heat flux measurements, we retain the range of 15 ± 3 mW m^{-2} for the mantle heat flux in stable continents. Outside of this range, the differences of average heat flux values between geological provinces cannot be accounted for by changes of mantle heat flux and hence must be attributed to changes of crustal heat production. The ranges of heat flux and heat production values are the same for all provinces between 200 My and 2.5 Gy, with a weak trend of decreasing average heat flux and heat production with age (Perry *et al.*, 2006). The range is narrower in Archean provinces where high heat flux values are not found, possibly because a very radioactive crust would have been too hot to be stabilized (Morgan, 1985). Averaging the heat production of the crust of different ages yields a range of 0.79–0.9 μW m^{-3} (Tables 8.4 and E.4).

Values of the mantle heat flux may sometimes be obtained by two independent methods. In the Vredefort structure in South Africa, the deep crustal section exposed allowed estimation of the crustal heat production and left a value of 18 mW m^{-2} for the Moho heat flux in the Kaapvaal craton (Nicolaysen *et al.*, 1981), very close to the value deduced from xenolith (P,T) data (Rudnick and Nyblade, 1999).

In the Abitibi province of the Archean Superior province, Canada, xenolith suites from the Kirkland Lake kimberlite pipe yield temperatures and pressures corresponding to a wide depth range (Figure 7.10). With an estimate of thermal conductivity in the mantle, Rudnick and Nyblade (1999) found a best-fit Moho heat flux ≈ 18 mW m^{-2}, within a range of 17–25 mW m^{-2}. Combining the variations of heat flux, gravity and crustal thickness from seismic data, Guillou *et al.* (1994) obtained a range of 7–15 mW m^{-2} for Q_m. For the Abitibi, the two independent estimates can be combined to reduce the uncertainty: values lower than 15 mW m^{-2} are inconsistent with the xenolith data and values higher than 18 mW m^{-2} are inconsistent with heat flux and heat production data.

There are now enough data to directly assess whether the mantle heat flux varies as a function of age or as a function of distance across a craton. In the Canadian Shield, xenoliths samples from the Jericho kimberlite pipe of the Archean Slave province, more than 2000 km northwest of the Abitibi, give a best-fitting value of 15 mW m^{-2} for mantle heat flux within a range between 12 and 24 mW m^{-2} (Russell *et al.*, 2001). This wide range is due to the interpretation technique, which leaves the temperature dependence of thermal conductivity as a free parameter. In the Lac de Gras kimberlite pipes, also in the Slave province, surface heat flux data are available and allow tighter constraints of 12–15 mW m^{-2} for Q_m (Mareschal *et al.*, 2004).

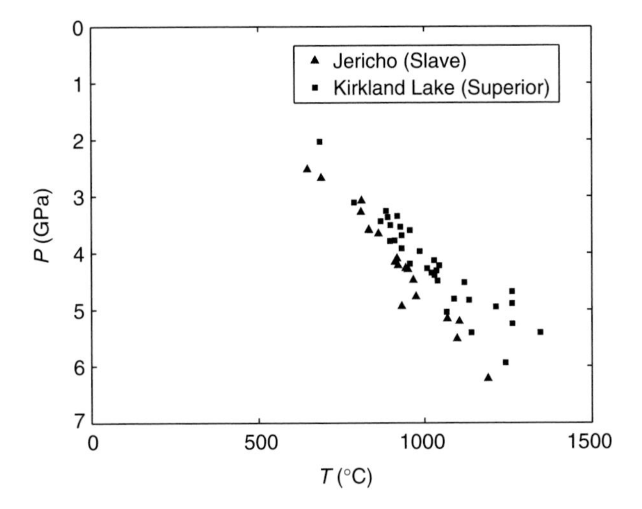

Figure 7.10. Mineral assemblages in mantle xenoliths record the pressure and temperature conditions at the time when they were erupted. Two suites, one from the Slave and one from the Superior Province of the Canadian Shield give temperature profiles in the same range as the thermal models.

Crustal models lead to mantle heat flux values of 10–15 mW m^{-2} for the *ca.* 1.8 Gy Trans-Hudson orogen (Rolandone *et al.*, 2002) and for the *ca.* 1.0 Gy Grenville province (Pinet *et al.*, 1991).

7.2.4 Regional variations of heat flux and lithospheric temperatures

Table 7.3 gives the average heat flux from different provinces grouped according to age. The trend of decreasing heat flux with age is weak at best and shows remarkable exceptions. In North America, variations of heat flux can be accounted for by changes of crustal heat production, as explained above. For stable continents, the very wide range of average heat fluxes within each age group (Archean, 36–50 mW m^{-2}; Proterozoic, 41–94 mW m^{-2}; Paleozoic: 30–57 mW m^{-2}) implies that age is not a proxy for heat flux. This range suggests that surface heat flux reflects the structure and composition of the continental crust, which vary due to the competing mechanisms of crustal extraction from the mantle and crustal recycling. In the Proterozoic provinces, high heat flux and crustal heat production (e.g. Wopmay orogen, Thompson Belt in the Trans-Hudson orogen, Gawler craton in Australia) are always associated with recycled (Archean) crust. By contrast, juvenile Proterozoic crust is characterized by low heat flux (e.g. all the juvenile belts of the Trans-Hudson, the Proterozoic rocks of the Kola peninsula).

Table 7.3. *Mean heat flux and heat production in major provinces*

	$<Q>^a$	N_Q^c	$<A>^a$	σ_A^b	N_A^c	References
	mW m^{-2}			μW m^{-3}		
Archean						
Dharwar (India)	36 ± 2.1	36	0.8			Roy and Rao (2000)
Kaapvaal basement†(S. Africa)	44	81	1.8	/		Ballard *et al.* (1987), Jones (1988)
Zimbabwe (S. Africa)	47 ± 3.5	10	1.34	/	/	Jones (1987)
Yilgarn (Australia)	39 ± 1.5	23	3.3	/	540	Cull (1991)
Superior (N. America)	41 ± 0.9	70	0.72	0.73	64	Mareschal *et al.* (2000)
Slave (N. America)	50 ± 3.5	3	2.3	1.0	20	Mareschal *et al.* (2004)
Ukrainian Shield	36 ± 2.4	12	0.9	0.2	7	Kutas (1984)
Wyoming (N. America)	48.3 ± 5.7	6	3.1	2.1	6	Decker *et al.* (1980)
Total Archean [e]	41 ± 0.8	188				Pollack *et al.* (1993)
Proterozoic						
Aravalli (India)	68 ± 4.9	13	7			Roy and Rao (2000)
Namaqua (S. Africa)	61 ± 2.5	20	2.3		10	Jones (1987)
Gawler (Australia)	94 ± 3	6	3.6	/	90	Cull (1991)
Sao Francisco craton (Brazil)	42 ± 5	3	1.5	0.6	3	Vitorello *et al.* (1980)
Braziliane mobile belt (Brazil)	55 ± 5	8	1.7	1.2	5	Vitorello *et al.* (1980)
Trans-Hudson (N. America)	42 ± 2.0	49	0.73	0.50	47	Rolandone *et al.* (2002)

Table 7.3. *(Continued)*

	$< Q >^a$	N_Q^c	$< A >^a$	σ_A^b	N_A^c	References
	mW m^{-2}			μW m^{-3}		
Wopmay (N. America)	90 ± 1.0	12	4.8	1.0	20	Lewis *et al.* (2003)
Grenville (N. America)[d]	41 ± 2.0	30	0.80		17	Mareschal *et al.* (2000)
Total Proterozoic[e]	48 ± 0.8	675				Pollack *et al.* (1993)
Paleozoic						
Appalachians (N. America)	57 ± 1.5	79	2.6	1.9	50	Jaupart and Mareschal (1999)
Basement United Kingdom	49 ± 4.4	6	1.3	0.5	6	Lee *et al.* (1987)
Urals	30 ± 2	40				Kukkonen *et al.* (1997)
Total Paleozoic	58.3 ± 0.5	2213				Pollack *et al.* (1993)

[a] Mean ± one standard error
[b] Standard deviation on the distribution
[c] Number of sites
[d] Area-weighted average value
[e] Total in the compilation by Pollack *et al.* (1993) excluding more recent measurements
† Heat flux in basement after accounting for heat production in the sediments

Within each province, there is regional (on a scale in the order of 400 km) variability which must be considered when calculating geotherms. This is illustrated in Table 7.2 which shows that regional variability is high within all provinces. Thus, there is no geotherm characteristic of a single province.

On the scale of a whole continent, the average heat production and the crustal differentiation index are positively correlated (Figure 7.8). Thus, regions with high heat production (and hence high surface heat flux) are systematically associated with an enriched upper crust. In most cases, this is due to highly radiogenic granites which do not extend very deep, as in the Appalachians province for example. All else being equal, Moho temperatures decrease with increasing differentiation index and increase with increasing heat production. With the correlation between heat production and differentiation index, variations of crustal temperatures are much smaller than for a single universal model for the vertical distribution of radioelements. From the standpoint of large-scale geophysical models, the relations between surface heat flux, Moho heat flux and temperature are non-linear and cannot be reduced to simple correlations.

Steady-state thermal conductive models are only valid if heat flux is less than about 90 mW m^{-2}. Higher heat flux leads to melting in the lower crust and to a weak lithospheric mantle that can deform easily, suggesting that other heat transport mechanisms are effective.

A low heat flux is not a sufficient condition for a steady-state thermal regime. In a thick lithosphere, with a thermal time constant in the order of 1 Gy, long-term thermal transients are inevitable.

7.3 Long-term transients: Stabilization and secular cooling of the continental lithosphere

The thickness of the continental lithosphere implies very long thermal relaxation times with many interesting consequences.

7.3.1 Archean conditions

There is no consensus on the mechanism of formation of cratonic roots. One popular model is inspired by sea-floor formation and invokes melting and melt extraction at the top of large mantle plumes. This case is analogous to that of oceanic ridges. According to an alternative model, the continental lithosphere is generated by melting above subduction zones (Carlson *et al.*, 2005). The key consequence is that the proto-root may remain hydrated, in which case it is intrinsically buoyant but not intrinsically more viscous. Geoid anomalies provide evidence for a negative intrinsic density contrast between Cratons and younger continents (Turcotte and McAdoo, 1979; Doin *et al.*, 1996).

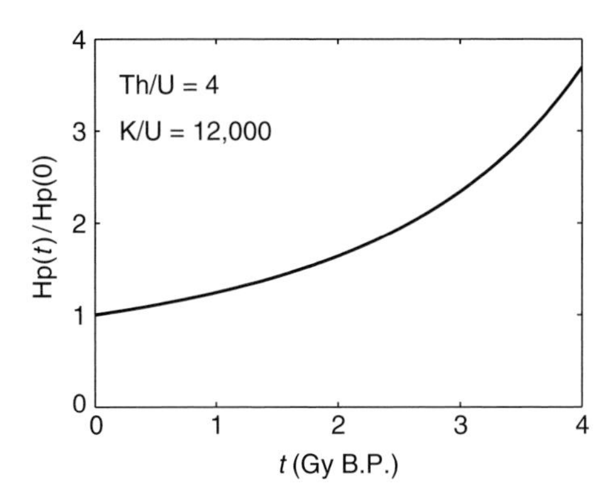

Figure 7.11. Past heat production relative to present as a function of time. Past heat production is calculated for present Th/U = 4 and K/U = 12,000.

The Archean era saw the stabilization of large cratons and the emergence of geological processes that are still active today. In the Archean, crustal metamorphism was biased towards high-temperature low-pressure conditions in contrast to more recent analogs, indicating that crustal temperatures were higher than today. In apparent contradiction, cratons achieved stability because they had strong lithospheric roots, indicating that temperatures in the lithospheric mantle were not much hotter than today. Proposed mechanisms of formation of lithospheric roots involve either stacking of subducted slabs, or melting of mantle over hot spots. These mechanisms result in different initial thermal conditions and evolution for the stabilized lithosphere.

In the Archean, heat production in the Earth was more than double the present (Figure 7.11), which might suggest higher temperatures in the crust and in the mantle as well as higher heat flux at the base of the young continental lithosphere. On average, however, the Archean crust today is associated with fewer heat-producing elements than its modern analogs (see Table E.4). When corrected for age, the total amount of crustal heat production in Archean times was close to that presently observed in Paleozoic provinces. Save for a few anomalous regions with high radioactivity, crustal heat production in the Archean is thus not sufficient to account for crustal temperatures that are higher than those of modern equivalents. The origin of the high-temperature low-pressure metamorphic conditions must thus be sought in other mechanisms, perhaps widespread magmatic perturbations.

Crustal radioactivity heats the crust in a geologically short time, but a much longer time is required to heat up the lower lithosphere. In Archean times, the thermal structure of the young continental lithosphere remained sensitive to initial conditions, i.e. conditions which led to the extraction of continental material from the mantle and to the stabilization of thick roots. If the lithospheric mantle is formed by the under-thrusting of subducted slabs beneath the crust, it will initially be colder

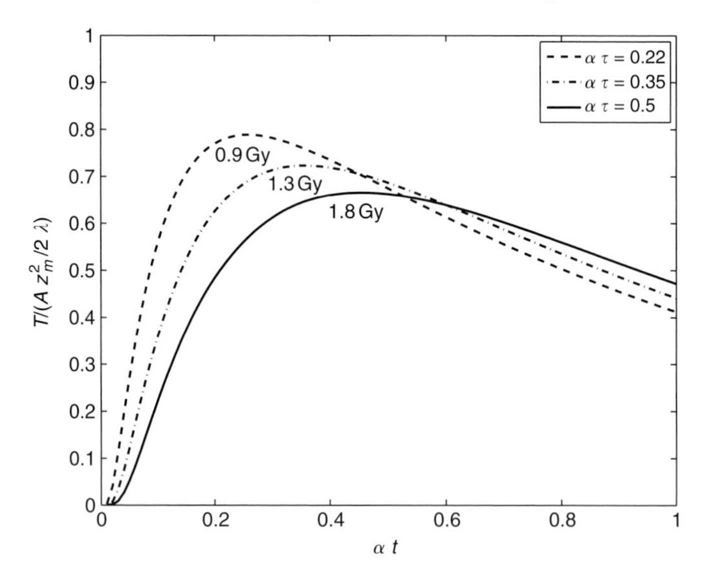

Figure 7.12. Temperature at the base of the lithospheric root after its stabilization beneath the crust. The temperature is divided by the maximum temperature increase due to crustal heat production at the time of stabilization, $Az_m^2/(2\lambda)$. α is the average decay constant of the radioelements (corresponding to a half-life of about 2.5 Gy). τ is the thermal relaxation time of the root, L^2/κ. The chosen values of $\alpha\tau$ correspond to root thicknesses 160, 200, and 240 km.

than in steady state. Mareschal and Jaupart (2005) have estimated the time needed for time-dependent crustal radioactivity to heat up the entire lithosphere. When the half-life of crustal radioactivity is of the same order as the thermal time of the lithosphere, lithospheric temperatures cannot adjust to the time-dependent radiogenic heat production and remain lower than predicted by steady-state calculations. Following isolation of a continental root from the convecting mantle, the "radiogenic" temperature component at the base of the lithosphere reaches a maximum after 1–2 Gy, depending on lithospheric thickness (Fig 7.12). The peak temperature is $\approx$ 70% of what one would infer from steady state models. Thus, temperatures in the crust and deep in the continental root are effectively decoupled for a long time. If the root forms with its initial temperature below steady state, the mantle temperature will always be below equilibrium with crustal heat production. Depending on the mechanism of root formation, the lithospheric mantle could well remain sufficiently cold and strong to preserve Archean features (van der Velden *et al.*, 2005).

7.3.2 Rundown of the heat producing elements. Secular cooling of the lithosphere

In thick continental lithosphere, the time scale for diffusive heat transport is comparable to the half-lives of uranium, thorium and potassium, implying that temperatures are not in equilibrium with the instantaneous rate of radiogenic heat generation. The lithospheric mantle undergoes secular cooling even when thermal conditions at the base of the lithosphere remain steady. The magnitude of transient effects depends on mantle heat production as well as on lithosphere thickness. Even large values of heat production do not introduce large transients in a shallow lithosphere. Conversely, even small values of heat production lead to significant transient effects in a thick lithosphere.

In a lithosphere that is thicker than 200 km, the geotherm is transient and sensitive to past heat generation. For the same parameter values, and in particular for the same values of present heat production, the deeper part of the temperature profile diverges from a steady-state calculation because of the long time to transport heat to the upper boundary. Depending on the amount of radioelements in the lithospheric mantle, the vertical temperature profile may exhibit significant curvature and may be hotter than a steady-state profile by as much as 150 K (Figure 7.13). For typical values of heat production in the lithospheric mantle, this secular cooling contributes about 3 mW m^{-2} to the total heat flow. Predicted cooling rates for lithospheric material are in the range 50–150 K Gy^{-1}, close to values reported recently for mantle xenoliths from the Kaapvaal craton, South Africa (Albarède, 2003; Bedini *et al.*, 2004).

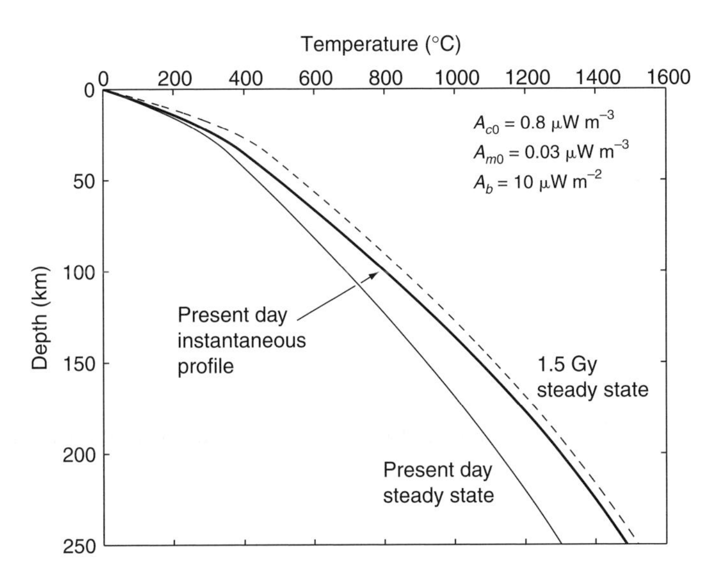

Figure 7.13. Transient geotherm with decaying heat sources in the lithospheric mantle. Two steady-state calculations corresponding to the present values 0.8 μW m^{-3} and 0.03 μW m^{-3} for heat production in the crust and mantle and to the same values of heat production back-calculated at 1.5 Gy. Due to the large relaxation time of thick lithosphere, temperatures are never in equilibrium with radioactive heat sources.

One important consequence of such long-term transient behavior stems from the shape of the vertical temperature profile. Applying a steady-state thermal model to xenolith (P,T) data leads to an overestimate of the mantle heat flux and an underestimate of the lithosphere thickness.

In order to calculate the effect on the heat flux of the rundown of the heat sources in the mantle lithosphere, we solve the heat equation,

$$\frac{1}{\kappa}\frac{\partial T}{\partial t} = \frac{\partial^2 T}{\partial z^2} + \frac{A}{\lambda}\exp(-\alpha t), \tag{7.8}$$

where $\alpha \approx 0.3 \times 10^{-9}\ \mathrm{y}^{-1}$. We assume that the temperature is 0 at the surface $z = 0$ and that the heat flow is 0 at the base of the lithosphere $z = L$. We shall assume that the effect of the initial conditions has been damped out and can be neglected, and a solution can be obtained by expanding the temperature in the following Fourier series:

$$T(z,t) = \sum_{k=0}^{\infty} c_k(t) \sin\left((2k+1)\frac{\pi z}{2L}\right), \tag{7.9}$$

with

$$1 = \frac{4}{\pi}\sum_{k=0}^{\infty}\frac{1}{2k+1}\sin\left((2k+1)\frac{\pi z}{2L}\right), \quad 0 < z < L. \tag{7.10}$$

The solution for temperature is obtained as,

$$\begin{aligned} T(z,t) = {} & \frac{16AL^2}{\pi\rho C}\sum_{k=0}^{\infty}\frac{\exp(-\alpha t) - \exp(-(2k+1)^2\pi^2\kappa t/4L^2)}{4\alpha L^2 - (2k+1)^2\pi^2\kappa} \\ & \times \frac{1}{(2k+1)}\sin\left((2k+1)\frac{\pi z}{2L}\right). \end{aligned} \tag{7.11}$$

Neglecting transient terms due to the poorly defined initial condition, we obtain the surface heat flow as,

$$\begin{aligned} \lambda\frac{\partial T}{\partial z} & = 8AL\exp(-\alpha t)\sum_{k=0}^{\infty}\frac{1}{4\alpha\tau - (2k+1)^2\pi^2} \\ & = AL\frac{\tan\sqrt{\alpha\tau}}{\sqrt{\alpha\tau}}\exp(-\alpha t), \end{aligned} \tag{7.12}$$

with $\tau = L^2/\kappa$.

Because the lithospheric thermal relaxation time is comparable to the half-life of the radioactive elements, the lithosphere can never be in equilibrium with its present heat production and the actual temperatures lie somewhere between the present and the 1.5 Gy steady-state geotherms (Figure 7.13).

7.3.3 Secular decrease in heat flux from the mantle

The secular decrease in heat flux from the mantle Q_m due to the rundown of radioactive elements can be approximated by,

$$Q_m(t) = Q_{m0} \exp(-\alpha t), \tag{7.13}$$

where t is time and the decay constant α is of the order of 3 Gy^{-1}. Because of the lithosphere thickness, the decrease in surface heat flux will lag behind the decrease in heat flux at the base of the lithosphere. From solution of the heat equation, with initial temperature 0, the surface heat flux is obtained as,

$$\frac{Q_0(t)}{Q_m 0} = \frac{\exp(-\alpha t)}{\cos\sqrt{(\alpha\tau)}} + \pi \sum_{n=0}^{\infty} \frac{(-)^n (2n+1) \exp[-(2n+1)^2 \pi^2 t/4\tau]}{\alpha\tau - (2n+1)^2 \pi^2/4}, \tag{7.14}$$

where $\tau = L^2/\kappa$. This solution is the sum of a long-term evolution and a transient which depends on the initial condition.

If $t/\tau > 1$, the surface heat flux is no longer sensitive to the poorly known initial condition and can be approximated by,

$$Q_0(t) = \frac{Q_m 0 \exp(-\alpha t)}{\cos\sqrt{\alpha\tau}} = \frac{Q_m(t)}{\cos\sqrt{\alpha\tau}}. \tag{7.15}$$

For $L = 250$ km, $\tau \approx 2$ Gy. For $\alpha \approx 3\ \mathrm{Gy}^{-1}$, the correction is not negligible.

For a secular decrease of temperature at the base of the lithosphere specified by,

$$T_m(t) = T_m(0) \exp(-\alpha t), \tag{7.16}$$

we find the long-term solution, independent of the initial condition:

$$Q_0(t) = \lambda \frac{T_m(t)}{L} \frac{\sqrt{\alpha\tau}}{\sin\sqrt{\alpha\tau}} = Q_m(t) \frac{\sqrt{\alpha\tau}}{\sin\sqrt{\alpha\tau}}, \tag{7.17}$$

The long-term solution has a similar expression to that for a decreasing basal heat flux. The mantle temperature decreases by $\approx 100\ \mathrm{K\,Gy}^{-1}$, which is a small fractional change. Thus, if one uses an exponential law to approximate this secular decrease, the "effective" decay constant α is smaller than in the previous calculation, and the correction is small.

7.3.4 Secular cooling and lithospheric thickening

In order to model the effect of the secular decrease in basal heat flux on the thickness of the lithosphere, the base of the lithosphere is defined by its viscosity. For a temperature- and pressure-dependent rheology, this may be approximated by a linear function relating basal temperature to depth, $T_m = T_0 + \gamma g \rho z$, where γ is the slope of the softening curve, ρ is the lithosphere density and g is the acceleration of gravity. The temperature and lithosphere thickness following a stepwise change in heat flux ΔQ at the base of the lithosphere can be approximated (Gliko and Mareschal, 1989). The lithosphere thickness $L(t)$ is given by:

$$L(t) = L_0 - \frac{\Delta Q}{\lambda(\beta - \gamma g \rho)} 2\sqrt{\frac{t}{\pi \tau}} + ..., \tag{7.18}$$

where L_0 is the initial lithosphere thickness, λ is the thermal conductivity, β is the temperature gradient, and $\tau = L_0^2/\kappa$. This relationship is valid for $\Delta Q << \lambda(\beta - \gamma g \rho)L$. For the secular decrease in mantle heat flux, $Q(t) = Q_{m0} \exp(-\alpha t)$, it yields for $\alpha t < 1$:

$$L(t) = L_0 \left(1 + \frac{4\alpha t}{3}\sqrt{\frac{t}{\pi \tau}} + \mathcal{O}((t/\tau)^{1/2})\right). \tag{7.19}$$

The lithospheric thickness determined from steady-state surface heat flux (equation 7.15) is

$$L_e(t) \approx L_0 (1 + \alpha(t - \tau/2) + ...). \tag{7.20}$$

Retaining only the low-order terms, we obtain the difference between the lithospheric thickness and the apparent thickness determined from the surface heat flux:

$$\Delta L = L_0 \left(\frac{4}{3}\alpha t \sqrt{\frac{t}{\pi \tau}} - \alpha(t - \tau/2)\right) \tag{7.21}$$

This equation is limited to the leading terms of the series expansion and is valid only for $\alpha\tau < 1$ and $\alpha t < 1$. With the same values for λ and τ as used previously, we find that for $t = 10^9$ years, the error on estimated lithospheric thickness $\Delta L/L \approx 10\%$.

7.4 Thermal perturbations in compressional orogens

In the following sections, our objectives are not to develop comprehensive thermo-mechanical models for the evolution of orogens, but to focus on some important factors that control their thermal evolution.

7.4.1 General features

In tectonically active regions, advection of heat usually dominates over conduction and temperatures are strongly time dependent. Thermal evolution models depend very much on the choice of boundary conditions at the base of the lithosphere and cannot be assessed against heat flux data for several reasons. One difficulty comes from the variable quality and density of heat flow data in active regions. In the western United States, the numerous heat flux data from the Basin and Range and Rio Grande rift are very noisy because of hydrological perturbations (Lachenbruch and Sass, 1978), a situation reminiscent of young sea floor. Furthermore, inclusion in the data set of measurements made for geothermal energy exploration has introduced a strong bias towards excessively high values. Far from thermal steady state, one may not use heat flux data to estimate lithospheric temperatures by downward extrapolation of shallow heat flux measurements. In a continent that is being deformed, heat flow and temperatures depend on the competing effects of crustal thickness changes, which imply changes of crustal heat production, and deformation, which affect the temperature distribution. Thus, erosion or crustal extension initially cause steeper geotherms and enhanced heat flux. After these transient effects decay, the reduced crustal thickness leads to a lower heat flux than initially. Conversely, crustal thickening causes the geothermal gradient and the heat flux to decrease at first and then to increase due to higher crustal heat production. In many cases, heat flux also records shallow processes such as the cooling of recently emplaced plutons. Because crustal composition is often affected by syn or post orogenic magmatism, there is no general rule to predict the final crustal thickness, composition, and heat production distribution.

Following cessation of tectonic and magmatic activity, one must distinguish between two types of transients. Crustal temperatures return to equilibrium with local heat sources in less than 100 My. This is followed by a much slower transient associated with re-equilibration of the lithospheric mantle. For a thick lithosphere, such transients may last as long as 500 My and result in negative or positive heat flow anomalies. Such slow thermal relaxation has two important features. First, it involves deep thermal anomalies whose lateral variations are efficiently smoothed out by heat conduction and which do not lead to spatial variations of surface heat flux over distances < 500 km. Second, it is linked to changes of thermal boundary layer thickness which may be detectable by other methods.

7.4.2 Compressional orogens

Unless the crust has anomalous composition, the total radiogenic heat production increases with crustal thickness. Steady-state conditions have not been reached in young orogens where heat flow is also enhanced by erosion. High heat flux values have been measured in Tibet and parts of the Alps (Jaupart *et al.*, 1985). These values imply high temperatures in the shallow crust. Because these variations are of short wavelength, they have been attributed to the cooling of shallow plutons. One should note that crustal melting and emplacement of granite intrusions in the upper crust modify the vertical distribution of radioelements. Thus, one should not use the same heat production model before and after orogenesis (Sandiford and McLaren, 2002).

Effect of overthrusting: stacking of two slabs

The effect of underthrusting of a cold slab under the lithosphere will first be cooling in the lithosphere overriding the slab, followed by thermal re-equilibration. A simple analytical solution can be obtained for simple geometry and boundary conditions.

With no heat sources, and fixed heat flux at the base $\lambda\Gamma$, the equilibrium temperature profile is a linear temperature profile across the two blocks. Assuming that the two blocks have the same thickness a, the initial condition for the transient temperature is $T(z,t=0)=0$ for $z<a$ and $T(z,t)=\Delta T=-\Gamma a$ for $a<z<2a$.

Using Laplace transform techniques, the solution can be obtained as:

$$
\begin{aligned}
T(z,t) &= \Delta T \sum_{n=1}^{\infty} \frac{(-)^n}{k_n} \sin(k_n z/a)\sin(k_n)\exp(-k_n^2 \kappa t/a^2) \quad z<a \\
&= \Delta T \sum_{n=1}^{\infty} \frac{(-)^n}{k_n} \cos(k_n(z-2a)/a)\cos(k_n)\exp(-k_n^2 \kappa t/a^2)
\end{aligned}
\tag{7.22}
$$

$a<z<2a,$

with $k_n=(2n-1)\pi/4$. These equations represent only the transient component of the temperature; to obtain the total temperature, one must add the steady state, $T(z,t=\infty)=\Gamma z$.

Thermal re-equilibration after the stacking of two slabs requires a time more than twice the thermal time constant $\tau=a^2/\kappa$. For a 100km thick slab with $\tau \approx 300$ My, thermal equilibration will take more than $2.5\times\tau$, 750 My (Figure 7.14). During that time the temperature in the overriding slab is lower than it was initially. The surface heat flux is given by:

$$
\frac{q(t)}{\lambda\Gamma} = \left(1-\sum_{n=1}^{\infty}(-)^n \sin(k_n)\exp(-\kappa k_n^2 t/a^2)\right). \tag{7.23}
$$

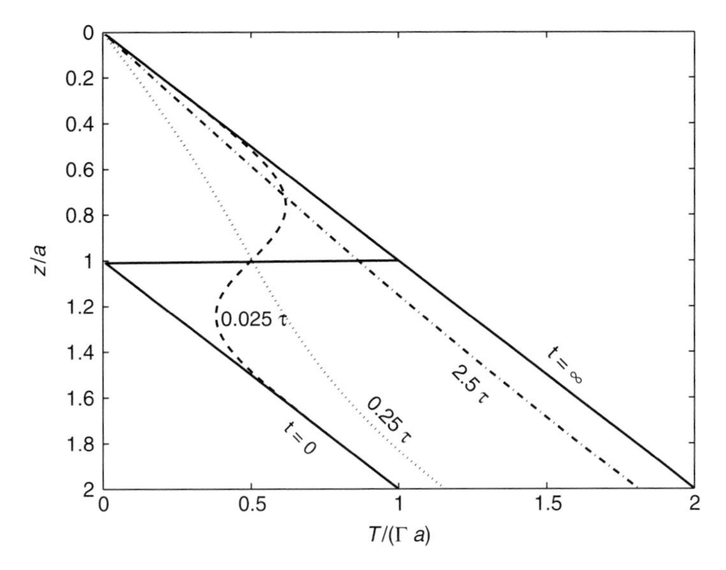

Figure 7.14. Temperature profiles at different times after stacking of two slabs of equal thickness a and each with linear temperature profiles initially. Temperature is normalized to the initial temperature at the base of the slabs, and depth to the slab thickness. The heat flux is constant at the lowermost boundary (equation 7.22).

This decreases and drops to half the initial value before it starts recovering $\approx 0.25\,\tau$. It has been speculated that this effect accounts for the low heat flow in a continent overriding an oceanic plate (e.g. the North or South American Cordillera) (Ziagos *et al.*, 1985).

Crustal scale thrusting During continental collision, one crustal block can be thrust over another. To determine the thermal effect of thrusting one crustal block over another, one must account for crustal heat production. Simplified models also allow for simple analytical solutions to be calculated. For uniform heat production A in the crust and heat flux from Moho q_m, the temperature at the base of the crust

$$T_m = \frac{Q_m a}{\lambda} + \frac{A a^2}{2\lambda}. \tag{7.24}$$

For constant heat production throughout the crust, crustal radioactivity increases the temperature at the base of the crust by $Aa^2/2\lambda$. If the average heat production is the same in both blocks, the crustal component to the heat flux doubles, and the crustal contribution to the temperature at the base of the crust is multiplied by four. The initial conditions depend on the mechanism of doubling crustal thickness. As a (crude) first approximation, we can assume that emplacement of one crust over the other is rapid and the pre-collision temperature profile is maintained in each block.

After one block overrides another one, both with the same thickness a, and with uniform heat generation A and assuming no heat flux at the base, the equilibrium temperature profile is,

$$T(z) = \frac{2Aaz}{\lambda} - \frac{Az^2}{2\lambda}. \tag{7.25}$$

Assuming that each layer was initially in steady state, the initial temperature perturbation is,

$$\Delta T(z, t=0) = -\Gamma z = \frac{-Aaz}{\lambda}, \; 0 < z < a \tag{7.26}$$

$$\Delta T(z, t=0) = -\frac{3\Gamma a}{2} = \frac{-3Aa^2}{2\lambda}, \; a < z < 2a. \tag{7.27}$$

The solution is again obtained with Laplace transforms. For constant heat flux at the base, this gives:

$$\begin{aligned} T(z,T) = {} & \frac{\Gamma a}{2} \sum_{n=1}^{\infty} \frac{(-)^n}{k_n} \sin(k_n) \sin(k_n z/a) \exp(-k_n^2 \kappa t/a^2) \\ & + \Gamma a \sum_{n=1}^{\infty} \frac{(-)^n}{k_n^2} \cos(k_n) \sin(k_n z/a) \exp(-k_n^2 \kappa t/a^2), \; 0 < z < a \end{aligned} \tag{7.28}$$

$$\begin{aligned} T(z,t) = {} & \frac{\Gamma a}{2} \sum_{n=1}^{\infty} \frac{(-)^n}{k_n} \cos(k_n) \cos(k_n(z-2a)/a) \exp(-k_n^2 \kappa t/a^2) \\ & + \Gamma a \sum_{n=1}^{\infty} \frac{(-)^n}{k_n^2} \sin(k_n) \cos(k_n(z-2a)/a) \exp(-k_n^2 \kappa t/a^2), \; a < z < 2a, \end{aligned} \tag{7.29}$$

with $k_n = (2n-1)\pi/4$.

The increase in temperature following the superposition of two crustal blocks can be very large (up to 800 K for a heat production of 0.8 μW m^{-3}), but it requires more than 25 My for the temperature to be well above the initial temperature (Figure 7.15). The initial and boundary conditions for the problem of crustal thickening rest on too simplified assumptions, and other processes (erosion, etc.) must be accounted for. These calculations demonstrate the importance of crustal heat production and that a much thickened crust can not be stable.

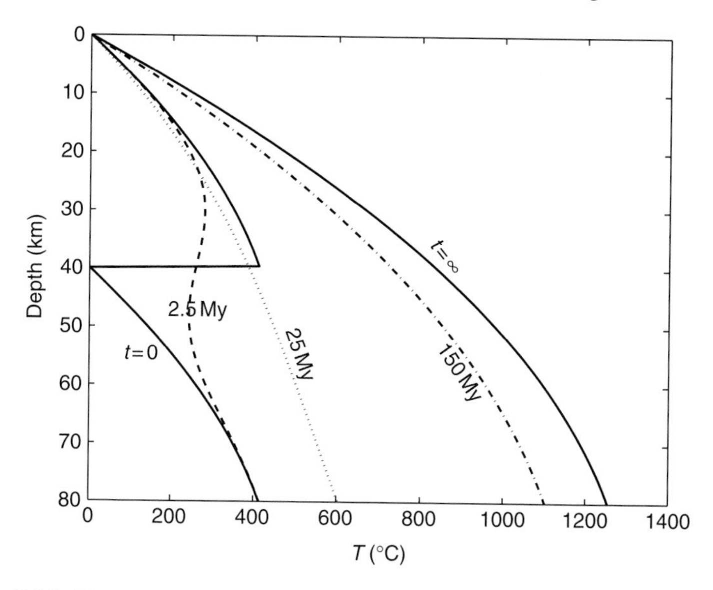

Figure 7.15. Temperature profiles at different times after superposing two 40 km thick crustal blocks with uniform heat production 0.8 μW m^{-3}. The solution superposes to the transient (equation 7.28) a steady-state term for a constant heat flux at the base of 15 mW m^{-2}.

7.4.3 Metamorphism. P-T-t paths

Schematically, an orogenic cycle consists of a phase of crustal thickening, followed by a phase of relaxation during which rocks are exhumed and brought near to the surface. Metamorphic reactions that occur during this cycle are controlled by pressure and temperature conditions. The important point is that the temperatures controlling mineral re-equilibration during metamorphic reactions are usually very far from the equilibrium geotherm.

How the pressure and temperature evolve with time depends on the mechanism of crustal thickening (e.g. thrusting, uniform crustal or lithospheric thickening). It also depends on crustal heat generation, and on possible heat inputs from the mantle (following delamination, for example). Other effects (e.g. shear heating along thrust faults, refraction of heat by lateral changes in thermal conductivity) must also be considered. During burial or erosion, pressure changes are almost instantaneous, while temperature changes are delayed. The path followed on a P-T-t diagram is thus characterized by high dP/dT. The characteristics of the P-T-t paths have been analyzed in great details by England and Thompson (1984).

To illustrate how pressure and temperature conditions change with time, we can use moving half-space solutions (equation 4.121). For a fast burial or erosion rate, the "apparent" dP/dT, i.e. the slope experienced by a rock parcel changes. In the

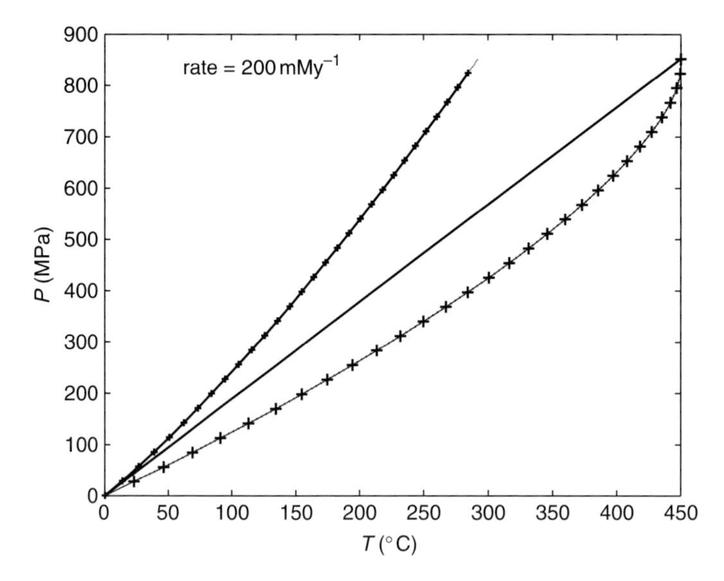

Figure 7.16. Evolution of pressure and temperature during burial and erosion, calculated with the moving medium solution of the heat equation (4.121). Curves represent the temperature and pressure of a rock parcel as it is buried (upper curve) or un-buried (lower curve) during deposition or erosion. For these calculations, the reference geotherm is assumed 15 K km^{-1} and the same rate of 0.2 km My^{-1} is assumed for deposition or erosion. Both paths are calculated assuming equilibrium initially. Crustal heat production is not included. Crosses represent 5 My intervals.

case of burial, this slope is always greater than the reference one; for erosion, the slope is steep at depth, and decreases when the parcel reaches the boundary layer near the surface (Figure 7.16).

Denudation rates are often calculated by geochronology from the difference in radiometric ages between minerals that have different closure temperatures. Depth is inferred from an average temperature gradient. Figure 7.17 illustrates the potential error on depth estimates resulting from assuming a constant gradient. Another uncertainty comes from crustal heat production which causes the reference temperature gradient to decrease with depth.

Evolution of the metamorphic (P, T) conditions is completely different for high temperature metamorphism, caused by contact with intrusions. In this case, the temperature may increase while pressure remains constant.

Temperature in moving half-space with heat production

It is possible to generalize equation 4.121 to include heat sources A that are turned on at $t = 0$, with the initial condition that the temperature gradient Γ is constant; in this case the initial condition is not in steady state before erosion/burial starts.

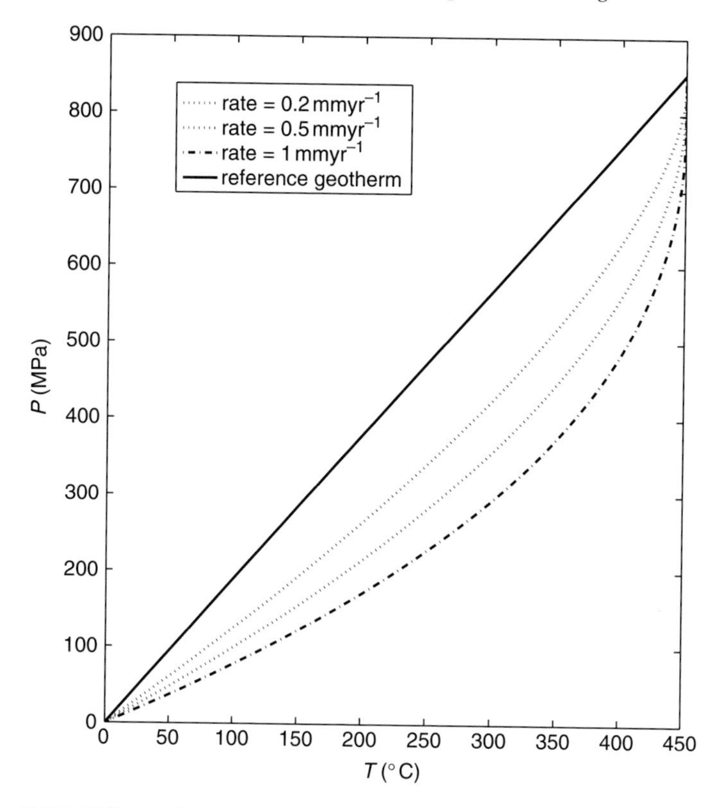

Figure 7.17. Effect of the erosion rate on the P-T-t path, and on the shallow temperature gradient. For these calculations, the reference geotherm is assumed 15 K km^{-1} and the erosion rate varies between 0.2 and 1 km My^{-1}. Crustal heat production is not included. The pressure corresponding to the 300 °C isotherm varies between 300 MPa for rapid decompression to 550 MPa for the reference geotherm

Temperature is obtained as:

$$T(z,t) = \Gamma(z - vt) + \frac{\kappa A t}{\lambda} + \left(\frac{\Gamma}{2} - \frac{\kappa A}{2v\lambda}\right)\left((z+vt)\exp\left(\frac{vz}{\kappa}\right)\right.$$
$$\left.\operatorname{erfc}\left(\frac{z+vt}{2\sqrt{\kappa t}}\right) + (vt - z)\operatorname{erfc}\left(\frac{z-vt}{2\sqrt{\kappa t}}\right)\right). \quad (7.30)$$

Numerical models of orogen evolution The simple one-dimensional thermal models (Figures 7.16 and 7.17) illustrate how crustal temperature changes during mountain building. Many competing mechanisms control the evolution of compressional orogens and their inclusion in thermal calculations requires the use of numerical models (Beaumont *et al.*, 2004; Jamieson *et al.*, 2004).

Thermal relaxation of thick continental lithosphere

Once active deformation and magmatism have ceased, the return to thermal equilibrium takes a long time. Thermal relaxation time depends on the lateral extent of the perturbed region, and on the boundary condition at the base of the lithosphere.

The large relaxation time of a thick continental lithosphere might lead one to conclude that all thermal perturbations decay slowly and leave a heat flow anomaly for a long time. In some cases where the thermal perturbation is narrow, a large thickness may in a sense be self-defeating as it enhances lateral heat transfer. Thus, thermal relaxation of some tectonic or magmatic perturbations may in fact be more sensitive to width than to thickness.

Gaudemer *et al.* (1988) and Huerta *et al.* (1998) have shown that temperatures in orogenic belts depend on belt width and on local values of heat production and thermal conductivity. One consequence is that (P,T,t) metamorphic paths may record belt width as well as other characteristics.

Another example is provided by flood basalt provinces where large volumes of magma rose through the lithosphere. For a laterally extensive thermal perturbation, one should detect a relict thermal signal for more than 200 My. There is no heat flux anomaly over the Deccan traps, India, which erupted at about 65 My (Roy and Rao, 2000). The same is true over the Parana basin in Brazil which saw the emplacement of large magma volumes at 120 Ma (Hurter and Pollack, 1996). It seems that, in both cases, eruptive fissures are localized in relatively small areas, suggesting that the zone affected by magma ascent may be a few km in width. In this case, thermal perturbations decay rapidly by horizontal heat transport. One consequence is that lithospheric seismic velocity anomalies that are associated with large magmatic events cannot be accounted for by thermal effects and hence reflect compositional variations. In several cases, it seems that the lithosphere has been modified over a large depth interval. For example, a pronounced low-velocity anomaly of narrow width (120 km) extends through the whole mantle part of the lithosphere beneath the south central Saskatchewan kimberlite field in the Trans-Hudson orogen, Canada (Bank *et al.*, 1998). Similar anomalies have been found beneath the Monteregian–White Mountain–New England hot spot track in northeastern America or beneath the Bushveld intrusion in South Africa (Rondenay *et al.*, 2000; James *et al.*, 2001).

7.5 Thermal regime in regions of extension

Intracontinental rifts and regions of extension, such as the Basin and Range, are always characterized by high heat flux. Extension itself causes the geotherms to become steeper and increases the heat flux, but the total extension in intracontinental rifts is small (10–15%) and cannot account for the elevated heat flux.

The high elevation in regions of extension is apparently inconsistent with the thinned crust and requires support from a hot and buoyant mantle. Another feature of zones of extension is their proximity to plateaux that have been uplifted without deformation (e.g. the Colorado plateau, the Tanzania craton). Chronological data on plateau uplifts are often ambiguous.

The thermal effects of extension are transient and after return to equilibrium, the heat flux at the surface of thinned crust will reflect the smaller amount of radioactive elements and hence will be lower than before extension. Further changes of crustal heat production may occur due to the injection of basaltic melts, which are depleted in radioelements with respect to average continental crust. This explains, for example, why heat flux is slightly lower in the 1,000 My Keweenawan rift than in the surrounding Superior province (Perry *et al.*, 2004).

In the Basin and Range province of the southwestern United States (Sass *et al.*, 1994; Morgan, 1983), high regional heat flux values (110 mW m^{-2}) are consistent with an extension rate of 100% (Lachenbruch and Sass, 1978; Lachenbruch *et al.*, 1994). High temperatures imply lower densities and should lead to an elevated topography. On the other hand, crustal thinning results in lower topography. The two effects compete in the Basin and Range province where the present high elevation cannot be accounted for only by extension and lithospheric thinning. According to Lachenbruch *et al.* (1994), the mantle lithosphere beneath the Basin and Range has been delaminated and not simply stretched.

A striking feature of zones of extension is that the transition between the region of elevated heat flux and the surroundings is as sharp as the sampling allows us to determine. This is observed across the boundaries of the Colorado Plateau and the Basin and Range in North America (Bodell and Chapman, 1982), between the East African rift and the Tanzanian craton (Nyblade, 1997), or between the Baikal rift and the Siberian craton (Poort and Klerkx, 2004). Where the sampling is sufficient, heat flux exhibits short wavelength variations. These variations are probably due to shallow magmatic intrusions, a hypothesis well justified by the numerous volcanic edifices that dot such areas; they also may reflect groundwater movement (Poort and Klerkx, 2004). For studies of lithospheric structure, one must separate between a high background heat flux due to extension and local anomalies reflecting shallow magmatic heat input, which requires measurements at close spacing.

7.5.1 Thermal models of rifts and zones of extension

The rate of deformation in zones of extension is controlled by the thermal regime. Zones of extension are characterized by a thinner crust, and their topography must be supported by a light and thus hot mantle. In order to provide the needed buoyancy, the mantle temperature anomaly must be very large (> 500 K). It has also been noted that uplift of zones of extension takes place in a short time and thus can not be accounted for by conductive heating of the lithosphere, but requires advective heating of the lithospheric mantle or its replacement by hot asthenospheric material.

Lithospheric extension and thinning will instantly result in a steeper temperature gradient and an increased heat flux. Thermal conduction cannot account for rapid thinning of the lithosphere and advective processes such as delamination, underplating of magmas at the base of the crust, or diapiric uprise of the asthenosphere are necessary to account for the rapid uplift of plateaux and the development of extension zones (Mareschal, 1983).

7.5.2 Underplating

For a layer of magma brought to the base of the crust, the thermal perturbation can be calculated by assuming that the initial temperature of the underplated layer is that of the asthenosphere. As the magma is molten or partially molten, its solidification will release latent heat (see Chapter 11) which must be included. In order to avoid solving the Stefan problem, the extra heat will be included in the specific heat of the layer by increasing the initial temperature by L^*/C_p, where L^* is the latent heat and C_p is the specific heat. If the layer is confined in the region $a < z < b$, the initial thermal perturbation is $T(z,t=0) = 0$ for $0 < z < a$ and for $b < z < L$ and $T(z,t=0) = \Gamma(L-z)$ for $a < z < b$ (Figure 7.18). The decay of this perturbation depends on the boundary condition at the base of the lithosphere. For constant temperature, it is

$$T(z,t) = \sum_{n=1}^{\infty} A_n \sin(n\pi z/L) \exp(-n^2\pi^2 t/\tau), \tag{7.31}$$

where the coefficients A_n are given by,

$$\begin{aligned} \frac{A_n}{2\Gamma} = & \frac{(L-a)(\cos(n\pi a/L) - (L-b)\cos(n\pi b/L))}{n\pi} \\ & + \frac{L^2}{n^2\pi^2}(\sin(n\pi a/L) - \sin(n\pi b/L)), \end{aligned} \tag{7.32}$$

where $\tau = L^2/\kappa$. The temperature perturbation decays quite rapidly and is never very large except near the intruding layer (Figure 7.19).

The surface heat flux $q(t)$ is given by,

$$q(t) = \frac{\lambda}{L} \sum_{n=1}^{\infty} A_n n\pi \exp(-n^2\pi^2 t/\tau). \tag{7.33}$$

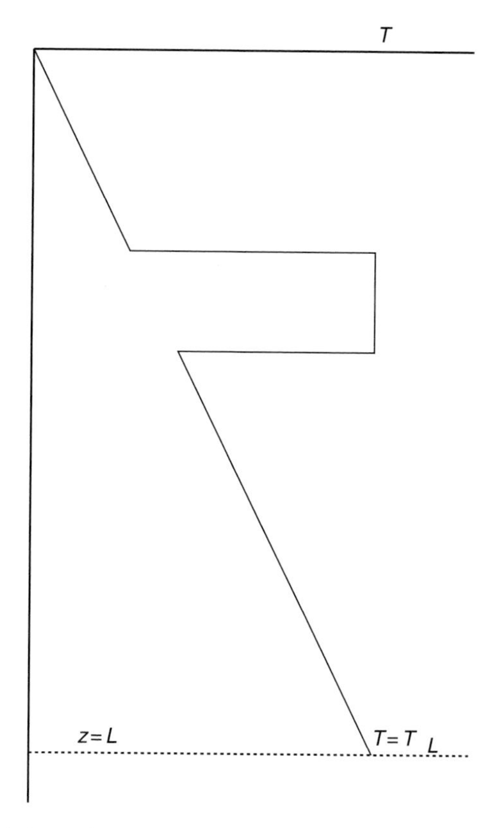

Figure 7.18. Initial condition for the lithospheric temperature following underplating at the base of the crust. A layer is emplaced at temperature T_L between depths $z = a$ and $z = b$.

The transient surface heat flux reaches its peak a short time after underplating occurs ($\approx$25 My, depending on the intrusion depth) and its amplitude can be significant, i.e. comparable to the basal heat flux (Figures 7.19 and 7.20).

The thermal uplift $h(t)$ can also be calculated from the average density variation over a column of constant thickness L;

$$h(t) = \alpha \int_0^L T(z,t)dz = 2\alpha L \sum_{n=1}^{\infty} \frac{A_{2n-1}}{2n-1} \exp(-(2n-1)^2\pi^2 t/\tau). \qquad (7.34)$$

The amplitude of uplift following underplating depends on the thickness of the layer but it will be modest. For a standard value of the thermal expansion coefficient $\alpha = 3 \times 10^{-5}$ K^{-1} and an excess temperature of 600 K, the surface uplift will 18×10^{-3} times the layer thickness. For example, the underplated layer must be at least 60 km thick to cause an uplift of the order of 1000 m.

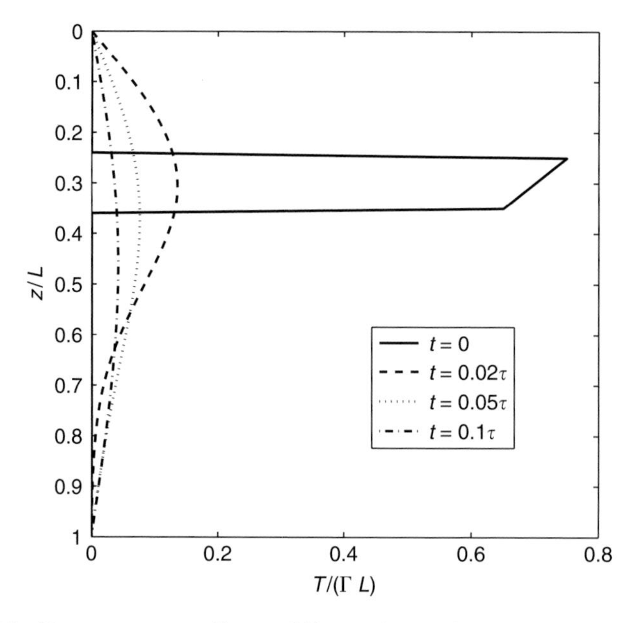

Figure 7.19. Temperature profiles at different times after underplating at the base of the crust. Temperature is normalized to that at the base of the lithosphere, depth to lithospheric thickness and time to $\tau = L^2/\kappa$, i.e. 500 My for $L = 150$ km.

7.5.3 Mantle delamination

Bird (1979) proposed that delamination of the mantle lithosphere could cause rapid plateau uplift. After delamination, i.e. the rapid peeling off of the mantle lithosphere and its replacement by asthenospheric material, the thermal perturbation in the mantle lithosphere is,

$$T(z,t=0) = (T_L - T_M)\frac{L-z}{L-z_M}, \; z_M < z < L, \tag{7.35}$$

where T_M is Moho temperature and z_M is Moho depth. Using a Fourier series expansion, we obtain for the transient temperature perturbation:

$$T(z,t) = (T_M - T_L)\sum_{n=1}^{\infty}\left(\left(\frac{z_M}{L}-1\right)\cos\frac{n\pi z_M}{L} - \frac{1}{(n\pi)}\sin\frac{n\pi z_M}{L}\right) \times \frac{1}{n\pi}\sin\frac{n\pi z}{L}\exp(-n^2\pi^2\kappa t/L^2), \tag{7.36}$$

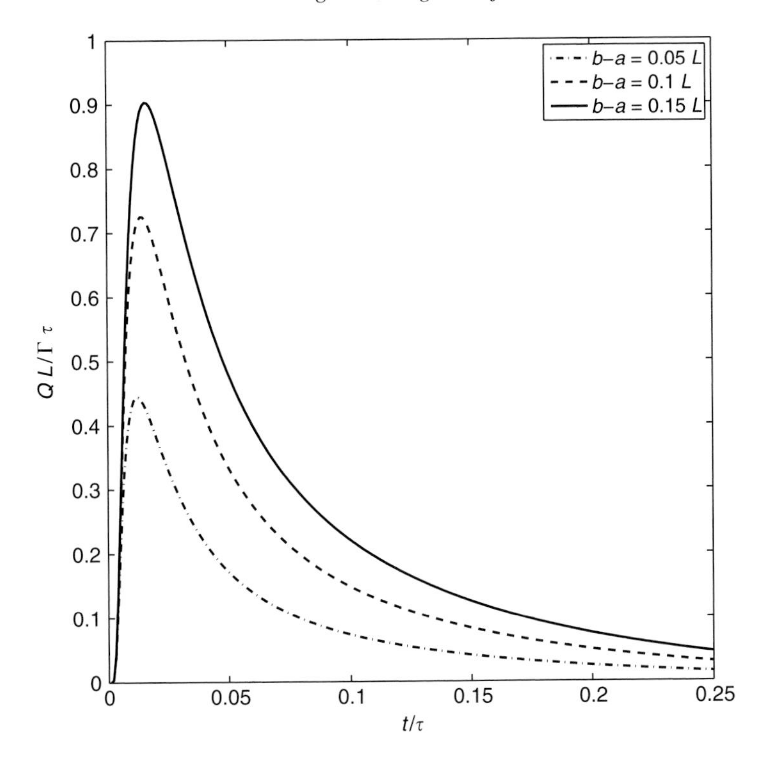

Figure 7.20. Surface heat flux variation after underplating at the base of the crust for different thicknesses of the underplating layer. Flux is relative to $\lambda\Gamma/L$, i.e. the steady-state heat flux across the lithosphere; time is normalized to $\tau = L^2/\kappa$, i.e. 500 My for $L = 150$ km.

and the surface heat flux,

$$q(z,t) = \frac{\lambda(T_M - T_L)}{L} \sum_{n=1}^{\infty} \left(\left(\frac{z_M}{L} - 1 \right) \cos \frac{n\pi z_M}{L} - \frac{1}{n\pi} \sin \frac{n\pi z_M}{L} \right) \times \exp(-n^2\pi^2\kappa t/L^2). \tag{7.37}$$

The average change in density of a lithospheric column is,

$$\frac{< \Delta\rho >}{\rho} = \frac{\alpha}{2}(T_L - T_M)\frac{(L - z_M)}{L} \sum_{n=1}^{\infty} \frac{2}{(2n-1)^2\pi^2} \times \left(\left(\frac{z_M}{L} - 1 \right) \cos \frac{(2n-1)\pi z_M}{L} - \frac{1}{(2n-1)\pi} \sin \frac{(2n-1)\pi z_M}{L} \right) \times \exp(-(2n-1)^2\pi^2\kappa t/L^2). \tag{7.38}$$

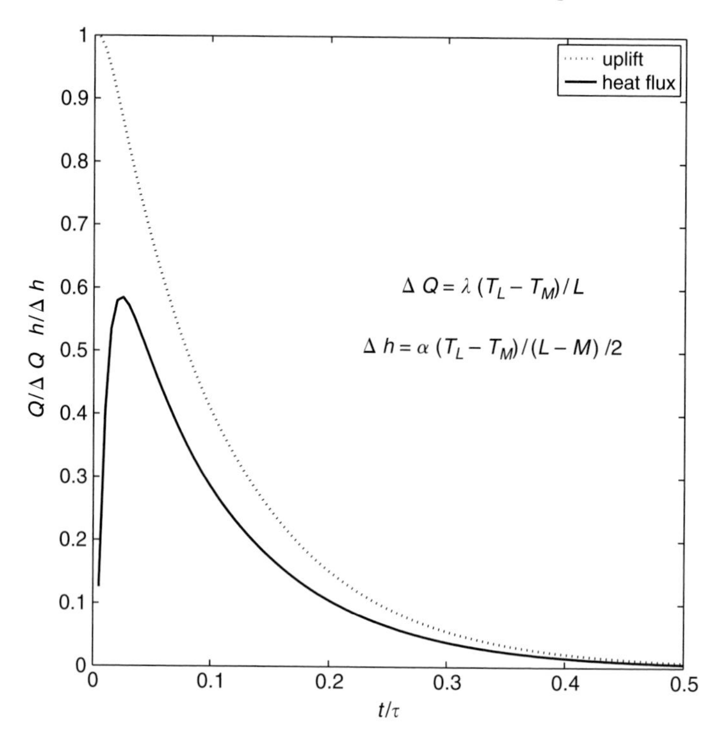

Figure 7.21. Surface heat flux variation and uplift after delamination of the lithospheric mantle. Crustal thickness is 1/4 lithospheric thickness. Heat flux amplitude $\Delta Q = \lambda(T_L - T_M)/L$ and uplift $\Delta h = \alpha(T_M - T_L)(L - z_M)/2$.

This gives an uplift $\Delta h = < \Delta\rho > L/\rho$ with amplitude $\alpha(T_L - T_M)(L - z_M)/2$. With low Moho temperature (600 °C) and a 150 km thick lithosphere, an uplift ≈ 1.1 km can be achieved almost instantly. Relaxation of the topography to half the initial level requires about 80 My (see Figure 7.21). The peak in surface heat flux perturbation lags 25 My behind the uplift and is $\approx 0.5 \times \lambda \times (T_M - T_L)/L$, i.e. 7.5 mW m^{-2}. This small value is consistent with the normal (60 mW m^{-2}) heat flux reported for the Colorado plateau. Similarly, the Tanzania craton between the two branches of the East African rift is characterized by elevations higher than 1000 m but low heat flux (40 mW m^{-2}) (Nyblade, 1997).

7.5.4 Delamination and extension

As discussed earlier, a simple model for the Basin and Range calls for crustal extension and delamination of the lithospheric mantle, i.e. its replacement by hot asthenospheric mantle. If we assume that instantaneous stretching immediately follows delamination, the initial thermal perturbation is,

$$\begin{aligned} T(z,t=0) &= \Gamma(\beta-1)z, \ 0<z<\zeta \\ &= \Gamma(L-z), \ \zeta<z<L, \end{aligned} \tag{7.39}$$

where β is the crustal stretching factor, ζ is the thickness of the extended crust, i.e. $\zeta \times \beta$ is the pre-extension crustal thickness. Assuming a fixed temperature at the base of the lithosphere (initial depth), the thermal perturbation can be developed in Fourier series as follows:

$$T(z,t) = \Gamma L \sum_{n=1}^{\infty} A_n \sin\left(\frac{n\pi z}{L}\right) \exp(-n^2\pi^2 t/\tau), \tag{7.40}$$

with the coefficients:

$$\begin{aligned} A_n = (\beta-1)&\left(\frac{L\sin(n\pi\zeta/L)}{n^2\pi^2} - \frac{\zeta\cos(n\pi\zeta/L)}{n\pi}\right) \\ &+ (1-\zeta/L)\left(\frac{\cos(n\pi\zeta/L)}{n\pi} + \frac{\sin(n\pi\zeta)/L}{n^2\pi^2(1-\zeta/L)}\right). \end{aligned} \tag{7.41}$$

The peak temperature decreases rapidly as the perturbation diffuses through the entire lithosphere (Figure 7.22).

The transient component of the surface heat flux is given by,

$$Q(t) = \lambda\Gamma \sum_{n=1}^{\infty} A_n(n\pi)\exp(-n^2\pi^2 t/\tau), \tag{7.42}$$

and the thermal component of the uplift is given by,

$$\delta h(t) = \alpha \int_0^L T(z,t)dz = \alpha\Gamma L^2 \sum_{n=1}^{\infty} \frac{2A_{2n-1}}{(2n-1)\pi} \exp(-(2n-1)^2\pi^2 t/\tau). \tag{7.43}$$

The peak value of the surface heat flux can be larger than the steady-state flux across the lithosphere (Figure 7.23). The amplitude of the thermal uplift depends both on crustal extension and the ratio of crustal to lithospheric thickness; it is proportional to $< \Delta T > \propto 0.5(\beta-1)\zeta^2 + 0.5(1-\zeta)^2$. In addition, we must account for crustal thinning which reduces the initial elevation by $(\beta-1)\zeta(\rho_m-\rho_c)/\rho_m$, where ρ_m and ρ_c are mantle and crustal densities respectively.

7.5.5 Extension and magma intrusions

Extension of the lithosphere can be accompanied by the ascent of low-density magmas in the mantle, simultaneously with crustal stretching. On the lithospheric

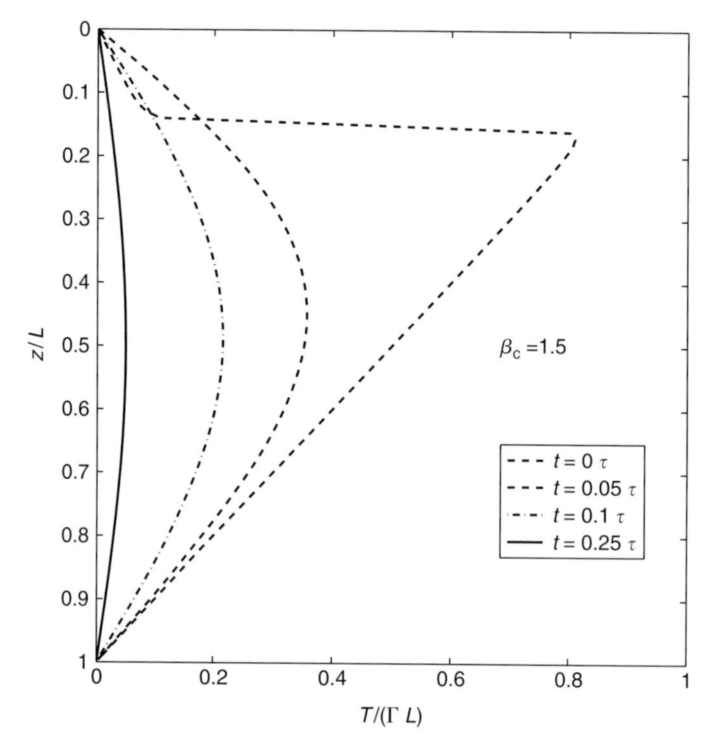

Figure 7.22. Transient temperature profiles following delamination of the lithospheric mantle and crustal extension. The temperature anomaly is relative to the temperature at the base of the lithosphere.

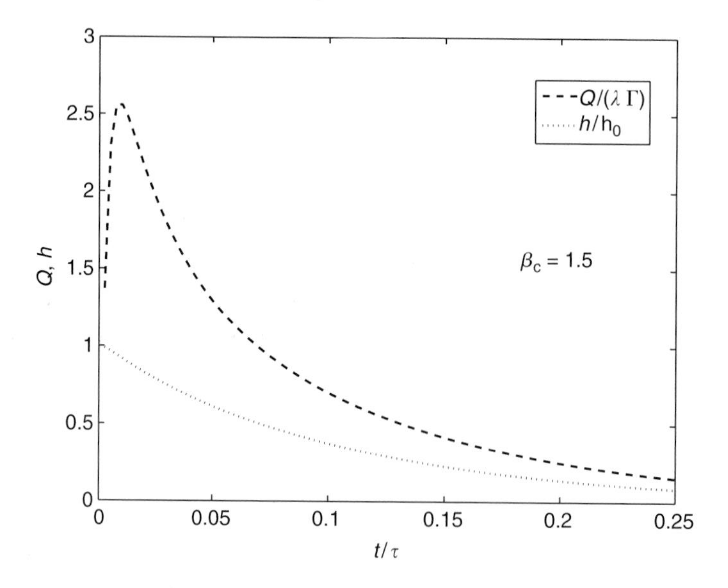

Figure 7.23. Excess surface heat flux and topography following delamination of the lithospheric mantle and crustal extension with the same parameters as in Figure 7.22.

scale, the thermal effect of these intrusions is equivalent to the addition of heat sources in the mantle. The amount of heat introduced in the lithospheric mantle is equal to the volume of intrusions times the specific and latent heat carried by the intrusions, i.e. $C_P \Delta T + L^*$. The equivalent heat production is obtained as,

$$A = \frac{\Delta \dot{V}}{V} \rho (C_P \Delta T + L^*) \tag{7.44}$$

where $\Delta \dot{V}/V$ is the intrusion rate. For $\Delta \dot{V}/V = 10^{-15}$ s^{-1}, $\Delta T = 300$ K, and $L^*/C_p = 300$ K, the amount of heat injected is equivalent to 1.8 μW m^{-3}. The temperature perturbation is then obtained by solving the heat equation with a source term in the lithospheric mantle, and initial condition $T = 0$:

$$\begin{aligned} \frac{\partial T}{\partial t} &= \kappa \frac{\partial^2 T}{\partial z^2}, \quad 0 < z < z_M \\ \frac{\partial T}{\partial t} &= \kappa \frac{\partial^2 T}{\partial z^2} + \frac{A}{\rho C_p}, \quad z_M < z < L. \end{aligned} \tag{7.45}$$

The temperature, surface heat flux and uplift due to the magma intrusions can be calculated with standard methods. For a constant flux at the base of the lithosphere, the excess surface heat flux due to a constant rate of magmatic intrusion starting at $t = 0$ is given by,

$$\begin{aligned} q(t) = A(L - z_M) - \frac{8AL}{\pi^2} \sum_{n=1}^{\infty} \frac{(-)^n}{(2n-1)^2} \sin\left((2n-1)\frac{\pi(L - z_M)}{2L}\right) \\ \times \exp\left(\frac{-(2n-1)^2 \pi^2 \kappa t}{4L^2}\right), \end{aligned} \tag{7.46}$$

where L and z_M are lithospheric and crustal thicknesses respectively. The surface uplift h(t) is:

$$\begin{aligned} h(t) = \frac{\alpha A L^3}{6\lambda} \Bigg(\left(1 - \frac{z_M}{L}\right)\left(2 + 2\frac{z_M}{L} - \frac{z_M^2}{L^2}\right) \\ - \frac{192}{\pi^4} \sum_{n=1}^{\infty} \frac{(-)^n}{(2n-1)^4} \sin\left((2n-1)\frac{\pi(L - z_M)}{2L}\right) \exp\left(\frac{-(2n-1)^2 \pi^2 \kappa t}{4L^2}\right) \Bigg). \end{aligned} \tag{7.47}$$

The input of heat can be substantial if intrusions penetrate throughout the lithospheric mantle: with the same parameters as above, intrusions penetrating a 50 km thick layer carry an additional 90 mW m^{-2} in the lithosphere. But the surface heat flux lags behind and increases by only 20% of that amount after a time $0.2\ L^2/\kappa$, ≈ 50 My.

7.5.6 Lithospheric extension. Discussion

Zones of extension are in a transient thermal regime. As extension proceeds, the geothermal gradient steepens and the surface heat flux increases. Topography is supported by a low density, hot mantle lithosphere. Complete replacement of the mantle lithosphere as a result of diapiric uprise of the asthenosphere or delamination leads to rapid uplift, but is followed by slow subsidence as the lithosphere cools off. Intrusion of magmas in the mantle lithosphere during extension brings sufficient heat to maintain the elevated heat flux.

7.6 Passive continental margins. Sedimentary basins

The subsidence of sedimentary basins and passive continental margins provides a good record of the relaxation of thermal perturbations in the lithosphere and is sensitive to lithosphere thickness. Such transients have been recorded in intracratonic basins located away from active plate boundaries and have generated a lot of interest (Haxby *et al.*, 1976; Nunn and Sleep, 1984; Ahern and Mrkvicka, 1984). Subsidence is also affected by tectonic, metamorphic and eustatic effects. In order to identify these effects, some authors have assumed that the continental lithosphere has a well-defined characteristic cooling time of 60 My (Bond and Kominz, 1991) and that subsidence phases that are significantly longer than this require other than thermal effects, such as renewed extension for example. These assumptions are not justified for a thick continental lithosphere with long thermal relaxation time.

Several authors have suggested that intracratonic basin subsidence can be explained by cooling of an initially hot lithosphere, similar to the cooling and subsidence of the oceanic lithosphere (Sleep, 1971; Sleep and Snell, 1976; Turcotte and Ahern, 1977; McKenzie, 1978). Different models proposed assume that: (1) the lithosphere is initially hot and: (2) cooling, thermal contraction and subsidence coincide with the return of the lithosphere to thermal equilibrium. These models differ by their initial conditions and by their basal boundary conditions (McKenzie, 1978; Hamdani *et al.*, 1991, 1994). The latter is an important factor determining the duration of thermal subsidence. The duration of the subsidence episode varies by a factor of three between various intracratonic basins of North America. For a fixed temperature at the base of the lithosphere, theory would imply that the continental lithosphere thickness is about 115 km and 270 km beneath the Michigan and Williston basins respectively (Haxby *et al.*, 1976; Ahern and Mrkvicka, 1984). For two basins of similar age located on the Precambrian basement of the same continent, such a large difference is surprising. The rate of subsidence depends on the boundary condition at the base of the lithosphere and is slower for a fixed flux than for a fixed temperature. Differences in subsidence rates may thus be related to the thermal processes at the base of the lithosphere (Hamdani *et al.*, 1994). These arguments rely on 1D thermal models which have recently been questioned (Kaminski and Jaupart, 2000). According to Haxby *et al.* (1976), for example, the initial perturbation beneath the Michigan basin has a radius of about 120 km, which is less than the thickness of the North American lithosphere. In this case, the assumption of purely vertical heat transfer is not tenable.

7.6.1 Cooling of the lithosphere after a heating event

The model of Sleep and Snell (1976) is similar to the cooling plate model with fixed temperature at the base for the oceanic lithosphere (see Section 6.2.4). More generally, we can consider an initial condition with a uniform temperature perturbation ΔT between depths a and b in the lithosphere.

With the initial condition, $T(z,t=0) = \Delta T$ for $a < z < b$, and $\Delta T = 0$ for $z < a$ and $z > b$, and the boundary conditions $T(z=0,t) = T(z=L,t) = 0$, the temperature perturbation $T(z,t)$ in the lithosphere $0 < z < L$ is given by the following Fourier series:

$$T(z,t) = \frac{2\Delta T}{\pi} \sum_{n=1}^{\infty} \frac{1}{n} \left(\cos\left(\frac{n\pi a}{L}\right) - \cos\left(\frac{n\pi b}{L}\right) \right) \times \sin\left(\frac{n\pi z}{L}\right) \exp(-n^2\pi^2\kappa t/L^2). \tag{7.48}$$

Temperature profiles are shown on Figure 7.24. The surface heat flux is:

$$q(t) = \frac{2\lambda\Delta T}{L} \sum_{n=1}^{\infty} \left(\cos\left(\frac{n\pi a}{L}\right) - \cos\left(\frac{n\pi b}{L}\right) \right) \exp(-n^2\pi^2\kappa t/L^2), \tag{7.49}$$

and the thermal subsidence

$$h(t) = \alpha\Delta T(a-b) - \frac{4\alpha\Delta T L}{\pi^2} \sum_{n=1}^{\infty} \frac{1}{(2n-1)^2} \times \left(\cos\left(\frac{(2n-1)\pi a}{L}\right) - \cos\left(\frac{(2n-1)\pi b}{L}\right) \right) \exp(-(2n-1)^2\pi^2\kappa t/L^2). \tag{7.50}$$

For $a = 0$ and $b = L$, the model is identical to the cooling plate model used for the oceanic lithosphere. The subsidence is accompanied by increased surface heat flux (Figure 7.25), but for the constant temperature boundary condition at the base, the subsidence rate is not proportional to surface heat flux.

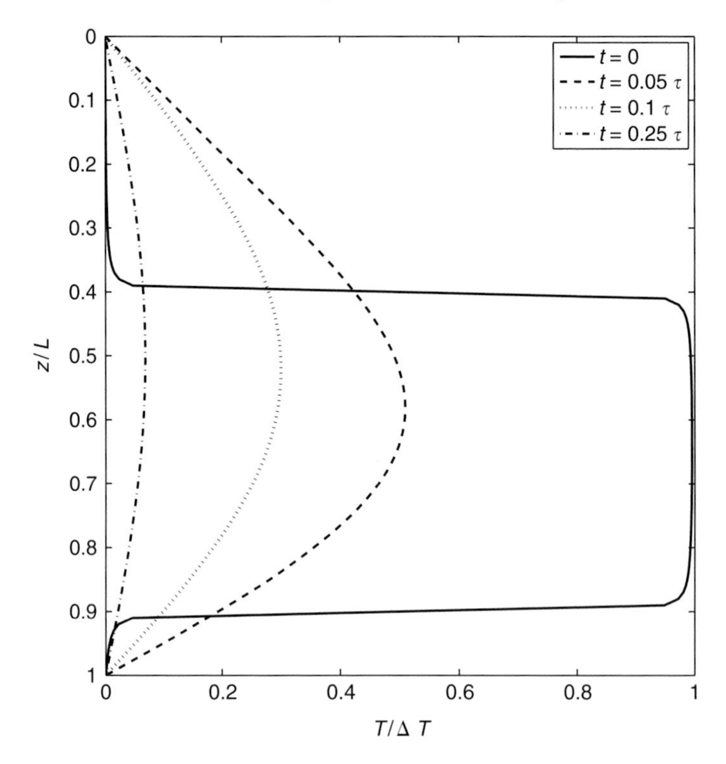

Figure 7.24. Temperature profiles across the lithosphere after the emplacement of a wide intrusion in its lowermost part ($0.4 < z/L < 0.9$). Cooling is almost completed after $t = 0.25L^2/\kappa$, i.e. $>$ 150 My for 150 km thick lithosphere.

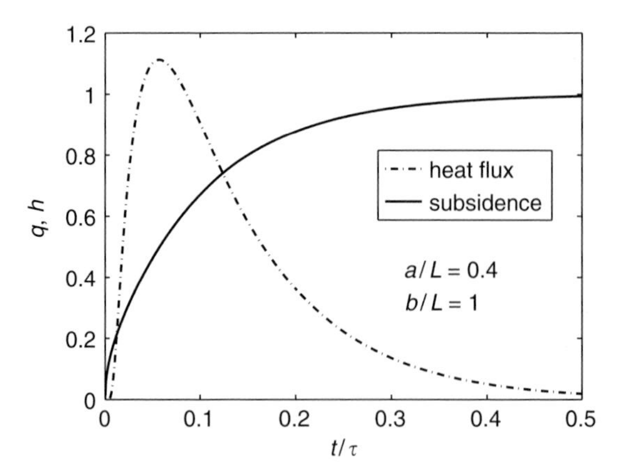

Figure 7.25. Subsidence and heat flux for the cooling intrusion model. The initial condition is a temperature perturbation ΔT between depths a and b. Temperatures are fixed at the surface and base of the plate. Total subsidence is $\alpha \Delta T(b-a)$. Heat flux is divided by $\lambda \Delta T/L$.

This represents the thermal subsidence and Isostatic adjustments must be added to have total subsidence. In the case of a basin, the isostatic amplification is due to the additional weight of the sediment added on top.

7.6.2 The lithospheric stretching models

McKenzie (1978) proposed that basin subsidence follows the stretching of the lithosphere. Stretching results in a thinner and hotter lithosphere. Its cooling and return to equilibrium causes the subsidence. Mass conservation requires lithospheric thinning to compensate for the extension. The stretching factor β is defined as the ratio of initial to final lithospheric thickness. The initial temperature perturbation $T(z,t)$, schematized in Figure 7.26, is given by,

$$\begin{aligned} T(z,t=0) &= T_L \frac{(\beta-1)z}{L}, \ z < L/\beta \\ T(z,t=0) &= T_L\left(1-\frac{z}{L}\right), \ \frac{L}{\beta} < z < L. \end{aligned} \tag{7.51}$$

The temperature profile is obtained by decomposing the initial condition in Fourier components:

$$\begin{aligned} T(z,t) &= \frac{2\beta T_L}{\pi^2} \sum_{n=1}^{\infty} \frac{1}{n^2} \sin\left(\frac{n\pi}{\beta}\right) \sin\left(\frac{n\pi z}{L}\right) \exp\left\{\frac{-n^2\pi^2\kappa t}{L^2}\right\} \\ q(t) &= \frac{2k\beta T_L}{\pi L} \sum_{n=1}^{\infty} \frac{1}{n} \sin\left(\frac{n\pi}{\beta}\right) \exp\left\{\frac{-n^2\pi^2\kappa t}{L^2}\right\}. \end{aligned} \tag{7.52}$$

Using the relationship,

$$\sum_{n=1}^{\infty} \frac{1}{(2n-1)^3} \sin\{(2n-1)z\} = \frac{\pi^2 z - \pi z^2}{8}, \tag{7.53}$$

the total subsidence is

$$\begin{aligned} h(t) = &\frac{\alpha(\beta-1)T_L L}{2\beta} - \frac{4\alpha T_L \beta L}{\pi^3} \\ &\times \sum_{n=1}^{\infty} \frac{1}{(2n-1)^3} \sin\left\{(2n-1)\frac{\pi}{\beta}\right\} \exp\left\{\frac{-(2n-1)^2\pi^2\kappa t}{L^2}\right\}. \end{aligned} \tag{7.54}$$

The amplitude of the thermal subsidence depends on the stretching factor but the shape of the subsidence curves does not vary much with it (Figure 7.27).

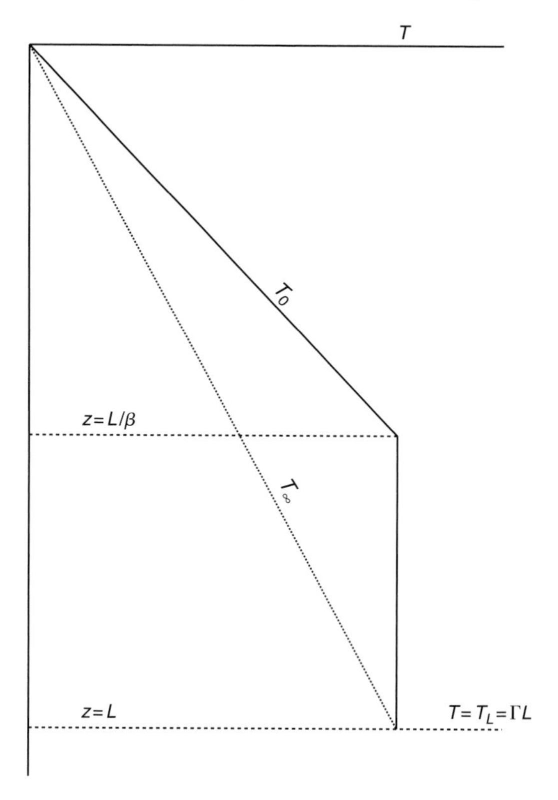

Figure 7.26. Temperature profile following uniform stretching of the lithosphere. After stretching by β, the lithospheric thickness is reduced to L/β, and the temperature gradient is increased by a factor β.

This model has been widely used because it provides a reasonable geological mechanism to explain the initial temperature condition preceding subsidence. It has been applied to several basins such as the North Sea or the Pannonian basin (Sclater and Christie, 1980; Sclater *et al.*, 1980b) and to stretched continental margins, such as those of eastern Canada or western Europe (Royden and Keen, 1980; Le Pichon and Sibuet, 1981). It is less clear that it is applicable to cratonic basins, such as the Michigan or Williston basins, where there is no evidence of stretching (such as crustal thinning) and where the almost circular shape is not consistent with the stretching model.

7.6.3 Horizontal transport of heat

Sedimentary basins have a radius that is comparable to lithospheric thickness. Consequently, horizontal transport of heat cannot be neglected in thermal models of basin subsidence. Also, isostatic compensation is not achieved locally and must

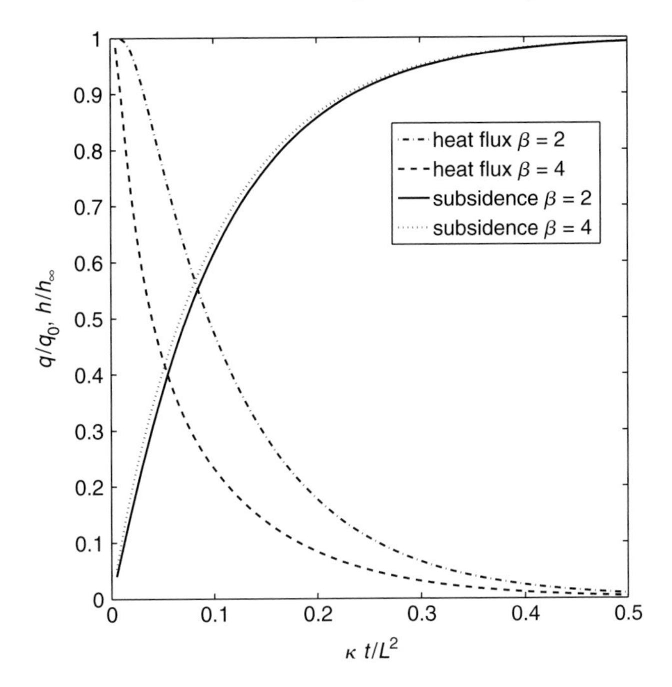

Figure 7.27. Transient heat flux and thermal subsidence for the stretching model. Initially, the transient heat flux is $q_0 = \lambda(\beta - 1)T_L/L$ and the total subsidence $h_\infty = \alpha(\beta - 1)T_L L/(2\beta)$. Both heat flux and subsidence increase with β.

be calculated from a flexural model. The effective elastic thickness of the lithosphere is controlled by temperature and increases with time.

Calculations are usually done using Fourier or Hankel transforms (see Appendix A). For example, the temperature in the lithosphere after a heating event with an initial temperature perturbation $T(x, y, z, t = 0) = \Delta T(x, y)$ for $a < z < b$, and 0 for $z < a$ and $z > b$:

$$
\begin{aligned}
T(k_x, k_y, z, t) &= 2\Delta T(k_x, k_y) \exp(-\kappa(k_x^2 + k_y^2)t) \\
&\quad \times \frac{1}{\pi} \sum_{n=1}^{\infty} \frac{1}{n} \left(\cos\left(\frac{n\pi a}{L}\right) - \cos\left(\frac{n\pi b}{L}\right) \right) \\
&\quad \times \sin\left(\frac{n\pi z}{L}\right) \exp(-n^2\pi^2\kappa t/L^2),
\end{aligned}
\tag{7.55}
$$

where $\Delta T(k_x, k_y)$ is the Fourier transform of the initial condition. One can immediately see that components with wave-vectors $k_x^2 + k_y^2 > \pi^2/L^2$ will be attenuated much more rapidly because of horizontal transport of heat. Duration of subsidence depends on the width of the thermal anomaly as well as on lithospheric thickness.

Accounting for horizontal heat transfer, a relationship can be found between the width of the thermal anomaly and the lithospheric thickness. For all the intracratonic basins in North America, the models can not fit the subsidence history for lithosphere thicknesses less than 170 km. Observations are best fitted for a model with fixed heat flux boundary condition at the base of the lithosphere (Kaminski and Jaupart, 2000).

7.7 Geophysical constraints on thermal structure

Most physical properties depend on temperature (among many other factors) and different geophysical data sets are available to test thermal models and/or complement the information from surface heat flux.

7.7.1 Constraints from seismology

The 3D seismic velocity structure of the upper mantle determined by seismic tomography shows a strong correlation with the geology and heat flux. In particular the presence of lithospheric roots beneath cratons is associated with higher seismic velocity and lower temperature than outside (van der Lee and Nolet, 1997; Poupinet *et al.*, 2003, for example). Horizontal differences in seismic velocities can be interpreted in terms of compositional and thermal differences. A vertically integrated picture of the variations in seismic velocities can be obtained by calculating the differences in shear wave travel time across a region of the mantle. In North America, travel time delays calculated between 60 and 300 km depth show large-scale features similar to those seen in the heat flow map (Figures 7.28 and 7.1). There are also differences within the stable continent that might be related to small differences in mantle heat flux and lithospheric thickness.

Direct calculation of the velocity profile

For a given composition of the lithospheric mantle, it is possible to determine the variation of seismic velocity with pressure and temperature. Laboratory measurements of the elastic moduli and density have been made for various mantle minerals and can be extrapolated to mantle temperatures and pressures by using the infinitesimal strain approximation (Goes *et al.*, 2000). The velocities for different mineral assemblages are calculated by using the Reuss–Voigt–Hill averaging method (see Appendix D). The variation of elastic parameters of minerals with temperature include an "an-harmonic", i.e. frequency independent, and an an-elastic components. The latter is due to attenuation of the seismic waves at high temperature. The an-elastic effects become important at high temperature but are not as well constrained experimentally as the an-harmonic component.

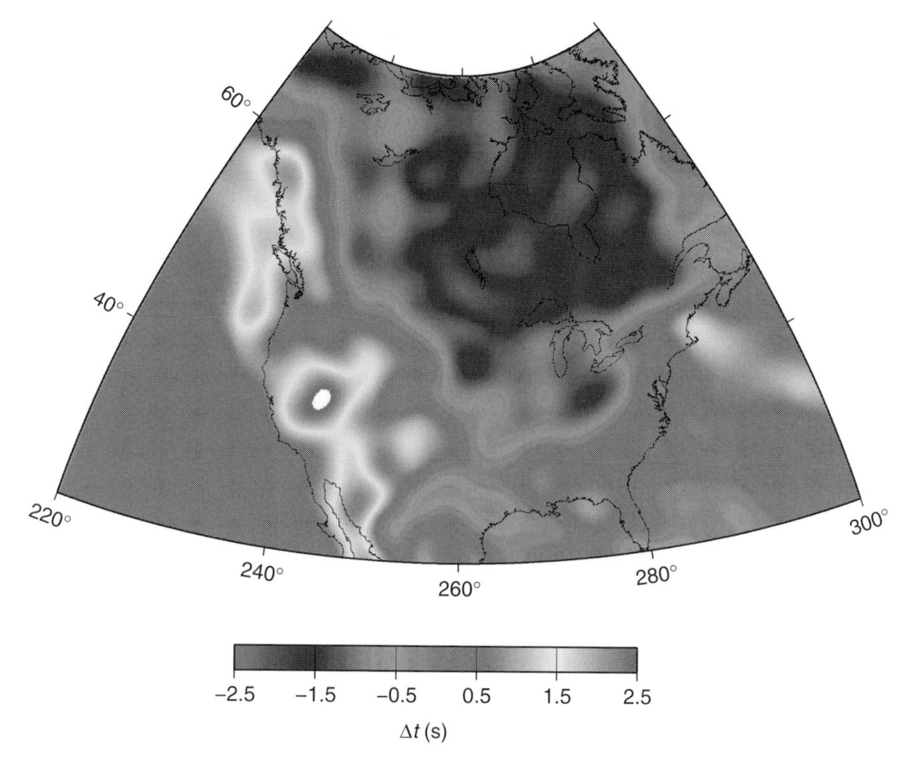

Figure 7.28. Shear wave travel time delays for North America. The delays in the arrival time of shear waves are calculated between 60 and 300 km from the surface wave tomography model of van der Lee and Frederiksen (2005). See plate section for color version.

For a given composition of the mantle, it is thus possible to calculate how seismic velocities vary with temperature and pressure. The effect of temperature on shear wave velocity is more important than that of composition, if kept within plausible range. For cratons with low surface heat flux, the shear wave velocity profiles depend very strongly on the Moho heat flux (Figure 7.29). Typical profiles show a decrease in velocity throughout the lithospheric mantle where the temperature gradient is conductive; velocities increase in the asthenosphere where the temperature gradient becomes isentropic. The depth of the velocity minimum depends strongly on the Moho heat flux. It varies between 200 and 300 km for Moho heat flux varying from 18 to 12 mW m^{-2}. The corresponding difference in shear wave travel time delays is ≈ 1.1 s.

Inverse calculations

Heat flux data and thermodynamic constraints can be used to narrow down the range of mantle temperatures consistent with seismic tomography models. Shapiro

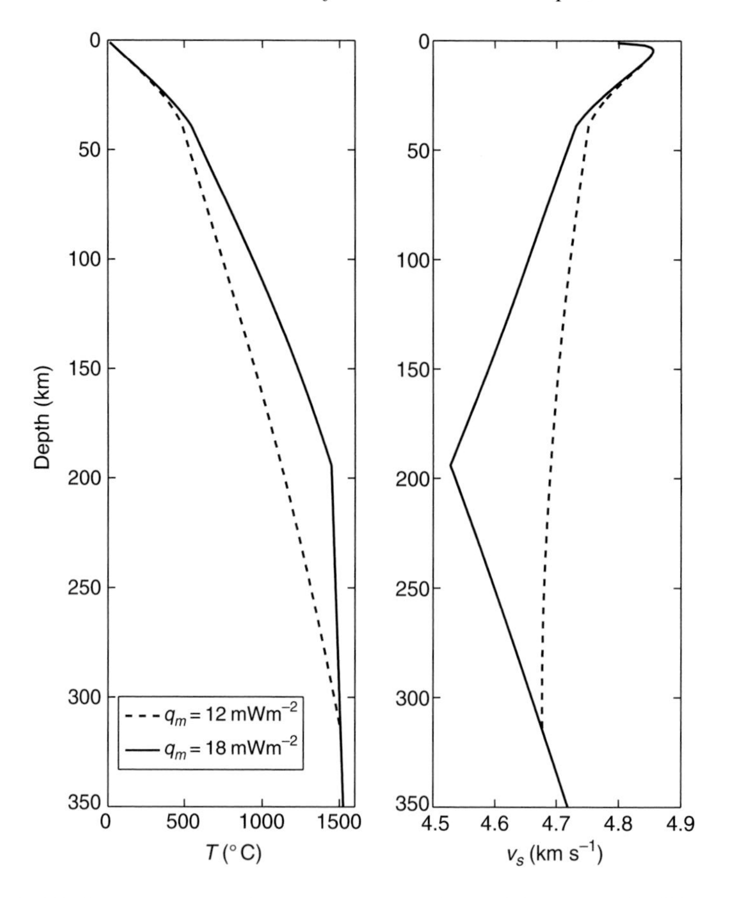

Figure 7.29. Temperature and shear wave velocity profiles across the lithosphere. The profiles are calculated for the same surface heat flux 45 mW m^{-2} and mantle heat flux values of 12 and 18 mW m^{-2}. Seismic velocities are calculated for the same cratonic mantle composition as that proposed by McDonough and Rudnick (1998) with a value of 0.086 for the Fe number.

and Ritzwoller (2004) inverted surface wave data to obtain vertical profiles of S-wave velocity through both continents and oceans. For a given compositional model, these data can be converted to temperature. In a given area, the solution domain allows for non-monotonous variations of temperature with depth, i.e. with zones where temperature decreases with depth, which are not physically realistic. Applying the constraint that temperature must always increase with depth leads to a narrower solution domain. The range can be further narrowed down by eliminating solutions outside the range of Moho temperatures allowed by surface heat flux data. One striking result is that this procedure gets rid of non-physical solutions with negative vertical temperature gradients (Figure 7.30).

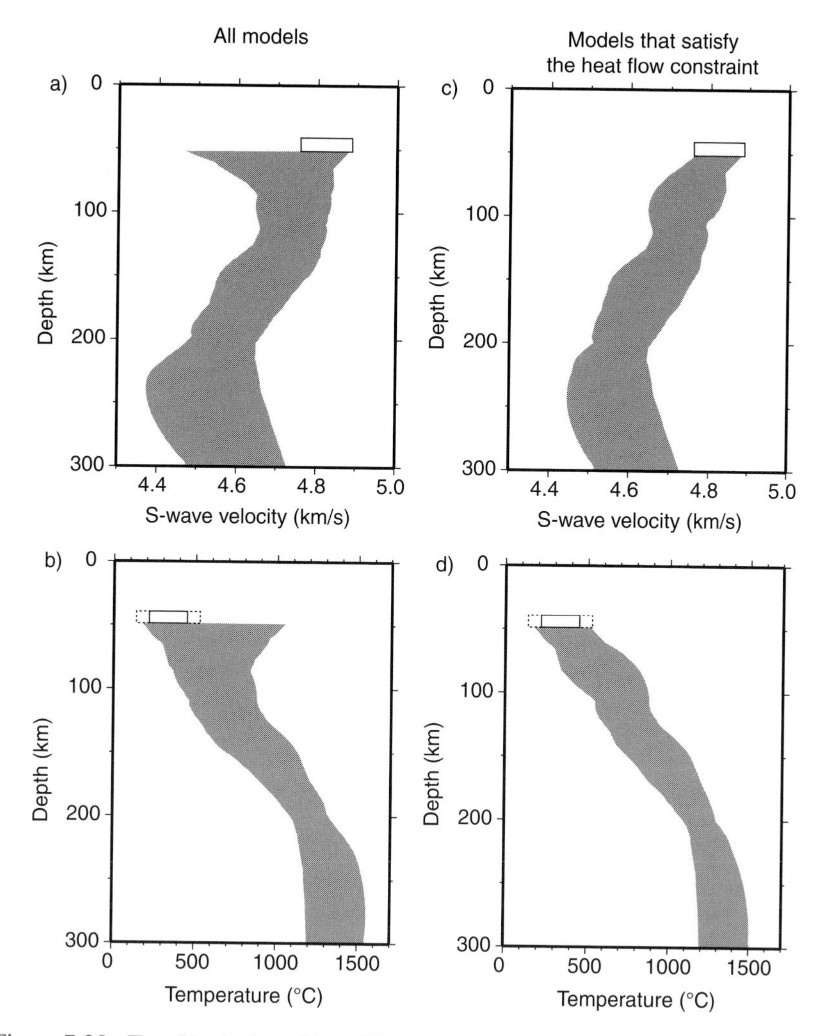

Figure 7.30. Top: Vertical profiles of S-wave velocity through the Canadian Shield obtained by diffraction tomography. Bottom: Vertical temperature profiles deduced from the velocity models. The left panel shows the whole solution domain, which includes non-physical temperature profiles such that temperature decreases with depth at shallow levels. The right panel shows the solutions that are consistent with the bounds of the Moho temperature deduced from heat flow studies. All the non-realistic temperature profiles have been eliminated.

7.7.2 Other geophysical constraints

Strength, seismicity, elastic thickness and thermal regime of the lithosphere

In the oceans, both the effective elastic thickness (T_e) and the maximum depth of earthquakes increase with the age of the oceanic lithosphere (see Section 6.4.4). The effective elastic thickness of the lithosphere is related to the yield strength envelope which is useful to understand how the temperature profile affects the strength of the lithosphere. The depth where the strength begins to decrease corresponds to the transition from brittle to ductile. In the oceans, this depth is strongly controlled by temperature i.e. by the age of the plate (McKenzie *et al.*, 2005).

In the continents, the strength profile is complicated by the rheological stratification in the lithosphere. A relationship between age, thermal regime and strength of the continental lithosphere was suggested by Karner *et al.* (1983). There are large differences in effective elastic thickness between active regions and cratons, for example between the Basin and Range and stable North America (Lowry and Smith, 1995), or between the cratons and the East African rift (Pérez-Gussinyé *et al.*, 2009). The effective elastic thickness is large over most cratons, such as the Canadian Shield (Audet and Mareschal, 2007) or the West African craton (Pérez-Gussinyé *et al.*, 2009). Within the Shields, there appear to be large variations in elastic thickness (from 40 to >100 km) that, so far, have been difficult to explain in terms of thermal structure (Audet and Mareschal, 2007). The seismogenic zone is usually shallow (<30 km) beneath the continents and most earthquakes are confined to the crust. The lack of earthquakes in the continental mantle may be attributed to combined effects of temperature and water on the mantle rheology (Maggi *et al.*, 2000).

Depth to the Curie isotherms

The main sources of lithospheric magnetic fields (magnetic anomalies) are in the crust and not in the mantle. The Curie isotherm for magnetite, $\approx 580\,^\circ$C, will normally be located in the upper mantle beneath continents and oceans. However, high surface heat flux and elevated temperatures in the lower crust will cause a shallow Curie isotherm with thinning of the magnetic crust and a local source of magnetic anomaly. Because of the dipolar character of the magnetic field, aeromagnetic anomalies maps are dominated by shallow sources, but satellite magnetic data are useful in estimating the total crustal magnetization and the depth to the Curie isotherm (Hamoudi *et al.*, 1998). Satellite magnetic data have been used to confirm elevated lower crustal temperatures beneath the Basin and Range (Mayhew, 1982) or to delineate the edge of the North American craton (Purucker *et al.*, 2002). Although recent satellite missions have acquired high quality magnetic data with sufficient resolution for very large-scale lithospheric studies (Maus *et al.*, 2006), they do not allow resolution of the detailed thermal structure.

Thermal isostasy

In the oceans, long-wavelength bathymetric variations are caused by density variations in the lithosphere. The depth of sea floor below sea level is directly related to the average lithospheric density and temperature. Note that the oceanic geotherm is not in steady state. Crough and Thompson (1976) have applied similar concepts to the continental lithosphere.

The uplift/subsidence Δh due to changes in lithospheric density is calculated from an isostatic balance condition:

$$\Delta h = \frac{< \delta\rho >}{\rho} L = \alpha < \Delta T > L, \tag{7.56}$$

where brackets indicate averaging. In stable regions, small regional variations in mantle heat flux (± 3 mW m^{-2}) lead to maximum differences in mantle temperature ± 150 K and elevation differences that could reach 450 m. The effect of thermal density variations is thus important but it is less than that of variations in crustal thickness and composition.

On a wider scale, low mantle temperatures beneath the cratons should increase the density of the mantle and keep the elevation much lower than the observed mean elevation. This observation led Jordan (1981) to propose that the cratonic mantle is made up of refractory residual mantle with lower density than the off-cratonic mantle. This compositional effect balances the thermal effect to give to the cratons their present elevation.

In continents, the component of the topography due to thermal isostasy is usually small, except in regions of extension. In the Basin and Range, where the typical crustal thickness is 30 km, the average elevation of 1,750 m requires the upper mantle density to be extremely low. Differences in temperature can account for only part of the elevation and low-density magma intrusions are thought to also contribute to the buoyancy of the mantle (Lachenbruch and Morgan, 1990). The high heat flux in the Canadian Cordillera suggests that thermal isostasy contributes to part of the elevation (Lewis *et al.*, 2003). The buoyancy of the mantle beneath the Colorado plateau is likely to be in part thermal, although the heat flux is not high in the Plateau, possibly because not enough time has elapsed to allow the effect of higher mantle temperature to be conducted to the surface (Bodell and Chapman, 1982; Hasterok and Chapman, 2007a,b). Active regions are in transient thermal regime and simple steady-state models relating elevation and average mantle temperature to surface heat flux remain approximative.

Electrical conductivity

Among the many factors that affect the electrical conductivity of mantle rocks, temperature is the most important. Others include oxygen fugacity, iron content, and the presence of hydrous minerals. The electrical conductivity of assemblages of different minerals can be calculated from different mixing laws, or bounds can be established. These bounds have been used to infer mantle temperature from electrical conductivity. Comparison between different cratons based on both electrical conductivity and seismic velocity confirms the temperature differences between the Kaapvaal and Slave craton for instance (Jones *et al.*, 2009).

Concluding remark

Many physical parameters depend on temperature: they provide constraints on the thermal structure of the continental lithosphere, but they should not be considered as proxies for heat flux which, combined with crustal heat production, remains the principal constraint on the thermal structure of the lithosphere.

Exercises

7.1 Calculate the temperature profile in a half-space when the surface heat flux is q_0, and the heat-producing elements are uniformly distributed in a layer of thickness D, $q_0 > A_0 \times D$. Keeping $A_0 \times D$ constant, determine how temperature varies with D at a fixed depth $z > D$.

7.2 Calculate the temperature profile in a half-space when the surface heat flux is q_0, and the heat-producing elements decrease exponentially with depth $A(z) = A_0 \exp(-z/D)$. Keeping $A_0 \times D$ constant, determine how temperature varies with D at a depth $z \gg D$.

7.3 Find a vertical distribution of HPEs, $A(z)$, such that the heat flux at any level $q(z)$ is proportional to the heat production at the same level: $q(z) = D \times A(z)$.

7.4 Consider a crust made up of two layers of thickness h_1 and h_2 with heat production A_1 and A_2. Show that the differentiation index $D_I = A_1(h_1+h_2)/(A_1h_1+A_2h_2)$. What is the maximum possible value of D_I? For given $h_1/(h_1+h_2)$, surface and Moho heat fluxes and thermal conductivity, find an analytical expression for Moho temperature in function of D_I.

7.5 The Basin and Range covers 800,000 km^2 and average heat flux is $\approx$ 40 mW m^{-2} above the continental average. Estimate the extension rate needed to sustain increased heat flux for the different mechanisms described in this chapter.

7.6 Deccan traps erupted $> 1.5 \times 10^6$ km^3 of basalt. Assuming that the specific heat of basalt is 1000 J kg^{-1} K^{-1}, that they were at their melting temperature, $\approx$1300 °C, that the latent heat is $\approx$ 300 kJ kg^{-1}, estimate the total energy brought to the surface. How does that compare to present heat loss of earth, if the traps were emplaced in 10^6 or 3×10^4 y?

7.7 To determine the effect of a volcanic plateau, determine the solution of the heat equation for a half-space when the initial temperature is $T(z, t=0) = T_m$ for $0 < z < a$ and $T(z, t=0) = \Gamma(z-a)$ for $z > a$ and the surface temperature $T = 0$ for $t > 0$.

7.8 Determine the temperature perturbation caused by the injection of magmas in the lithospheric mantle (see Section 7.5.5) and derive the surface heat flux and the uplift. This requires solving the heat equation with sources between $M < z < L$, initial condition $T = 0$ and boundary conditions $T = 0$ at $z = 0$, $\partial T/\partial z = 0$ at $z = L$. Why is was a flux condition preferred to a temperature condition at the lower boundary?

7.9 Consider the following initial temperature distribution: $T(z, t = 0) = 0$ for $0 < z < a$ and $T(z, t = 0) = -\Delta T$ for $z > a$. Determine how this temperature perturbation will change with time. Also determine how the heat flux at the surface $z = 0$ will change with time. This temperature perturbation represents the transient following the thrusting of an infinite layer of thickness a over a half-space when both the layer and half-space have the same temperature gradient. Compare with the solution for two layers (equation 7.22).

7.10 Determine the cooling of an intrusion in the lower lithosphere when the initial temperature perturbation is the same as given in Section 7.5.5, but for a boundary condition of constant flux at the base of the lithosphere, i.e. $\partial T/\partial z = 0$ for $z = L$.

8

Global energy budget. Crust, mantle and core

Objectives of this chapter

We establish the thermodynamic framework for studying how the Earth cools down and contracts. The energetics of the Earth cooling involve effects of both slow contraction and fast convection. We review all terms in the energy budget and elucidate internal energy transfers that do not contribute to the bulk energy loss and to secular cooling. We examine how the energy loss is balanced by internal heat generation and cooling of the Earth and derive the equation for the average temperature of the mantle. The heat flux integrated over the Earth's surface represents the energy loss of the Earth. Heat flux measurements are distributed very unevenly and may not record all the heat that is lost through some areas. We assess different procedures that may be used to integrate these data and evaluate the uncertainty of the final result. We review current knowledge on the Earth's composition and calculate the total amount of heat produced by radioactive decay in the planet.

8.1 Thermodynamics of the whole Earth

8.1.1 The global energy budget

We assume that the Earth's surface is stress-free, such that there is no work due to external shear stresses, and set up the global energy balance for the Earth as follows:

$$\frac{d\left(U+E_c+E_g\right)}{dt} = -\int_S \mathbf{q}\cdot\mathbf{n}\,dS + \int_V H dV - p_a\frac{dV}{dt} + \int_V \Phi dV, \tag{8.1}$$

where $\mathbf{n}$ is the unit normal vector, S is the Earth's outer surface and V is the Earth's total volume. The left-hand side collects the different energy components, internal energy U, kinetic energy E_c and gravitational potential energy E_g. The right-hand side lists all the processes that contribute to energy changes: surface heat flux $\mathbf{q}$, internal heat generation H, the work of atmospheric pressure p_a as the planet contracts and finally Φ, which accounts for energy transfers to or from external systems, such as tidal dissipation. Dissipation induced by internal convective motions is not included because it is involved in internal energy transfers and does not act to change the total energy of the system (see Chapter 5).

The dominant terms on the right are the Earth's rate of heat loss and internal heat generation, which are inferred from field measurements and chemical Earth models. The other terms are evaluated theoretically and shown to be negligible. We shall show that changes of gravitational energy are compensated by changes of strain energy E_S, which is the energy required to compress matter to its present local pressure.

The gravitational energy is the energy required to bring matter to infinity. For the Earth, assuming spherical symmetry, it is calculated as follows:

$$E_g = -\int_0^R \rho(r) g(r) r 4\pi r^2 dr, \tag{8.2}$$

with ρ and g the spatially varying density and gravity. This energy is negative because the accretion process releases energy. This energy can be computed for the present Earth and an upper limit can be obtained for a sphere with uniform density and with the same mass as Earth:

$$E_g = -\frac{3}{5}\frac{GM^2}{R}, \tag{8.3}$$

where G is the gravitational constant.

Kinetic energy may be broken down into three different components:

$$E_c = E_{\mathrm{rot}} + E_{\mathrm{contr}} + E_{\mathrm{conv}}, \tag{8.4}$$

corresponding to the bulk rotation of our planet, radial contraction induced by secular cooling and internal convective motions respectively. One may easily show that the latter two are negligible compared to the first one.

Estimates for gravitational, kinetic and internal energy components are listed in Table 8.1. The smallest component is kinetic energy and the largest by far is gravitational energy, which is larger than internal energy by at least one order of magnitude. In a constant-mass planet, gravitational energy changes due to thermal contraction, chemical differentiation, and vertical displacements of the Earth's surface (tectonic processes and erosion deposition). These various processes are associated with different energy transport mechanisms and must be dealt with separately.

Table 8.1. *Energy components*

	Value	Units
Rotational energy	2.1×10^{29} †	J
Internal energy (for 2500 K average temperature)	1.7×10^{31}	J
Gravitational energy (uniform sphere)	2.2×10^{32} †	J
Rotation angular velocity	7.292×10^{-5}	rad s^{-1}
Polar moment of inertia	8.036×10^{37}	kg m^2
Total mass	5.974×10^{24}	kg
Total volume	1.08×10^{21}	m^3
Mass mantle	$\approx 4.0 \times 10^{24}$	kg
Mass crust	$\approx 2.8 \times 10^{22}$	kg
Mass core	$\approx 1.95 \times 10^{24}$	kg

† See Section 2.2.

8.1.2 Changes in gravitational energy: Contraction due to secular cooling

The Earth contracts as it cools down, which implies a change of gravitational energy:

$$\frac{\Delta E_g}{E_g} = -\frac{\Delta R}{R}. \tag{8.5}$$

Contraction induces a change in gravity which in turn induces changes in pressure and density in the Earth's interior. Jaupart *et al.* (2007) have shown that the most important effect is that of temperature and that, for uniform cooling by an amount ΔT,

$$\frac{\Delta R}{R} \approx \frac{\langle\alpha\rangle\, \Delta T}{3}, \tag{8.6}$$

where $\langle\alpha\rangle$ is an average value for the coefficient of thermal expansion and ΔT is negative. For $\langle\alpha\rangle \approx 2 \times 10^{-5}$ K^{-1} and a secular cooling rate of 100 K Gy^{-1}, the contraction velocity is $dR/dt \approx -10^{-13}$ m s^{-1}, which is much less than the typical convective velocity of $\approx 10^{-9}$ m s^{-1}. The induced change of gravitational energy is $\approx$ 4 TW, which, as we shall see, corresponds to 10% of the total energy loss of the Earth. This cannot be considered as negligible but, as shown below, is converted to strain energy, and not to heat (Lapwood, 1952; Flasar and Birch, 1973).

Thermal contraction also affects the moment of inertia C and hence the Earth's rotation:

$$\Delta C/C = 2\Delta R/R \tag{8.7}$$

and, with conservation of angular momentum $\Delta\omega/\omega = -\Delta C/C$, the change in rotational energy:

$$\Delta E_{\rm rot}/E_{\rm rot} = \Delta\omega/\omega = -2\Delta R/R, \tag{8.8}$$

where ω is the Earth's rotation velocity. Thus, some of the gravitational potential energy goes into rotational energy. Rotational energy is smaller than gravitational energy by three orders of magnitude, however, and this change may be neglected in the global energy budget.

To elucidate energy transfer processes, we consider thermodynamics at the local scale. Changes of internal energy per unit mass u are such that (see Chapter 3):

$$\rho\frac{Du}{Dt} = -\nabla\cdot\mathbf{q} + H + \Phi + \psi - p\nabla\cdot\mathbf{v}. \tag{8.9}$$

In this equation, ψ stands for viscous dissipation, which does not appear in the bulk energy budget for the planet:

$$\psi = -\nabla\cdot[\boldsymbol{\sigma}\cdot\mathbf{v}] + \mathbf{v}\cdot(\nabla\cdot\boldsymbol{\sigma}). \tag{8.10}$$

The velocity field is decomposed into a component due to contraction, $\mathbf{v_c}$, and a convective component, $\mathbf{w}$. One key difference between these two is that the azimuthal average of the radial convective velocity, $\overline{w}_r$, is zero, in contrast to that of contraction. Purely radial contraction involves no deviatoric stress in a viscous fluid, and proceeds with negligible acceleration. In this case, the azimuthal average of the momentum equation reduces to a hydrostatic balance:

$$0 = -\nabla\overline{p} + \overline{\rho}\mathbf{g}, \tag{8.11}$$

where the bar denotes azimuthal averages. We consider separately the effects of contraction, which act on the average density and pressure, and the effects of convection, which involve departures from these averages. In the internal energy equation (3.28), we identify the work done by pressure as a change of strain energy,

$$-\overline{p}\nabla\cdot\mathbf{v_c} = \overline{\rho}\frac{De_D}{Dt}, \tag{8.12}$$

and break internal energy down into heat content u_T and strain energy u_s,

$$\overline{\rho}\frac{Du}{Dt} = \overline{\rho}\left(\frac{Du_T}{Dt} + \frac{Du_D}{Dt}\right). \tag{8.13}$$

In the total energy balance (8.9), the last term on the right is the work done by gravity. By definition, this term can be written as the change of gravitational potential energy when it is carried over to the left-hand side of the balance,

$$\overline{\rho}\frac{De_g}{Dt} = -\overline{\rho}\mathbf{g}\cdot\mathbf{v_c}. \tag{8.14}$$

Collecting all terms, the energy balance (8.9) is written as,

$$\overline{\rho}\frac{D\left[e_T + e_D + e_g + e_c\right]}{Dt} = -\nabla\cdot\mathbf{q} + H + \psi - \nabla\cdot(\overline{p}\mathbf{v_c}), \tag{8.15}$$

where we have not accounted for terms associated with convective motions. Kinetic energy is also negligible and, by inspection, one deduces from this equation that

$$\overline{\rho}\frac{D\left[e_D + e_g\right]}{Dt} = -\nabla\cdot(\overline{p}\mathbf{v_c}). \tag{8.16}$$

This can also be demonstrated by recalling the identity,

$$\nabla\cdot(\overline{p}\mathbf{v_c}) = \overline{p}\nabla\cdot\mathbf{v_c} + \mathbf{v_c}\cdot\nabla\overline{p}. \tag{8.17}$$

Using the hydrostatic balance (equation 8.14), the right-hand side of this equation can be recast as,

$$\begin{aligned}\overline{p}\nabla\cdot\mathbf{v_c} + \mathbf{v_c}\cdot\nabla\overline{p} &= \overline{p}\nabla\cdot\mathbf{v_c} + \overline{\rho}\mathbf{g}\cdot\mathbf{v_c} \\ &= -\overline{\rho}\frac{De_D}{Dt} - \overline{\rho}\frac{De_g}{Dt},\end{aligned} \tag{8.18}$$

which is indeed equation (8.16). Integrating equation (8.16) over the whole Earth, we obtain,

$$\frac{dE_g}{dt} + \frac{dE_D}{dt} = -p_a\frac{dV}{dt}, \tag{8.19}$$

where E_D is the total strain energy of the Earth. The work of the atmospheric pressure on the right-hand side is very small so that this equation states that the change of gravitational energy is compensated for by one of strain energy, implying that no heat is generated.

8.1.3 Other energy sources: Tidal heating, crust–mantle differentiation

We review a number of minor energy sources that were lumped together as variable Φ in the bulk energy budget (8.1) and show that they can be neglected.

Earth's rotation is accelerating because of post-glacial re-adjustments, and is slowing down because of tidal interaction with the Moon. The torque exerted on the Moon is due to the lag between the tidal potential and the tidal bulge which is 2.9°

ahead of the potential. In the Earth–Moon system, angular momentum is conserved, but there is a net loss of the rotational and gravitational potential energy due to frictional dissipation. With laser ranging, the changes in the Earth–Moon distance have been measured accurately (3.7 cm y^{-1}) and the slowing down of Earth's rotation due to tidal interaction can be calculated exactly. The effect of the solar tidal potential on Earth rotation is $\approx$20% that of the Moon and it must be included in the calculations. The slowing down of Earth's rotation is 5.4×10^{-22} rad s^{-1} leading to a 0.024 ms y^{-1} increase in the length of the day. The accompanying energy loss is 3 TW, which must be accounted for by dissipation in the oceans, in the solid Earth and in the Moon. According to Lambeck (1977), 90–95% of the energy from tidal friction gets dissipated in the seas and oceans. The contribution of the solid Earth tide depends on the quality factor Q (see Section 2.4.2). For the values of Q in the mantle suggested by seismology, dissipation by the solid Earth tide accounts for < 0.1TW (Zschau, 1986). Such a low value has been confirmed by satellite observations of the lag between the solid Earth tide (0.16°) and the lunar potential. This observation implies that dissipation by the solid Earth is 0.083 TW (Ray *et al.*, 1996).

Some gravitational potential energy is released by the extraction of continental crust out of the mantle and the associated changes of the density structure of the Earth. In contrast to the change of gravitational energy due to contraction, this energy contributes to the bulk energy balance as compositional energy (Braginsky and Roberts, 1995). A rough estimate of the loss in potential energy is given by,

$$\delta E_g = \int_c^R g(R) dm, \tag{8.20}$$

where c is the core–mantle boundary, R the radius of the Earth and dm is the mass difference. With total mass of crust 2.6×10^{22} kg, $\delta\rho/\rho = 0.1$, $g(R) \approx 10$ m s^{-2}, we get $\delta E_g \approx 3 \times 10^{28}$ J. For a constant rate of crustal growth during 3 Gy, the contribution to the energy budget is small, 0.3 TW. If the crust differentiated during two or three short episodes, each episode may have added 1–2 TW to the mantle budget.

8.1.4 Secular cooling equation

To derive an equation for temperature, we return to local variables. Introducing variables of state, we write,

$$\rho \frac{De_T}{Dt} = \rho T \frac{Ds}{Dt} = \rho C_p \frac{DT}{Dt} - \alpha T \frac{Dp}{Dt}, \tag{8.21}$$

where s, the entropy per unit mass, has been expressed as a function of temperature and pressure and α is the coefficient of thermal expansion. From (3.28), we deduce that,

$$\rho C_p \frac{DT}{Dt} = \alpha T \frac{Dp}{Dt} - \nabla \cdot \mathbf{q} + H + \psi. \tag{8.22}$$

By definition,

$$\frac{Dp}{Dt} = \frac{\partial p}{\partial t} + \mathbf{v} \cdot \nabla p = \frac{\partial p}{\partial t} + \mathbf{v_c} \cdot \nabla p + \mathbf{w} \cdot \nabla p. \tag{8.23}$$

The first two terms on the right of this equation are responsible for what has been called "adiabatic heating", which is the only remaining contribution of contraction. This may be safely neglected for the solid Earth because of the small contraction velocity and small expansion coefficient. The last term on the right is not negligible and contributes a key term in the energy budget, the buoyancy flux.

To elucidate the role of the buoyancy flux, we decompose temperature into an azimuthal average and a perturbation, such that $T = \overline{T} + \theta$. We then average equation 8.22 over a spherical shell of radius r and obtain (see Chapter 4),

$$\rho C_p \frac{D\overline{T}}{Dt} = -\rho \alpha g \overline{w_r \theta} - \frac{\partial \overline{q}_r}{\partial r} + H + \overline{\psi}, \tag{8.24}$$

where w_r is the radial convective velocity component such that $\overline{w}_r = 0$. The first term on the right is the buoyancy flux (Chapter 4). Integrating over the planet, we finally obtain,

$$\begin{aligned} \int_V \rho C_p \frac{D\overline{T}}{Dt} dV &= M \langle C_p \rangle \frac{d \langle T \rangle}{dt} \\ &= - \int_V \rho \alpha g \overline{w_r \theta}\, dV - \int_S \overline{q}_r\, dS + \int_V H\, dV + \int_V \overline{\psi}\, dV, \end{aligned} \tag{8.25}$$

where M is the mass of the Earth and $\langle C_p \rangle$ and $\langle T \rangle$ are its average heat capacity and average temperature, respectively. As shown in Chapter 4,

$$- \int_V \rho \alpha g \overline{w_r \theta}\, dV + \int_V \overline{\psi}\, dV = 0, \tag{8.26}$$

which states that viscous dissipation is balanced by the bulk buoyancy flux. This explains why internal viscous dissipation does not appear in the bulk energy budget.

With this final step, we have reduced the bulk energy balance to,

$$M \langle C_p \rangle \frac{d \langle T \rangle}{dt} = - \int_A \overline{q}_r\, dA + \int_V H\, dV, \tag{8.27}$$

which is the secular cooling equation.

8.2 Heat loss through the ocean floor

The energy budget of the Earth is calculated from the same inputs that are used in constructing the global heat flow map of the Earth (Figure 8.1). That the global heat flow map is so perfectly similar to the sea-floor age map (Figure 2.14) is not a coincidence: sea-floor ages are the main input used to determine oceanic heat flux. The heat flow map is based on a combination of heat flux measurements from continents and their margins, and on the cooling model combined with sea-floor ages for the oceans. We explain below why we prefer to rely on a model than on the raw data to calculate the oceanic heat flow, and how we remove the sampling bias that affects continental heat flow data.

8.2.1 Oceanic heat flux data

For the purposes of calculating the rate at which the Earth is losing heat, the most direct and unbiased method is to integrate individual measurements of heat flux over the surface. This method fails in the oceans because the heat flux data are

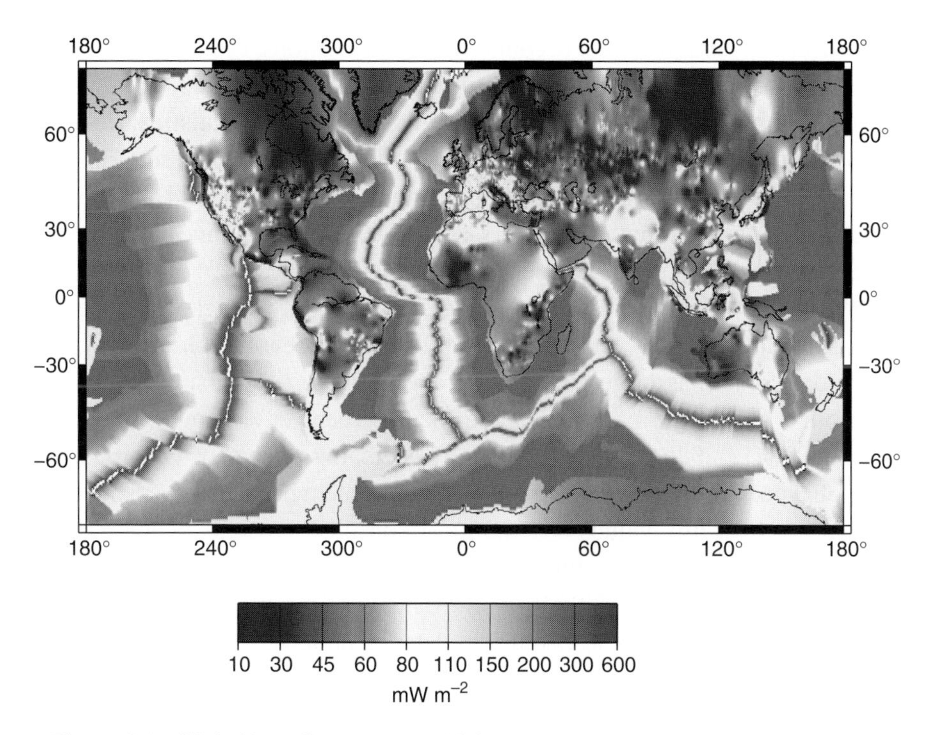

Figure 8.1. Global heat flow map combining land heat flux measurements and the cooling plate model for the oceans, where heat flux is calculated as the maximum of 48 mW m^{-2} and $490/\sqrt{\tau}$ mW m^{-2}, where τ is age in My. See plate section for color version.

perturbed by hydrothermal circulation (Chapter 6). Such circulation affects large volumes of oceanic rocks as shown by the extent of alteration in ophiolite massifs (Davis and Elderfield, 2004).

Heat transport through permeable rock and sediment involves two mechanisms: conduction through the solid and static matrix and water flow through pores and fractures into the sea. Measuring the latter cannot be done over large areas, so that the available data account only for heat conduction. The vast majority of marine heat flux determinations rely on the probe technique, such that a rigid rod carrying a thermistor chain is shoved into sediments (see Appendix C). Obviously, this requires thick sedimentary cover, a systematic bias in the measurement environment that has important consequences. A better technique is to measure temperatures in deep-sea drill-holes because it involves a large depth range through crystalline basement, but it is time consuming and feasible in a small number of sites only. Measurements made with both techniques at a few selected locations show that shallow probes provide reliable results (Erickson *et al.*, 1975).

High heat flux values near oceanic ridges were one of the decisive observations confirming sea-floor spreading (Von Herzen and Maxwell, 1959; Langseth *et al.*, 1965). These early surveys made it very clear that heat flux data exhibit large scatter and that heat conduction cannot account for all observations. Figure 6.4 shows heat flux data from the compilation by Stein and Stein (1992) binned in 2 My age intervals. Binning data for all oceans by age group is done for statistical reasons, in the hope that measurement errors cancel each other in a large data set. This is not valid if measurement errors are not random, which is the case here. Data reliability depends in part on environmental factors, such as basement roughness and sediment thickness. A rough basement/sediment interface leads to heat refraction effects and focusing of hydrothermal flows. At young ages, sea floor roughness depends in part on spreading velocity and varies from ocean to ocean. Such effects explain why the data scatter is largest for young ages (Figure 6.4). Using the raw data turns a blind eye to the problem of the measurement environment (Harris and Chapman, 2004) and does not deal with the systematics of scatter. The first age bin in the global oceanic data set presents a specific problem because it is characterized by the largest heat flux values and the most conspicuous signs of hydrothermal activity. A proper average for this age bin requires data at very young ages, less than 1 My, say, which are virtually non-existent. Accounting for the cooling of oceanic lithosphere at young ages requires detailed understanding of hydrothermal flows (Davis and Elderfield, 2004). In areas of hydrothermal circulation, sediment ponds are frequently zones of recharge, such that downward advection of cold water lowers the temperature. Recharge tends to occur over wide regions in contrast to discharge, which is usually focused through basement outcrops. By design, heat flux measurements require sedimentary cover and are systematically biased towards

Table 8.2. *Potential temperature of the oceanic upper mantle*

T, °C	Reference	Method
1315	McKenzie *et al.* (2005)	Depth + heat flux with cooling model
1280	McKenzie and Bickle (1988)	Average basalt composition
1315–1475	Kinzler and Grove (1992)	Basalt composition

anomalously low heat flux areas. This systematic error cannot be eliminated by a large number of measurements. We shall give a specific example below. Far from oceanic ridges, sediment cover is thicker and hydraulically resistive, so that hydrothermal circulation is confined to the crystalline crust in closed systems. In such conditions, heat flux varies spatially but the integrated value is equal to the heat extracted from the lithosphere.

Systematic biases in oceanic heat flux measurements are accounted for by detailed heat flux surveys in specific environments combined with theoretical estimates of the amounts of heat that are lost by the oceanic lithosphere. Below the superficial sediment cover and the shallow permeable basement, heat can only be transported by conduction. The simplest thermal model for the oceanic lithosphere is such that cooling proceeds unhampered over the entire age span of the oceans. Due to the large separation of horizontal and vertical scales, one may safely assume that conductive heat transport occurs over the vertical. This is in fact the standard boundary layer model that has been used in Chapter 4, for which the surface heat flux at age τ is

$$Q(\tau) = C_Q \tau^{-1/2}, \tag{8.28}$$

where C_Q is a constant and which is valid for arbitrary temperature-dependent physical properties (see Chapter 6). For constant properties, $Q(\tau) = \lambda T_M / \sqrt{\pi\kappa\tau}$, where T_M is the average temperature in the well-mixed convective mantle. The value for the mantle temperature T_M remains subject to some uncertainty (Table 8.2), but the value of the constant C_Q may be determined empirically from the heat flux data.

An alternative model is the plate model, in which a boundary condition is applied at the base of the "plate" (Chapter 6). In this case, the heat flux tends to a constant value at large times, when the plate is in thermal equilibrium with heat brought from below. Simple boundary conditions at the base of the plate are adopted for mathematical convenience and must be regarded as approximations. For example, the fixed temperature boundary condition requires infinite thermal efficiency for heat exchange between the plate and the mantle below. For very short time, the surface heat flux does not depend on the lower boundary condition and is the same as for the cooling half-space (equation 8.28). The details are provided in Chapter 6. Thus, it is better to use a half-space model because it relies on a reduced set of hypotheses and because it does fit the oceanic data (see Chapter 6). At the very least, the theoretical heat flux model provides a reference function for a best-fit relationship between heat flux and age. This simple model breaks down at ages larger than about 80 My for reasons that are still debated. For older ages, we need not invoke the questionable plate model because heat flux data exhibit small scatter: the heat flux is approximately constant $q_{80} \approx 48$ mW m^{-2} (see also Chapter 6). Deviations from this value are ± 3 mW m^{-2} and exhibit no systematic trend as a function of age. The mean is determined with an uncertainty of 1 mW m^{-2}, which represents 1% of the average oceanic heat flux. This has little impact on the total heat loss estimate which is dominated by the young sea floor contribution.

8.2.2 Estimating the total oceanic heat loss

The total oceanic heat loss is determined by integration:

$$Q_0 = \int_0^{\tau_{\max}} q(\tau) \frac{dA}{d\tau} d\tau, \tag{8.29}$$

where $A(\tau)$ is the distribution of sea floor with age, which is deduced from maps of magnetic anomalies of the ocean floor. One must also account for the contribution of marginal basins, whose heat flux conforms to the standard oceanic heat flux model, as demonstrated by Sclater *et al.* (1980a). Table 8.6 compares the various estimates that have been obtained in the past. The heat loss estimate of Pollack *et al.* (1993) was based on $C_Q = 510$ mW m^{-2} My$^{1/2}$, which is clearly an upper bound. This specific value was taken from the analysis of Stein and Stein (1992), which itself was based on the plate model with constant basal temperature. One feature of this model is that $T_M = 1725$ K, a high value that is not consistent with the average ridge axis temperature derived from the compositions of mid-ocean ridge basalts (Table 8.2).

Sea-floor ages are determined from the pattern of marine magnetic anomalies and the well-established chronometry of geomagnetic reversals. The areal distribution of sea-floor ages (Figure 8.2) can be approximated by a linear relationship:

$$\frac{dA_1}{d\tau} = C_A(1 - \tau/\tau_m). \tag{8.30}$$

Different authors agree that $\tau_m = 180$ My but have proposed slightly different values of the coefficient C_A. Small differences are due to the different methods of analysis but a more important one arises from the exclusion of marginal basins in

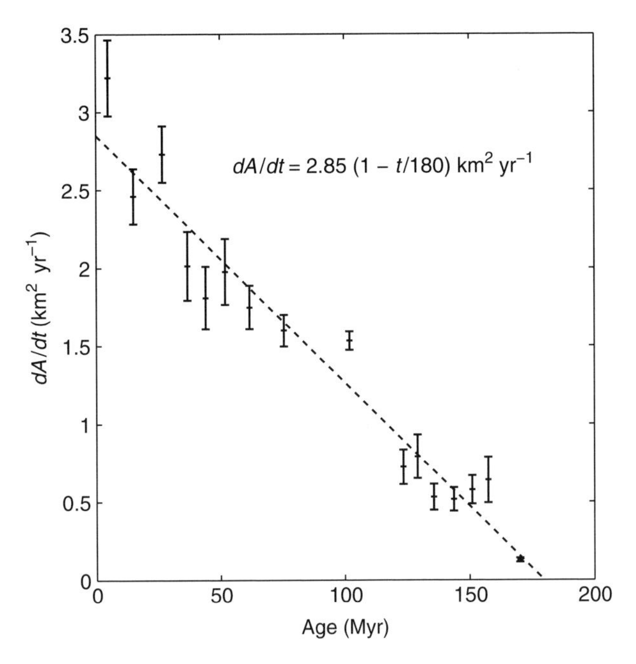

Figure 8.2. Distribution of sea-floor area as a function of age, from Cogné and Humler (2004). The distribution does not include the marginal basins and yields an area of 257×10^6 km^2 for the oceans.

some of the analyses. Interestingly enough the lower estimate of the coefficient implies a value too low for the total surface area of the oceans.

One independent constraint is brought by the total area of ocean floor, which is sensitive to the exact location of the continent–ocean boundary. A detailed analysis of continental margins leads to a total continental area of 210×10^6 km^2 (Cogley, 1984). Consequently, the total sea-floor surface is 300×10^6 km^2. For a triangular age distribution with a maximum age $\tau_m = 180$ My, this implies that $C_A = 3.34$ km^2 y^{-1}. Subtracting the contribution of marginal basins, this corresponds exactly to the Rowley (2002) estimate. This discussion illustrates that uncertainties may come from unexpected variables, the area of the sea floor in this particular instance.

Uncertainties come from estimates of the total area of ocean floor, or, more precisely, the total area of continental shelves as well as departures from simple triangular age distribution. The former is less than 3% and implies a much smaller uncertainty on the global heat loss estimate because the total surface of the Earth is known very precisely: a change in the area of oceans is compensated by an opposite change in the area of continents. Considering the difference between the average oceanic and continental heat flux values, the resulting uncertainty on the global heat loss estimate is only 1%. One could integrate equation 8.29 with the exact distribution of sea-floor age rather than the approximate triangular distribution. The difference is negligible (<0.3%) provided the two distributions include the same total area.

Integrating separately sea floor younger and older than 80 My gives,

$$
\begin{aligned}
Q_{80-} &= \int_0^{80} C_Q \tau^{-1/2} C_A (1 - \tau/180) d\tau = 24.3 \text{ TW} \\
Q_{80+} &= q_{80} \int_{80}^{180} C_A (1 - \tau/180) dt' = 4.4 \text{ TW} \\
Q_{\text{oceans}} &= 29 \pm 1 \text{ TW},
\end{aligned}
\tag{8.31}
$$

where the uncertainty comes mostly from that of coefficient C_Q. The present estimate is slightly less than earlier estimates because of the slightly lower ridge temperature (or equivalently, the slightly smaller value of coefficient C_Q in the heat flux versus age relationship) and because of the revised estimate for the mean accretion rate at zero age C_A. For $C_Q = 510$ mW m^{-2} My$^{1/2}$ and $C_A = 3.45$ km^2 y^{-1}, the heat loss would be 31 TW. Sea-floor bathymetry provides an independent measure of the heat lost by the cooling plate for ages less than 80 My but cannot be used in old basins because it is almost flat. The total oceanic heat loss depends on the distribution of sea-floor ages, i.e. on the rate of sea-floor creation and destruction in subduction zones (Figure 8.3).

These estimates of oceanic heat loss do not account for the contribution of hot spots which are areas of enhanced heat flux (Von Herzen *et al.*, 1982). The heat flux from hot spots can be calculated from the buoyancy of bathymetric swells (see Chapter 6). These estimates are in the range 2–4 TW and must be added to the heat loss due to plate cooling. This is an upper limit because the plate can be subducted before all the heat has been evacuated (see Chapter 6). We discuss below the relationship between the hot spot component and core heat loss.

8.3 Heat loss through continents

8.3.1 Estimating the continental heat loss

In the oceans, a simple theoretical cooling model predicts how the surface heat flux varies and is used to determine the total heat loss in young ocean basins even when there are no data. In the continents, no such model exists and estimating the total heat loss can only be done with heat flux data. The measurements and their corrections are discussed in Appendix C. Continental heat flux measurements are made in boreholes that usually penetrate a few hundred meters. The quality of the measurements can be controlled by comparing temperature profiles from neighboring boreholes and/or by repeat measurements. This procedure allows detection of errors and environmental

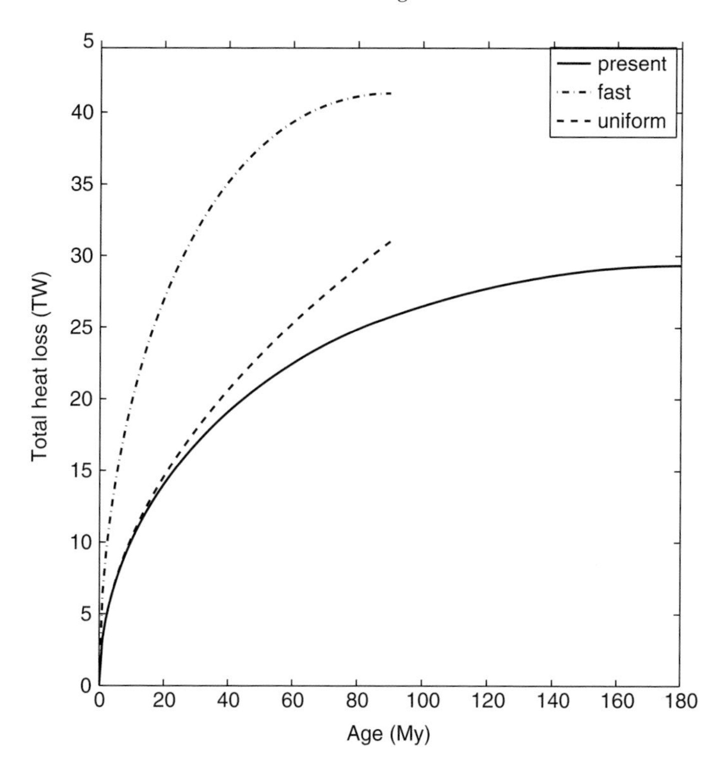

Figure 8.3. Cumulative total heat loss of the sea floor for three different distributions of sea-floor age: the present triangular distribution, triangular distribution with twice the present spreading rate and oldest sea floor 90 My, uniform distribution of ages from 0 to 90 My with the same spreading rate as present. The total oceanic area is 300×10^6 km^2, and $C_Q = 490$ mW m^{-2} My$^{1/2}$; for ages > 80 My, heat flux is 48 mW m^{-2}.

perturbations and provides data that are far more reliable than in the oceans. For the present discussion, we must note the shortcomings of the data set: (1) the quality of continental heat flux measurements, as measured for example by the depth of the boreholes that have been used, is variable; (2) there is no complete repository of heat flux (and related physical properties) data; and (3) there is a strong bias in the geographical and geological distribution of the data. Figure 8.4 shows the location of the continental heat flux measurements that were used to construct the heat flow map (Figure 8.1). This data set is larger than that used by Pollack *et al.* (1993) which is missing almost 50% of the data measured in India or in Canada. The distribution remains uneven with most of the data located between 30 and 60 N. It contains more than 3,000 entries for the United States, but only 600 entries for Africa, 120 for Australia, 9 for Antarctica. The mean of all the continental heat flux data is 80 mW m^{-2}. The global data sets include a large number of entries from the United States. The mean heat flux for the United States data subset is 112 mW m^{-2}.

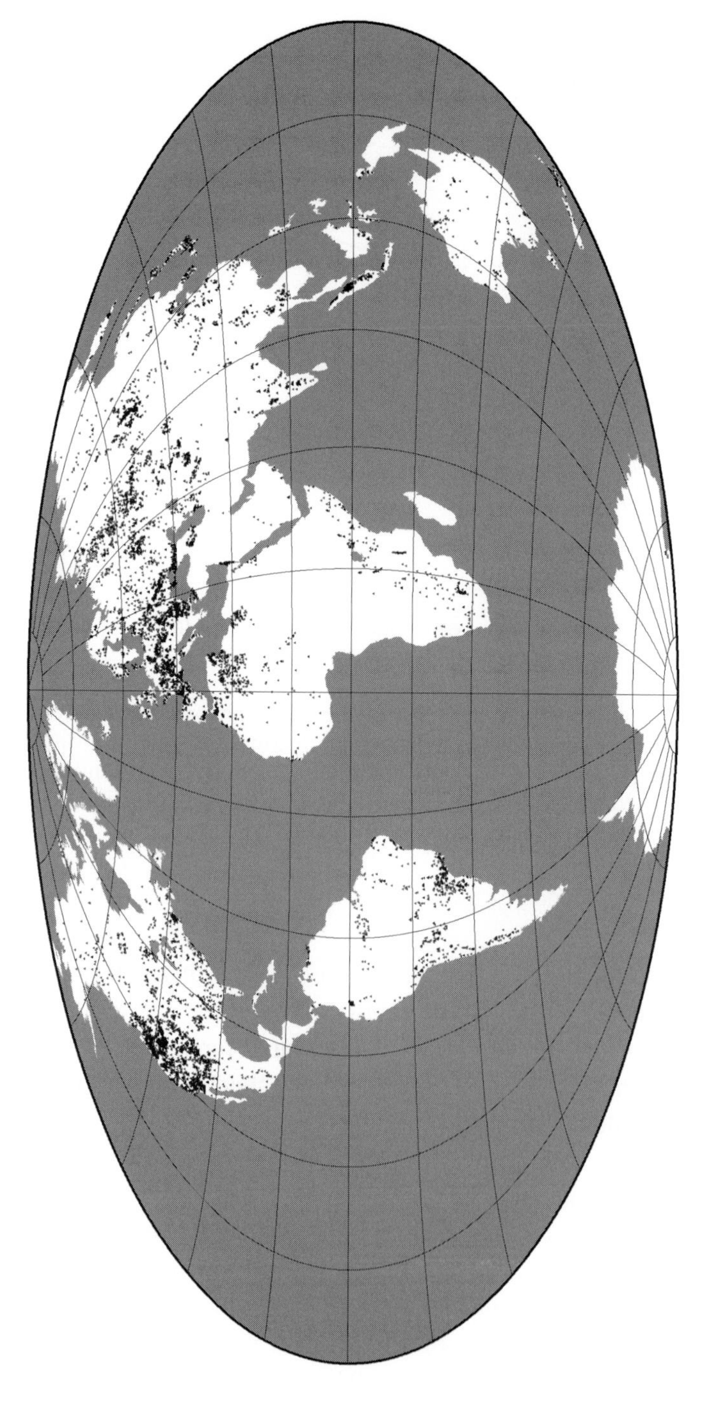

Figure 8.4. Location of land heat flux measurements. This compilation includes approximately 22,000 sites.

Table 8.3. *Continental heat flux statistics* †

	$\mu(Q)$ mW m^{-2}	$\sigma(Q)$ mW m^{-2}	$N(Q)$
World			
all values	79.7	162	14123
averages 1° × 1°	65.3	82.4	3024
averages 2° × 2°	64.0	57.5	1562
averages 3° × 3°	63.3	35.2	979
USA			
all values	112.4	288	4243
averages 1° × 1°	84	183	532
averages 2° × 2°	78.3	131.0	221
averages 3° × 3°	73.5	51.7	128
without USA			
all values	65.7	40.4	9880
averages 1° × 1°	61.1	30.6	2516
averages 2° × 2°	61.6	31.6	1359
averages 3° × 3°	61.3	31.3	889

† μ is the mean, σ is the standard deviation and N is the number of values

The tectonically active Basin and Range province that covers a large area of the United States, has very high flux (Figure 7.1). Many measurements were made for geothermal energy exploration, resulting in over-sampling of high heat flux regions. Excluding the values from the United States, the mean continental heat flux drops to only 65 mW m^{-2}.

The sampling bias can also be removed by weighting the data by area as demonstrated in Table 8.3. Using average heat flux values over 1° × 1° windows yields a mean continental heat flux of 65.3 mW m^{-2}. This mean value does not decrease significantly when averaging is done over wider windows. The histograms of heat flux values or averages over 1° × 1° windows have identical shapes, except for the extremely high values (>200 mW m^{-2}) that bias the data. The same analysis on the US data alone gives a mean that converges to a value (74 mW m^{-2}) higher than the word average, reflecting the contribution of tectonically active regions.

Another approach has been to bin heat flux values by geological age and to estimate the mean heat flux within each age group by analogy with oceanic heat flux. The total continental heat loss can then be calculated as an area-weighted average of the mean heat flux of each age group (Pollack *et al.*, 1993). The mean continental heat flux obtained by this method is 65 mW m^{-2}, the same as that obtained by averaging over spatial windows.

For a mean continental heat flux value of 65 mW m^{-2}, the contribution of all the continental areas (i.e. 210×10^6 km^2) to the energy loss of the Earth represents $\approx$14 TW. This number includes the submerged margins and continental areas with active tectonics, where higher than normal heat flux values are associated with thick radiogenic crust and/or shallow magmatic activity. Uncertainty in this number is due to lack of adequate data coverage in Greenland, Antarctica and large parts of Africa. To estimate the uncertainty, we assume that heat flux in those areas is equal to either the lowest or the highest average heat flux recorded in well-sampled geological provinces (36 and 100 mW m^{-2} respectively). This procedure allows departures of $\pm$1.5 TW from the estimate of 14 TW. The uncertainty must be less than 1.5 TW because the poorly sampled regions are vast and must encompass geological provinces of various ages and geological histories; for instance, both Antarctica and Greenland are known to include high and low heat flux regions. For the sake of simplicity, we retain an estimate of 1 TW for the uncertainty.

8.3.2 Various contributions to the surface heat flux in continental areas

Determining the heat loss from the mantle through the continental lithosphere requires accounting for the crustal heat production. In stable continents, for ages greater than about 500 My, continents are near thermal steady state such that surface heat flux is the sum of heat production in the lithosphere and of the heat supply at the base of the lithosphere. Recently active regions are in a transient thermal regime and the high surface heat flux reflects cooling of the continental lithosphere. After removing the crustal heat production, it is possible to estimate the transient component of the heat flux, which originates in mantle cooling.

Stable continents

The contributions of heat production in the crust and in the mantle lithosphere have been discussed at length in Chapter 7. It has been shown that, for stable regions, crustal heat production makes the dominant contribution and the heat flux from the mantle is low.

Recently active regions and continental margins

Submerged and recently active (i.e. during the past 200 My) continental areas cover 92×10^6 km^2 , $\approx$45% of the total continental surface (Table 8.5). These regions are not in thermal steady state and are characterized by higher heat flux than the continental average. Because of the long thermal relaxation time of the continental lithosphere, present surface heat flux samples the inputs of heat from the mantle of the past 100–200 My. The crustal component can now be calculated from crustal thickness and average heat production. After accounting for crustal heat production, the heat from the mantle (some of which is included in the transient component) can be estimated.

Compressional orogens In compressional orogens, crustal and lithospheric thickening result in reduced temperature gradients and heat flux, but the total heat production in the thick crust is high. These two competing effects lead to a complex transient thermal structure and few generalizations can be made on the surface heat flux. For instance, very high heat flux values (> 100 mW m^{-2}) have been measured on the Tibetan plateau (Francheteau *et al.*, 1984; Jaupart *et al.*, 1985; Hu *et al.*, 2000). They have been attributed to shallow magma intrusions and yield little information on mantle heat flux. In contrast, present surface heat flux remains low in the Alps, and after removing crustal heat production, heat flux at the Moho is estimated to be as low as 5 mW m^{-2} (Vosteen *et al.*, 2003). Heat flux from the mantle is also low beneath the North American Cordillera (Brady *et al.*, 2006), and beneath the South American Cordillera, at least where it has not been affected by back-arc extension (Henry and Pollack, 1988).

Zones of extension and continental margins In rifts, recently extended regions, continental margins and basins, heat flux is high because of a large transient component, which ultimately represents additional inputs of heat from the mantle. Crustal extension and lithospheric thinning will initially result in steepening the temperature gradient and increasing the heat flux. Long-term thermal relaxation depends on heat supply at the base of the lithosphere. The heat flux is high (75–125 mW m^{-2}) in zones of extension and in continental rifts (Morgan, 1983). A typical feature is a sharp transition over a few tens of kilometers between a region of elevated heat flux and surrounding areas, i.e. Colorado Plateau–Basin and Range in North America (Bodell and Chapman, 1982), East African Rift–Tanzanian craton (Nyblade, 1997), Baikal Rift–Siberian craton (Poort and Klerkx, 2004). The absence of lateral diffusion of heat shows that the enhanced heat flux in the extended area is not due to changes of heat supply at the base of the lithosphere but is the direct result of extension and lithospheric thinning. Where the sampling is sufficient, heat flux exhibits short wavelength variations. These variations are partly due to the cooling of shallow magmatic intrusions and to groundwater movement. The actual heat loss may be higher than indicated by conductive heat flux measurements because of the heat transport by hot springs and volcanoes. The contribution of wide regions of extension is more significant than that of rifts. In the Basin and Range Province in the southwestern United States, 100% extension accounts for the high average heat flux (105 mW m^{-2}) (Lachenbruch and Sass, 1978). In this province, the heat delivered by volcanoes is negligible and the integrated effect of heat transport by groundwater is small (Lachenbruch and Sass, 1978). This interpretation depends on assumptions made about pre-extensional heat flux and on crustal heat production. The average heat production of the crust is the same as in stable regions and yields a total crustal heat production of $\approx$33 mW m^{-2}(Ketcham, 1996). This implies that the transient component due to cooling and mantle heat flux in the Basin and Range contribute a total heat flux of $\approx$70 mW m^{-2}. Detailed models to account for this heat loss involve delamination of the lithospheric mantle or stretching of the lithosphere and heat supply by magma intrusions (see Chapter 7). Regardless of the mechanism, however, at least $\frac{2}{3}$ of the heat flux in this region comes from the mantle.

In the Yellowstone system, the heat flux is locally $>$ 40 Wm^{-2} but the total heat loss remains modest: including both conductive and convective processes, it is $\approx$5 GW (Fournier, 1989). It would thus require 200 "Yellowstones" to increase the continental heat loss by 1 TW. The effect of continental hotspots on the budget seems presently negligible. Values for the total heat loss through geothermal systems in the East African rift are comparable to those of Yellowstone (Crane and O'Connell, 1983). In continental as well as in oceanic rifts, the heat loss is underestimated because of hydrothermal heat transport. However, because continental rifts are narrow and their total surface area is small, the error will not significantly affect the continental heat flux budget. For instance, the total heat loss for the Gregory rift, in Kenya, is $\approx$20 GW (Crane and O'Connell, 1983). Similar values have been inferred for Baikal (Poort and Klerkx, 2004).

Large igneous provinces testify to periods of enhanced volcanic activity in the continents, but their effect on the heat flow budget is also negligible. In the Deccan, where 500,000 km^3 of basalts were deposited about 60 My ago, there is no heat flow anomaly, suggesting that the magmas did not heat up the lithosphere over large volumes. Assuming that the lavas were deposited over 1 My, the heat that they carried to the surface contributed less than 0.1 TW to the energy budget.

Continental margins account for an important fraction ($\approx$30%) of the continental surface. Margins are characterized by gradual crustal thinning towards oceanic basins, which implies a lateral variation of the crustal heat flux component. The average heat flux of the margins (78 mW m^{-2}) is higher than in stable regions despite the thinner crust. This higher heat flux is explained by the cooling of the stretched lithosphere and is reflected in thermal subsidence (Vogt and Ostenso, 1967; Sleep, 1971). Where detailed information on crustal thickness is available, the input of heat from the mantle can be calculated.

Table 8.4. *Estimates of bulk continental crust heat production from heat flux data (Jaupart and Mareschal, 2003)*

Age group	Heat production μW m^{-3}	Total (40 km crust) mW m^{-2}	% Area †
Archean	0.56–0.73	23–30	9
Proterozoic	0.73–0.90	30–37	56
Phanerozoic	0.95–1.21	37–47	35
Total Continents	0.79–0.99	32–40	100

† Fraction of total continental surface, from Model 2 in Rudnick and Fountain (1995).

Table 8.5. *Surface area and heat flux in oceans and continents*

	Area	Total heat flux
Oceans		
Oceanic	273×10^6 km^2	
Marginal basins	27×10^6 km^2	
Total oceans	300×10^6 km^2	32 TW
Continents		
Precambrian	95×10^6 km^2	
Paleozoic	23×10^6 km^2	
Stable continental	118×10^6 km^2	
Active continental	30×10^6 km^2	
Submerged (margins and basins)	62×10^6 km^2	
Total continental	210×10^6 km^2	14 TW

8.3.3 Mantle heat loss through continental areas

Different methods lead to a value of 14 TW for the integrated heat flux from continental areas. Accounting for geothermal and volcanic transport has no significant impact on this value. The estimated average heat production of the continental crust ranges between 0.79 and 0.99 μW m^{-3} (Table 8.4) and its total volume is $\approx 0.73 \times 10^{10}$ km^3, which gives a total heat production between 6 and 7 TW for the continental crust.

Little is known about the amounts of radiogenic elements in the lithospheric mantle. Direct estimates rely on a few exposures of peridotite massifs, which are typically depleted (Rudnick *et al.*, 1998), and on mantle xenoliths from kimberlite pipes, which are usually enriched (Russell *et al.*, 2001). Considerations on the thermal stability of continental roots and consistency with heat flux measurements as well as with petrological temperature estimates lead to the conclusion that enrichment must be recent and associated with metasomatic infiltrations (Jaupart and Mareschal, 1999; Russell *et al.*, 2001). This enrichment process is probably limited in both area and volume and our best estimate of radiogenic heat production in the lithospheric mantle comes from peridotite massifs. For the sake of completeness, we take a value of 0.02 μW m^{-3} from (Rudnick *et al.*, 1998) and consider an average lithosphere thickness of 150 km. The total heat thus generated in the subcontinental lithospheric mantle is about 0.5 TW, which is only accurate within a factor of about two.

Subtracting the contribution of radioactive sources from the total heat loss out of continents, we thus arrive at an estimate of the heat input from the mantle of 6–7 TW, most of which is brought through the tectonically active regions and the continental margins.

8.4 Heat loss of the Earth

Table 8.6 compares the various heat loss estimates that have been obtained. Over more than 35 years, therefore, heat loss budgets for the Earth have not changed much because they are dominated by the oceanic contribution. Oceanic heat loss is due to a single process, cooling of the lithosphere that is generated at a mid-ocean ridge, and is not affected by poor knowledge of radiogenic heat production: neither the oceanic crust nor the oceanic lithospheric mantle contain significant amounts of U, Th and K. As a consequence, oceanic heat loss estimates depend only on the distribution of sea-floor ages and on the scaling constant in the heat flux model for young sea floor. Neither of these have been modified significantly by the increasingly accurate age maps and heat flux data that have been obtained.

Table 8.6. *Estimates of the continental and oceanic heat flux and global heat loss*

	Continental mW m^{-2}	Oceanic mW m^{-2}	Total TW
Williams and von Herzen (1974)	61	93	43
Davies (1980)	55	95	41
Sclater *et al.* (1980a)	57	99	42
Pollack *et al.* (1993)	65	101	44
Jaupart *et al.* (2007)†	65	94	46

† The average oceanic heat flux does not include the contribution of hot spots. The total heat loss estimate does include 3 TW from oceanic hot spots.

The continental estimates have changed by larger amounts due solely to increasing data coverage and better corrections for sampling bias.

8.5 Radiogenic heat sources in the mantle

The bulk composition of the Earth cannot be measured directly for lack of direct samples from the lower mantle and the core, and hence has been estimated using various methods. All approaches rely on two different kinds of samples: meteorites, which represent the starting material, and samples of today's upper mantle. Both show rather extensive variations of composition due to their different histories. Processes in the early solar nebula at high temperature contribute one type of compositional variation. Processes within the Earth, which occur at lower temperatures, contribute another type of compositional variation. Stated schematically, one has a range of compositions from the early solar system and a range of compositions for the present Earth, and one must devise a procedure to correct for the chemical evolution of these two different systems.

Chondrites represent samples of undifferentiated silicate material from the solar system prior to melting and metallic core segregation. Different families of chondrites with different compositions were generated from the solar composition by processes in the early nebula. Perturbations were essentially brought in the gas state and elemental behavior must be classified according to volatility (or condensation temperature). For our energy budget, the important elements are uranium, thorium and potassium. The first two are associated with very high condensation temperatures and called "refractory lithophile" elements. These two elements had the same behavior in the early solar system because they have the same ratio in all types of chondritic meteorites. Potassium is a "moderately volatile" element with a lower condensation temperature. The best match with solar concentration ratios is achieved by CI carbonaceous chondrites. We do know, however, that CI chondrites have larger amounts of volatiles, including water and CO_2, than the Earth. As regards the Earth's mantle, one seeks to establish a systematic compositional trend through the available samples and to identify the most primitive (and least differentiated) end-member. Four methods have been used to derive estimates of Earth composition from the two types of materials.

The first method relies only on direct samples from the mantle (Ringwood, 1962). Basalts and peridotites are complementary rocks, because the latter is the solid residue of the partial melting event that led to basalt genesis. Thus, mixing them back together in the appropriate proportions yields the starting material, which is named "pyrolite". Clearly, one has to choose samples which have not been affected by leaching and low-temperature alteration. Unfortunately, this procedure is not efficient for uranium, which is very mobile, and hence one should not use pyrolite models for heat-producing elements. In a second method, one selects a specific type of chondrite and works through the processes that turn this material into Earth-like material: devolatilization (loss of water, CO_2, and other volatile elements) followed by reduction (loss of oxygen). Many authors use CI chondrites (e.g., Hart and Zindler (1986)). Errors come from mass loss estimates, but also from the starting CI chondrite composition since this group of meteorites is quite heterogeneous. Javoy (1999) argued in favor of a different type of meteorite on the basis of the oxidation state of the solar nebula as it started to condense. The only meteorites with the right oxidation state are enstatite chondrites, which are therefore close to the material which went into the protoplanets. These chondrites are largely degassed, save for sulfur, so the volatile loss correction is small.

A third method tries to avoid a specific choice for the starting composition and aims at determining it. Hart and Zindler (1986) defined the compositional trends of chondritic meteorites and mantle peridotites, which are not parallel to one another. Each trend records the effects of the two different sets of processes operating in the primitive solar nebula and in the Earth, hence the intersection can only be the starting Earth material. In this case, the error comes from the scatter around the two compositional trends. A fourth method relies on elemental ratios. For refractory lithophile elements, such as uranium and thorium, the concentration ratio is independent of chondrite type and hence is a property of the starting Earth material. Once these ratios have been determined, two procedures can be used to determine primitive abundances from measurements on peridotite samples. In one procedure, one starts with one specific element for which a reliable bulk Earth content can be determined and then work sequentially to all the others using elemental ratios. The element of choice is Mg because, although it is not the most refractory element, its behavior during melting and alteration is well understood (Palme and O'Neill, 2003). The other procedure is to study the relationship between abundance and elemental ratios: the primitive abundance is that which corresponds to the chondritic ratio (McDonough and Sun, 1995). Variations in abundances and elemental ratios observed can be accounted for by melting and melt extraction, which act to deplete the solid peridotite. Depletion effects are intrinsically non-linear, so that one can not apply simple linear regression tools to the data. With a realistic treatment of depletion effects and statistical analysis of the highly scattered data, Lyubetskaya and Korenaga (2006) have obtained a model for the bulk silicate Earth (BSE) that is more depleted than previous ones.

Table 8.7 lists different estimates of uranium, thorium and potassium abundances in the bulk Earth. Uncertainties are large for the individual concentrations but small for concentration ratios. The uncertainty on each estimate ($\approx 15\%$) is consistent with the range spanned by the other estimates. Unfortunately, one cannot separate uncertainties due to starting chemical data from those of the calculation algorithm,

Table 8.7. *Radioelement concentration and heat production in meteorites, in the bulk silicate Earth, in the Earth's mantle and crust*

	U (ppm)	Th (ppm)	K (ppm)	A* (pW kg^{-1})
CI chondrites				
Palme and O'Neill (2003)	0.0080	0.030	544	3.5
McDonough and Sun (1995)	0.0070	0.029	550	3.4
Bulk silicate Earth				
From CI chondrites				
Javoy (1999)	0.020	0.069	270	4.6
From EH chondrites				
Javoy (1999)	0.014	0.042	385	3.7
From chondrites and lherzolites trends				
Hart and Zindler (1986)	0.021	0.079	264	4.9
From elemental ratios				
and refractory lithophile elements abundances				
McDonough and Sun (1995)	0.020 ± 20%	0.079 ± 15%	240 ± 20%	4.8 ± 0.8
Palme and O'Neill (2003)	0.022 ± 15%	0.083 ± 15%	261 ± 15%	5.1 ± 0.8
Lyubetskaya and Korenaga (2006)	0.017 ± 0.003	0.063 ± 0.011	190 ± 40	3.9 ± 0.7
Depleted MORB source				
Workman and Hart (2005)	0.0032	0.0079	25	0.59
Average MORB mantle source				
Su (2000); Langmuir *et al.* (2005)	0.013	0.040	160	2.8
Continental crust				
Rudnick and Gao (2003)	1.3	5.6	1.5×10^4	330
Jaupart and Mareschal (2003)	/	/	/	293–352

because each author uses his own data and method. With these values of U, Th and K concentrations, we have derived estimates of the radiogenic heat production rate in the Earth. We have used the revised decay constants listed in Table E.1 (Rybach, 1988), which differ slightly from the earlier values given by Birch (1965). Heat production values vary within a rather large range, from 3.7 to 5.1 pW kg^{-1}. The highest value corresponds to the BSE model of Palme and O'Neill (2003) and the lowest one to the EH chondrite model of Javoy (1999). The latter model implies important differences at early ages because its Th/U and K/U ratios differ strongly from the others.

From the different models for the bulk silicate Earth, which includes continental crust, we find a total rate of heat production of 20 TW, with an uncertainty of 15%. After removing the contributions of the continental crust (6–7 TW) and the lithospheric mantle (≈ 1 TW), heat production in the mantle amounts to a total of 13 TW, with an uncertainty of 20%.

The uncertainty on the total heat production of the mantle is large. So far, it has not been possible to measure directly the heat production of the mantle. This situation has changed with the development of underground observatories to detect neutrinos (Fiorentini *et al.*, 2005). A fraction of the neutrino flux comes from geoneutrinos, more precisely antineutrinos, that are produced by radioactive decays in the Earth. The decay of U, Th and K in the Earth produces antineutrinos that have a distinctive energy spectrum. The sensitivity of detectors so far is not sufficient for detecting K decays, but sufficient for U and Th. The number of geoneutrino events is small and include some neutrinos from nuclear power plants. An estimate of the total radioactive heat production of the Earth was made from geoneutrino observations at the Kamland observatory in Japan (Enomoto *et al.*, 2007). So far, the error bars are too large for these results to be of practical value, but the limits bracket the heat production of bulk silicate Earth. Another difficulty is that the local crust contributes up to 80% of the total flux, and must be accounted for (Perry *et al.*, 2008). With the increasing number of observations of geoneutrinos at different observatories, and good control on the flux of geoneutrinos from the local crust, it will become possible to determine the abundance of radioactive elements in the mantle.

Independent constraints can be derived from the compositions of mid-ocean ridge basalts (MORB), which have been sampled comprehensively. MORBs span a rather large compositional range and common practice has been to identify different chemical reservoirs from various end-members. In this framework, depleted basalts come from a depleted reservoir whose complement is enriched continental crust. Enriched basalts can be attributed to primitive mantle tapped by deep mantle plumes or to secondary enrichment processes, for example infiltrations of low-degree melts and metasomatic fluids in subduction zones (Donnelly *et al.*, 2004). The heat production rate of the depleted MORB mantle source is very small, ≈ 0.6 pW kg^{-1}: at this rate, the entire mantle would generate only 2.4 TW. This source, however, does not provide an exact complement of average continental crust (Workman and

Table 8.8. *Various estimates of the global budget*

	Stacey and Davis (2008)	Davies (1999)	Jaupart *et al.* (2007)
Total heat loss	42	41	46
Continental heat production	8	5	7
Upper mantle		1.3	
Lower mantle		11–27	
Mantle heat production	19	12–28‡	13
Latent heat–core differentiation	1.2	<1	
Mantle differentiation	0.6	0.3	0.3
Gravitational (thermal contraction)	2.1		
Tidal dissipation		0.1	
			0.1
Core heat loss	3	5	10
Mantle cooling	10	9†	16§

† Mantle cooling is fixed.
‡ Lower mantle heat production is variable and calculated to fit the mantle cooling rate.
§ Mantle cooling is adjusted to fit the other terms in the budget.

Hart, 2005). An alternative approach avoids the separation of hypothetical mantle reservoirs and determines the average composition of all the mantle that gets tapped by mid-ocean ridges (Su, 2000; Langmuir *et al.*, 2005). Composition of the average MORB mantle source is then derived from a well-constrained melting model. This reservoir is a mixture of different components and is the average mantle lying below oceanic ridges. It is depleted in compatible elements and represents a complement of continental crust. Thus, it may be interpreted as the mantle reservoir that has been processed to form continents (Langmuir *et al.*, 2005). There may be a volume of primitive undepleted mantle lying at depth that has never been sampled by mid-oceanic ridges. Therefore, one obtains a lower bound on the total amount of radioelements in the mantle by assuming that the average MORB source extends through the whole mantle. From Table 8.7, this leads to a lower bound of 11 TW for the total mantle heat production. Adding radioelements from the continental crust and lithospheric mantle, which contribute 7–8 TW (Table 8.8), we obtain a lower bound of 18 TW for the total rate of heat production in the Earth. This is consistent with the BSE models and their uncertainties.

8.6 Heat flux from the core

The outer core is made of molten iron and hence has very low viscosity, contrary to the deep mantle which is much more viscous. Thus, the heat flux out of the core is controlled by the efficacy of mantle convection and cannot be considered as an independent input. Nevertheless, the thermal evolution of the core controls the energy available to drive the geodynamo and one may thus deduce constraints on the heat flux at the core–mantle boundary (CMB) using thermodynamics. An energy balance can be written for the core, in much the same way as it is done above for the mantle. The main differences come from electromagnetic processes and chemical buoyancy due to inner core crystallization. Low viscosity maintains the convective core very close to the reference (radially symmetric) state.

The energy balance of the core equates the heat flux at the CMB to the sum of secular cooling, Q_C, latent heat from inner core crystallization, Q_L, compositional energy due to chemical separation of the inner core (often called gravitational energy, but see Braginsky and Roberts (1995)), E_ξ, and, possibly, radiogenic heat generation, Q_H. Secular cooling makes the inner core grow, which releases latent heat and compositional energy and the first three energy sources in the balance can be related to the size of the inner core and its growth rate (Braginsky and Roberts, 1995). The growth rate of the inner core is too small (about 300 km Gy^{-1}) to be determined by observation. Thus, one has to resort to indirect means. Energy requirements for the geodynamo do not appear directly in the bulk energy balance for the core because they are accounted for by internal energy transfers, just like viscous dissipation in the mantle. The entropy balance, however, depends explicitly on dissipation (Φ_c), which is achieved mostly in the form of ohmic dissipation. Combining the energy and the entropy balances, an efficiency equation can be written, which is to leading order (Braginsky and Roberts, 1995; Labrosse, 2003; Lister, 2003):

$$\begin{aligned}\Phi_c + T_\Phi \Delta S_{\text{cond}} = {}& \frac{T_\Phi}{T_{CMB}}\left(1 - \frac{T_{CMB}}{T_{ICB}}\right) Q_L + \frac{T_\Phi}{T_{CMB}}\left(1 - \frac{T_{CMB}}{T_C}\right) Q_C \\ &+ \frac{T_\Phi}{T_{CMB}}\left(1 - \frac{T_{CMB}}{T_H}\right) Q_H + \frac{T_\Phi}{T_{CMB}} E_\xi,\end{aligned} \tag{8.32}$$

where T_i is the temperature at which heat due to process (i) is released and where

$$\Delta S_{\text{cond}} \equiv \int \lambda \left(\frac{\nabla T}{T}\right)^2 dV \tag{8.33}$$

is the entropy production due to heat conduction. The efficiency equation (8.32) shows that heat is less efficiently transformed into ohmic dissipation than compositional energy.

All the source terms on the right-hand side of the efficiency equation (8.32), save for radiogenic heating, are linked to inner core growth and are proportional to its growth rate. If ohmic dissipation Φ_c and radiogenic heat production Q_H can be estimated, therefore, one can calculate the inner core growth rate and the heat flux across the CMB. Ohmic dissipation in the core is dominated by small-scale components of the magnetic field, which cannot be determined directly because they are screened by crustal magnetic sources (Hulot *et al.*, 2002). High-resolution numerical models of the geodynamo suggest that $\Phi_c = 0.2-0.5$ TW (Christensen *et al.*, 2004). These models rely on the Boussinesq approximation, and hence neglect the isentropic temperature gradient in both the heat and entropy balances. As a rough correction, one may add an estimate of the associated dissipation $\Delta S_{\text{cond}} \sim 1$ TW, which would bring the total dissipation estimate to $\Phi_c = 1$–2 TW, close to that of Roberts *et al.* (2003). According to Labrosse (2003), this implies a heat flux of 4–9 TW at the CMB.

One must add to this estimate heat that is conducted along the isentropic gradient in the well-mixed liquid core. Estimates for the thermal conductivity of the core are in the range 40–60W m^{-1} K^{-1}, and the isentropic gradient in the core is ≈ 0.8K km^{-1}, resulting in a heat flux of 32–48 mW m^{-2}, i.e. $\approx 3-5$ TW (Buffett, 2007). The total heat flux across the CMB would thus be in the range 7–14 TW.

Studies of hot spot swells have led to estimates of $\approx$2–4 TW for the heat flux carried by mantle plumes (Sleep, 1990). The relevance of this heat flux to the cooling of the core is difficult to assess, in part because some of the plumes may not come from the CMB and in part because plumes account for only part of the core heat loss. As explained in Chapter 5, the convective heat flux includes a component due to cold downwellings. In the Earth, convection is mostly driven by internal heating and secular cooling and this breaks the symmetry between up- and downwelling currents, at the expense of the former. In such a situation, heat transport at the CMB is dominated by the spreading of cold material from subducted slabs (Labrosse, 2002; Zhong, 2005; Mittelstaedt and Tackley, 2006). The contribution of hot spots only provides a lower bound to the total core heat loss.

8.7 Mantle energy budget

We have obtained estimates for all the terms in the energy budget and hence can estimate the amount of energy due to mantle cooling directly (Table 8.8). Comparing this procedure to those adopted by Stacey and Davis (2008) and Davies (1999) sheds light on important aspects. For Stacey and Davis (2008), the total heat production (27 TW) is significantly higher than the value of 20 TW for BSE. It seems that Stacey and Davis have added the crustal heat production to BSE. The core heat loss is low because it is assumed identical to the heat carried by hot spots. Stacey and Davis further assumed that all the gravitational energy released by thermal contraction (2.1 TW in his estimate) goes to heat. In Davies (1999), the secular cooling of the mantle is derived from petrological data and the lower mantle heat production is the variable that is adjusted to balance the budget when all the other variables are fixed. Core cooling is also assumed to be identical to the total heat flux from hot spots. Individual lines of budget shown in Table 8.8 present the average value for each term. The uncertainty on several of the components of this budget are much larger than the uncertainty on the total heat loss of the Earth.

Exercises

8.1 Calculate and compare the total heat loss through the sea floor for a linearly decreasing (triangular) and a uniform age distribution of sea floor, assuming the same spreading rate, for a cooling half-space.

8.2 Calculate how the total heat loss through the sea floor for a linearly decreasing age distribution varies with the spreading rate, assuming the half-space heat flux age relationship $Q = C_Q t^{-1/2}$, with $C_Q = 490$ mW m^{-2} My$^{1/2}$, and the total area of the sea floor is 300×10^6 km^2.

9

Mantle convection

Objectives of this chapter

Mantle convection involves many processes and physical effects that are seldom found in other convective systems. We evaluate the most important ones and discuss the impact of each one on the heat loss properties and thermal structure of the mantle. We develop scalings for the heat flux and typical velocity and demonstrate that they can all be reduced to simple statements on the dynamics of a thermal boundary layer. We aim at an understanding of each process and its control variables, and do not attempt to build an all-encompassing physical model of mantle convection.

9.1 Introduction

Compared to Rayleigh–Benard convection that has been studied in Chapter 5, convection in the Earth's mantle involves a series of processes that all act to enhance the impact of the upper thermal boundary layer on heat transport. The surface heat flux evacuates heat released by radioactive decay within the mantle as well as sensible heat due to secular cooling. The presence of continents over part of the Earth's surface restricts the efficiency of heat loss to the atmosphere and hydrosphere, and enhances heat flux through oceanic areas. Separation of the oceanic and continental domains with different heat transport characteristics at the Earth's surface generates large-scale horizontal temperature variations in the shallow mantle. Mantle rheology is highly sensitive to temperature, implying that cold material in the upper thermal boundary layer deforms less readily than the interior. All these effects act to increase the temperature difference across the upper thermal boundary layer with respect to that of the boundary layer just above the core. The upper thermal boundary layer of the Earth's mantle not only sets the rate of heat loss of our planet, but is also the most important part of the mantle convection system because it is associated with the largest buoyancy forces.

In our analysis of convection, we rely on the tools that have been developed in Chapter 5. We use the dissipation equations and a few assumptions on the flow structure derived from both laboratory experiments and numerical calculations. The Prandtl number of Earth's mantle is extremely large ($\approx 10^{24}$!), so that we consider convection in the infinite Pr limit with no Pr dependence to worry about.

9.2 Elongated convection cells

Oceanic plates can be identified as the upper parts of convective cells fed by upwellings beneath ocean ridges and feeding downwellings in subduction zones. Here, we investigate the influence of their horizontal dimensions on the heat flux. For Rayleigh–Benard convection in a fluid layer of large horizontal dimensions, the stable convection regime takes the form of cells that are about as wide as the layer thickness h as long as the Rayleigh number does not exceed a threshold value of about 10^5. In contrast, the convective cells of the Earth's mantle extend over horizontal distances that are much larger than the thickness: the Pacific plate, for example, stretches over about 10,000 km. Elongated cells of this kind can be generated in different ways due to geometrical constraints on convection in a finite volume, and have been observed in both laboratory and numerical simulations. The basic principle is that the fluid layer must contain a finite number of cells because of its finite size, but can rarely fit cells of the same width. Cells of different widths are therefore generated, including some that are more elongated than others. Such cells may be transient and involved in merge-and-split cycles. Dietsche and Muller (1985) have shown that it takes an aspect ratio of at least seven to achieve convective cells that are unaffected by the side-walls in the central region of an experimental tank. In the Earth, there are no side-walls and several effects effectively delineate oceanic domain boundaries, including the thick and stable roots of old continents.

We consider Rayleigh–Benard boundary conditions, with temperatures maintained at T_o and $T_o + \Delta T$ at the top and bottom. The convective system is therefore characterized by two dimensionless numbers: the Rayleigh number and the dimensionless width of the cell L/h. We further consider that the upper and lower horizontal surfaces are free to deform and move in the horizontal direction with velocities equal to U and $-U$ at the top and bottom, respectively. By symmetry, velocity is zero in the middle of the cell. For a convection cell extending over width L, the main effect is that the average heat flux across the cell decreases as L increases. The cell is bounded by one upwelling and one downwelling, and fluid motions are essentially horizontal in the central region (Figure 9.1). Vertical motion is most important at the sides and is developed over a width that scales with depth h. Kinetic dissipation may be split into two different components. One component is generated by the horizontal shear flow with a vertical gradient of horizontal velocity

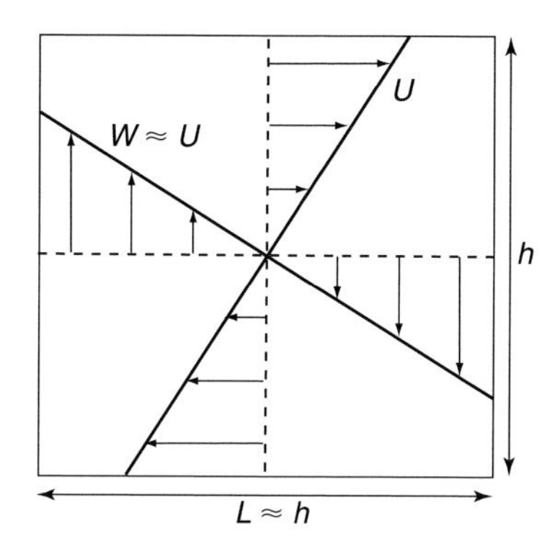

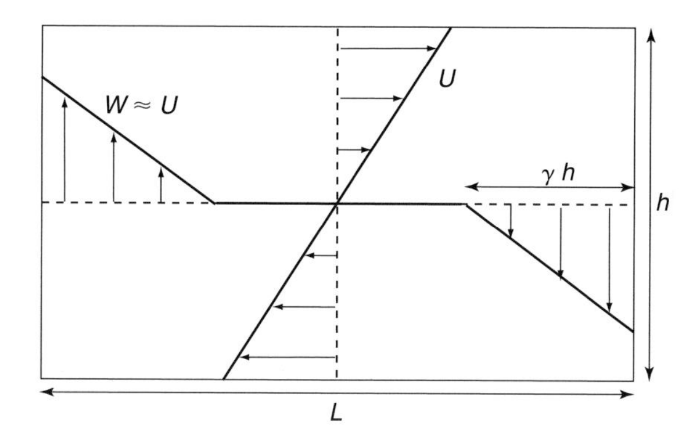

Figure 9.1. Schematic illustration showing the flow field in a convection cell. Top: convection cell of aspect ratio ≈ 1, which is appropriate for Rayleigh–Benard convection. With free boundaries at large Prandtl number, a velocity gradient extends over the whole layer. Bottom: convection cell of aspect ratio larger than 1, showing the difference between the horizontal and vertical velocity fields. The horizontal velocity field stretches over the whole layer thickness, whereas the vertical velocity field extends over kinetic boundary layers of width γh.

$\sim 2U/h$. The other component is associated with the vertical flow, which is such that velocity drops from the maximum value W over horizontal distance $\delta = \gamma h$, where γ is some proportionality constant. In Rayleigh–Benard convection, convective cells have an aspect ratio of 1, such that $\gamma \approx 0.5$. The structure of the upwellings and downwellings does not depend on the width of the cell if it is large enough, and we therefore expect that γ takes the same value of ≈ 0.5 in elongated cells. This is confirmed by numerical simulations (Figure 9.2). For this flow configuration the kinetic dissipation equation (5.90) in Chapter 5 can be written as follows:

$$\mu\left[\left(\frac{2U}{h}\right)^2 hL + 2\left(\frac{W}{\gamma h}\right)^2 \gamma hL\right] = \frac{\alpha g Q}{C_p} hL, \tag{9.1}$$

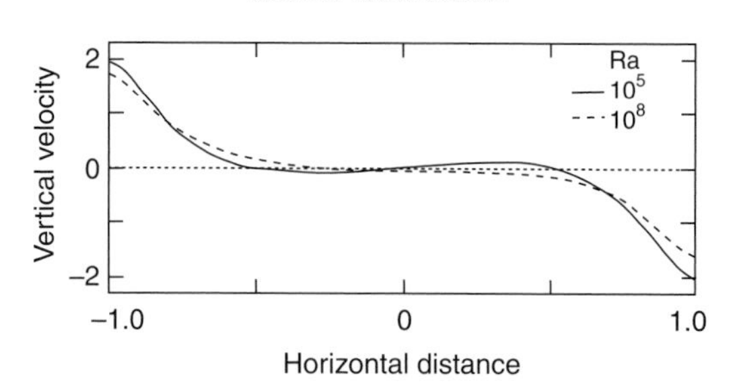

Figure 9.2. Horizontal profile of vertical velocity at $z = h/2$ in a large aspect-ratio convection cell, at two different values of the Rayleigh number, from Grigné *et al.* (2005). Note that the momentum boundary layer extends to a horizontal distance of 0.5 for all values of Rayleigh number.

where we allow for two dissipation components due to the horizontal and vertical velocity fields. As illustrated by Figure 9.1, such a distinction is not warranted for a cell of aspect ratio ≈ 1.

In the convection cell, the core region is at temperature $T_i = T_o + \Delta T/2$, by symmetry. The upper thermal boundary layer thickness scales with δ such that $Q = \lambda \Delta T / 2\delta$. Using the dominant balance between vertical diffusion and horizontal advection that has been discussed in Chapter 5, we obtain,

$$\delta = \sqrt{\frac{\kappa L}{U}}. \tag{9.2}$$

Terms U and W are related to one another through mass conservation: $Uh/2 \sim W\gamma h$. Substituting for W and U into equation 9.1 and rearranging, we obtain,

$$\frac{\delta}{h} \sim \left[\frac{L}{h}\left(\frac{L}{h} + \frac{1}{8\gamma^3}\right)\right]^{1/3} \mathrm{Ra}^{-1/3}, \tag{9.3}$$

where we can see that, by setting $L \approx h$, we recover the scaling for normal convection cells, $\delta \sim h\mathrm{Ra}^{-1/3}$. This can be recast in the form of an equation for the Nusselt number,

$$\mathrm{Nu} \sim \frac{Q}{\lambda \Delta T / h} \sim \frac{h}{\delta} \sim \frac{1}{\left[\left(\frac{L}{h}\right)^2 + \frac{L}{8\gamma^3 h}\right]^{1/3}} \mathrm{Ra}^{1/3}. \tag{9.4}$$

An explicit solution to the heat balance equation in the upper boundary layer leads to (Grigné *et al.*, 2005),

$$\mathrm{Nu} = \left(\frac{1}{2\pi}\right)^{2/3} \frac{1}{\left[\left(\frac{L}{h}\right)^2 + \frac{L}{8h\gamma^3}\right]^{1/3}} \mathrm{Ra}^{1/3}$$

$$= \left(\frac{1}{2\pi}\right)^{2/3} \frac{1}{\left[\left(\frac{L}{h}\right)^2 + \frac{L}{h}\right]^{1/3}} \mathrm{Ra}^{1/3} \text{ for } \gamma = 1/2. \tag{9.5}$$

Figure 9.3 shows that this approximate solution is in good agreement with numerical calculations if one uses $\gamma = 1/2$, as expected. Equation 9.2 can be used to derive an expression for the flow velocity U.

These results show that the Nusselt number and the heat flux decrease with the width of the convection cell. The control factors that set the size of convection cells on Earth have not yet been determined and there is no reason to suppose that their size remains constant. Without such an understanding, one may not evaluate how the Earth has evolved through geological time. As discussed later, mantle convection cells change size as new oceans appear and old ones disappear. The Earth today has cells of aspect ratio ≈ 3, such that its rate of heat loss is about half that for cells of aspect ratio 1. The significance of this is best appreciated by comparing it to the current imbalance between the Earth's rates of heat loss and

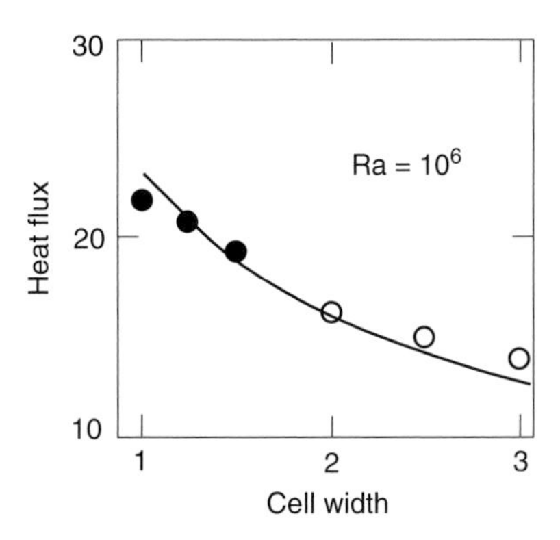

Figure 9.3. Dimensionless heat flux Nu as a function of the width of a convective cell in Rayleigh–Benard convection with free boundaries, from the numerical calculations of Grigné *et al.* (2005). Solid symbols correspond to stable cells. Open symbols stand for transient cells that are involved in slow merge-and-split cycles with neighbouring cells. The solid line is from equation 9.5.

internal heat production, which amounts to a factor of about 2. The Earth may well regulate its energy balance by adjusting the size of its convection cells.

9.3 The impact of continents on convection

We know by observation that continents are regions of low heat flux at the Earth's surface and we also know that this is due to their thick and rigid lithospheric roots. In contrast, oceanic plates lose large amounts of heat. Thus, the upper boundary condition for mantle convection does not belong to the usual categories of fixed temperature or fixed heat flux. The presence of continents has two consequences for mantle convection. One is that the lateral change of heat flux that occurs between continents and oceans induces a horizontal temperature variation that affects the flow characteristics. In other words, the ocean–continent dichotomy introduces a new length scale into the problem. The other consequence is that the cooling of the Earth is less efficient than in a continent-free planet and depends on the area occupied by continents, which has probably changed with geological time. Thus, one must allow for continental growth explicitly in an evolutionary model for the Earth. From a mechanical point of view, continents perturb the convective flow pattern as their lithospheric roots divert mantle flow.

Continental heat flux and a new boundary condition

Here, we shall focus on purely thermal effects to evaluate the dependence of surface heat flux on the size of continents. This allows simple derivations that illustrate clearly the factors at play and their incidence on flow characteristics. We begin by discussing the boundary condition imposed by a rigid lid of finite thickness, where, by construction, heat is transported by conduction. In a thin lid of large aspect ratio, such that its width is much larger than its thickness, vertical heat transport dominates and hence heat flux through the lid is

$$q = +\lambda_L \frac{T - T_o}{d}, \tag{9.6}$$

where λ_L is the thermal conductivity of the lid, T_o is the temperature at the top of the lid, which can be assumed to be constant with little error, and T is the temperature at the base of the lid. The basal temperature is allowed to vary laterally because of the interaction between the moving fluid and the lid material. If we neglect the lid thickness in comparison with the fluid layer depth, we may recast the above equation in the form of a boundary condition involving the fluid temperature and the heat flux at the top of the fluid, i.e. at $z = 0$,

$$\left(\lambda \frac{\partial T}{\partial z}\right)_{z=0} = \lambda_L \frac{T - T_o}{d}, \tag{9.7}$$

where λ is the thermal conductivity of the convecting fluid below the lid. This is best written in dimensionless form by scaling temperature with ΔT, the temperature difference across the fluid layer, and depth by the layer thickness h. In such a form, the boundary condition can now be recast as follows:

$$\left(\frac{\partial T}{\partial z}\right)_{z=0} - \mathrm{B}T = 0, \tag{9.8}$$

where variables are dimensionless and T_0 has been set to 0. B is a parameter called the Biot number:

$$\mathrm{B} = \frac{\lambda_L}{\lambda}\frac{h}{d}, \tag{9.9}$$

which may be interpreted as a thermal impedance. An insulating lid may be achieved with either a large thickness or a small thermal conductivity: for such conditions, B is small, so that $\mathrm{B}T \ll 1$ and the boundary condition reduces to one of zero heat flux. For a highly conducting lid, corresponding to large values of λ_L or small values of d, the boundary condition is completely different and is one of zero temperature. For a lid with an aspect ratio that is not much larger than 1, we may not neglect horizontal heat transport within the lid and one must solve for the temperature fields in both the convecting fluid and the lid simultaneously.

In the Earth, the conductivity ratio is about unity, $d \approx 300$ km and $h \approx 3000$ km for whole-mantle convection, so that $\mathrm{B} \approx 10$. As we shall see, this may be considered small because it leads to heat fluxes that are a small fraction of those of free convective cells. Indeed, the heat flux through the base of a cratonic lithosphere may be as small as 12 mWm^{-2} (see Chapter 7), which must be compared to the mean oceanic heat flux of 100 mWm^{-2}.

Main features of sub-continental convection

We again consider Rayleigh–Benard boundary conditions (free boundaries and fixed temperatures) and do not allow for internal heat generation. Two dimensionless numbers characterize the flow, the Rayleigh number and the dimensionless lid width $2a/h$. In all cases, the presence of a poorly conducting lid at the upper boundary of a fluid has three important consequences. One is that the fluid beneath the lid gets heated and becomes involved in an upwelling centered on the continent (Figure 9.4). Because of the difference in the efficiency of heat transport at the upper and lower boundaries, the average temperature of the fluid layer is larger than $T_o + \Delta T/2$ (Figure 9.5). A third consequence is that the sub-continental upwelling feeds a large-scale horizontal flow that sustains an elongated convective cell structure. The heat flux is less than that through a layer with no lid despite the higher average temperature.

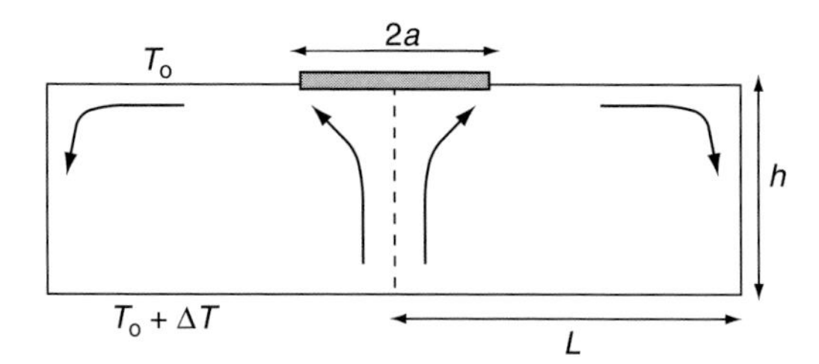

Figure 9.4. Illustration introducing geometrical parameters for convection beneath a conducting lid. The base of the lid must be at a higher temperature than the top of the fluid away from the lid, which is fixed at some value T_o. Thus, the lid generates an upwelling beneath itself which feeds an elongated cell of width L.

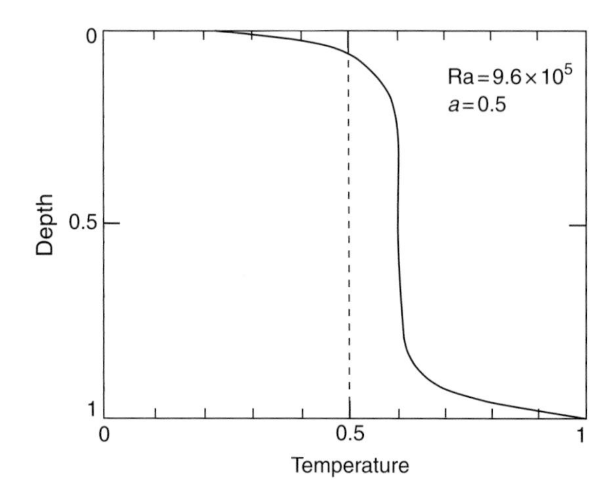

Figure 9.5. Vertical profile of temperature beneath a conductive lid in the laboratory experiments of Guillou and Jaupart (1995). Ra and a stand for the Rayleigh number and the dimensionless lid width. Temperature is not zero at the top surface, as in the adjacent oceanic-type domains, and is larger than 1/2 in the interior.

For simplicity, we shall assume that the continental lid is completely insulating, so that the heat flux out of the convecting mantle is set to zero in continental regions. Calculations and laboratory experiments show that continents generate elongated cells with an upwelling beneath the continent that feeds horizontal flows extending to large distances. The aspect ratio of these continental cells depends on the width of the lid and slightly on the Rayleigh number, and may be as large as 3. For Rayleigh numbers that are larger than about 10^6, the flow can no longer be described by a simple cell configuration and involves small-scale instabilities that are superimposed on a larger-scale circulation. This is indeed documented in the oceans, as discussed in Chapter 6. A large-scale cell-like structure is still present in the flow and the convective pattern can be characterized by a set of small plumes carried and distorted by a horizontal shear flow (Guillou and Jaupart, 1995).

We restrict the discussion to Rayleigh numbers smaller than 10^6, for which the cellular flow structure is marked and boundary layer instabilities are either absent or subdued. A simple argument allows an estimate of the average temperature in continental convective cells, denoted T_m. We have already explained that in such conditions T_m is not the average of the two boundary temperatures, as in a normal Rayleigh–Benard configuration. We use the fact that the flow is dominantly horizontal away from the vertical edges of the cell and is characterized by velocity U. The velocity is the same at both horizontal boundaries because of mass conservation. For the bottom boundary, we may solve the heat equation,

$$U\frac{\partial T}{\partial x} = \kappa \frac{\partial^2 T}{\partial z^2}, \tag{9.10}$$

where we have again neglected horizontal heat conduction which is effected over the large width L of the cell. This equation can be recast as a 1D diffusive heat transport problem if one uses time t instead of horizontal distance x: $t = x/U$. With the boundary conditions that $T = T_m$ at $x = 0$ and $T = T_o + \Delta T$ at $z = L$, we recover the classical solution of Chapter 4 involving the error function such that,

$$T(x,z) = T_m + (T_o + \Delta T - T_m)\text{erfc}\left(\frac{z}{2\sqrt{\kappa x/U}}\right). \tag{9.11}$$

The heat flux is therefore

$$q_b(x) = \lambda \frac{T_o + \Delta T - T_m}{\sqrt{\pi \kappa x/U}}. \tag{9.12}$$

The total heat supplied to the layer through the base is readily obtained by integrating this heat flux over width L of the cell,

$$Q_b = \frac{2\lambda}{\sqrt{\pi\kappa}}(T_o + \Delta T - T_m)\sqrt{UL}. \tag{9.13}$$

The same analysis can be applied to the horizontal flow near the top boundary with one key difference: cooling to the boundary temperature T_o is imposed over a smaller distance $\delta x = L - a$ due to the insulating lid that extends from $x = 0$ to $x = a$. Thus, the heat flux through the top is

$$Q_t = \frac{2\lambda}{\sqrt{\pi\kappa}}(T_m - T_o)\sqrt{U(L-a)}. \tag{9.14}$$

In steady state, the heat fluxes through the top and bottom boundaries must be equal and hence,

$$(T_m - T_o)\sqrt{U(L-a)} = (T_o + \Delta T - T_m)\sqrt{UL}, \tag{9.15}$$

which leads to,

$$\frac{T_m - T_o}{\Delta T} = \frac{1}{1 + \sqrt{1 - a/L}}. \tag{9.16}$$

This equation predicts that $(T_m - T_o) = \Delta T2$, if $a = 0$, we therefore recover the normal Rayleigh–Benard result if there is no insulating lid at the top. In the other limit, such that $a = L$, this equation correctly predicts that $T_m - T_o = \Delta T$, but this is not relevant to geological conditions. In reality, continents do not extend over the whole surface of the Earth and the cell width L increases with the width of the continent. Even if continents did cover the whole surface of the planet, this equation would not apply because it relies on approximation of a zero continental heat flux, which is only meaningful if there are oceanic plates.

The predictions of equation (9.16) are quite close to numerical calculations (Figure 9.6). They illustrate the important principle that continental insulation acts to increase the average mantle temperature. The numerical results have been obtained for a cell size L that has been set independently of the other parameters, corresponding to the situation of the previous section where some lateral boundary prevents free development of the cell. In reality, cell size depends on the width of the lid (Guillou and Jaupart, 1995; Grigné *et al.*, 2007).

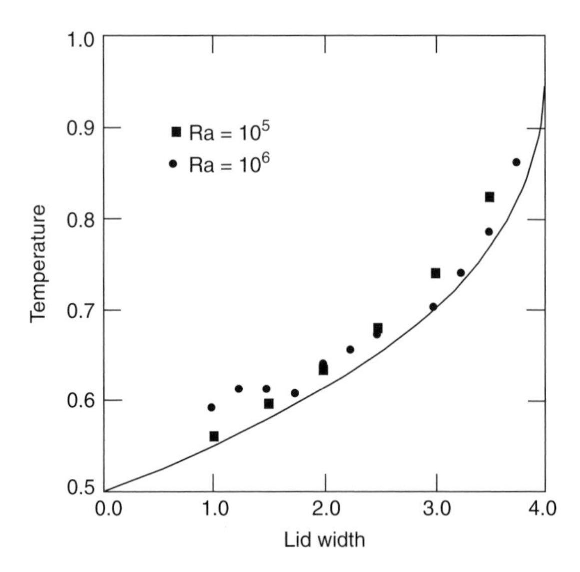

Figure 9.6. Mean temperature of a convective fluid layer with an insulating lid at the top, as a function of the width of the lid (scaled to the width of the convective cell), for two values of the Rayleigh number, from the numerical calculations of Grigné *et al.* (2007). In all cases, the elongated continental cell extends to $L = 4 \times h$. The solid line is the prediction of equation 9.16.

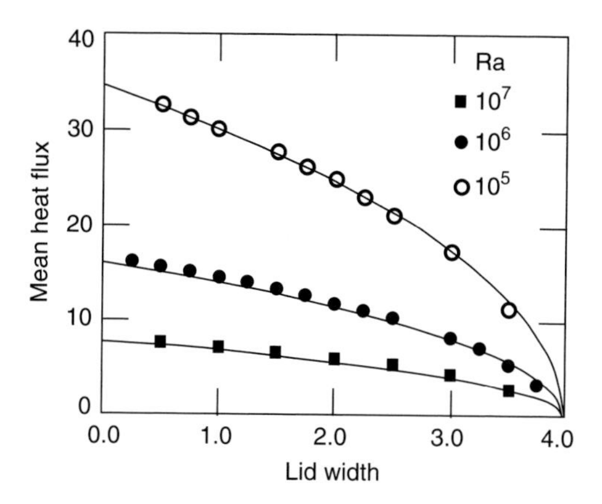

Figure 9.7. Mean dimensionless heat flux in a convecting fluid with an insulating lid at the top boundary, from Grigné *et al.* (2007). The convective cell is in all cases such that $L = 4 \times h$. Solid lines have been obtained from empirical equation 9.17 for the relevant values of Rayleigh number.

For a full solution, one needs to specify velocity. When an insulating lid is present at the upper boundary, an upwelling develops beneath the lid over a lateral distance that no longer scales with the layer depth h. In this case, simple scaling arguments do not apply, due to the multiplicity of length scales. An empirical relationship fits numerical results over wide ranges of continental sizes and Rayleigh numbers (Figure 9.7):

$$\mathrm{Nu} = \frac{1}{(2\pi)^{2/3}} \mathrm{Ra}^{1/3} \sqrt{1 - a/L} \left(\frac{L^2}{h^2} + \frac{L}{h} \right)^{1/3}, \tag{9.17}$$

which is simply the same expression as that for an elongated Rayleigh–Benard cell (equation 9.5) multiplied by a scale factor involving the lid width, $\sqrt{1 - a/L}$.

9.4 Convection with internal heat sources

Convection in the Earth's mantle is driven in large part by internal heat generation due to the decay of uranium, thorium and potassium. In this case, as shown in Chapter 5, there are two dimensionless numbers: the Rayleigh number and the dimensionless heat generation.

9.4.1 Pure internal heating: no heat supplied from below

Consider first a fluid layer that is not heated from below and within which radioactive sources generate heat at rate H per unit volume. In this case, one cannot define a Rayleigh number on the basis of an externally imposed temperature difference. In steady state, in contrast to the Rayleigh–Benard cases studied in Chapter 5, the heat flux is known because it is equal to the amount of heat generated in the layer and it is the temperature difference across the layer that must be solved for.

As shown in Chapter 5, one may introduce a temperature scale for internal heat generation:

$$\Delta T_H = \frac{Hh^2}{\lambda}, \tag{9.18}$$

and the relevant Rayleigh number is

$$\mathrm{Ra}_H = \frac{\rho_o g \alpha H h^5}{\lambda \kappa \mu}. \tag{9.19}$$

In steady state, the thermal structure of the convecting layer can be split into an upper boundary layer and a convective interior: there is no basal thermal boundary layer because no heat is supplied to the fluid layer from below (Figure 9.8). An important feature is that the fluid interior is not well mixed, with a negative temperature gradient whose magnitude decreases with increasing Ra_H.

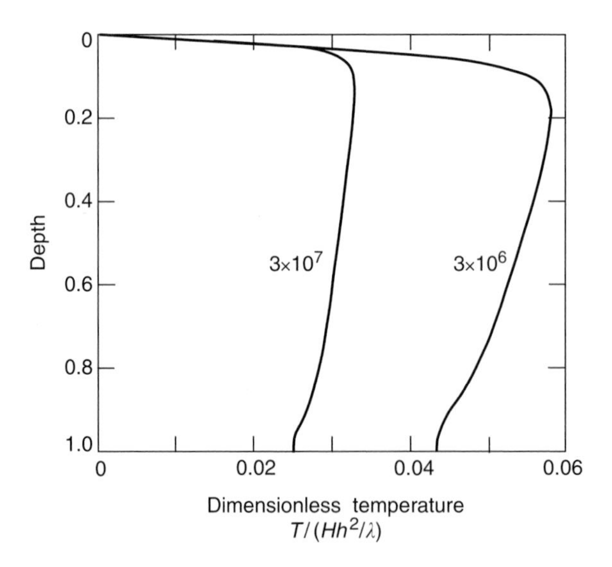

Figure 9.8. Vertical profile of the horizontally averaged temperature for internally heated convection, from the 3D numerical calculations of (Parmentier *et al.*, 1994). The Rayleigh numbers (3×10^6 and 3×10^7) are defined in equation 9.19. Note the absence of a lower thermal boundary layer and the small vertical temperature gradient that exists in the fluid interior despite vigorous convection.

The heat flux through the top of the layer is known because it is equal to the rate of heat generation in the layer:

$$Q = Hh, \tag{9.20}$$

and the unknown is the temperature contrast across the upper thermal boundary layer, denoted ΔT_i. Dimensional analysis along the same lines as those in Chapter 5 shows that the dimensionless temperature contrast, $\Delta T_i/\Delta T_H$, is a function of two dimensionless numbers Ra_H and Pr. For the Earth's mantle in the infinite Pr limit, one seeks to determine the following relationship:

$$\frac{\Delta T_i}{\frac{Hh^2}{\lambda}} = f(\mathrm{Ra}_H), \tag{9.21}$$

where $f(\mathrm{Ra}_H)$ is some function of the Rayleigh number.

We proceed as in Chapter 5 and use the kinetic dissipation equation 5.90. The major difference with the Rayleigh–Benard case, i.e. convection due to heating from below, is in the distribution of the convective heat flux, which determines that of the buoyancy flux. To specify how convective heat flux varies in the layer, we use the horizontally averaged heat equation 5.70 in Chapter 5:

$$0 = -\frac{d}{dz}\left(-\lambda\frac{d\overline{T}}{dz} + \rho_o C_p \overline{w\theta}\right) + H, \tag{9.22}$$

where $\overline{T}$ is the horizontal average of temperature, θ the temperature fluctuation and w the vertical velocity. Integrating between $z = 0$ (the base of the fluid layer) and $z = h$, we obtain the convective heat flux at any depth z:

$$+\rho_o C_p \overline{w\theta} = \lambda\frac{d\overline{T}}{dz} + Hz, \tag{9.23}$$

where we have used the fact that temperature gradient and vertical velocity drop to zero at $z = 0$ (no heat is brought into the layer from below). Neglecting the small temperature gradient in the fluid interior, we see that the convective heat flux decreases linearly with increasing depth in the layer, which reflects that fluid parcels get heated by radioactive decay as they go down. The buoyancy flux integral is therefore,

$$\int_0^h \rho_o \alpha g \overline{w\theta} dz = \int_0^h \frac{\alpha g}{C_p}(\rho_o C_p \overline{w\theta}) dz \tag{9.24}$$

$$= \frac{\alpha g}{C_P}\int_0^h \left[\lambda\frac{d\overline{T}}{dz} + Hz\right] dz \tag{9.25}$$

$$= \frac{\alpha g}{C_P}\left(-\lambda \Delta T + H\frac{h^2}{2}\right). \qquad (9.26)$$

For large Rayleigh numbers, one may neglect $\lambda \Delta T$ compared to $Hh^2/2$. With this choice of coordinates, the buoyancy flux is positive, corresponding to an upward flux of heat. To scale the amount of kinetic dissipation, we again rely on a velocity scale U and on the fact that flow develops over distances that scale with the layer depth h. The kinetic dissipation equation is therefore:

$$\mu \left(\frac{U}{h}\right)^2 hS \sim H\frac{h^2}{2}S \qquad (9.27)$$

where S is the area of fluid in the horizontal plane. We also use the same equations as in Chapter 5 for the thickness of the upper thermal boundary layer, denoted δ, and the surface heat flux Q,

$$Q \sim \lambda \frac{\Delta T_i}{\delta}, \; \delta \sim \sqrt{\frac{\kappa h}{U}}, \qquad (9.28)$$

where one should note that ΔT_i is not equal to the temperature difference across the layer, ΔT, due to the negative temperature gradient that exists in the fluid interior (Figure 9.8). To close the problem, we use $Q = Hh$ and obtain,

$$\Delta T_i \sim \frac{Hh^2}{\lambda} \mathrm{Ra}_H^{-1/4}. \qquad (9.29)$$

Experiments on thermal convection in internally heated fluids are difficult to achieve in the laboratory and have been limited to a narrow range of Rayleigh numbers and hence do not allow a test of this simple prediction. Three-dimensional numerical calculations confirm this scaling law at surprisingly small values of the Rayleigh number ($\mathrm{Ra}_H \sim 10^6$, Figure 9.9).

Equation 9.29 can be recast as an equation for the heat flux $Q = Hh$:

$$Q = C_Q \lambda \left(\frac{\rho_o g \alpha}{\kappa \mu}\right)^{1/3} \Delta T_i^{4/3}, \qquad (9.30)$$

where we recover the local boundary layer scaling law for the convective heat flux (equation 5.82). The constant of proportionality that enters this scaling law is given in Table 9.1 and is close to that for Rayleigh–Benard convection.

One might be tempted to write the local heat flux scaling law in dimensionless form involving a Nusselt number Nu. For purely internal heating, however, there is no externally imposed temperature difference and one can only use ΔT_H as

Table 9.1. *Constant for the local heat flux scaling law (equation 5.82) for convection between free boundaries*†

Heating mode	C_Q	Reference
Internally heated	0.302	Parmentier and Sotin, 2000
Mixed §	0.346	Sotin and Labrosse, 1999
Heated from below	0.378	Hansen *et al.*, 1992

† From numerical calculations in the infinite Pr limit.
§ The fluid layer is heated from below and from within.

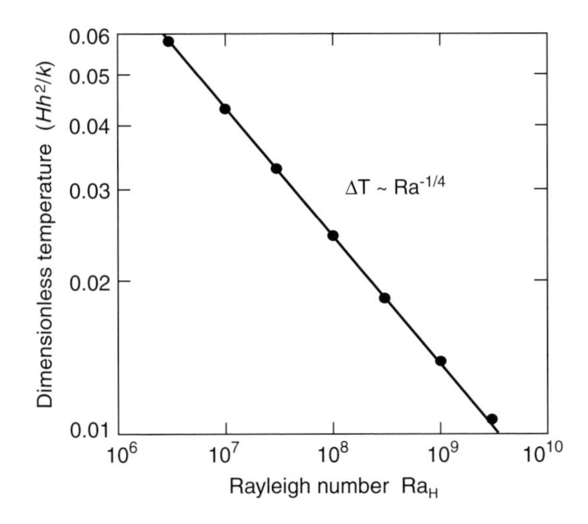

Figure 9.9. Temperature difference across the thermal boundary layer at the top of an internally heated fluid layer, from the 3D numerical calculations of (Parmentier and Sotin, 2000). The Rayleigh number is defined in equation 9.19. The data are consistent with the simple scaling law 9.29 over the whole Rayleigh number range.

temperature scale. In this case, one obtains a trivial and useless result:

$$\mathrm{Nu} = \frac{Q}{\lambda \Delta T_H / h} = 1, \tag{9.31}$$

due to the global heat balance $Q = Hh$.

9.4.2 Layer heated from below and from within

In the Earth's mantle, one must allow for heating by radioactive decay and by the core. In this case, the thermal structure is made of two different thermal boundary layers and a region with negative temperature gradient near the bottom

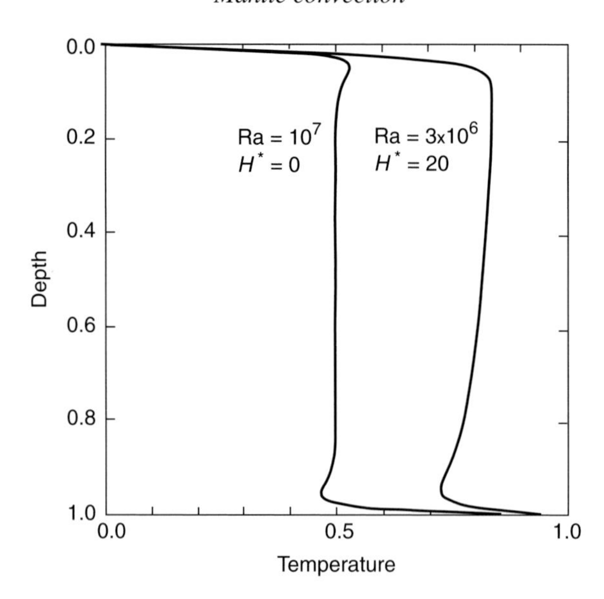

Figure 9.10. Vertical profiles of the horizontally averaged temperature in a convecting layer for two different types of heating: pure heating from below with no internal heat production, and mixed heating with heat supplied from below as well as from within due to radioactive decay, (from Sotin and Labrosse (1999)). Note the differences between the two profiles and the thick region of negative temperature gradient at the base of the layer with internal heating.

(Figure 9.10). One can interpret the latter feature as due to the ponding of cold plumes coming from the upper boundary. No theory has been developed yet for such a complicated structure. We consider a total temperature difference ΔT across the layer and heat production H and seek scaling laws for two variables, surface heat flux Q_T and the temperature difference ΔT_i across the top boundary layer. In dimensionless form, these two variables are the Nusselt number Nu, which has already been defined, and $\Theta = \Delta T_i / \Delta T$. As explained in Chapter 5, two dimensionless numbers describe convection, the Rayleigh number Ra and the dimensionless heat generation $H^* = (Hh^2)/(\lambda \Delta T)$.

We rely again on the local scaling law for heat flux through the upper thermal boundary layer, so that we need only determine Θ. Dimensional analysis dictates that Θ depends on H^* and Ra. We first note that $\Theta = 1/2$ if $H^* = 0$, and further that internal heat generation acts to increase Θ with respect to that value. Thus, $\Theta = 1/2 + f(H^*, \mathrm{Ra})$, where f is a function to be determined. We further know that this function must increase with H^*, as just explained, and must decrease with Ra: for given H^*, increasing Ra enhances the impact of bottom heating compared to that of internal heating, such that $\Theta \to 1/2$ as $\mathrm{Ra} \gg 1$. We assume that the two heating mechanisms contribute independently to the temperature contrast across the top boundary layer. We use the previous result for pure internal heating and write in dimensional form,

$$\Delta T_i = \frac{\Delta T}{2} + C^* \frac{Hh^2}{\lambda} \left(\frac{\rho g \alpha H h^5}{\lambda \kappa \mu} \right)^{-1/4}, \tag{9.32}$$

where C^* is some proportionality constant. We note that, as required, $\Delta T_i = \Delta T/2$ when $H = 0$. This relationship can be recast as a function of H^* and Ra:

$$\Delta T_i = \frac{\Delta T}{2} + C^* \Delta T H^{*3/4} \text{Ra}^{-1/4} \tag{9.33}$$

or, equivalently,

$$\Theta = \frac{1}{2} + C^* H^{*3/4} \text{Ra}^{-1/4}. \tag{9.34}$$

Numerical calculations for a large range of Rayleigh numbers and dimensionless internal heating rates are consistent with this prediction and yield $C^* = 1.24$ (Figure 9.11). The Nusselt number is (Table 9.1 and equation 9.30)

$$\text{Nu} = \frac{Q_T}{\lambda \Delta T/h} = 0.346 \text{Ra}^{1/3} \Theta^{4/3}, \tag{9.35}$$

where we have introduced the heat flux at the upper boundary, Q_T, which is not equal to the heat flux through the lower boundary due to internal heat production. By construction, this expression for the Nusselt number is consistent with the local heat flux scaling law in the upper boundary layer (equation 5.82). These results are only valid for a fluid layer that is heated from below, which requires that $\Delta T_i < \Delta T$ (or $\Theta < 1$). The proportionality constant in this heat flux scaling law is not identical to that for no internal heat generation (Table 9.1), which is probably due to slightly

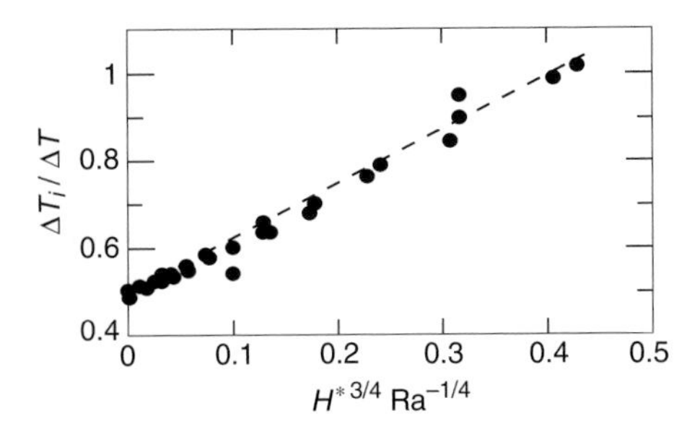

Figure 9.11. Temperature difference across the upper thermal boundary layer for a fluid layer that is both heated from below and heated from within by radioactive decay as a function of $H^{*3/4}\text{Ra}^{-1/4}$, (from Moore, 2008). The data are consistent with the scaling law given by equation 9.34.

different dynamics in the upper boundary layer. We comment on this interesting aspect in the next section.

9.4.3 *Dynamics of the upper thermal boundary layer*

In a convecting layer, the dynamics of the upper thermal boundary layer depend in part on its intrinsically unstable thermal stratification and in part on the effect of plumes (or, less specifically, convective elements) coming from the lower boundary. One can evaluate the impact of the latter by considering how constant C_Q in the local heat flux scaling law (equation 5.82) depends on the heating mode for the layer. With only internal heating, there is no lower thermal boundary layer and hence the only external influence on the upper boundary layer is the slow and diffuse return flow of the downgoing plumes that detach from it. In the mixed heating mode, the thin lower boundary layer emits weak plumes. Finally, with no internal heat generation, the lower thermal boundary layer is as thick as the upper one and the ascending plumes are stronger. We observe from Table 9.1 that constant C_Q increases in a systematic fashion as the strength of basal heating increases: it is 0.302 for an internally heated fluid, 0.346 for mixed heating and 0.378 for basal heating only.

9.5 Temperature-dependent viscosity

The very large temperature differences that characterize geological systems imply large variations of physical properties and in particular viscosity. In such cases, one must re-evaluate the scaling laws that have been derived because they depend on a single viscosity value. We consider a viscosity function of the form:

$$\mu = \mu_o \exp\left(-\frac{T - T_o}{\Delta T_R}\right), \tag{9.36}$$

such that $\mu = \mu_o$ for $T = T_o$ (the temperature of the upper boundary) and where ΔT_R can be called a "rheological" temperature scale. For Rayleigh–Benard convection with temperature difference ΔT across the layer, three dimensionless parameters appear in the governing equations: the Prandlt number, a Rayleigh number calculated at some reference temperature and the viscosity ratio $\mu_o/\mu(T_o + \Delta T)$ or, equivalently, the temperature ratio $\Delta T/\Delta T_R$. The new dimensionless number, $\Delta T/\Delta T_R$, characterizes the viscosity variation that is generated in the convecting fluid. For $\Delta T/\Delta T_R \sim 1$, one expects no significant differences with the constant viscosity case. Indeed, all the scaling laws that have been obtained on thermal convection in real fluids have been achieved with a finite temperature difference, which necessarily implies the existence of small viscosity variations. The success of the simple isoviscous scaling laws is due to the thermal structure of the fluid layer, which is such that temperature contrasts are essentially confined to thin boundary layers at the top and bottom. As regards kinetic dissipation, for example, the largest amount is achieved in the fluid interior, over thickness h, where the average temperature is almost constant. For $\Delta T/\Delta T_R \gg 1$, however, there is a large viscosity difference between the upper and lower boundaries and one expects profound changes in the convective flow pattern as well as on the magnitude of the convective heat flux.

For mantle material in the infinite Pr limit, we need only deal with two dimensionless numbers: a Rayleigh number calculated at some reference temperature and the viscosity ratio $\mu(T_o)/\mu(T_o+\Delta T)$ or, equivalently, the temperature ratio $\Delta T/\Delta T_R$.

9.5.1 Moderate viscosity contrasts

If the viscosity ratio is larger than one order of magnitude, we expect that the cold upper boundary deforms more slowly than the lower boundary. In the scaling analysis for a constant viscosity, we have written that kinetic dissipation occurs in the bulk fluid because there is no momentum boundary layer against stress-free boundaries in a large Prandtl number fluid. With the larger viscosity that exists in the upper thermal boundary layer, however, dissipation may be larger there than in the interior fluid. We introduce two thicknesses δ_T and δ_B for the upper and lower thermal boundary layers, respectively, such that $\delta_T > \delta_B$, and two temperature differences ΔT_T and ΔT_B, such that $\Delta T_T + \Delta T_B = \Delta T$. We also expect that the cold and viscous top layer moves at a velocity that is less than that of the hotter interior, and introduce two velocities U_T and U_B. The heat flux constraint leads to a relationship between these various variables:

$$Q = \lambda \frac{\Delta T_T}{\delta_T} = \lambda \frac{\Delta T_B}{\delta_B}, \tag{9.37}$$

where,

$$\Delta T_T = \Delta T \frac{\delta_T}{\delta_T + \delta_B} \tag{9.38}$$

and a similar equation for ΔT_B. As before, the heat balance equation in the upper boundary layer leads to,

$$\delta_T \sim \sqrt{\frac{\kappa h}{U_T}}. \tag{9.39}$$

One can also write an analogous equation for the lower boundary layer. Note that we have assumed that the flow in both boundary layers develops over the same horizontal distance, so that this theory refers to a cell-like convection pattern.

In the cold upper boundary layer, dissipation is associated with the bending of a viscous layer as it is entrained into a downwelling current. For bending along a circular trajectory, the strain rate is U_T/δ_T. Dissipation is achieved in the small circular quadrant sector where bending occurs, $\delta V \approx \delta_T^2 h$. Thus,

$$\epsilon_{U,T} \sim \mu_o \left(\frac{U_T}{\delta_T}\right)^2 \delta_T^2 h. \tag{9.40}$$

Dissipation in the hot interior is achieved with velocity gradients that are distributed through the layer, implying that,

$$\epsilon_{U,B} \sim \mu_i \left(\frac{U_B}{h}\right)^2 h^3, \tag{9.41}$$

where μ_i is the interior viscosity (which is close to that of the fluid at the lower boundary). The kinetic dissipation equation therefore involves two different types of deformation, bending in the viscous upper boundary layer and shearing in the fluid interior. In such conditions, the scaling constants for the two dissipation components may not be equal, so that one may not assume that the total dissipation scales as the sum of $\epsilon_{U,T}$ and $\epsilon_{U,B}$. By construction, however, both types of dissipation must scale with the buoyancy flux, so that,

$$\mu_o \left(\frac{U_T}{\delta_T}\right)^2 \delta_T^2 h \sim \mu_i \left(\frac{U_B}{h}\right)^2 h^3 \sim \frac{\alpha g Q}{C_P} h^3. \tag{9.42}$$

The ratio between the two components of kinetic dissipation is therefore,

$$\frac{\epsilon_{U,T}}{\epsilon_{U,B}} \sim \frac{\mu_o U_T^2}{\mu_i U_B^2} \sim \frac{\mu_o \Delta T_B^4}{\mu_i \Delta T_T^4}, \tag{9.43}$$

where we have used equations 9.37 and 9.39. One could invoke a principle of equipartitioning for these two dissipation components but, as we shall see, such a principle cannot hold for any value of the viscosity contrast. At large viscosity contrasts, part of the upper boundary layer remains stagnant and does not participate in convective motions. Thus, the dissipation ratio (9.43) cannot be specified. For given deformation mechanisms, we know that this ratio must be constant, however, and this is sufficient to derive useful relations that illustrate the differences between the two thermal boundary layers:

$$\frac{\Delta T_T}{\Delta T_B} \sim \left(\frac{\mu_o}{\mu_i}\right)^{1/4}, \quad \frac{U_T}{U_B} \sim \left(\frac{\mu_o}{\mu_i}\right)^{-1/2}, \quad \frac{\delta_T}{\delta_B} \sim \left(\frac{\mu_o}{\mu_i}\right)^{1/4}. \tag{9.44}$$

Figure 9.12 illustrates the growth of the upper thermal boundary layer as the viscosity contrast increases. One consequence, of course, is that the interior temperature is

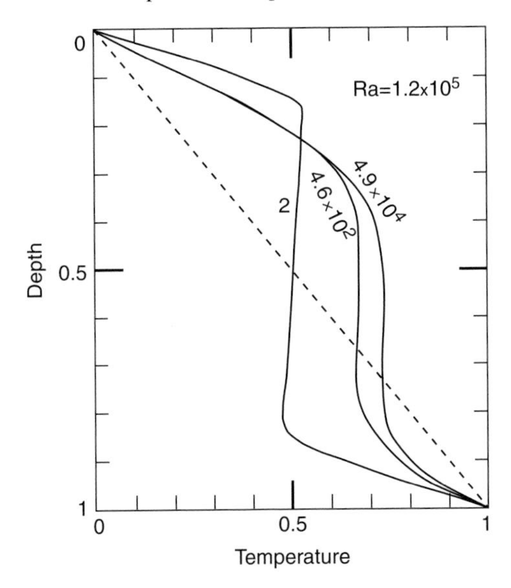

Figure 9.12. Vertical profile of the horizontally averaged temperature in a convective fluid with temperature-dependent viscosity at a Rayleigh number of about 10^5, from the laboratory experiments of Richter *et al.* (1983). Numbers along the curves indicate the viscosity ratio between top and bottom. With increasing viscosity ratio, the upper thermal boundary layer grows at the expense of the lower one, and the temperature of the well-mixed interior gradually increases.

no longer the average of the two boundary temperatures and increases with increasing viscosity contrast. The temperature profile therefore becomes increasingly asymmetrical.

The relationships in (9.44) ensure that local Rayleigh numbers calculated in the two boundary layers are of similar magnitudes:

$$\frac{\mathrm{Ra}_T}{\mathrm{Ra}_B} = \frac{\dfrac{\rho_o g \alpha \Delta T_T \delta_T^3}{\kappa \mu_o}}{\dfrac{\rho_o g \alpha \Delta T_B \delta_B^3}{\kappa \mu_i}} \sim 1. \tag{9.45}$$

Thus, the two boundary layers lie in the same critical dynamic state. We could have developed this argument in reverse, starting from the Rayleigh numbers and working our way to the kinetic dissipation estimates. From these, we obtain the temperature difference across the upper boundary layer and Nusselt number,

$$\Delta T_T \sim \Delta T \frac{(\mu_o/\mu_i)^{1/4}}{1+(\mu_o/\mu_i)^{1/4}} \tag{9.46}$$

$$\mathrm{Nu} \sim \mathrm{Ra}_o^{1/3} \frac{(\mu_o/\mu_i)^{1/3}}{\left[1+(\mu_o/\mu_i)^{1/4}\right]^{4/3}}, \tag{9.47}$$

where Ra_o is a Rayleigh number calculated with the viscosity at the top of the fluid,

$$\mathrm{Ra}_o = \frac{\rho_o g \alpha \Delta T h^3}{\kappa \mu_o}. \tag{9.48}$$

The thickness of the upper boundary layer is

$$\delta_T \sim h Ra_o^{-1/3} \left[\frac{(\mu_o/\mu_i)^{1/4}}{\left[1 + (\mu_o/\mu_i)^{1/4}\right)} \right]^{4/3}. \tag{9.49}$$

One can also obtain explicit relations for the other variables, including the temperature difference across the lower boundary layer.

The theoretical arguments rely on simple scalings which implicitly require that dissipation be scaled with a single viscosity value in each region of the fluid layer. This is appropriate as long as the viscosity contrast is not too large. We note that these solutions predict that, in the limit of very large viscosity ratio (i.e. $\mu_o/\mu_i \gg 1$), $\Delta T_T \approx \Delta T$. In this limit, the lower boundary layer accounts for a negligible part of the overall temperature difference, but must still supply the heat flux that is lost through the top. In such conditions, the flow structure changes drastically and the analysis of this section no longer applies.

9.5.2 Large viscosity contrasts

With a large viscosity ratio, such that $\Delta T/\Delta T_R \gg 1$, the top of the fluid is so viscous that it does not participate in convective motions: a stagnant lid develops at the upper boundary. Thus, only part of the total temperature difference is available for convection, implying that ΔT is not the relevant temperature scale for the active part of the fluid. In this case, the appropriate temperature scale is ΔT_R and the flow characteristics depend neither on the total temperature difference across the layer nor on the viscosity ratio. We use the fact that viscosity variations in the actively convecting interior are small and use scalings for constant viscosity convection:

$$Q \sim Q_R = \lambda \left(\frac{\rho_o g \alpha}{\kappa \mu_i} \right)^{1/3} \Delta T_R^{4/3}, \tag{9.50}$$

where Q_R is the "rheological" heat flux scale and μ_i is the viscosity at the temperature of the well-mixed fluid interior. Laboratory experiments at large viscosity contrasts support this scaling law (Figure 9.13) and indicate that temperature variations within the convective interior scale with the rheological temperature difference ΔT_R.

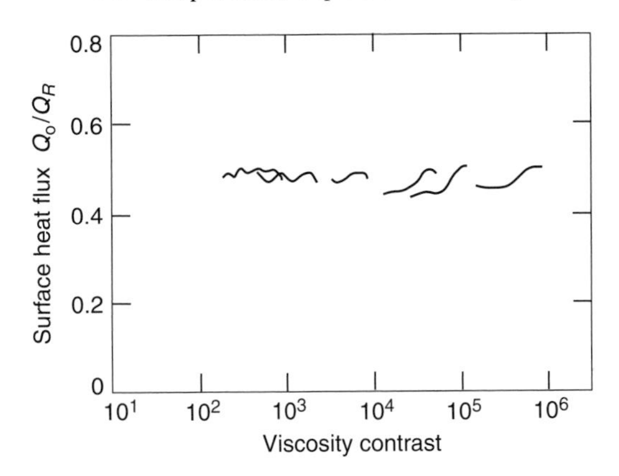

Figure 9.13. Heat flux through the top of a convecting fluid at high values of the viscosity ratio between top and bottom, from the laboratory experiments of Davaille and Jaupart (1993b). Heat flux values have been normalized by the local heat flux scale in the active part of the thermal boundary layer which involves the rheological temperature scale ΔT_R (equation 9.50). Each segment represents data from one transient experiment, in which a fluid layer above an adiabatic base is cooled from the top.

The temperature differences across the top and bottom boundary layers are ΔT_T and ΔT_B, respectively, and the associated boundary layer thicknesses are δ_T and δ_B. As argued above, $\Delta T_T \gg \Delta T_B$ because the upper thermal boundary layer is made of a thick stagnant lid and a thin unstable sub-layer. Also, $\Delta T_B \sim \Delta T_R$. We deduce that,

$$\mathrm{Nu} \sim \mathrm{Ra}_i^{1/3} \left(\frac{\Delta T_R}{\Delta T} \right)^{4/3}, \quad \delta_B \sim h \mathrm{Ra}_i^{-1/3} \left(\frac{\Delta T_R}{\Delta T} \right)^{-1/3}, \tag{9.51}$$

where Ra_i is a Rayleigh number calculated with the interior viscosity μ_i,

$$\mathrm{Ra}_i = \frac{\rho_o g \alpha \Delta T h^3}{\kappa \mu_i}. \tag{9.52}$$

A scaling for the thickness of the upper boundary layer can only be obtained in the limit of $\Delta T \gg \Delta T_R$. In this case, $\Delta T_T \sim \Delta T$ and hence,

$$\delta_T \sim h \mathrm{Ra}_i^{-1/3} \left(\frac{\Delta T}{\Delta T_R} \right)^{4/3} \tag{9.53}$$

$$\sim \delta_B \frac{\Delta T}{\Delta T_R}. \tag{9.54}$$

Table 9.2. *Rheological temperature scale for mantle rheologies* †

Creep regime	E (kJ mole^{-1})	V (cm^3 mole^{-1})	ΔT_R (K) §
Dry diffusion	261	6	92
Wet diffusion	387	25	62
Dry dislocation	610	13	39
Wet dislocation	523	4	46

† Representative values from (Korenaga and Karato, 2008).
§ Calculated from equation 9.57 with $T_i = 1700$ K at a pressure of 6 GPa.

9.5.3 Arrhenius viscosity dependence on temperature

In reality, mantle deformation mechanisms involve Arrhenius activation mechanisms, so that the temperature dependence of viscosity takes the following form:

$$\mu = a \exp\left(\frac{E + pV}{RT}\right), \tag{9.55}$$

where a is some constant, E is the activation energy, p is pressure, V is the activation volume and R is the perfect gas constant. One may expand this equation for temperatures that are close to that of the well-mixed interior region, T_i:

$$\mu \approx \mu_i \exp\left[-\frac{(E + pV)(T - T_i)}{RT_i^2}\right], \tag{9.56}$$

where μ_i is again the viscosity at the interior temperature T_i. This shows that the rheological temperature scale is

$$\Delta T_R = \frac{RT_i^2}{E + pV}. \tag{9.57}$$

This approximation is valid for the stagnant lid regime because there are only small variations of viscosity in the actively convecting region. Table 9.2 lists values of the rheological temperature scale for various deformation mechanisms that operate in the Earth's mantle. For large-scale mantle convection where viscosity variations span many orders of magnitude, the convenient approximation of an exponential viscosity law is less accurate.

9.6 Non-Newtonian rheology

The Earth's mantle is a polycrystalline aggregate that may deform in two different regimes depending on grain size, temperature, pressure and water content. The rheological law takes the following form:

$$\dot{e} = b^{-1}\sigma^n \exp(-H/RT), \tag{9.58}$$

where $\dot{e}$ and σ are strain rate and deviatoric stress, respectively, b is a proportionality constant, n is an exponent and $H = E + pV$ is an activation enthalpy. For diffusion creep, such that vacancies move through the crystalline lattice by diffusion, $n = 1$, so that the Newtonian solutions developed above apply. For dislocation creep, due to the migration of imperfections through the crystalline lattice, material exhibits shear-thinning behavior with viscosity that decreases with increasing applied stress. Current experimental studies suggest that exponent n varies within a range of about 3–5 and it is important to evaluate its influence on the characteristics of convection. Scaling laws for such rheological laws can be derived using the same methods as before, but one must treat n as yet another dimensionless number. We expect that the scaling relationships are more complex than before because of the complicated dimensions of constant b.

The key feature of non-Newtonian materials is that viscosity, $\mu \sim \sigma/(2\dot{e})$, is not an intrinsic material property and depends on the applied stress (or on the applied deformation rate). The scalar quantity that provides a measure of stress is the second invariant of the stress tensor, whose numerical value is independent of the coordinate system,

$$\sigma_2 = \sqrt{(\sigma_{ij}\sigma_{ij} - \sigma_{ii}\sigma_{jj})/2}. \tag{9.59}$$

For simplicity, we shall first ignore the pressure dependence of the rheological law and work with the same exponential temperature dependence as before, i.e. $\exp[-(T - T_o)/\Delta T_R]$. Although both n and ΔT_R are specified for each deformation mechanism, and hence are related to one another, it is useful to treat them as independent parameters. The constitutive relationship between the deviatoric stress and strain rate tensors may be written as follows:

$$\dot{e}_{ij} = \frac{1}{2b}\sigma_2^{n-1} \exp\left(\frac{T - T_o}{\Delta T_R}\right)\sigma_{ij}. \tag{9.60}$$

It is useful to introduce viscosity,

$$\mu = \frac{\sigma_{ij}}{2\dot{e}_{ij}} = \frac{b}{\sigma_2^{n-1}} \exp\left(-\frac{T - T_o}{\Delta T_R}\right) = \frac{b^{1/n}}{\dot{e}_2^{\frac{n-1}{n}}} \exp\left(-\frac{T - T_o}{n\Delta T_R}\right), \tag{9.61}$$

where $\dot{e}_2$ is the second invariant of the strain-rate tensor, which is calculated using the same formulation as σ_2 (equation 9.59).

For some applied temperature difference ΔT, the magnitude of the stress that is achieved in the convecting fluid is not specified a priori and depends on the rheology. One must separate variations of viscosity that are due to changes of stress, which depend on n, from those that are due to temperature, whose magnitudes are determined by temperature ratio $\Delta T/\Delta T_R$, as before. We determine a scale for convective stresses valid for all rheologies, denoted as σ, using the kinetic dissipation equation. We use the fact that velocity gradients develop throughout the fluid layer, such that $\dot{e} \sim U/h$, where U is a typical velocity. The rate of kinetic dissipation scales as $\sigma\dot{e}$. Thus,

$$\sigma \frac{U}{h} hS \sim \frac{\alpha g}{C_p} QhS, \tag{9.62}$$

where Q is the heat flux through the layer, hS is the volume of fluid in a layer of thickness h. Using the same scalings as before for the thermal boundary layer thickness δ and heat flux Q, $Q \sim \lambda \Delta T/\delta$ and $\delta \sim \sqrt{\kappa h/U}$, we obtain,

$$\sigma \sim \rho_o g \alpha \Delta T \delta. \tag{9.63}$$

This stress scale corresponds to the horizontal pressure difference generated by temperature variations within the thermal boundary layer. In this stress scale, δ is unknown and must be solved for. One may surmise, however, that, with increasing ΔT, σ increases, and viscosity decreases in a non-Newtonian fluid (such that $n \geq 1$). Thus, one expects that changes of convective vigor are more important in a non-Newtonian fluid than in a Newtonian one, and that they increase with increasing n. As ΔT increases, viscosity also varies due to temperature, but such variations involve different processes and depend on a different dimensionless number, temperature ratio $\Delta T/\Delta T_R$.

9.6.1 Small viscosity contrasts due to temperature: $\Delta T/\Delta T_R \ll 1$

For fluid at the interior temperature T_i, the proper viscosity scale is:,

$$\mu_i = \frac{b^{1/n}}{(U/h)^{\frac{n-1}{n}}} \exp\left(-\frac{T_i - T_o}{n\Delta T_R}\right). \tag{9.64}$$

This suggests that the rheological temperature scale is no longer ΔT_R, but $n\Delta T_R$, which is larger. This has important consequences which will be discussed below. We now solve for velocity scale U. Using μ_i and ΔT_R as viscosity and temperature scales, respectively, the dissipation equation yields:

$$\sigma\dot{e}h^3 \sim \frac{b^{1/n}}{(U/h)^{\frac{n-1}{n}}} \exp\left(-\frac{T_i - T_o}{n\Delta T_R}\right)\left(\frac{U}{h}\right)^2 h^3 \sim \frac{\alpha g}{C_p} Qh^3. \tag{9.65}$$

In this regime with small viscosity contrasts, and hence small temperature differences, the interior temperature is, by symmetry, $T_i \approx T_o + \Delta T/2$ and the horizontally averaged temperature profile is again characterized by two thermal boundary layers with equal temperature differences $\Delta T/2$ at the top and bottom. Introducing the boundary layers scaling relationships for δ and Q, we obtain the following results:

$$\delta \sim h \mathrm{Ra}_n^{-\frac{n}{n+2}}, \ \mathrm{Nu} \sim \mathrm{Ra}_n^{\frac{n}{n+2}}, \ U \sim \frac{\kappa}{h} \mathrm{Ra}_n^{\frac{2n}{n+2}} \tag{9.66}$$

where Ra_n is a modified Rayleigh number:,

$$\mathrm{Ra}_n = \frac{\rho_o \alpha g \Delta T h^{\frac{n+2}{n}}}{\kappa^{1/n} b^{1/n} \exp\left(-\dfrac{T_i - T_o}{n \Delta T_R}\right)}. \tag{9.67}$$

Substituting for $n = 1$ in these equations, one recovers the Newtonian scalings, of course. These scalings involve two independent parameters, n and ΔT_R.

To illustrate how the theory accounts for data, it is useful to study the influence of each parameter separately. For this purpose, Solomatov (1995) introduced two different Rayleigh numbers:

$$\begin{aligned} \mathrm{Ra}_o &= \frac{\rho_o \alpha g \Delta T h^{(n+2)/n}}{\kappa^{1/n} b^{1/n}} \\ \mathrm{Ra}_i &= \frac{\rho_o \alpha g \Delta T h^{(n+2)/n}}{\kappa^{1/n} b^{1/n} \exp\left(-\dfrac{T_i - T_o}{\Delta T_R}\right)} = \mathrm{Ra}_o \exp\left(\frac{T_i - T_o}{\Delta T_R}\right). \end{aligned} \tag{9.68}$$

The scaling laws can be written as a function of these two numbers. For the Nusselt number, for example,

$$\mathrm{Nu} \sim \mathrm{Ra}_o^{\frac{n-1}{n+2}} \mathrm{Ra}_i^{\frac{1}{n+2}}. \tag{9.69}$$

Figure 9.14 shows how the Nusselt number varies as a function of these two Rayleigh numbers and compares the predictions with the numerical results of Christensen (1985a). The theory also accounts for the average velocity in the convecting interior (Figure 9.15).

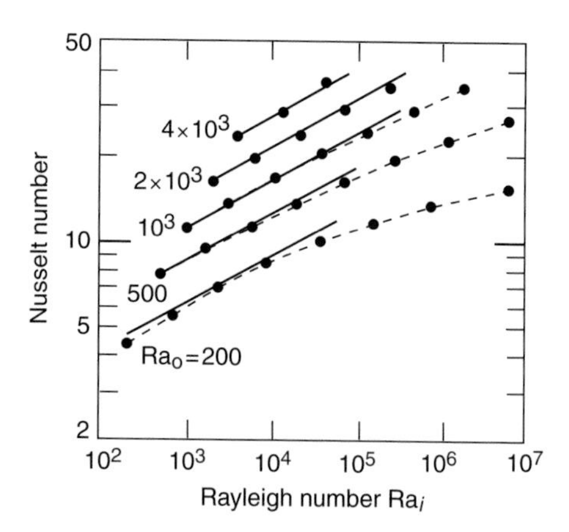

Figure 9.14. Nusselt number as a function of Rayleigh number Ra_i for convection in a non-Newtonian fluid with temperature-dependent viscosity, from Solomatov (1995). Curves are drawn for different values of the other Rayleigh number, Ra_o, and are compared to numerical results by Christensen (1985a). The scaling theory provides values of Nu to an unspecified proportionality constant, which has been adjusted to fit the data. The theory is not valid for large viscosity contrasts, which are attained for large values of Ra_i.

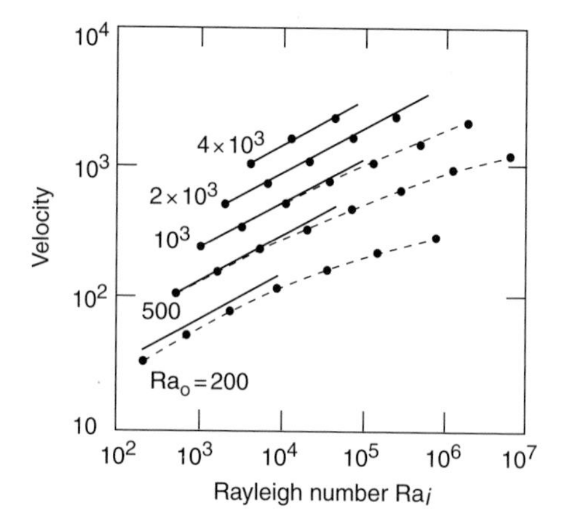

Figure 9.15. The same as Figure 9.14 for the average velocity in a non-Newtonian fluid with temperature-dependent viscosity, from Solomatov (1995).

The modified Rayleigh number that enters these scaling laws, Ra_n (equations 9.66), can be written in terms of a reference viscosity μ_n:

$$\mathrm{Ra}_n = \frac{\rho_o g \alpha \Delta T h^3}{\kappa \mu_n}, \tag{9.70}$$

where $$\mu_n = \frac{b^{1/n}}{(\kappa/h^2)^{\frac{n-1}{n}}} \exp\left(-\frac{T_i - T_o}{n\Delta T_R}\right). \tag{9.71}$$

The reference viscosity, μ_n, is calculated for a strain-rate of κ/h^2, which does not depend on the applied temperature difference. A more meaningful viscosity scale can be deduced from the heat flux expressed in dimensional form, which may be derived from equations 9.66:

$$Q \sim \lambda \left(\frac{\rho_o g \alpha}{\kappa^{1/n} b^{1/n} \exp\left(-\dfrac{T_i - T_o}{n\Delta T_R}\right)} \right)^{\frac{n}{n+2}} \Delta T^{\frac{2(n+1)}{n+2}}. \tag{9.72}$$

We note that Q does not depend on layer depth h, and hence that it is determined by local dynamics in the thermal boundary layer independently of the total fluid thickness. We also note that heat flux dependence on ΔT is stronger than for a Newtonian fluid (with $n = 1$), as expected. This key property is illustrated in a different form at large viscosity ratios, as discussed in a separate section below.

The complicated equation for Q (9.72) can be reduced to a familiar form,

$$Q \sim \lambda \left(\frac{\rho_o g \alpha}{\kappa \mu_i} \right)^{1/3} \Delta T^{4/3}, \tag{9.73}$$

where μ_i is the interior viscosity. This interior viscosity corresponds to the deformation rate specified by (9.66):

$$\mu_i = \frac{\left[b^{1/n} \exp\left(-\dfrac{T_i - T_o}{n\Delta T_R}\right)\right]^{\frac{3n}{n+2}}}{\kappa^{\frac{n-1}{n+2}} (\rho_o g \alpha \Delta T)^{\frac{2(n-1)}{n+2}}}. \tag{9.74}$$

We may verify that the interior viscosity reduces to, $\mu_i = b\exp[-(T_i - T_o)/\Delta T_R]$ for a Newtonian fluid with $n = 1$. In 9.73, the heat flux scaling is identical to that for a Newtonian fluid. One must keep in mind that, for $n > 1$, the interior viscosity μ_i is not an intrinsic fluid property and depends on the deformation rate, which itself depends on the driving temperature difference. Nevertheless, if one may measure the material viscosity *in situ*, this scaling law allows straightforward calculations.

The key result is that μ_i, which is the appropriate viscosity value for the convective flow, decreases with increasing temperature difference. This is a direct consequence of shear-thinning behavior: as one increases the temperature difference across the layer, convective vigor increases and so does the shear rate. This also implies that the convective heat flux increases more rapidly with increasing ΔT than in the Newtonian case, as already noted. We also note that, for a constant applied temperature difference, μ_i decreases with increasing n: in such conditions, shear-thinning behavior is enhanced, which leads to a larger viscosity decrease.

Using this viscosity value, one recovers scalings for other important variables. For example, using $Q \sim \lambda \Delta T / \delta$ and $U \sim \kappa h / \delta^2$, we obtain,

$$\delta \sim \left(\frac{\rho_o g \alpha \Delta T}{\kappa \mu_i} \right)^{-1/3}. \tag{9.75}$$

This allows a fully explicit expression for the convective stress scale σ:

$$\sigma \sim \rho_o g \alpha \Delta T \delta \tag{9.76}$$

$$\sim \mu_i \frac{U}{h} \tag{9.77}$$

$$\sim \left[b\kappa \left(\rho_o g \alpha \Delta T \right)^2 \exp\left(-\frac{T_i - T_o}{\Delta T_R} \right) \right]^{\frac{1}{n+2}}. \tag{9.78}$$

This is the stress value that leads to viscosity μ_i according to (9.61). This convective stress scale indeed increases with the applied temperature difference ΔT, as predicted. We also note that it depends on n.

9.6.2 *Large viscosity contrasts due to temperature:* $\Delta T / \Delta T_R \gg 1$

In this case, part of the upper thermal boundary layer remains stable and effectively behaves as a rigid material. Convection only affects a lower sub-layer with a driving temperature difference, denoted as ΔT_δ, that is smaller than the total temperature difference across the layer, ΔT, and that is proportional to the rheological temperature scale, ΔT_R. From the preceding section, we expect that ΔT_δ increases as n increases, due to stronger shear-thinning behaviour. Variables n and ΔT_R are independent, and hence we can write ratio $\Delta T_\delta / \Delta T_R$ as a function of the one remaining dimensionless parameter, which is n. The dissipation equation can be used as before, but dimensional analysis cannot resolve the scaling relationship between the temperature ratio and the rheological exponent.

One can determine the driving temperature difference in different ways. One can rely on the characteristics of the unstable part of the boundary layer. Solomatov and Moresi (2000) have defined the base of the rigid lid from vertical velocity profiles. An alternative method relies on convective heat flux. From the result of the previous section, the driving temperature difference scales with ΔT_R and heat flux can be written as,

$$Q = C(n) \lambda \frac{\left(\rho_o g \alpha \right)^{\frac{n}{n+2}}}{\left[\kappa b \exp\left(-\frac{T_i - T_o}{\Delta T_R} \right) \right]^{\frac{1}{n+2}}} \Delta T_R^{\frac{2(n+1)}{n+2}}, \tag{9.79}$$

Table 9.3. *Scaling constants for convection in non-Newtonian fluids with temperature-dependent viscosity* †

n	$\Delta T_\delta / \Delta T_R$	$C(n)$ ‡
1	2.4	0.528
2	3.6	0.755
3	4.8	0.971

† From (Solomatov and Moresi, 2000) .
‡ Constant in the local heat flux scaling law (equation 9.79).

where $C(n)$ is a constant that depends on the power-law exponent n (Table 9.3). As before, one should rewrite this scaling law in terms of the temperature difference across the unstable boundary layer, ΔT_δ:

$$Q = C_o \lambda \left(\frac{\rho_o g \alpha}{\kappa \mu_i} \right)^{1/3} \Delta T_\delta^{4/3}, \tag{9.80}$$

where viscosity μ_i is calculated as before and where C_o is a constant that does not depend on the power-law exponent. This constant is the same as that for Newtonian fluids, by construction. The unstable part of the boundary layer lies below material that behaves rigidly, so that we must use the scalings for rigid boundaries, i.e. $C_o \approx 0.16$ (Table 9.3). From the values of $C(n)$ that have been obtained by numerical calculations (Table 9.3), we derive the relationship between ΔT_δ and the rheological temperature scale ΔT_R:

$$\Delta T_\delta = \left(\frac{C(n)}{C_o} \right)^{3/4} \Delta T_R. \tag{9.81}$$

Figure 9.16 compares two different determinations of $\Delta T_\delta / \Delta T_R$. We note that, for $n = 1$, both are identical and very close to 2.4, a value derived from laboratory experiments (Davaille and Jaupart, 1993b). The two determinations differ slightly at larger values of n, but they exhibit the same tendency: they both increase with n, as predicted.

9.7 Mantle plumes as part of a large convective system

Oceanic shield volcanoes such as Kilauea, Hawaii, and Piton de la Fournaise, Reunion Island, have been attributed to isolated mantle plumes. Their relative fixity with respect to oceanic plates shows that they are weakly affected by the large-scale

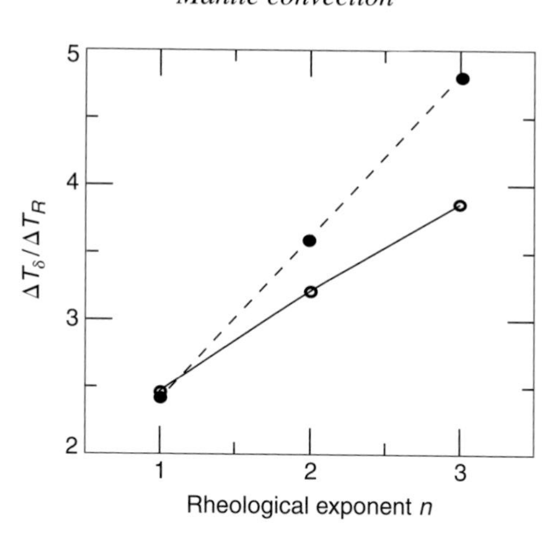

Figure 9.16. Temperature difference across the unstable part of the thermal boundary layer in a convecting with large variations of viscosity, as a function of n the exponent in the non-Newtonian rheological law, from the numerical calculations of Solomatov and Moresi (2000). The temperature difference has been scaled to the rheological temperature difference. Solid dots correspond to the definition of Solomatov and Moresi (2000), which is based on the velocity profile in the boundary layer. Open circles are deduced from the heat flux, as specified by equation 9.81.

mantle circulation associated with subduction zones and oceanic ridges. Furthermore, the source material for their lavas differs from that of oceanic ridges, which suggests that they come from a different source region (a different chemical reservoir). As a consequence, they have been studied as isolated plumes, with special emphasis on how they mix with surrounding mantle as they rise. In reality, such plumes are part and parcel of convection. At large Rayleigh numbers, we have seen that convective motions take the form of individual plumes that rise out of thin thermal boundary layers. With small viscosity variations, one deals with both downgoing and upgoing plumes of similar strengths. These plumes belong to a population and hence interact with one another. Thus, one may wonder whether their characteristics are similar to those of an isolated plume rising in otherwise stagnant fluid, as studied in Chapter 5. Here, we evaluate the relationship between the characteristics of plumes and those of the bulk convection in the fluid layer, as measured by the Rayleigh number for example.

9.7.1 Plumes in Rayleigh–Benard convection

We begin with the simplest configuration, that of Rayleigh–Benard convection, i.e. a fluid layer heated from below. Theory developed in Chapter 5 for large Rayleigh numbers predicts that the convective velocity scales as:

$$U \sim \frac{\kappa}{h} \mathrm{Ra}^{2/3}. \tag{9.82}$$

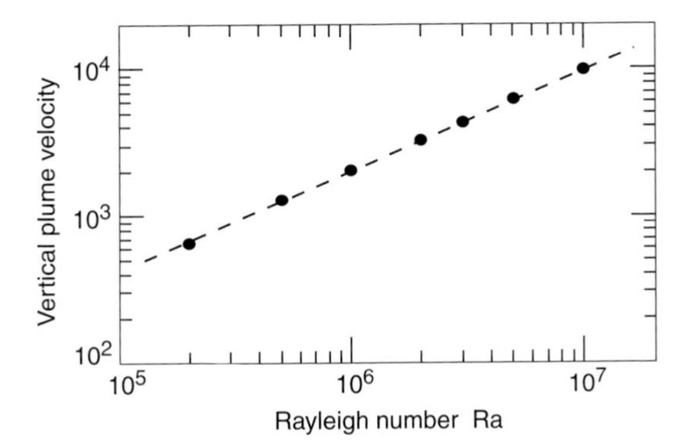

Figure 9.17. Vertical plume velocity as a function of Rayleigh number, from the Rayleigh–Benard numerical calculations of Galsa and Lenkey (2007). The dashed curve corresponds to the scaling law in (9.82). Calculations are made in 3D in a large aspect ratio domain in the limit of infinite Pr.

Numerical calculations for high Rayleigh numbers show that plume velocities are indeed proportional to this convective velocity scale (Figure 9.17).

It is illuminating to rewrite this velocity scale as follows:

$$U \sim \left[\frac{\kappa^2}{h^2} \left(\frac{\rho_o g \alpha \Delta T h^3}{\kappa \mu} \right)^{4/3} \right]^{1/2} \tag{9.83}$$

$$\sim \left[h^2 Q \frac{g\alpha}{\mu C_p} \right]^{1/2}, \tag{9.84}$$

where Q is the heat flux through the fluid layer, evaluated using the local boundary layer scaling law (equation 5.82). One recalls that the typical length scale for convective motions is the layer depth h. Thus, the plume spacing scales with h and each plume draws heat from an area $\sim h^2$. Thus, the power input into each plume is $P \sim Qh^2$ on average. Rewriting the plume velocity scale U as a function of P, we find that,

$$U \sim \sqrt{\frac{g\alpha P}{\mu C_p}}, \tag{9.85}$$

and recognize the scaling law for the velocity of an isolated plume (equation 5.3). The strong assumption that has been made in this argument is that the plume distance scales with h. This is assessed in Figure 9.18, which shows the number of plumes per unit area, N_p, in the numerical calculations of Galsa and Lenkey (2007). Each plume draws fluid from the thermal boundary layer over an area $\sim h^2/N_p$. If each plume draws fluid from an area that scales with h^2, N_p should not depend on the

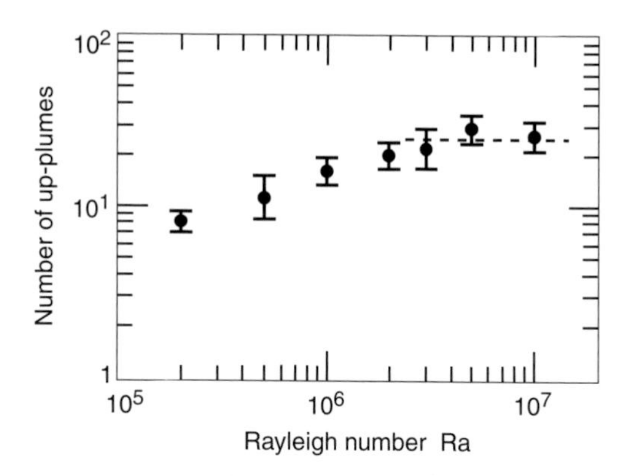

Figure 9.18. Number of plumes per unit area as a function of Rayleigh number, from the numerical calculations of Galsa and Lenkey (2007). For each value of Ra, the number of plumes that can be detected varies as a function of time and depth in the layer and the vertical bars indicate the spread of values. For $\mathrm{Ra} > 5 \times 10^6$, variations are not significant. At lower values of Ra, the number of plumes increases approximately as $\mathrm{Ra}^{1/3}$.

Rayleigh number. This is approximately verified for values of the Rayleigh number that are sufficiently large (in practice for $\mathrm{Ra} > \approx 5 \times 10^6$, Figure 9.18). This is also the validity domain for the heat flux scaling law that has been used.

One should not carry the analogy between the two plume velocity scalings further, however, because the plumes emerge from a laterally extensive thermal boundary layer and change shape as they rise. They are best characterized as sheet-like close to source and evolve towards almost cylindrical upwellings with increasing distance from source. Such plumes are intrinsically transient and their vertical velocity is not exactly constant as they rise or fall through the fluid layer. Thus, their dynamics are not the same as those of an isolated steady-state plume. Focusing on the source of a plume, one must allow for unsteadiness in both location, as the source drifts in the horizontal direction, and time, as the flux of heated fluid into the plume varies. Indeed, as shown in Figure 9.18, N_p varies with time and simple behaviour as a function of Ra is only valid on average. For the Earth, such issues are important when dealing with the apparent fixity of mantle plumes.

To proceed further, one must specify the dimensions of the plumes themselves. Using simple arguments, we derive a few basic controls on the collective dynamics of a plume population. Each plume grows out of a thermal boundary layer whose thickness δ is $\sim \mathrm{Ra}^{-1/3}$. By continuity, the initial plume width close to source is $\sim \delta$, corresponding to a cross-section $\Delta S \sim \delta^2 \sim \mathrm{Ra}^{-2/3}$. This is consistent with the numerical results of Galsa and Lenkey (2007) (Figure 9.19). From this, we deduce that the volume flux of hot or cold fluid in each plume, $U \Delta S$, is $\sim \mathrm{Ra}^{2/3}\mathrm{Ra}^{-2/3}$, i.e. does not depend on Ra. The plume heat flux is $\rho_o C_p U \Delta S \theta_p$, where θ_p is the

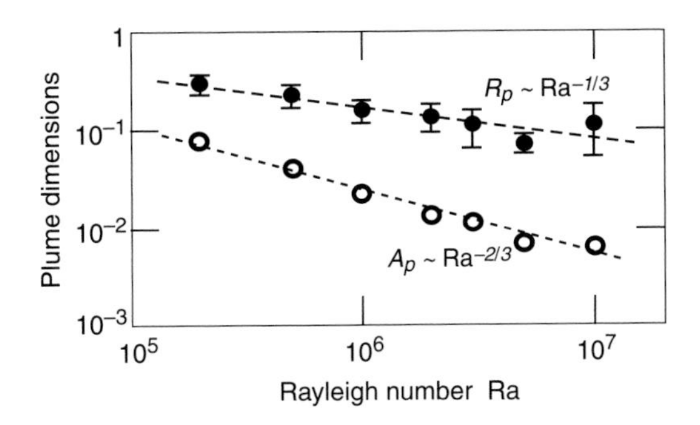

Figure 9.19. Area and radius of plumes as a function of Rayleigh number, from the convection calculations of Galsa and Lenkey (2007). Symbol size and vertical bar indicate the spread of values.

average plume temperature anomaly. On average, the plumes are responsible for convective heat flux, so that,

$$N_p U \Delta S \theta_p \sim \mathrm{Nu} \sim \mathrm{Ra}^{1/3}. \tag{9.86}$$

We now use the fact that $U \Delta S$ does not depend on Ra and we obtain a simple relationship between the number of plumes and their thermal anomalies. For large Ra ($> \approx 5 \times 10^6$), N_p is approximately constant, and hence $\theta_p \sim \mathrm{Ra}^{1/3}$. In this case, the plume temperature anomaly increases with the Rayleigh number, and with the power input that sustains them, as for an isolated plume. For the low Ra regime ($\mathrm{Ra} < 10^6$), $N_p \sim \mathrm{Ra}^{1/3}$ approximately (Figure 9.18), so that θ_p depends neither on the Rayleigh number nor on the heat flux. This behavior contrasts with that of an isolated plume.

9.7.2 *Heat flux carried by plumes*

The Earth's mantle is cooling down and is heated by *in situ* radioactive decay, which generates different thermal boundary layers at the top and bottom. In such conditions, the simple arguments developed above for Rayleigh–Benard convection no longer apply. The convective heat flux $Q = \rho_o C_p \overline{w\theta}$ sums up the positive contributions of both upwellings and downwellings. Thus, focusing on rising plumes only, for example, leads to an under-evaluation of the heat flux.

In a Rayleigh–Benard regime, such that there is no internal heating, downgoing and upgoing plumes account for an equal proportion of the convective heat flux, by symmetry. With internal heating, temperature contrasts are much larger in the upper thermal boundary layer than in the lower one and hence downgoing plumes account for a large part of the convective heat flux near the top of the fluid layer (Figure 9.20). As one goes deeper into the fluid layer, the contribution of upwellings

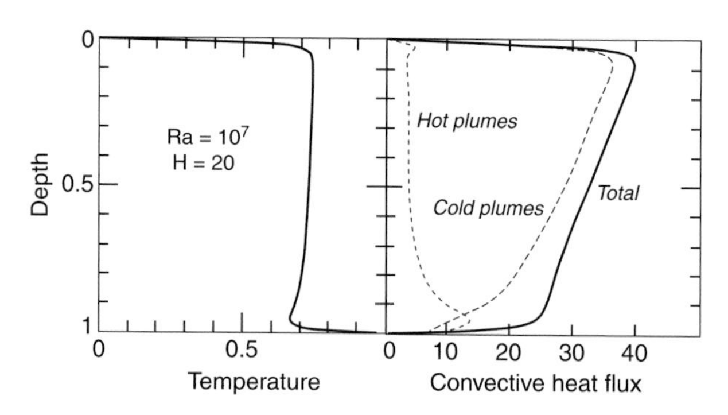

Figure 9.20. Vertical profile of the horizontally averaged temperature and of the convective heat flux in a fluid layer that is heated from below and internally heated, from the 3D numerical calculations of Labrosse (2002). The convective heat flux has been broken down into the contributions of upwellings and downwellings. Note that the convective heat flux decreases with depth in the layer due to internal heating.

to the bulk heat flux decreases, because cold descending fluid parcels get warmed up by local heat production. In the calculation shown in Figure 9.20, upwellings and downwellings contribute approximately equal fractions of the bulk heat flux at the top of the basal boundary layer.

For the Earth, these results have important implications for the cooling of the core. Mantle plumes have been attributed to instabilities of the thermal boundary layer at the core–mantle boundary and their cumulative heat flux has been calculated. This heat flux, however, must not be identified with the rate at which the core is losing heat. Core cooling is effected by both hot spots, which carry hot mantle drawn from the core–mantle boundary layer, and cold subducted mantle, which ponds at the top of the core.

9.8 Two scales of convection

Mantle convection involves a large number of processes and physical properties that depend on temperature and strain rate. Scaling arguments are made complicated by the existence of two scales of convection.

9.8.1 Large-scale convective motions

Mantle convection involves oceanic plates that span horizontal distances that are much larger than the thickness of the mantle, which is at odds with classical Rayleigh–Benard systems. We have mentioned above several physical effects that may account for this peculiar feature and that affect the rate at which Earth loses its internal heat. The other peculiar feature of mantle convection is the very large viscosity contrasts that are involved. Oceanic plates are more viscous than the mantle interior by many orders of magnitude. In fact, they do not deform in a purely viscous regime and are able to support the permanent loads of volcanic edifices. This is at odds with Rayleigh–Benard systems that have been reviewed above, in which convection is restricted to an interior region where viscosity variations do not exceed a few orders of magnitude. In principle, Earth should be in the stagnant lid regime with no subduction. This large discrepancy between physical expectations and geological reality has not yet been resolved.

9.8.2 Small-scale convection beneath the lithosphere

Heat flux measurements demonstrate that heat is supplied to the base of both oceanic and continental lithosphere. This is the situation that has been studied above, with convective motions that develop beneath a cold and viscous lid. In this case, the driving temperature difference for convection is proportional to the rheological temperature scale. Local thermal equilibrium between the rigid lithosphere and the actively convecting fluid below is achieved if the convective heat flux is evacuated by conduction through the lithosphere over thickness d. Denoting the convective heat flux by q_c and the temperature at the base of the lithosphere by T_b, this balance is such that,

$$\lambda \frac{T_b - T_o}{d} = q_c = \lambda \frac{\Delta T_\delta}{\delta}. \tag{9.87}$$

For an interior mantle temperature T_i, one has:,

$$T_b + \Delta T_\delta = T_i. \tag{9.88}$$

Using the scaling law for q_c derived above, these two equations can be solved for both T_b and d. For an Arrhenius rheological law, both the interior viscosity μ_i and the rheological temperature scale ΔT_R depend on pressure, i.e. depth. In this case, the local convective heat flux decreases with increasing lithosphere thickness. So does the conductive heat flux through the lithosphere, which decreases as the lithosphere thickens. The pressure dependences of the two types of heat fluxes are different and allow two solutions for the heat flux balance equation, one for small depth, corresponding to high heat flux and thin lithosphere, and another for a larger depth with a larger viscosity, corresponding to low heat flux and thick lithosphere (Figure 9.21) (Doin *et al.*, 1997). It is tempting to associate these two equilibrium states with oceanic and continental lithospheres, respectively. However, one must evaluate how each state is achieved in practice and whether or not it is stable to

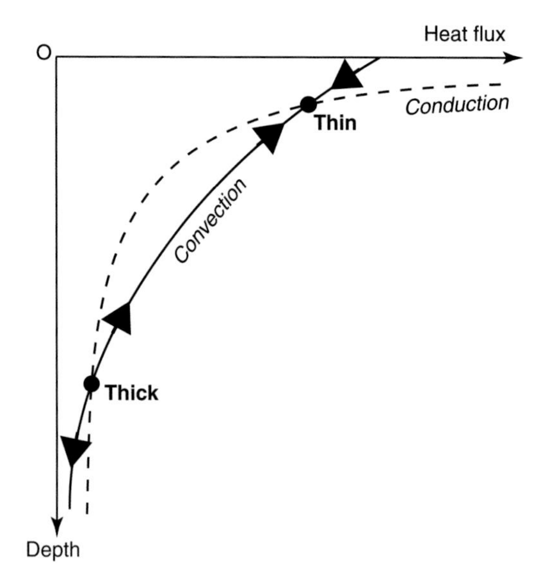

Figure 9.21. Schematic plot of the local convective heat flux beneath a stagnant lid as a function of depth for a viscosity that depends on pressure and temperature. The heat flux sustained by convection beneath a lid depends on the local viscosity, which depends on pressure. With increasing lid thickness, the convective heat flux decreases due to the pressure dependence of viscosity. The dashed line shows the heat flux that is carried by conduction across the stagnant lid, which also decreases with depth. A stable thermal state requires equality of the convective and conductive heat fluxes. There are two such states, but only one of them is stable to perturbations. Arrows denote how the convective heat flux varies when the stagnant lid thickness is not at an equilibrium value.

perturbations. This is best carried out by considering how the conductive heat flux changes when it is not in equilibrium with the convective one.

If we assume that there is no compositional difference between the lithospheric mantle and convecting mantle below, the lithosphere behaves rigidly because it is cold: this is the "pure" stagnant lid regime. In this case, the lithosphere thickness can change as temperatures evolve. Let us imagine that the convective heat flux is not equal to the conductive heat flux at the base of the lithosphere. On the one hand, if the convective heat flux is larger than the conductive one, the lithosphere heats up and thins, and hence the conductive heat flux increases. On the other hand, if the convective heat flux is less than the conductive one, the lithosphere cools down and hence thickens, so that the conductive heat flux decreases. This shows how lithosphere thickness and heat flux adjust to bring the system to an equilibrium state. As shown by Figure 9.21, however, the thickness changes in two different ways in the vicinity of the two equilibrium states. The shallow equilibrium state may be attained from both a thin or thick lithosphere. In contrast, the deep equilibrium situation can only be achieved if the lithosphere is created with precisely the right thickness. Starting from lithosphere that is either too thick or too thin, one cannot reach the deep equilibrium state and it is not a stable state.

One may evaluate this model further through a quantitative calculation of heat flux values for the two steady states. The solutions depend strongly on the activation volume, which is only known within large error bounds (Korenaga and Karato, 2008). The activation volume controls the magnitude of changes due to pressure, and the local heat flux beneath the lithosphere would not decrease significantly with depth if it is very small. For lithosphere that is made of depleted mantle which is both buoyant and more viscous than the asthenosphere, these arguments must be re-evaluated because thinning of the lithosphere due to an imbalance between the convective and conductive heat fluxes is unlikely.

9.9 Conclusion

This chapter has covered many different aspects of mantle convection but falls short of a comprehensive treatment in which all the pieces are assembled in a single model. One robust result emerges, however, in the form of the scaling law for convective heat transport. We have found that, in all cases, the convective heat flux depends only on the local dynamics of a thin thermal boundary layer, independently of large-scale flow. We have also shown how one can use the heat flux to determine the effective viscosity of mantle convection.

10

Thermal evolution of the Earth

Objectives of this chapter

The global energy budget drawn in Chapter 8 and geological evidence indicate that the Earth has been cooling down for several billions of years. We have many reasons to believe that the Earth was very hot after the giant impact and formation of the core. We discuss how the very young Earth might have cooled down rapidly, allowing continents to become stable and grow. We evaluate how to relate the present rate of energy loss to long-term thermal evolution of the mantle, and we discuss the impact of continental growth and the super-continent cycle on the Earth's thermal evolution.

10.1 Initial conditions

The Earth is presently in a regime which can be described as one of sub-solidus convection, such that motions predominantly occur in the solid state and depend on the rheological properties of mantle rocks. Melting only occurs at shallow levels and is thought to be a passive process that does not affect large-scale mantle dynamics. In the early stages of Earth's evolution, large amounts of energy were available and probably led to a planet that was almost entirely molten. A host of processes with different dynamics were active then, which may be separated into three categories: accretion, core formation and magma ocean crystallization.

10.1.1 Accretion of the Earth. Differentiation of the core

The accretion process of Earth brought together matter which was originally dispersed in the proto-solar nebula, thereby releasing gravitational energy. The total energy released can be estimated easily by taking the difference between the total gravitational energy before and after. Determining the fate of this energy is not straightforward and depends on how it is dissipated. The thermal consequences of core differentiation are quite different from those of accretion. Most of the processes involved remain speculative to some extent and we restrict ourselves to the points that are directly relevant to the thermal structure of early Earth. Two books (Newsom and Jones, 1990; Canup and Righter, 2000) deal with many of these issues.

During accretion, the gravitational energy of impactors is first transformed into kinetic energy and then dissipated in the form of heat at the impact. One may define two limit-cases. If no energy is lost to space, the temperature of the whole Earth is raised by an amount equal to

$$\Delta T = \frac{-E_g}{MC_P} \sim 3.75 \times 10^4 \text{ K}, \tag{10.1}$$

which would be sufficient to vaporize the whole planet. In the other limit, all the energy is released at shallow depth and lost to space by radiation. In this case, accretion would raise the temperature of the Earth by less than 70 K relative to that of the nebula (Stevenson, 1989). The actual evolution lies somewhere between these two limiting cases, involving partial dissipation of the impact energy within the planet and radiative heat transfer through the primordial atmosphere. One key variable is the ratio between time for energy transport to the surface and the time between two impacts, which depends on the size of the impactors. The larger the impactor, the larger is the depth of energy release, and the longer the time for energy transport to the surface. After the planetary embryo stage, impactors grew progressively in size (Melosh and Ivanov, 1999). The evolution towards larger and fewer impactors has two opposite effects on heat release: energy gets buried at greater depth whilst the time between two impacts increases, which enhances heat loss to the atmosphere. Assuming heat transport by diffusion and a typical accretion sequence, Stevenson (1989) found that a large part of the gravitational energy is stored within the planet. An extreme case is that of the giant impactor thought to be at the origin of the Moon. Calculations suggest that the whole Earth temperature was raised to as high as 7000 K (Cameron, 2001; Canup, 2004). In such conditions, the whole Earth melted and parts of it were vaporized to form a thick atmosphere. The question of whether or not previous impacts had been able to melt the Earth becomes irrelevant.

More gravitational energy was released by going from a uniform composition to a stratified iron core–silicate mantle system. For the Earth, the total amount of energy available corresponds to a global temperature rise of about 1700 K (Flasar and Birch, 1973). In contrast to the accretionary sequence, kinetic energy plays no role and gravitational potential energy is directly dissipated by viscous heating in both the iron and silicate phases. Our current understanding of this process suggests three mechanisms: iron droplets “raining” through a magma ocean, diapirs generated by Rayleigh–Taylor instability at a rheological interface and interstitial flow across a solid permeable matrix (Stevenson, 1990; Rushmer *et al.*, 2000). All three mechanisms may have been active at different times and have different implications for dissipation. Iron–silicate differentiation occurred very early in the solar system and affected planetesimals (Kleine *et al.*, 2002). The occurrence of a giant impact at a late stage does not reduce uncertainty about the dominant segregation mechanism. On the one hand, according to Canup (2004), large parts of the cores of the two proto-planets merged without remixing with silicates. On the other hand, Rubie *et al.* (2003) have argued in favor of iron emulsification in the molten silicate due to the large stresses involved.

For a Newtonian rheology, the amount of viscous heating is $\psi \sim \mu (\nabla \mathbf{v})^2 \sim U^2/L^2$, where μ is the viscosity of the fluid phase, which may be silicate or iron, and U and L are characteristic scales for velocity and length. In the case of an iron diapir, the velocity and length scales are the same for the metal and silicate phases, but the viscosity of the former is several orders of magnitudes lower. Viscous heating is thus concentrated in the silicate phase with little diffusion of heat into the iron phase because the descent is rapid compared with the diffusion time scale. This would differentiate a core that is initially colder than the lower mantle. In the case of interstitial flow, the small size of iron veins makes heat diffusion very effective and one expects molten iron to be near thermal equilibrium with the surrounding silicate phase. In the case of iron droplets raining down through molten silicate, the droplet size is set by equilibrium between surface tension and viscous drag and is typically 1 cm. As for iron veins, thermal equilibrium is achieved between the two phases and the core should initially be at the temperature of the lower mantle.

10.1.2 Magma ocean evolution

Both the giant impact and the formation of an iron core generated temperatures that were high enough to melt the whole silicate Earth and led to formation of a deep magma ocean. Cooling and crystallization of this magma involved heat transfer through the primordial atmosphere, convection, rotation and crystal-melt separation. Models have been aimed mostly at determining the extent of chemical stratification at the end of crystallization (Abe, 1997; Solomatov, 2000). The low viscosity of the melt and the size of Earth imply highly turbulent convective flows and rapid cooling, such that the lower parts of the magma ocean solidify in a few thousands of years. Two rheological transitions, from pure magma to slurry and from slurry to mush, affect the convective regime and the cooling rate. One important fact is that, in a convecting region, the isentropic temperature gradient is less than the gradients of liquidus and solidus, which causes more crystallization at the bottom (Figure 10.1).

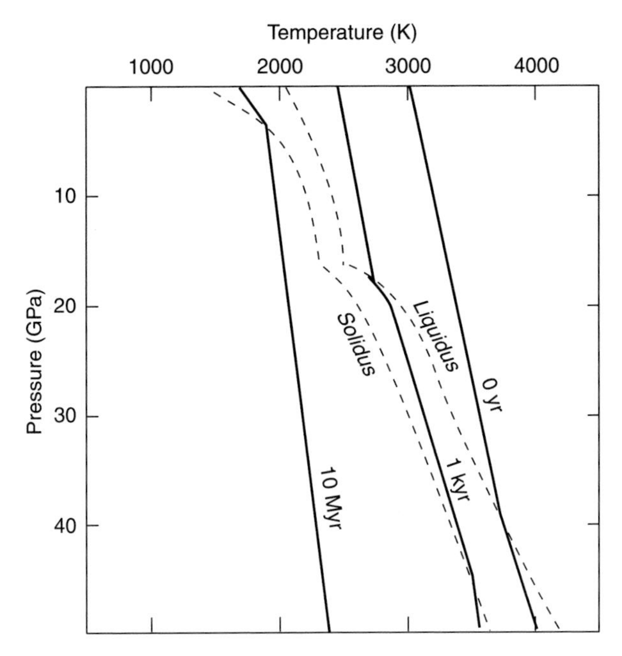

Figure 10.1. Three geotherms at different times in the early Earth, from Abe (1997). Note the different slopes of the geotherms, which are approximated by isentropic profiles, and the liquidus boundary, which imply that solidification proceeds from the bottom up. At about 10 My, the solid content in the partially molten upper mantle layer reaches the threshold value of 60%, which marks the cessation of liquid behavior. After that time, convection is in the sub-solidus regime controlled by solid behavior which still prevails today.

Starting from a superheated magma ocean (i.e. at temperatures above the liquidus), the initial phase has a fully molten upper layer which becomes thinner as cooling proceeds. A first transition occurs when temperatures drop below the liquidus. At this stage, there is no longer a fully liquid magma ocean and the Earth is made of a partially crystallized shell which may lie over already fully solidified mantle. The radial temperature profile is tied to the solidus which is steeper than the isentropic profile, which leads to convective overturn in the solid layer (Elkins-Tanton *et al.*, 2003). The two layers evolve with vastly different time scales because of their different rheologies. The bulk cooling rate is set by heat loss through the Earth's surface, which is controlled by the dynamics of the partially crystallized superficial magma. In this second phase, heat transport occurs mostly by melt–solid separation and solidification proceeds from the bottom up. The fully solidified layer at the base of the magma ocean grows rapidly and eventually becomes unstable. Convective overturn is slower than cooling of the magma and may be considered as a separate event which leads to decompression melting and the formation of a secondary magma ocean at the surface. The process of cooling and solidification of a deep magma ocean then repeats itself. This regime prevails until the shallow magma layer reaches the rheological threshold between liquid and solid behavior, which probably occurs when the solid fraction reaches about 60%. At this stage, the shallow partially crystallized layer becomes strongly coupled to the solid mantle below and cooling proceeds through bulk convection everywhere. According to (Abe, 1993, 1997), this phase of rapid cooling of the magma ocean was completed in a few 10 My and sets the initial conditions for secular cooling models of the solid Earth. From recent phase diagrams (Herzberg and Zhang, 1996; Litasov and Ohtani, 2002), this final rheological transition corresponds to a potential temperature of about 1800 ± 100 K for a mantle composed of dry pyrolite (Figure 10.2).

The magma ocean models of Abe (1997) depend on the phase diagram and composition of the mantle. The time scale for thermal evolution, however, is set by heat loss at the upper boundary, which can be described by the robust scaling laws for heat transport which decreases rapidly with temperature (see Chapters 5

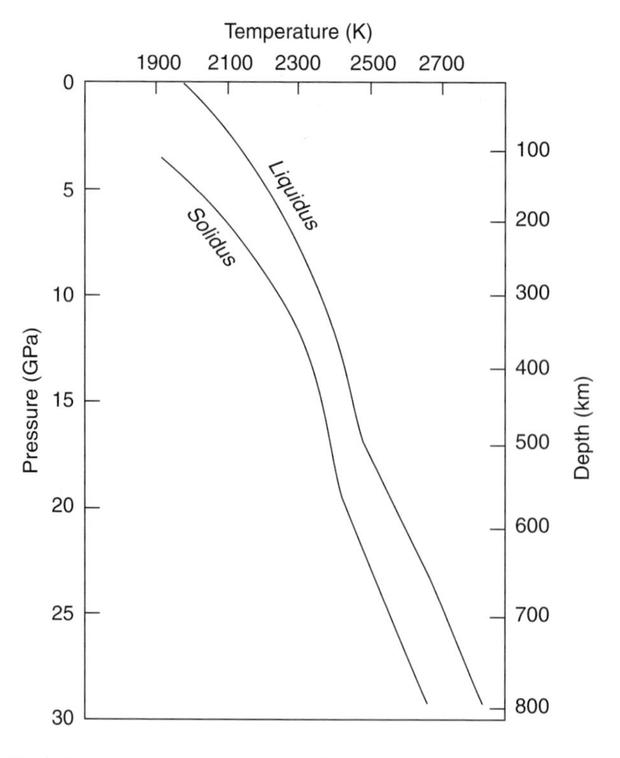

Figure 10.2. Solidus and liquidus for dry pyrolite as a function of pressure. Adapted from Litasov and Ohtani (2002).

and 9). Thus, there can be no doubt that the magma ocean phase was short-lived, which is confirmed by many geochemical studies. At 4.4 Gy, crustal material had already been extracted from the mantle according to Caro *et al.* (2003), and there was liquid water at the Earth's surface at 4.3 Gy according to Mojzsis *et al.* (2001).

10.1.3 Average secular cooling rate

The present-day mantle potential temperature is fixed at ≈ 1600 K by a fit to heat flux and bathymetry data regardless of the water content of mantle rocks (McKenzie *et al.*, 2005). Sub-solidus convection began at a mantle potential temperature of about 1800 ± 100 K, about 200 K higher than present. Even if the timing is not known precisely, this constrains the average cooling rate of the Earth to 50 ± 25 K Gy^{-1}. If the mantle contained significant amounts of water at the end of the magma ocean phase, the phase diagram must be shifted to lower temperatures. In this case, the starting potential temperature at the beginning of sub-solidus convection was even less than 1800 K, and the average cooling rate of the Earth must be lower than 50 K Gy^{-1}.

Other estimates of the Earth's cooling rate can be derived from the chemical composition of ancient rocks. Continental crustal material was different in the Archean than it is today. Basaltic lavas exhibit systematic compositional trends with time, including a secular decrease in average MgO content. MgO-rich ultramafic lavas named komatiites are common in the Archean and are almost absent from today's rock record. If komatiites were generated by deep mantle plumes, involving mantle that is essentially dry, one deduces that mantle plume temperatures have decreased by about 300 K in 3 Gy (Nisbet *et al.*, 1995). We have seen, however, that plumes are hotter than the average mantle, so that one cannot infer a cooling rate for the whole planet from data on a few selected upwellings. According to an alternative hypothesis, komatiites were generated in a subduction environment, involving mantle hydrated by downgoing plates. In that case, one concludes that this part of the mantle was only slightly hotter (≈ 100 K) in the Archean than today (Grove and Parman, 2004). In both cases, komatiites do not sample "average" mantle and it is not clear how to extrapolate these temperature estimates to the entire mantle.

Mid-ocean ridge basalts are suited to studies of the mantle's average temperature because they can be sampled over very large areas. They are a compositionally heterogeneous group, however, which translates into a wide temperature range ($\approx$200 K) (Klein and Langmuir, 1987; Kinzler and Grove, 1992). The liquidus temperatures of Phanerozoic MORBs and Archean MORB-like greenstones follow a trend of decreasing temperature as a function of time (Figure 10.3). The liquidus temperature can be used to infer the potential temperature that generated these

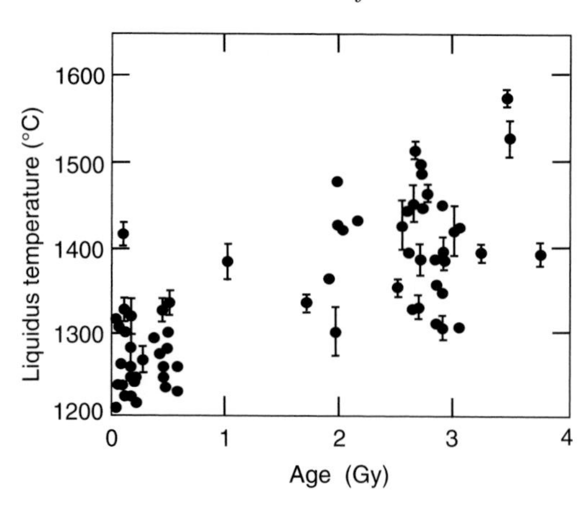

Figure 10.3. Liquidus temperatures of basalts from ophiolites and greenstone belts with MORB characteristics as a function of age. Adapted from Abbott *et al.* (1994).

melts by decompression. According to Abbott *et al.* (1994), the mantle temperature change since the late Archean (2.5 Gy) is 137 ± 8 K (from temperature ranges) to 187 ± 42 K (from temperature means). This leads to a cooling rate in the range $\approx 50 \pm 20$ K Gy^{-1}, which is consistent with estimates based on mantle rheology at the end of the magma ocean.

10.1.4 Convective heat flux

The Earth is cooling down, implying that its thermal structure is not in steady state. In this case, the horizontally averaged heat balance equation is,

$$\rho_o C_p \frac{\partial \overline{T}}{\partial t} + \frac{\partial}{\partial z}\left(\rho_o C_p \overline{w\theta}\right) = \lambda \frac{\partial^2 \overline{T}}{\partial z^2} + H. \tag{10.2}$$

In the mantle, as explained in Chapter 3, one should add an isentropic gradient to this temperature. Laboratory experiments and numerical calculations show that, in this case, the interior is well mixed and the horizontal average of temperature does not vary significantly with depth. Thus, one can write that, away from the thermal boundary layers at the top and bottom, $\overline{T}(z,t) = T_i(t)$, and the heat balance equation can be written as,

$$\frac{\partial}{\partial z}\left(\rho_o C_p \overline{w\theta}\right) = \left[H - \rho_o C_p \frac{dT_i}{dt}\right]. \tag{10.3}$$

This shows that bulk cooling can be treated as internal heat generation.

10.2 Thermal evolution models

10.2.1 The Urey ratio

Over the Earth's history, heat sources have decreased by a factor of about four. The decay time of bulk radiogenic heat production, which is such that heat production decreases by a factor e, and which is the weighted average of the individual decay times of the four relevant isotopes (Table 8.7), is 3 Gy. The efficiency of Earth's convective engine in evacuating this heat may be measured by the Urey ratio, Ur, which is the ratio of heat production over heat loss:

$$\mathrm{Ur} = \frac{\int_V H dV}{\int_A \mathbf{q}\cdot\mathbf{n}\, dA} = \frac{H_T}{Q}, \tag{10.4}$$

where Q is the total rate of heat loss and H_T is the rate of heat production. For $\mathrm{Ur} = 1$, heat loss and heat production balance one another exactly and the average mantle temperature remains constant. There is some ambiguity in thermal evolution models depending on the role assigned to continental heat sources. One may argue that continental heat sources are stored in the continental lithosphere and hence are not available to drive mantle convection. In this case, the present budget (Table 8.8) implies that $\mathrm{Ur} = 0.33$, with a total range of 0.21–0.49. Alternatively, one may argue that such separation between active and passive heat sources is rather arbitrary and that it may not be valid for all of Earth's history. Geochemical data indicate that the continental crust has been extracted continuously from the Earth's mantle. In this case, $\mathrm{Ur} \approx 0.43$. In both cases, the Urey number is less than 0.5, which provides a constraint on the mantle convection regime as well as a test of our understanding of mantle convection processes.

From the present-day energy budget, the cooling rate is about 120 K Gy^{-1}, which is less than the average value of about 50 K Gy^{-1} deduced from rheological and petrological arguments. This shows that the cooling rate has increased with time and provides a constraint on how the rate of heat loss has evolved. The global heat balance is

$$M\langle C_p\rangle \frac{d\langle T\rangle}{dt} = -Q + H_T, \tag{10.5}$$

where M is the mass of the Earth, $\langle C_p\rangle$ is an "effective" heat capacity which accounts for the variation of temperature with depth. Integrating over the age of the Earth, one deduces that

$$\frac{Q_{av} - H_{Tav}}{Q - H_T} = \frac{(dT/dt)_{av}}{d\langle T\rangle/dt}, \tag{10.6}$$

where Q_{av} and H_{Tav} are the time-averaged values of heat loss and heat production and $(dT/dt)_{av}$ is the average cooling rate. From available constraints on the cooling rate, we deduce that the ratio in equation 10.6 is less than 1 and probably as small as 0.4. This implies that the rate of heat loss varies less rapidly than that of heat production.

10.2.2 "Parameterized" cooling models

To evaluate the efficiency of mantle convection, we consider a heat flux scaling law of the following form:

$$Q = C_2 T^{1+\beta} \nu_m^{-\beta}, \tag{10.7}$$

where constant C_2 depends on fluid properties that depend weakly on temperature, β is a power-law exponent and ν_m is the viscosity in the well-mixed fluid below the thermal boundary layer. We have established in Chapters 4 and 8 that, in an asymptotic limit at very large values of Rayleigh number, the heat flux out of a fluid layer does not depend on the large-scale convective flow and is determined by the dynamics of a thin thermal boundary layer. In this case, the heat flux conforms to equation 10.7 with $\beta = 1/3$. In this section, however, we shall consider β as an unknown parameter to be determined from observation. Cooling models of this kind have been termed "parameterized" because they collapse all the physics of mantle convection into a single equation involving only temperature and two parameters, C_2 and β.

The mantle viscosity ν_m depends strongly on temperature. Temperature changes in the Earth are small compared to the absolute temperature (i.e., ≈ 200 K for a present day temperature of $T_0 \approx 1600$ K). In this case, one can approximate the Arrhenius law for mantle viscosity by an equation of the form $\nu = \nu_0 (T/T_0)^{-n}$, with $n \sim 35$ (Davies, 1980; Christensen, 1985b). The thermal evolution equation then takes the following form:

$$M \langle C_p \rangle \frac{dT}{dt} = -Q_0 \left(\frac{T}{T_0} \right)^{1+\beta(1+n)} + H_T(t), \tag{10.8}$$

where Q_0 is the heat loss at the reference potential temperature T_0. One may further linearize this equation by considering small temperature variations around T_0:

$$T = T_0 + \Theta, \text{with} \Theta \ll T_0 \tag{10.9}$$

$$M \langle C_p \rangle \frac{d\Theta}{dt} = -Q_0 \left[1 + \frac{\Theta}{T_0}(1 + \beta + \beta n) \right] + H_T(t). \tag{10.10}$$

The heat sources can be approximated as decreasing exponentially with time, $H_T(t) = H_{T0} \times \exp(-t/\tau_r)$, where $\tau_r \approx 3000$ My. The solution of equation 10.10 is,

$$\Theta = \Theta_0 \times \exp(-t/\tau_p) + \frac{Q_0 \tau_p}{M \langle C_p \rangle} \left(\exp(-t/\tau_p) - 1\right)$$

$$+ \frac{H_{T0} \tau_p \tau_r}{M \langle C_p \rangle T_0 (\tau_r - \tau_p)} \left(\exp(-t/\tau_r) - \exp(-t/\tau_p)\right), \tag{10.11}$$

where the relaxation time constant τ_p is given by

$$\tau_p = \frac{M \langle C_p \rangle T_0}{(1 + \beta + \beta n) Q_0}. \tag{10.12}$$

From equation 10.11, the Urey ratio as a function of time can be obtained. One finds that,

$$\lim_{t/\tau_p \to \infty} Ur = \frac{\tau_r - \tau_p}{\tau_r}. \tag{10.13}$$

The relaxation time of mantle convection is not known a priori but we can guess that it must be smaller than the age of the Earth. In this case, the above limit summarizes neatly the constraint brought by the Urey ratio. Within the framework of this simple parametrization scheme, the Urey ratio provides a value of relaxation time τ_p and hence of exponent β. A key point is that model predictions are not sensitive to initial conditions, which has two implications. One is that failure of a model to reproduce the present-day Urey ratio cannot be blamed on the poorly known initial condition. The other implication is that "backward" thermal calculations starting from the present become unreliable for old ages.

Using standard values for the parameters and variables involved, $n = 35$, $\beta = 1/3$, $M = 6 \times 10^{24}$ kg, $Q_0 = 30$ TW, $T_0 = 1300$ K (the temperature jump across the boundary layer is the relevant parameter here) and $\langle C_p \rangle = 1200$ J kg^{-1} K^{-1}, we find that $\tau_p \approx 800$ My and hence that Ur ≈ 0.75. This is larger than observed and reveals a fundamental flaw in the model set-up. The flaw may be in the physics of convection and, within the framework of this calculation, in the value for β. An alternative possibility may be that current heat loss estimates may be biased and are not representative of the long-term evolution of the Earth. This is addressed in a separate section below and we now briefly discuss the β value. In order to meet the constraint of the Urey ratio, one must increase the adjustment time of mantle convection, which implies small values of β. According to Conrad and Hager (1999), the resistance to bending of oceanic plates at subduction zones leads to $\beta \approx 0$. Larger degrees of melting in the past due to higher mantle temperature enhance dehydration of the residual solid and hence generate stiffer plates (Sleep, 2000). In these conditions, one may even obtain a negative value for β, such that the Earth's heat loss has in fact been increasing with time as the mantle was cooling (Korenaga, 2003). There is no evidence that resistance to bending actually limits subduction on Earth, however.

10.3 Fluctuations of the mantle heat loss

The heat loss of the Earth is effected in large part through the ocean floor and depends on the number, velocity and configuration of oceanic plates. One must ask whether the plate characteristics remain steady through time and also whether today's plates are truly representative, in a time-average sense, of the global mantle convection pattern. In turbulent flows, one frequently invokes the principle of *ergodicity* such that the spatial average of a variable, such as temperature or velocity, is equal to its time average. This requires a system that is large enough to include a large number of flow realizations. In a time-dependent system, this also requires a large separation of time scales, such that the characteristic time for the flow to sample the whole domain, or the mantle overturn time in our context, is small compared to the secular evolution time scale. In other words, one must ask whether today's oceanic plates allow a reliable heat loss average.

10.3.1 Vagaries of sea-floor spreading

The Earth's convecting mantle exhibits several features which make it very distinctive. One of them is the triangular age distribution of the sea floor (Figure 8.2, and Parsons, 1982). As shown by Labrosse and Jaupart (2007), this is at odds with laboratory convecting systems as well as numerical simulations, which illustrates current limitations in reproducing mantle convection processes. A few other peculiar features of mantle convection are worth mentioning. Heat loss is unevenly distributed at the Earth's surface. The Pacific Ocean alone accounts for almost 50% of the oceanic total, and 34% of the global heat loss of the planet. This is due in part to the large area of this ocean and in part to its high spreading rate. Oceanic plates are transient, such that changes of oceanic heat loss may occur when a new ridge appears or when one gets subducted. For example, the heat flux out of the Atlantic Ocean is about 6 TW, 17% of the oceanic total (Sclater *et al.*, 1980a). This ocean has almost no subduction and started opening only at 180 My. At that time, the generation of a new mid-ocean ridge led to an increase of the area of young sea floor at the expense of old sea floor from the other oceans, and hence to enhanced heat loss. The triangular age distribution may well be a consequence of this relatively recent plate reorganization.

From the standpoint of mantle dynamics, the most challenging features of mantle convection are perhaps the large variations in plate speeds and dimensions. With the small number of plates present, average values of spreading velocity and plate size may well be meaningless, and heat loss cannot be related to average velocity. In order to identify the key control variables, we derive a general equation for heat loss through the ocean floor. We use the half-space cooling model which is sufficiently accurate (equation 8.28). The distribution of sea-floor age f is a function of dimensionless age τ/τ_m such that,

$$\frac{dA}{d\tau} = C_A f\left(\frac{\tau}{\tau_m}\right), \tag{10.14}$$

where C_A is the plate accretion rate and where $f(0) = 1$. We obtain the total oceanic heat loss:

$$Q_{oc} = A_o \frac{\lambda T_M}{\sqrt{\pi \kappa \tau_m}} \frac{\displaystyle\int_0^1 \frac{f(u)}{\sqrt{u}} du}{\displaystyle\int_0^1 f(u) du} = A_o \frac{\lambda T_M}{\sqrt{\pi \kappa \tau_m}} \gamma(f), \tag{10.15}$$

where A_o is the total ocean surface and $\gamma(f)$ a coefficient which depends on the dimensionless age distribution. This equation introduces a new variable, the dimensionless age distribution. For the present-day triangular age distribution,

$$Q_{oc}^T = \frac{8A_o}{3} \frac{\lambda T_M}{\sqrt{\pi \kappa \tau_m}}. \tag{10.16}$$

This may be compared to that for a rectangular age distribution such that $f(\tau/\tau_m) = 1$, corresponding to 2D convection cells:

$$Q_{oc}^R = 2A_o \frac{\lambda T_M}{\sqrt{\pi \kappa \tau_m}}. \tag{10.17}$$

This shows that, all else being equal, changing the sea-floor age distribution may change the oceanic heat loss by significant amounts: going from a rectangular to a triangular age distribution, for example, increases the oceanic heat loss by about 30%. Variations of sea-floor spreading are constrained by the total oceanic area,

$$A_o = C_A \tau_m \int_0^1 f(u) du. \tag{10.18}$$

If we assume that the oceanic area and the age distribution remain constant, changes of τ_m imply changes of C_A, the average rate of sea-floor production. Conversely, changes of C_A imply changes of the maximum plate age and/or of the age distribution. Over a few tens of My, the oceanic area may change, however. Such changes are not related to rigid plate tectonics and occur due to diffuse deformation zones in plate interiors or at plate margins. These zones are found in both oceans and continents and presently account for $\approx 15\%$ of the Earth's surface (Gordon, 1998). In continents, extension occurs at the expense of oceans whereas shortening increases the oceanic area.

The present-day age distribution may well be a snapshot of a continuously evolving plate system. Examining individual ocean basins allows an understanding of how it arises (Sclater *et al.*, 1981). In the Atlantic, the age distribution is rectangular up to 80 My, and then the area per unit age decreases for older ages, due to the late opening of the North Atlantic and in part to subduction in the Antilles and South Sandwich arcs. The Indian Ocean has almost the same age distribution as the Atlantic except at very young ages (less than 4 My), where the area per unit age increases abruptly. In contrast, the older and larger Pacific Ocean exhibits no simple age distribution even over small age intervals. Thus, sea-floor spreading seems to proceed with a rectangular age distribution in ocean basins that do not exceed a certain size, as in ideal convection cells. The triangular age distribution develops out of a rectangular distribution and may be due to geometrical constraints on subduction that cannot be avoided when several spreading systems become active at different times and compete for space. Using geological reconstructions of oceanic plates, (Becker *et al.*, 2009) have proposed that the rate of sea-floor spreading has decreased by 25–50% during the past 140 My. They have also suggested that the age distribution function changed at 110 My from near rectangular to the present-day triangular distribution. Such reconstructions are based on the probability of subduction as a function of age and it is difficult to assess their accuracy.

Fluctuations in sea-floor spreading may occur on a longer time scale. Subduction of young sea floor occurs mostly at the edge of continents and may be due to the complex geometry of ocean–continent boundaries. With all continents assembled in a single landmass, the large continuous oceanic area imposes fewer constraints on spreading and subduction. In other words, the present-day distribution of subduction zones may be a transient feature associated with the breakup of Gondwana. The assembly and breakup of supercontinents occurs over characteristic time τ_W. Allegre and Jaupart (1985) have related this time to the "mean free path" of continents, such that continents sweep the whole surface of the Earth and necessarily run into one another. They obtained $\tau_W \approx 400$ My for present-day spreading rates and distribution of continents. τ_W varies as a function of continental area and drift velocity and was probably larger in the past when continents accounted for a smaller fraction of the Earth's surface. Geological data support such an increase of τ_W (Hoffman, 1997). Note that this observation runs against the intuitive notion that plates moved faster in the past. If the rate of heat loss of the Earth depends on the distribution of continents, it oscillates on a time scale of $\approx \tau_W$ over a long-term decreasing trend.

10.3.2 Heat flow out of the core

A similar uncertainty affects estimates for the heat loss of the Earth's core. This has been estimated from the core side using energy requirements for the dynamo. The core loses heat to an unstable boundary layer which grows at the base of the mantle. We can estimate the energy contained in this boundary layer, which we identify as the D$''$ layer. The temperature difference across this 200 km thick layer can be estimated to be about $\delta T = 1000$ K (e.g. Lay *et al.*, 1998). Assuming a linear temperature profile, the energy content of this layer is

$$U = \rho C_P 4\pi b^2 h \frac{\delta T}{2} \simeq 7.5\ 10^{28} \mathrm{J}, \tag{10.19}$$

with $b = 3480$ km the radius of the CMB, $\rho \simeq 5 \times 10^3$ kg m^{-3} density, $h = 200$ km and $C_P = 1000$ J kg^{-1} K^{-1}. This energy is transferred to the mantle when the boundary layer goes unstable. The time scale is that of conductive thickening of the layer i.e. $h^2/\pi\kappa \simeq 400$ My for $\kappa = 10^{-6}$ m^2 s^{-1}. One may argue that parts of the boundary layer are always unstable at any given time, so that there are only local fluctuations of heat flux but no time variations of the integrated heat loss. The paleomagnetic record indicates long periods of anomalous field behaviour, however, such as the Cretaceous Long Normal Superchron which saw no reversal of the magnetic pole for about 40 million years. These anomalous periods may be due to changes of cooling conditions in the Earth's core, which could be related to modifications of the mantle basal boundary layer. Dividing the total energy in the thermal boundary layer by the time scale of 400 My indicates that time variations of heat input at the base of the mantle could be as large as 5 TW.

10.3.3 Time-dependent fluctuations in mantle temperature?

The complex evolution of the oceanic plates probably lead to significant short-term variations of heat loss, with two consequences. One is that the present-day heat loss estimate may be biased by a transient plate configuration. Another consequence is that the mantle temperature may not decrease in a monotonous fashion and may undergo fluctuations due for example to supercontinent cycles. The large heat capacity of the Earth acts as a strong thermal buffer, however. An intrinsic time scale for the cooling of the Earth through the oceans may be defined as follows:

$$\tau_C = \frac{MC_pT_M}{Q_{oc}} = \frac{MC_p\sqrt{\pi\kappa\tau_m}}{\lambda A_o\gamma(f)}, \tag{10.20}$$

which is about 10 Gy. This time scale is the time required for both temperature and heat flux to drop by a factor e if all heat sources are instantly suppressed. Thus, fluctuations of heat loss must be spread over time intervals that are not short compared to τ_C to affect the average mantle temperature. This can be investigated using thermal calculations. The data indicate that the bulk rate of heat loss has changed less rapidly than heat production. Thus, for the purpose of example, we assume that the heat loss can be decomposed into two parts: a secular part which varies little through time, and a fluctuating part which reflects plate reorganizations and supercontinent cycles. Figure 10.4 shows that, for fluctuations over a supercontinent time

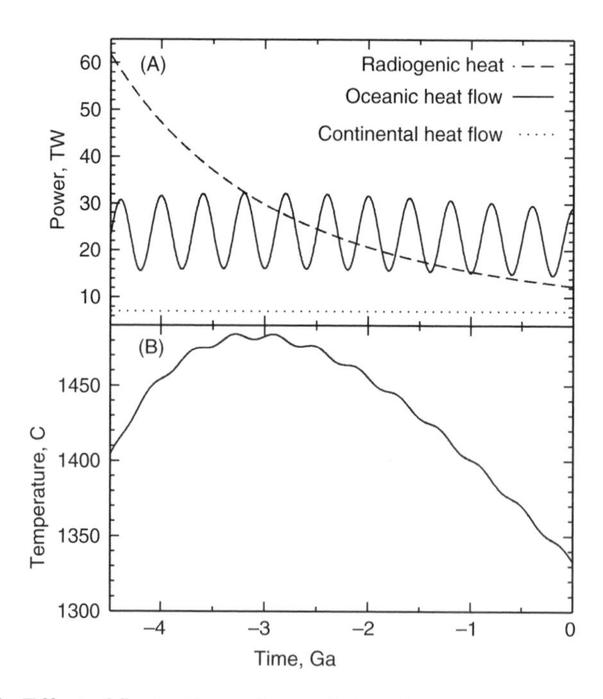

Figure 10.4. Effect of fluctuations of oceanic heat loss over a supercontinent time scale of $\tau_W = 400$ My on the secular thermal evolution of the mantle, from Labrosse and Jaupart (2007). Model calculations rely on a simple parametrization of sea-floor spreading that leads to a weak secular variation of oceanic heat loss. Heat loss fluctuations due to changes of sea-floor spreading are superimposed on the secular trend. These changes occur on a time scale that is short compared to that of the secular trend and leave almost no trace on mantle temperatures. (A): the various contributions to the Earth's energy budget that are used in the calculation. (B): predicted evolution of the average mantle temperature through time.

scale $\tau_W = 400$ My, the net effect on the mantle temperature is small and that the secular cooling trend is not affected significantly by fluctuations of heat loss.

10.4 Continental growth and cooling of the Earth

The growth of continents has two different effects on mantle convection. One is that it increases the size of the continental domain at the expense of the oceanic one, where large heat fluxes are achieved. This lowers the cooling rate. The other effect is the extraction of radioelements from the mantle and their storage in the continental crust, which reduces the amount of internal heat generation driving convection. To evaluate the former, one needs reliable scalings for the dimensions of oceanic plates, which are not yet available. We have suggested, however, that the widths of oceanic convective cells that are observed today are affected by the presence of continents. One consequence is further reduction of the Earth's cooling rate: free convection cells are limited to part of the Earth's surface and stretch over large horizontal distances, implying a lower heat flux than with the shorter cells typical of the Rayleigh–Benard configuration. One may discuss the second effect with the scaling laws that have been developed (Chapters 5 and 9).

Continental crust is enriched in radioactive elements that have been extracted from the Earth's mantle. These elements are stored near the surface in material that does not deform easily and does not participate in convection. Thus, continental growth acts to reduce the amount of internal heat generation driving mantle convection. We evaluate the consequences of this fundamental geological process for the cooling of the Earth. We determine the internal temperatures that are required to evacuate the amount of heat produced in the planet by two mechanisms, convection in a fluid with viscosity μ over thickness h, and conduction in a rigid crust enriched in radioactive elements over thickness d_c. The total amount of heat generated is the same in both cases, so that $H_{\mathrm{conv}} h = H_{\mathrm{cond}} d_c$, where we have introduced the two different rates of heat generation that are involved. In order to focus on internal heat generation, we consider that no heat is brought into either the convecting fluid or the rigid crust from below. Thus, the surface heat flux is the same in both cases, and is equal to $Q = H_{\mathrm{conv}} h = H_{\mathrm{cond}} d_c$. The temperature differences between the surface and the Earth's interior in the two cases are,

$$\Delta T_{\mathrm{conv}} = C_Q^{-3/4} \left(\frac{H_{\mathrm{conv}} h}{\lambda} \right)^{3/4} \left(\frac{\kappa \mu}{\rho_o g \alpha} \right)^{1/4} \tag{10.21}$$

$$\Delta T_{\mathrm{cond}} = \frac{H_{\mathrm{cond}} d_c^2}{2\lambda}, \tag{10.22}$$

where C_Q is the constant in the local heat flux scaling law (equation 9.30). For a comparison, it is best to use the ratio between the two,

$$\frac{\Delta T_{\rm cond}}{\Delta T_{\rm conv}} = \frac{C_Q^{3/4}}{2}\frac{d_c}{h}\left(\frac{\rho_o g \alpha H_{\rm conv} h^5}{\lambda \kappa \mu}\right)^{1/4} = \frac{C_Q^{3/4}}{2}\frac{d_c}{h}{\rm Ra}_H^{1/4}, \tag{10.23}$$

where we recognize the Rayleigh number Ra_H for internal heating (equation 9.19) and where the proportionality constant is $C_Q^{3/4}/2 \approx 0.2$. In the Earth, $d_c \approx 30$ km and $h \approx 3000$ km, and hence $d_c/h \approx 10^{-2}$. This result shows that it takes very large Rayleigh numbers, i.e. extremely vigourous convection, for $\Delta T_{\rm cond}$ to be larger than $\Delta T_{\rm conv}$. In practice, this requires $\mathrm{Ra}_H > 5 \times 10^{10}$, which was only possible in early stages of Earth's evolution. For reference, this Rayleigh number is about 10^9 for today's mantle viscosity, so that $\Delta T_{\rm cond}/\Delta T_{\rm conv} \approx 0.4$. In other words, conduction in a thin enriched crust evacuates radioactive heat with a lower internal temperature than solid-state convection. This simple calculation demonstrates that the most efficient cooling mechanism for a planet is to exhaust the energy source for mantle convection by growing a radioactive crust at the top.

10.5 Conclusion

Simple thermal models relying on scaling laws for the convective heat flux do not allow realistic predictions of mantle temperatures through time. The thermal evolution of the Earth depends on two variables that remain poorly understood, the distribution of sea-floor ages and the rate of continental growth. Relating the velocities and subduction behavior of oceanic plates to mantle convection processes is a difficult challenge.

11

Magmatic and volcanic systems

Objectives of this chapter

In this chapter, we consider only the cooling of magmas and do not investigate melt generation. Magmas intrude the crust and may accumulate in large reservoirs. Many processes occur in these reservoirs, involving replenishment by melts with compositions that may change with time, crystal settling, compositional convection as well as late-stage equilibration with meta-somatic fluids percolating through already solidified cumulates. We shall focus on the thermal aspects of crystallization. We begin with an analysis of latent heat release due to solidification. We evaluate how long magma reservoirs can remain active and feed eruptions.

11.1 A few features of crustal magma reservoirs

11.1.1 Dimensions and time scales

Crustal magma reservoirs can be studied in the field in two different ways: by using erupted lavas on the one hand and studying plutonic bodies brought to the surface by erosion on the other. There is a lingering controversy about the relationship between the two because it is not clear that all plutons were once volcanic reservoirs feeding eruptions. There is no doubt, however, that most volcanic systems involve at least one storage zone, such that large volumes of crystallized magma must be left at depth in order to account for the changes of lava composition that occur. We can therefore deduce from age determinations on lavas over what length of time magmatic systems remain active. For example, Mount Adams, state of Washington, erupted lavas for more than 500,000 years (Hildreth and Lanphere, 1994). The Bishop Tuff system, California, has been active for more than one million years (Halliday *et al.*, 1989; Hildreth and Wilson, 2007). There can be little doubt that magma reservoirs are open systems that evolve due to repeated replenishment and withdrawal events. Igneous complexes that have been unroofed by erosion demonstrate that magma reservoirs can grow to very large dimensions (Table 2.2). The Bushveld complex, in South Africa, for example, extends over a thickness of at least 7 km and over an area of about 6×10^4 km^2. Thus, when dealing with magmas, one has to consider bodies that may be as thin as a few tens of centimeters (dykes) and as thick as ten kilometers or more. We shall see that a range of cooling and crystallization behaviors are possible depending on size.

11.1.2 *Evolution of magma in a reservoir*

Large igneous complexes are usually compositionally and petrologically stratified, with a large-scale upward trend towards silicic compositions that is interrupted by a roof sequence. Changes of composition appear as changes of mineral assemblages, mineral compositions as well as trace element concentrations. Many petrological studies have focused on chemical and mineralogical variations, leaving temperature as the variable to be adjusted from phase diagram constraints. Yet, it is the thermal evolution that dictates how cooling and solidification proceed. Cooling is not only the driving mechanism for crystallization and hence differentiation, but it also determines how, where and how quickly crystals nucleate and grow.

Crystal settling has long been the favored explanation for igneous differentiation and layering (Wager and Brown, 1968). This model requires that settling occurs at a faster rate than cooling and crystallization. Yet, most crystals are generated in the cold thermal boundary layers that develop at the margins of a magma body, where there are strong gradients of temperature, crystal content and viscosity. Direct observations on lava lakes, which will be reviewed in more detail below, reveal that the major crystal phases which form in the cooling lava do not settle to any appreciable extent (Wright *et al.*, 1976; Helz, 1980). Thus, an assessment of crystal settling must begin with a sound understanding of the structure and evolution of thermal boundary layers. This is also required for a study of convective motions, which may act to transport melt and possibly suspended crystals from one cooling interface to another.

Rocks that are encasing a magma reservoir also deserve study. When attempting to determine their thermal evolution, one may wonder to what extent one must account for the intricate processes affecting the interior of the reservoir.

11.1.3 *Structure of magmatic boundary layers*

Upon emplacement at temperature T_i, magma loses heat to the colder country rock. The thermal boundary layer must be split in three different parts (Figure 11.1). That part which is below the solidus is fully solid and may include a chill phase that formed upon emplacement. Between the solidus and liquidus, magma is partially

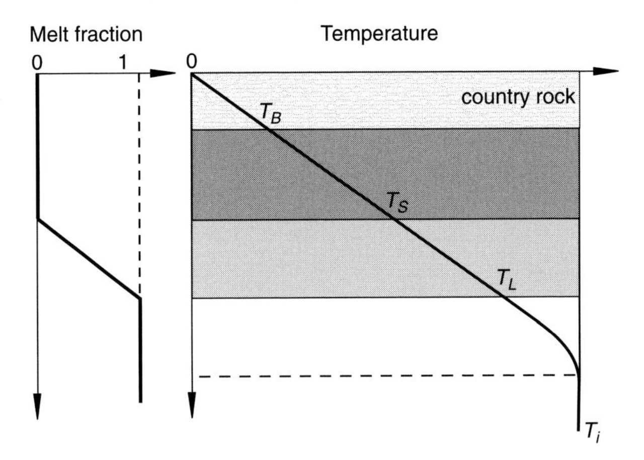

Figure 11.1. Schematic diagram illustrating the thermal structure at the margin of a magma body. The top of the magma body is at temperature T_B and is underlain by a thermal boundary layer that extends to the molten interior at temperature T_i. A partially crystallized layer exists between the solidus and liquidus temperatures. In that layer, the melt fraction goes from zero to 1.

crystallized and the behavior of such a two-phase system cannot be approximated by that of a pure liquid. The third part of the boundary layer is made of superheated magma, which may or may not exist depending on the initial conditions. The partially crystallized zone may be subdivided further into two different domains separated by a rheological transition that occurs for a solid fraction of about 60%. Above that threshold value, crystals are interconnected and cannot move with respect to one another. Below the threshold value, crystals are suspended within a continuous liquid phase. The interconnected crystal framework, which grades into the fully solidified roof region continuously, is immobile and cannot participate in convective instabilities. Lower down in the thermal boundary layer, the melt + crystals mixture may be treated as an equivalent liquid with properties that depend on the crystal content and the composition of the interstitial liquid. Depending on the initial melt composition, the interstitial liquid may be less dense or denser than the uncrystallized magma in the interior of the magma body, so that no general statement can be made on stability. A finite undercooling is required for nucleation of the crystals. Thus, another subdivision must be made at the external edge of the thermal boundary layer at temperatures slightly below the liquidus, where the melt may be devoid of crystals because of kinetic limitations on crystallization.

Drilling in Hawaiian lava lakes has provided a wealth of information on the structure of a natural magmatic system in the process of cooling and crystallizing. The observations dealt mostly with the upper boundary layer, but were complemented by studies of fully solidified lakes. The initial erupted magmas were olivine tholeiites

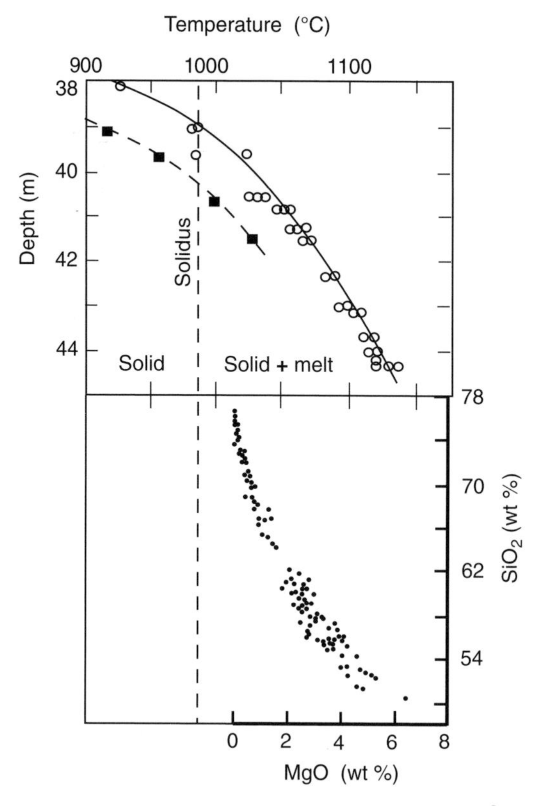

Figure 11.2. Profiles of temperature and interstitial melt composition in the upper boundary layer of Kilauea Iki lava lake. Within the layer, the relationship between composition and temperature is single valued, so that one can plot the concentrations of the major oxides (SiO_2 and MgO) as a function of temperature. Open circles stand for temperatures calculated for the composition of the interstitial melt in thermodynamic equilibrium conditions. Black squares correspond to *in situ* thermocouple temperature measurements. Redrawn from (Helz, 1980, 1976). Note that the interstitial melt composition spans a very large composition range.

with various amounts of olivine phenocrysts.[1] In the boundary layer, the proportions of mineral phases and interstitial melt varied continuously (Figure 11.2). Silica-rich melts were sampled including rhyolites with as much as 75–76% SiO_2. Thus, compositions representing the whole line of descent were present at any one time. Using laboratory experiments on Hawaiian olivine tholeiites, Helz and Thornber (1987) reproduced the same mineral assemblages as those found in the lakes. For a given mineral assemblage, the difference in temperature between the laboratory experiments and the *in situ* partially crystallized lava was less than $\approx$ 15 K,

[1] Phenocrysts denote crystals that grew from the magma that carries them, in contrast to xenocrysts which belong to other melts or to encasing rocks and which have been entrained into the magma.

showing that, to a good approximation, crystals and coexisting melt were in local thermodynamic equilibrium.

11.1.4 Convection

A magmatic thermal boundary layer includes crystallized magma that is almost solid and hence only a fraction of the temperature difference that exists across it is available to drive convection. The smallest part of the boundary layer that can participate in convective motions is made of crystal-free magma and may include both superheated and undercooled melt. Partially crystallized magma from deeper in the thermal boundary layer can also become unstable, but this cannot be assessed simply as both thermal and compositional effects on density must be taken into account.

To evaluate the possibility that convection occurs in a magma reservoir, we focus on crystal-free magma, such that cooling generates a negative density contrast that is potentially unstable regardless of the magma compositions. For simplicity, we assume a simple equation of state for the magma, such that,

$$\Delta\rho = -\rho_i \alpha \Delta T, \tag{11.1}$$

where α is the coefficient of thermal expansion. The Rayleigh number must exceed a value of about 10^3 for convection to occur (see Chapter 5). For convective heat transport to be significant in the heat budget, we require the Nusselt number to exceed a value of 2, such that the convective heat flux is at least twice the conductive heat flux. In transient cooling conditions, one must focus on the upper thermal boundary layer because there can be no thermal convection from the stable lower boundary layer. We use typical values of 3×10^{-5} K^{-1} for α and 10^{-6} m^2 s^{-1} for κ and fix ΔT at a small value 10 K. For basaltic magma with $\mu = 10$ Pa s, we find that the upper boundary layer must extend over at least 10 cm for convection to occur. For a Nusselt number of 2, the magma body must be four times as thick; i.e. it must be thicker than 40 cm for convection to be significant. Alternatively, we consider a 1 km thick basaltic magma body and ask what temperature difference is required to generate convection. We find that the Rayleigh number is largely supercritical with even a tiny temperature difference of 0.01 K. These simple arguments indicate that it is impossible to enforce stability in a large liquid body and that thermal convection cannot be avoided in all magma reservoirs of significant size. That convection operates with small temperature differences has been used to argue that it does not play an important role because it does not control the heat budget and crystallization history, but we shall show that this is erroneous.

The occurrence of thermal convection has been detected in Hawaiian lava lakes. In these lakes, temperature measurements through the upper thermal boundary

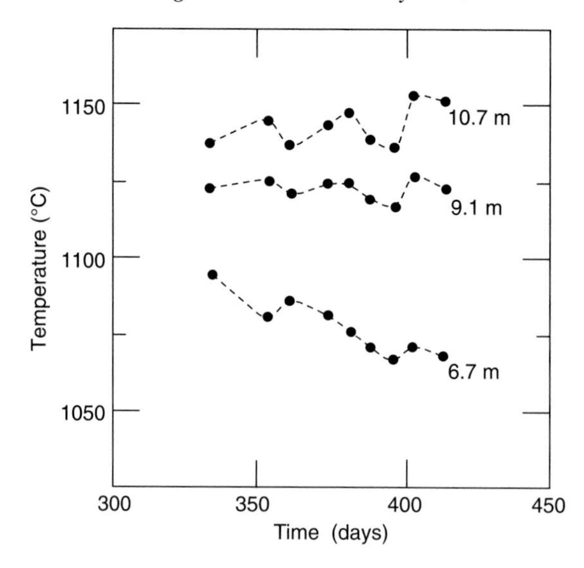

Figure 11.3. Time variation of temperature in the Makaopuhi lava lake, Kilauea volcano, Hawaii, at three different depths. The magnitude of temperature fluctuations increases with increasing depth, i.e. towards the lower edge of the thermal boundary layer, which is indicative of convection. Adapted from (Wright and Okamura, 1977).

layer are in good agreement with purely conductive cooling calculations (Wright and Okamura, 1977). Lava lakes degas for a long time and the upward motion of bubbles in the lake interiors prevents thermally driven instabilities. In the Makaopuhi lake, degassing was limited to an early phase and several observations indicate that thermal convection did develop eventually. For the first seven months, as the crust grew to about 9 m thickness, its composition did not change (Wright and Okamura, 1977) and temperatures decreased in a monotonic fashion. Both observations indicate that there was no convection within the lake. After that initial phase, however, significant temperature fluctuations were recorded (Figure 11.3). Furthermore, the olivine phenocrysts that had been retained in the growing crust of the initial phase were no longer observed in the new crust. From these two facts, Wright *et al.* (1976) deduced that convection had set in. The magnitude of temperature fluctuations that were recorded is about 20 K, which is consistent with an effective temperature-dependent viscosity law and the scalings developed in Chapter 5 (Davaille and Jaupart, 1993a). In magma reservoirs, thermal convection does not modify temperature profiles greatly in comparison with those for purely conductive cooling. That this is so in a lava lake can be understood easily when one considers that the conductive heat flux through a 30 m thick crust with a temperature difference of 10^3 K between magma and the atmosphere is very large, $\approx 10^2$ W m^{-2}, which is of the same order of magnitude as the convective heat flux.

In magma reservoirs, we shall see that convection has a weak impact on the heat balance at the roof also, but that it nevertheless is an important process for igneous differentiation.

11.2 Initial conditions: Super-heated magma?

The evolution of a melt body after emplacement depends on the initial conditions. Measurements and observations in lavas that erupt at the Earth's surface indicate that they usually bear phenocrysts and are close to their liquidus temperature. This has led to a wide consensus that magmas are not superheated when they get intruded in the Earth's crust. One exception is provided by some obsidian flows which have developed thick homogeneous glassy margins completely devoid of crystals. Here, we recapitulate a few important processes that affect the temperature of ascending magma. One point that will emerge is that there are major differences between magmas that are emplaced at depth and those that erupt at the Earth's surface, so that it may be misleading to use the latter to infer intrusion conditions in the Earth's crust.

Upon its formation at depth in the crust or mantle, magma is at its liquidus temperature, by definition. It may entrain some fragments of the source rock, but it is in thermodynamic equilibrium with them. These fragments are not phenocrysts that have grown from the magma, and hence have no bearing on the issue of superheat. As magma rises away from its source, pressure decreases which implies a change of liquidus temperature. There are marked differences between dry magmas and those that contain volatile species in solution. Figure 11.4 shows that the liquidus increases with pressure in anhydrous tholeiite basalt. For this magma, the slope of the liquidus at pressures larger than 800 MPa is $\approx 100\ \mathrm{K\,GPa^{-1}}$ or $\approx 3\ \mathrm{K\,km^{-1}}$. It is smaller at pressures below 800 MPa, corresponding to depths less than about 25 km (Figure 11.4), in association with a change of mineral phases on the liquidus. For common melt compositions, it is typically in a 1–2 $\mathrm{K\,km^{-1}}$ range in the continental crust.

Volatile-rich magmas behave in different ways below and above the depth of volatile saturation. Below that depth, the phase diagrams are similar to those for dry magmas, with liquidus temperatures that decrease as pressure is relieved during ascent. Above the depth of saturation, the opposite behavior is observed, with liquidus temperatures that now increase as pressure is relieved further (Figure 11.5), due to the ever-decreasing amounts of volatile species that remain dissolved in the melt. With rare exceptions, magmas contain some volatiles in solution, and hence become water saturated at shallow crustal levels. One may split the ascent path into two different parts: a deep one where the melt is not water saturated and hence becomes superheated as it rises, and a shallow part where magma follows

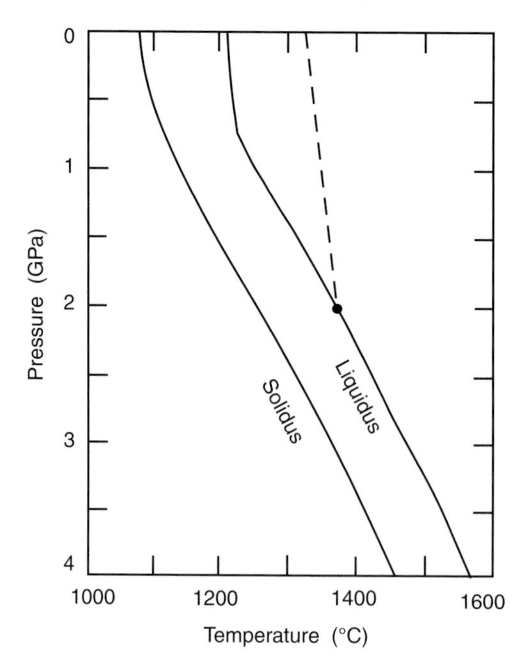

Figure 11.4. Solidus and liquidus for anhydrous tholeiite basalt. The dashed line shows a hypothetical decompression path for magma generated at 2 GPa pressure, or about 70 km. Adapted from Green (1982).

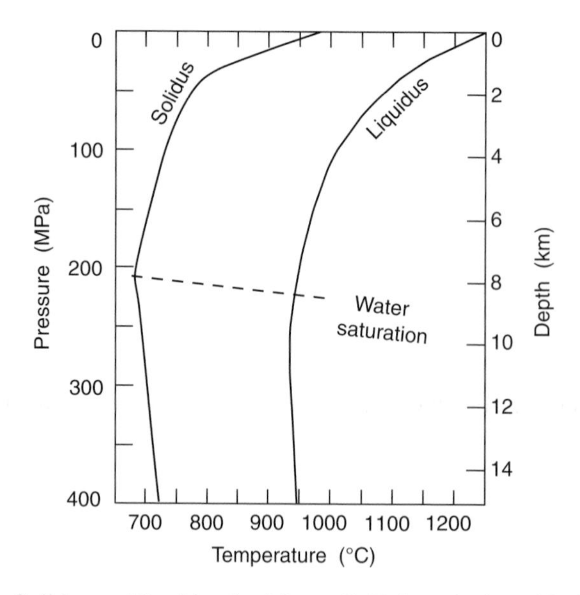

Figure 11.5. Solidus and liquidus for Mount St Helens dacite with 6 wt % water. The dashed line shows the saturation curve for the liquid. Above the saturation level, the water content of magma is dictated by the solubility law and varies along the liquidus. Adapted from Blundy and Cashman (2001).

the saturation curve and sees its superheat decrease. Save for heat losses to country rock, the amount of superheat therefore increases and reaches a maximum at some depth. The final amount of superheat upon emplacement depends on the amount of heat lost to cold country rock during ascent and on the initial water content. This emphasizes the difference between erupted lavas, which have almost always reached volatile saturation, and magmas that intrude the deep crust. We expect that the amount of superheat is much larger in the latter than in the former.

If it rises along an isentrope, anhydrous basaltic magma generated at a depth of 70 km ($\approx$2 GPa) may be superheated by more than 100 K when it intrudes a shallow crustal reservoir (Figure 11.4). We use a simple heat balance to estimate the amount of cooling that may occur during ascent. The main transport mechanism of magmas through the shallow mantle and the crust is hydraulic fracturing. Melt that rises in a dyke of length a and width d loses heat to the walls, such that

$$\rho C_p w a d \frac{dT}{dz} \approx 2qa, \tag{11.2}$$

where w is the average magma velocity and q is the heat flux through the walls. The heat flux through the static country rock is $q = \lambda \Delta T/\delta$, where λ is thermal conductivity, ΔT the temperature difference between the wall and the far field and δ the width of the thermal boundary layer that develops. We deduce that,

$$\frac{dT}{dz} \approx \frac{\kappa}{dw} \frac{\Delta T}{\delta}. \tag{11.3}$$

The width of the thermal boundary layer depends on the duration of the flow t, and is such that $\delta \approx 2\sqrt{\kappa t}$. For $\kappa = 10^{-6}$ m^2 s^{-1} and a short flow duration of one month, $\delta \approx 1$ m. Appropriate values for the other variables in the case of basaltic magmas are $\Delta T = 10^3$ K, $d = 2$ m and $w = 1$ m s^{-1}. We therefore estimate that $dT/dz \approx 1$ K km^{-1}, which is below the liquidus slope in crustal pressure conditions (1–3 K km^{-1}).

This simple balance has been designed to err on the high side with a large temperature contrast between the dyke walls and the far field. For intrusion events that last for several months, or several years, as for example during the Laki eruption in Iceland (Thordarson and Self, 1993), the amount of cooling would be much smaller. Some dykes in the Isle of Mull, Scotland, have developed chilled margins, showing that they were emplaced in cold country rock that was able to absorb large amounts of heat. Others have no such margins, which has been interpreted as evidence that they fed flows for several months through rocks that had been heated (Holness and Humphreys, 2003).

Volatile undersaturated magma is likely to become and remain superheated during ascent. In contrast, hydrous magmas that stall at pressures that are less than the saturation threshold may have lost their superheat and may have started to crystallize. Through a thick magma reservoir, the liquidus varies by a significant amount, $\approx$10 K over 5 km thickness, for example. Thus, even if magma entering the base of the reservoir is at its liquidus, the top of the reservoir may be superheated. We conclude that magma is likely to be above the liquidus when it intrudes the continental crust at depth. We also note that large melt sheets that were generated by meteorite impacts, such as at Sudbury, Ontario, were almost certainly superheated. Thus, we shall treat the amount of superheat as a variable. Many authors have assumed that it is zero, partly on practical grounds because it gets rid of one unknown, and partly on the grounds that convection would suppress it so rapidly that it does not matter. We shall show that such reasoning is erroneous and that a significant amount of crystallization occurs in the superheated phase if there is one.

11.3 Cooling and crystallization of magma sheets: Conduction

We first investigate cooling and crystallization in a diffusive regime. As discussed above, this is valid for thin magma bodies where thermal convection is not likely to develop. Another reason is that, even if convection does develop, the thermal boundary layer at the reservoir floor evolves in a regime dominated by conduction. Finally, we use the diffusive regime to illustrate the effects of latent heat release and to discuss how solidification proceeds in relation to the boundary conditions. We focus on the structure of the thermal boundary layer that develops against the cold surface (the country rock contact) and assume that the melt layer is very thick so that cooling proceeds independently of the other boundary. We start with a pure substance, such that all melt turns to solid at the freezing point, and then study multi-component melts with a finite crystallization interval. We shall show that, as regards rocks encasing the reservoir, solutions for a pure substance are close to those for a multi-component melt and hence one need not deal with the extra complexities associated with a finite crystallization interval.

We study cooling conditions, but the solutions can also be used for melting due to heating at a boundary with appropriate sign changes. We consider a layer of melt with liquidus temperature T_L at initial temperature T_i and set the temperature at the base to T_B at time $t = 0$ (Figure 11.6). A measure of the importance of latent heat release on the global thermal budget is provided by the Stefan number,

$$\mathrm{St} = \frac{L}{C_p\,(T_L - T_B)}, \qquad (11.4)$$

which compares the amount of heat released per unit mass of liquid (i.e. L) to the amount of heat that must be extracted to cool material down to the temperature at the boundary. If $\mathrm{St} \ll 1$, latent heat contributes a negligible amount to the global

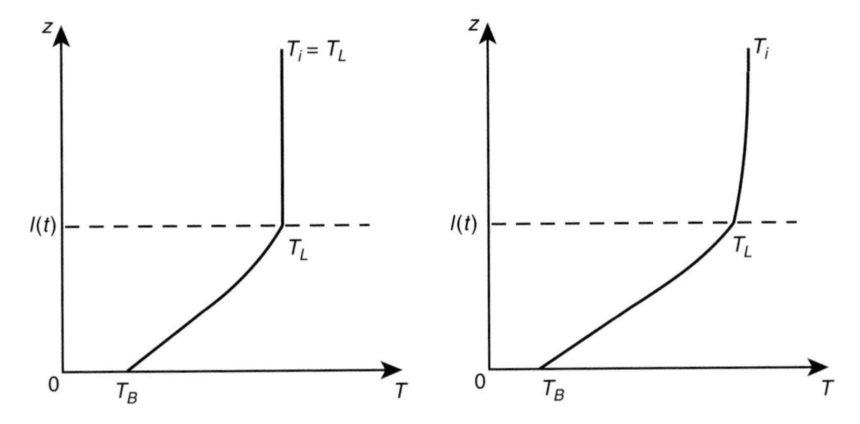

Figure 11.6. Schematic diagram illustrating the cooling of a pure melt from a horizontal boundary. The crystallization front lies at $z = l(t)$. Left panel: melt is initially at its melting temperature T_L. Right panel: melt is superheated upon emplacement, such that its initial temperature is above the freezing point. Note the jump in heat flux that occurs at the crystallization front, which is due to latent heat release (equation 11.6).

thermal budget and hence need not be accounted for. For typical geological cases (Appendix D.1.6), $L \approx 3 \times 10^5$ J kg^{-1}, $T_L - T_B \approx 800$ K and $C_p \approx 10^3$ J kg^{-1} K^{-1}, implying that St ≈ 0.4. We conclude that latent heat cannot be neglected and investigate the consequences for cooling and solidification.

11.3.1 Pure substance

For a pure substance, solidification occurs at $z = l(t)$ where $T = T_L$. In the absence of convection, temperature obeys the diffusion equation both above and below the phase change boundary,

$$\frac{\partial T}{\partial t} = \kappa \frac{\partial^2 T}{\partial z^2}. \tag{11.5}$$

One should in principle allow for different thermal properties in the melt and solid phases, but this leads to complicated mathematical expressions which cloud the important features. For most geological materials, thermal properties vary within relatively narrow ranges, so that changes upon solidification do not have important thermal consequences.

One way to derive the governing equation for the motion of the phase change boundary is to work in a moving reference frame attached to the boundary, so that motion by infinitesimal amount dl is achieved by the solidification of a mass $dm = \rho dl$ per unit area, and hence is accompanied by the release of an amount of heat $dQ = \rho L dl$ per unit area. The phase change boundary has no mass, and hence cannot accumulate or lose heat. For a pure substance, the solidification temperature is constant, save for the negligible effect of pressure. The thermal budget for the moving boundary therefore dictates that,

$$0 = -\lambda \left(\frac{\partial T}{\partial z} \right)_{l(t)^-} + \lambda \left(\frac{\partial T}{\partial z} \right)_{l(t)^+} + \rho L \frac{dl}{dt}. \tag{11.6}$$

This equation shows that the heat flux into the solid (at $z = l(t)^-$) is larger than that from the melt due to latent heat release. The equation can also be written as a jump condition on the heat flux,

$$\left[-\lambda \frac{\partial T}{\partial z} \right]_{l(t)} = \rho L \frac{dl}{dt}, \tag{11.7}$$

where the expression in brackets stands for a jump in the enclosed quantity. The right-hand side of that equation can also be understood as involving a jump condition on the solid (or liquid) fraction, which goes from 0 to 1 across the boundary.

The equation (11.6) is known as the Stefan boundary condition. One important feature of this boundary condition is that it is non-linear and solutions cannot be superposed. The non-linearity is not apparent in equation 11.6. To demonstrate it, we consider the condition of continuity of temperature at the boundary:

$$T^+(l^+(t), t) = T^-(l^-(t), t) = T_L. \tag{11.8}$$

Differentiating this equation, we find that,

$$\frac{dl}{dt} = -\frac{\dfrac{\partial T^+}{\partial t}}{\dfrac{\partial T^+}{\partial z}} = -\frac{\dfrac{\partial T^-}{\partial t}}{\dfrac{\partial T^-}{\partial z}}, \tag{11.9}$$

and the Stefan boundary condition is written in the form:

$$\lambda \left(\frac{\partial T}{\partial z} \right)_{l(t)^-} - \lambda \left(\frac{\partial T}{\partial z} \right)_{l(t)^+} = \rho L \frac{\partial T^- / \partial t}{\partial T^- / \partial z} = \rho L \frac{\partial T^+ / \partial t}{\partial T^+ / \partial z}, \tag{11.10}$$

which makes the non-linearity evident.

We consider that melt is emplaced at a uniform temperature T_i. If we suppose for simplicity that the melt body extends over a very large distance, initial stages of cooling and solidification proceed in a thin boundary layer independently of what happens at the other boundary. In such conditions, there are no scales for length and time. One can search for relations between distance and time, however. Denoting by d and τ distance and time, the heat diffusion equation imposes that $d \sim \sqrt{\kappa \tau}$, as usual. What is more interesting is that the phase change heat balance (equation 11.6) introduces a similar relationship. The main balance is between the heat flux through the solid and the amount of heat released by solidification so that one has

$$\rho L \frac{dl}{dt} \approx \lambda \frac{T_L - T_B}{l(t)}, \tag{11.11}$$

which implies that

$$l(t) \sim \mathrm{St}^{-1/2} \sqrt{\kappa t}. \tag{11.12}$$

The full result involves an unknown constant of proportionality which is expected to be of order 1. This argument indicates that the phase boundary moves as the square root of time, which will be verified by the full solutions developed later. In such conditions, mathematical solutions can be sought in terms of a similarity variable combining length and time, $\eta = z/(2\sqrt{\kappa t})$. This implies that the phase change lies at

$$\eta = l(t)/(2\sqrt{\kappa t}) = \Lambda_L, \tag{11.13}$$

where Λ_L is a constant to be determined.

Zero super heat. Fixed temperature at the boundary

We consider that melt is emplaced at the melting temperature T_L, i.e. $T_i = T_L$ (Figure 11.6). In this case, no heat is transported from the melt to the solid and the thermal balance at the phase change boundary reduces to

$$\lambda \left(\frac{\partial T}{\partial z} \right)_{l(t)^-} = \rho L \frac{dl}{dt}. \tag{11.14}$$

We seek a solution of the form:

$$T = T_B + (T_L - T_B)\theta(\eta), \tag{11.15}$$

where η is the similarity variable defined in (11.13). Substituting for this expression in the heat equation, we find,

$$\ddot{\theta} + 2\eta\dot{\theta} = 0, \tag{11.16}$$

with the following boundary conditions:

$$\theta(0) = 0, \theta(\Lambda_L) = 1 \tag{11.17}$$

$$\dot{\theta}(\Lambda_L) = 2\mathrm{St}\Lambda_L, \tag{11.18}$$

where we recognize the Stefan number St defined above. In this form, the problem involves only one dimensionless parameter, St, which was introduced above on the basis of a simple physical argument. Alternatively, solutions can be sought as a function of parameter Λ_L, which depends on St:

$$\theta = \frac{\operatorname{erf}(\eta)}{\operatorname{erf}(\Lambda_L)}, \quad \text{for } \eta \leq \Lambda_L \tag{11.19}$$

$$\theta = 1, \quad \text{for } \eta \geq \Lambda_L, \tag{11.20}$$

where erf is the error function introduced in Chapter 4. We shall see that the temperature profile for the solid (for $\eta \leq \Lambda_L$) is the same in many different cases.

Constant Λ_L is obtained using the conditions at the solidification front, which lead to a transcendental equation:

$$\sqrt{\pi}\Lambda_L \exp(\Lambda_L^2) \operatorname{erf}(\Lambda_L) = \mathrm{St}^{-1}. \tag{11.21}$$

Figure 11.7 shows how Λ_L varies as a function of St. One may see that Λ_L decreases as St increases and $\Lambda_L \to 0$ as St $\to \infty$. In the limit of infinite St, the amount of heat that must be extracted through the base of the layer for solidification to proceed must be infinite, which cannot be achieved by conduction with a finite temperature contrast. Thus, in this limit, crystallization is prevented, so that $l(t) = 0$ at all times: the phase change boundary remains stuck at the base of the layer. For large values

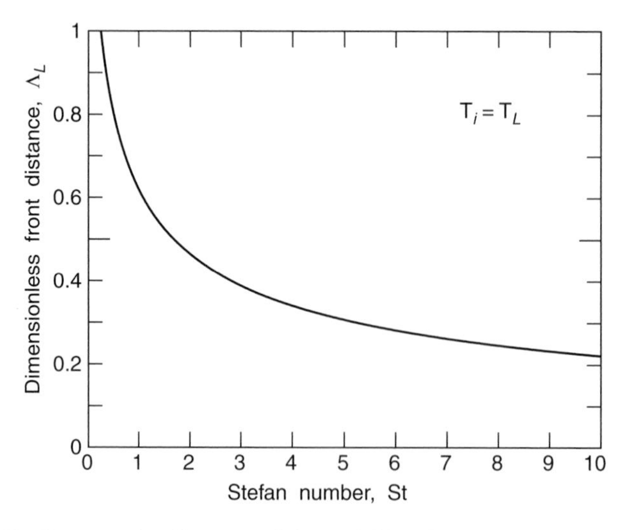

Figure 11.7. Constant for distance of the solidification front for a pure substance emplaced at its freezing point, Λ_L in equation 11.13, as a function of the Stefan number.

of St, Λ_L is small and one can expand equation 11.21. To leading order, we obtain,

$$\Lambda_L^2 = (2\ \mathrm{St})^{-1} \text{ for St} \gg 1 \tag{11.22}$$

This is such that $\Lambda_L \sim \mathrm{St}^{-1/2}$, which had been predicted above with a simple heat balance argument.

A final result of interest is the heat flux through the base of the layer,

$$-\lambda \left(\frac{\partial T}{\partial z} \right)_{z=0} = -\frac{1}{\operatorname{erf}(\Lambda_L)} \left(\lambda \frac{T_L - T_B}{\sqrt{\pi \kappa t}} \right). \tag{11.23}$$

For reference, the term in brackets on the right-hand side of this equation corresponds to a layer which is cooled from below in exactly the same thermal conditions but with no phase change. The error function is by construction such that it is positive and always less than 1, so that the constant of proportionality $1/\operatorname{erf}(\Lambda_L)$ is larger than 1. This shows that latent heat release enhances the amount of heat that must be extracted out of the layer. As St increases, the enhancement factor also increases and tends to ∞ as $\mathrm{St} \to \infty$. We also see that the basal heat flux decreases with time as the solidification front moves away from the cooling interface.

Zero superheat. Phase change moving at a constant velocity

In the previous solution, the solidification front migrates through the layer according to $l(t) \sim \sqrt{\kappa t}$, which is due to the boundary condition of a fixed temperature. We have calculated the heat flux that must be extracted out of the layer to maintain the basal temperature at a constant value T_B. This basal heat flux decreases as $t^{-1/2}$ and is very large at small times. This may not be realistic because, in practice, this heat flux is limited by the efficiency of heat transport in the underlying medium. We now consider another solution which illustrates the relationship between the basal heat flux and the solidification rate. In many problems, it is useful to consider the ideal case of a crystallization front that moves at constant velocity because this allows simple mass balance calculations for chemical differentiation. In this case, the basal heat flux is initially small but must increase with time, as we now show.

Consider that the phase change boundary moves at a constant speed V, such that $l(t) = Vt$. The temperature of the basal boundary cannot be prescribed a priori and will be found as part of the solution. It does not remain constant and decreases with time. The introduction of velocity V implies a relationship between distance and time which fundamentally changes the problem. We scale distance by κ/V, time by κ/V^2 and temperature by L/C_p.

For a constant velocity, the proper similarity variable is $\eta = z - Vt$, or, in dimensionless form,

$$\eta = \frac{V}{\kappa}(z - Vt). \tag{11.24}$$

Temperature is written as,

$$T(z,t) = T_L + \frac{L}{C_p}\theta(\eta), \tag{11.25}$$

where θ is the solution of

$$\ddot{\theta} + \dot{\theta} = 0, \tag{11.26}$$

with the following boundary conditions:

$$\theta(0) = 0 \text{ and } \dot{\theta}(0) = 1. \tag{11.27}$$

We deduce that,

$$\theta = 1 - \exp(-\eta) \tag{11.28}$$

$$T(z,t) = T_L + \frac{L}{C_p}\left(1 - \exp\left[-\frac{V}{\kappa}(z - Vt)\right]\right). \tag{11.29}$$

From this, we calculate the basal heat flux,

$$q(0,t) = -\lambda\left(\frac{\partial T}{\partial z}\right)_{z=0} = -\rho LV \exp\left[\frac{V^2 t}{\kappa}\right]. \tag{11.30}$$

At small times (for $t \ll \kappa/V^2$), the solutions behave in the following manner:

$$T(0,t) \approx T_L - \frac{L}{C_p}\frac{V^2 t}{\kappa}$$

$$q(0,t) \approx -\rho LV\left(1 + \frac{V^2 t}{\kappa}\right), \tag{11.31}$$

which can be deduced from the simple heat balance written above (equation 11.11). As the solidification front moves away, the basal temperature decreases and the basal heat flux increases, in contrast to the preceding solution for a constant basal temperature.

Finite superheat. Fixed temperature at the boundary

We now consider that the melt is above its solidification temperature upon emplacement, such that $T_i > T_L$ (Figure 11.6). We treat the problem to shed further light onto the controls on solidification and also to pave the way for the analysis of multi-component melts with a finite crystallization interval. In this situation, the phase change boundary receives heat from the overlying melt and loses heat to the underlying solid. The extra heat flux due to the superheat acts to decrease the rate of solidification.

As before, temperature obeys the diffusion equation and we seek solutions of the following form:

$$T(z,t) = T_B + B_1 \mathrm{erf}\left(\frac{z}{2\sqrt{\kappa t}}\right) \quad \text{for } z \le l(t)$$
$$T(z,t) = A_2 + B_2 \mathrm{erf}\left(\frac{z}{2\sqrt{\kappa t}}\right) \quad \text{for } z \ge l(t), \tag{11.32}$$

such that the phase change boundary lies at $l(t) = 2\Lambda_L\sqrt{\kappa t}$, where Λ_L is again a constant to be solved for. We note that the temperature profile in the solid (i.e. for $z \le l(t)$) takes exactly the same form as before:

$$T(z,t) = T_B + (T_L - T_B)\frac{\mathrm{erf}\left(\frac{z}{2\sqrt{\kappa t}}\right)}{\mathrm{erf}(\Lambda_L)} \quad \text{for } z \le l(t). \tag{11.33}$$

The temperature profile in the melt is also found as a function of the as yet unknown constant Λ_L:

$$T(z,t) = T_i - (T_i - T_L)\frac{1 - \mathrm{erf}\left(\frac{z}{2\sqrt{\kappa t}}\right)}{1 - \mathrm{erf}(\Lambda_L)} \quad \text{for } z \ge l(t). \tag{11.34}$$

Substituting for these expressions into the heat balance at the moving phase change boundary (equation 11.6), we obtain an equation for Λ_L,

$$\sqrt{\pi}\Lambda_L \exp(\Lambda_L^2)\mathrm{erf}(\Lambda_L) = \mathrm{St}^{-1}\left[1 - \frac{T_i - T_L}{T_L - T_B}\frac{\mathrm{erf}(\Lambda_L)}{1 - \mathrm{erf}(\Lambda_L)}\right], \tag{11.35}$$

which reduces to the previous result in the limit of zero superheat ($T_i = T_L$).

The solution involves two dimensionless numbers, the Stefan number St and the dimensionless superheat $(T_i - T_L)/(T_L - T_B)$. With increasing superheat, Λ_L decreases (Figure 11.8) and the basal heat flux increases (in proportion to $1/\mathrm{erf}\,\Lambda_L$), due to the extra heat supplied by the overlying melt. One can see, of course, that it takes large amounts of superheat to affect the results significantly.

11.3.2 Multi-component melts: Mushy layers

Latent heat release has so far been treated for a pure substance such that solidification occurs at a single temperature. This is not valid for magmas, which crystallize over a large temperature interval. The structure, evolution and dynamics of partially crystallized (or partially molten) regions has received a lot of attention (Roberts and Loper, 1983; Hills *et al.*, 1983; Huppert and Worster, 1985; Worster, 1986; Tait and Jaupart, 1992). For the cooling of very thick magma bodies such as the

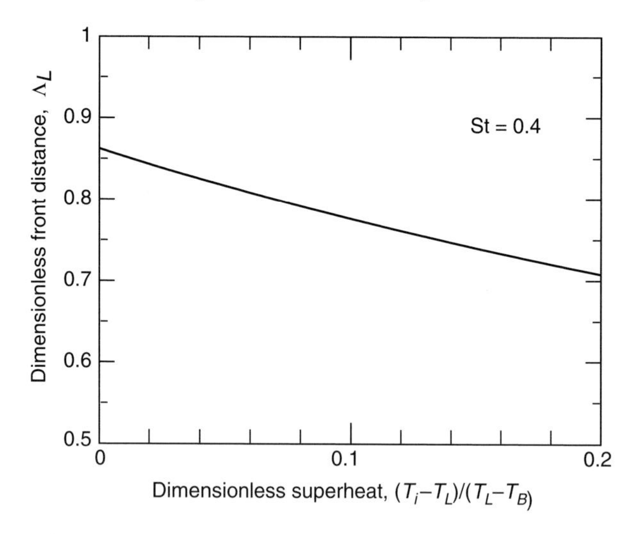

Figure 11.8. Constant for distance of the solidification front for a pure substance that is superheated upon emplacement, Λ_L in equation 11.13, as a function of dimensionless superheat for a fixed Stefan number of 0.4.

Earth's core, the effects of pressure variations, cannot be neglected. For crustal magma reservoirs, one may usually ignore such complicating factors. Crystals and melt have different densities and hence are likely to undergo relative motions with respect to one another. When the solids form a rigid network that remains attached to the cooling interface, the partially molten region is called a mush. When the solid crystals are isolated and are able to move in an absolute frame of reference, one refers to a slurry. For simplicity, however, we shall refer to partially crystallized melt as a mush. The simplest case is such that relative motions between crystals and melt are negligible, which provides a benchmark for more complicated situations.

Structure and properties of a mushy layer

We consider for simplicity a binary mixture whose composition C is specified by the proportion of one end-member in the solid solution. The phase diagram is specified by the liquidus and solidus curves,

$$T_L(C) \text{ and } T_S(C). \tag{11.36}$$

Denoting by H_s and H_l the enthalpies per unit mass of the solid and liquid phases, respectively, the bulk enthalpy per unit mass of a local mush parcel is

$$\rho H = \Phi \rho_s H_s + (1 - \Phi) \rho_l H_l, \tag{11.37}$$

where Φ is the solid fraction. We shall assume, as before, that the solid and liquid phases have the same physical properties, which does not affect the behavior of the solutions (i.e. such as the time dependence of the solidification rate, for example). In many natural magmas, the physical properties of all the phases are not known precisely and it is futile to develop elaborate models that cannot be used in practice. Additional difficulties arise for transport properties such as thermal conductivity, for example, which depend on the geometrical arrangement and shape of the coexisting phases. As cooling proceeds, the local enthalpy change is

$$\frac{\partial}{\partial t}\rho H = \Phi\rho\frac{\partial H_s}{\partial T} + (1-\Phi)\,\rho\frac{\partial H_l}{\partial T} - \rho\,(H_l - H_s)\,\frac{\partial\Phi}{\partial t}. \tag{11.38}$$

Introducing C_p, the heat capacity of the solid and liquid phases,

$$C_p = \frac{\partial H_s}{\partial T} \text{ and } C_p = \frac{\partial H_l}{\partial T}, \tag{11.39}$$

we obtain:

$$\frac{\partial}{\partial t}\rho H = \rho C_p\frac{\partial T}{\partial t} - \rho L\frac{\partial\Phi}{\partial t}, \tag{11.40}$$

where

$$L = H_l - H_s \tag{11.41}$$

is the latent heat.

To close the system of governing equations, we specify how the solid fraction Φ evolves as cooling proceeds. In the mushy layer, a useful assumption is that of local thermodynamic equilibrium between melt and solid, which has been tested in the field for basaltic magmas (Figure 11.2). If one assumes further that melt and solid do not move with respect to one another, so that the local bulk composition of the mixture remains equal to that of the starting melt, solid fraction Φ may be written as a function of temperature in the crystallization interval. Thus,

$$\frac{\partial\Phi}{\partial t} = \frac{d\Phi}{dT}\frac{\partial T}{\partial t}, \tag{11.42}$$

where $d\Phi/dT$ is negative. Thus, the bulk enthalpy change can be expressed as,

$$\frac{\partial}{\partial t}\rho H = \rho\left(C_p - L\frac{d\Phi}{dT}\right)\frac{\partial T}{\partial t}, \tag{11.43}$$

where we see that latent heat release may be understood as augmenting the heat capacity of the mushy material. Function $\Phi(T)$ describes how the solid fraction varies as a function of temperature and may be quite complicated for some natural magmas. We shall use two simple models in order to illustrate the relevant physical principles in the clearest manner.

In calculations where we are not interested in changes of melt composition, we shall assume that the solid fraction depends linearly on temperature in the crystallization interval, such that,

$$\Phi = \frac{T_L - T}{T_L - T_S} = \frac{T_L - T}{\Delta T_c}, \tag{11.44}$$

where $\Delta T_c = T_L - T_S$. As shown in Figure 11.9, this is a reasonable approximation when dealing with a basaltic mushy layer as a whole. In this case, $d\Phi/dT$ is constant, such that latent heat is released evenly through the crystallization interval.

In other models, we shall be interested in changes of magma composition due to crystallization. In such cases, we shall use a simple binary eutectic diagram with a constant composition for the solid phase that precipitates. The melt composition can be written as a fraction of the solid in solution, which facilitates the analysis (as it introduces a few dimensionless numbers) and allows comparisons with laboratory experiments on salt–water systems. The phase diagram of the diopside–anorthite system, a very well-characterized silicate binary solution, is shown on Figure 11.10.

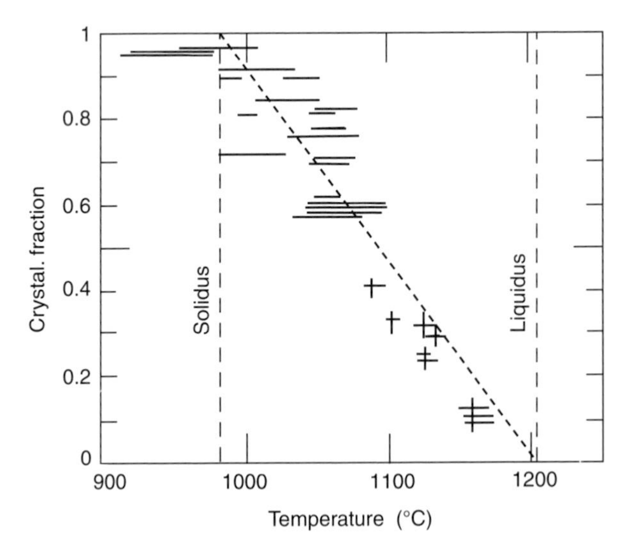

Figure 11.9. Crystallized fraction as a function of temperature for Hawaiian basalts, from (Wright *et al.*, 1976). The data are based on *in situ* measurements in lava lakes. The dashed line is a linear approximation through the crystallization interval.

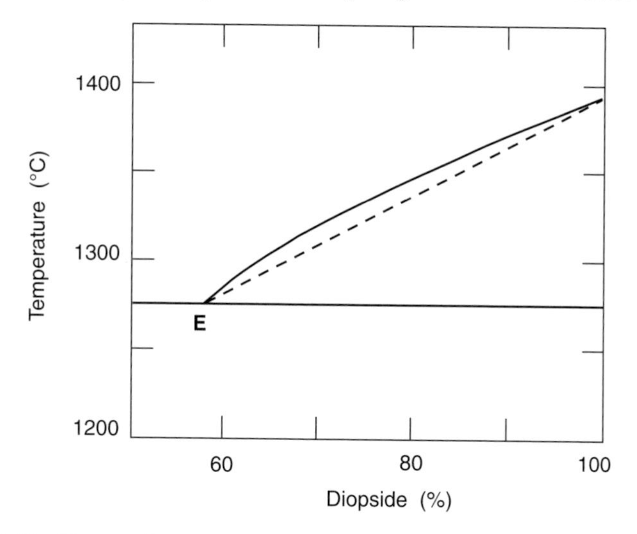

Figure 11.10. Phase diagram for the diopside–anorthite system (continuous curve), from Morse (1980). This system is not a true eutectic but implications are only important for detailed studies of mineral precipitates. E is the eutectic point. The phase diagram can be approximated by a straight line (dashed) with little error.

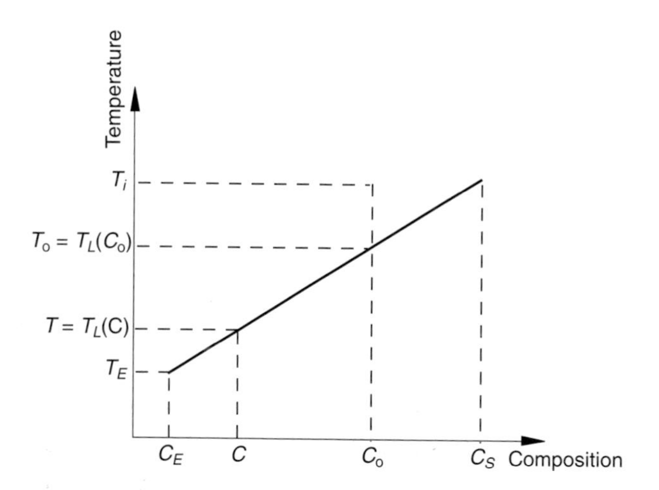

Figure 11.11. Diagram for a binary eutectic showing the main variables. C_s, C_E, C_0 solid, eutectic and initial compositions, respectively.

We can approximate the liquidus by a straight line such that,

$$T_L(C) - T_L(C_S) = \gamma\,(C - C_S), \tag{11.45}$$

where $T_L(C)$ is the liquidus temperature for melt with composition C, and C_S is the composition of the solid (Figure 11.11). Without separation of melt and crystals, the total amount of material with composition C_S is conserved, such that

$$\Phi C_S + (1 - \Phi) C = C_o, \tag{11.46}$$

where C_o is the initial melt composition. Within the mushy layer, the assumption of local thermodynamic equilibrium implies that $T = T_L(C)$. From the two equations above, we deduce that,

$$\Phi = \frac{T - T_L(C_o)}{T - T_L(C_S)} \text{ for } s(t) \leq z \leq l(t). \tag{11.47}$$

One important difference from the previous model is that $d\Phi/dT$ is not constant and depends on the initial melt composition:

$$\frac{d\Phi}{dT} = \frac{T_L(C_o) - T_L(C_S)}{[T - T_L(C_S)]^2}. \tag{11.48}$$

Note that, in this case, latent heat is not released evenly through the crystallization interval.

With the assumption of local thermodynamic equilibrium, crystallization sets up a composition gradient within the mushy layer which induces chemical diffusion. Diffusion of chemical species through magmas is very inefficient, however, so that it may be neglected in a first approximation. We return to this aspect later in this section.

Cooling by conduction

We consider again the 1D problem and consider for simplicity a melt that is cooled from below, so that thermal stratification is dynamically stable. This applies to the base of a magmatic intrusion, for example. Starting from a pure melt at some initial temperature T_i, we consider that the lower boundary is maintained at a constant temperature T_B. The top of the mushy layer lies at $z = l(t)$ and its base at $z = s(t)$ (Figure 11.12). The problem involves three regions labelled 1,2,3 with temperature distributions T_1, T_2 and T_3, respectively, and one must specify what happens at the melt–mush and mush–solid interfaces. There is no reason to assume that the solid fraction is continuous across these interfaces. For example, the heat balance at the moving lower interface is,

$$\rho L (\Phi_s - 1) \frac{ds}{dt} = \lambda \left(\frac{\partial T_2}{\partial z} \right)_{s^+} - \lambda \left(\frac{\partial T_1}{\partial z} \right)_{s^-}, \tag{11.49}$$

which relates a jump in the solid fraction to a jump in the heat flux. For a pure substance, we had written a similar balance at the phase change boundary (such that $\Phi_s = 0$). To close the problem, one must add that temperature is continuous and

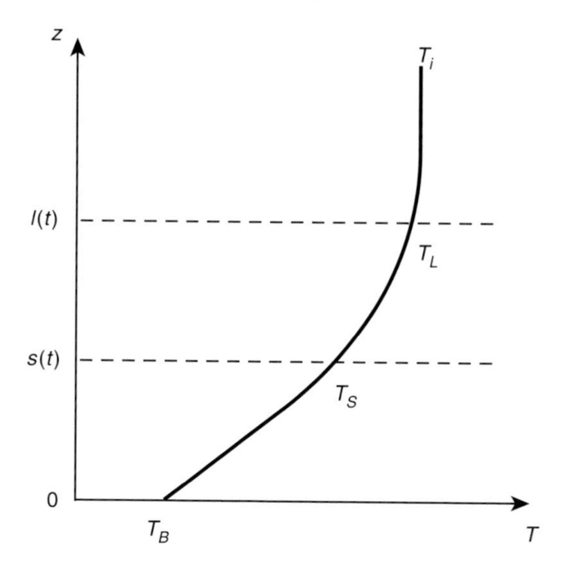

Figure 11.12. Diagram showing the growth of a partially crystallized layer above a boundary that is kept at a constant temperature. The heights of the crystallization front at the liquidus and of the solidification front at the solidus are denoted by $l(t)$ and $s(t)$ respectively.

some statement on the nature of the mush boundaries. In practice, it may be shown that, save for peculiar circumstances (Worster, 1986b) there is no jump, so that the heat flux is also continuous. Field data on a natural basaltic magmatic mushy layer support this. In this case, one writes that

$$T\left[s(t),t\right] = T_S \text{ and } T\left[l(t),t\right] = T_L. \tag{11.50}$$

The bulk enthalpy balance in the mushy layer is then,

$$\frac{\partial}{\partial t}\rho H = \rho\left(C_p - L\frac{d\Phi}{dT}\right)\frac{\partial T}{\partial t} = \nabla\cdot(\lambda\nabla T). \tag{11.51}$$

Here, we focus on thermal evolution and assume that the solid fraction varies linearly as a function of temperature in the crystallization interval (equation 11.44). Thus,

$$\rho\left(C_p + \frac{L}{\Delta T_c}\right)\frac{\partial T}{\partial t} = \lambda\frac{\partial^2 T}{\partial z^2}. \tag{11.52}$$

This defines an effective thermal diffusivity :

$$\kappa^* = \frac{\lambda}{\rho\left(C_p + L/\Delta T_c\right)} = \frac{\kappa}{1 + \dfrac{L}{C_p\Delta T_c}}. \tag{11.53}$$

With standard values for the parameters, $L/C_p\Delta T_c \approx 2$; the adjustment is not negligible. To complete the model, we add heat diffusion equations for the pure melt above the mushy layer and for the fully solidified magma below it.

As for the case of a pure substance, we seek solutions of the form:

$$\begin{aligned} T_1 &= T_B + B_1 \mathrm{erf}\left(\frac{z}{2\sqrt{\kappa t}}\right) && \text{for } z \leq s(t) \\ T_2 &= A_2 + B_2 \mathrm{erf}\left(\frac{z}{2\sqrt{\kappa^* t}}\right) && \text{for } s(t) \leq z \leq l(t) \\ T_3 &= A_3 + B_3 \mathrm{erf}\left(\frac{z}{2\sqrt{\kappa t}}\right) && \text{for } z \geq l(t), \end{aligned} \tag{11.54}$$

so that the upper and lower mush boundaries are such that,

$$\begin{aligned} s(t) &= 2\Lambda_S\sqrt{\kappa t} \\ l(t) &= 2\Lambda_L\sqrt{\kappa t}, \end{aligned} \tag{11.55}$$

where Λ_S and Λ_L are two constants to be determined. Applying the boundary and interface conditions, we find two equations for these two constants:

$$\begin{aligned} &\frac{T_S - T_B}{\mathrm{erf}\,(\Lambda_S)} \exp(-\Lambda_S^2) \\ &= \sqrt{\frac{\kappa}{\kappa^*}} \frac{T_L - T_S}{\mathrm{erf}\left[\sqrt{\frac{\kappa}{\kappa^*}}\Lambda_L\right] - \mathrm{erf}\left[\sqrt{\frac{\kappa}{\kappa^*}}\Lambda_S\right]} \exp(-\frac{\kappa}{\kappa^*}\Lambda_S^2) \\ &\frac{T_i - T_L}{\mathrm{erfc}(\Lambda_L)} \exp(-\Lambda_L^2) \\ &= \sqrt{\frac{\kappa}{\kappa^*}} \frac{T_L - T_S}{\mathrm{erf}\left[\sqrt{\frac{\kappa}{\kappa^*}}\Lambda_L\right] - \mathrm{erf}\left[\sqrt{\frac{\kappa}{\kappa^*}}\Lambda_S\right]} \exp(-\frac{\kappa}{\kappa^*}\Lambda_L^2). \end{aligned} \tag{11.56}$$

For numerical results, we consider that $L/(C_p\Delta T_c) = 2$ and use the definition for the Stefan number, i.e. $\mathrm{St} = L/C_p(T_S - T_B)$. For a given value of the Stefan number, chosen here to be 0.4, the only free parameter is the amount of superheat. The mush thickness, $\delta(t) = l(t) - s(t)$, decreases as the superheat increases because the superheat dictates the temperature gradient that develops through the mush. High superheat produces a steep gradient and a thin mush. It is instructive to compare the results for the pure substance and for a mushy layer (Figures 11.8 and 11.13) because the former corresponds to a mushy layer of zero thickness. For the pure

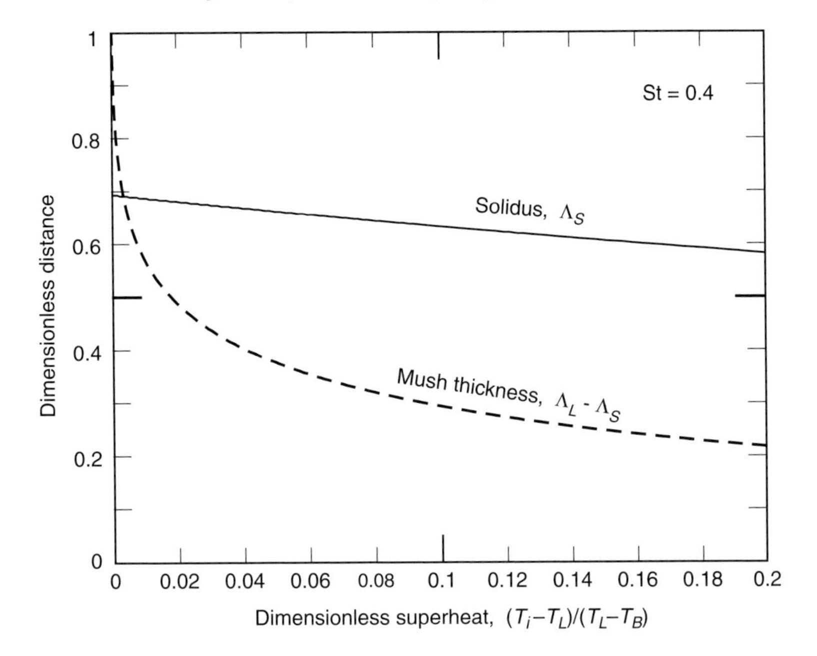

Figure 11.13. Constant for distance of the solidification front $s(t)$, Λ_S, and mush thickness $\delta(t) = l(t) - s(t)$, $\Lambda_L - \Lambda_S$, as a function of superheat for a Stefan number of 0.4. See text for explanations.

and multicomponent melts, the position of the solidification front is proportional to Λ_L and Λ_S, respectively. They do not vary much. As explained above, they specify the temperature profile in the fully solidified region (equation 11.19) as well as the basal heat flux. We conclude that allowing for a crystallization interval does not affect the thermal regime of encasing rocks significantly. We also see that, for given superheat, the solidification front for the pure substance lies between the top and bottom of the mush, which may be expected on intuitive grounds.

In this model for a mush, there is no jump of the temperature gradient at the solidification front, in contrast to the case of a pure substance (Figure 11.6). One consequence is that the mush thickness, $\delta(t) = l(t) - s(t)$, tends to infinity as the superheat goes to zero (Figure 11.8). Thus, in the limit of zero superheat, crystallization proceeds through the whole melt layer even at small times. This is not realistic and can be corrected in two different ways. One way is to account for the kinetics of crystallization, such that the nucleation of crystals can only occur in undercooled melt. In this case, the true crystallization front is not exactly at the liquidus and no crystals grow in the melt away from the basal thermal boundary layer. This is addressed below. Another process is chemical diffusion, which generates a thin chemical boundary later ahead of the crystallization front, leading to what has been called "constitutional supercooling". In the context of mushy layers, this effect has been discussed by Worster (1986).

Contact temperature at the boundary of a magma body

Save for the case of a solidification front that moves at constant velocity, all solutions were obtained for a fixed temperature at the contact between magma and country rock and we briefly discuss the values that this temperature may take in practice. For simplicity purposes, we assume that magma and country rock have the same thermal properties. Without crystallization, we have shown in Chapter 4 (equation 4.47) that, if magma and country rock are initially at temperatures T_i and T_{ro}, respectively, the contact temperature is $T_B = (T_i + T_{ro})/2$ and stays constant as the two materials exchange heat by diffusion. For tholeiitic basalt at its liquidus temperature of 1200 °C and $T_{ro} < 200$ °C, corresponding to emplacement in continental crust at depths less than about 10 km, $600 < T_B < 700$ °C. This prediction is valid when magma gets chilled and does not release its latent heat of solidification, which occurs if the temperature contrast between magma and country rock is very large.

With crystallization in the magma, the contact temperature is enhanced because of latent heat release. As shown above (equation 11.43), one may treat latent heat as augmenting the effective heat capacity of the cooling magma, such that,

$$C_p^* = C_p + \frac{L}{T_L - T_S} = C_p\left(1 + \mathrm{St}^*\right), \tag{11.57}$$

where St* is a Stefan number. For typical values of latent heat (3×10^5 J kg^{-1}), heat capacity (10^3 J kg^{-1}K^{-1}) and of the crystallization interval (≈ 200 K), $\mathrm{St}^* \approx 1.5$. In this case, equation 4.47 for the contact temperature now gives

$$T_B \approx \frac{T_{ro} + \sqrt{1 + St^*}\, T_i}{1 + \sqrt{1 + St^*}}. \tag{11.58}$$

For $\mathrm{St}^* = 1.5$ and for the same conditions as above, we find that $730 \leq T_B \leq 810$ °C. Temperatures in the upper part of this range are above the solidi of most crustal assemblages, implying partial melting of the wall rocks. If this occurs, one must also allow for the latent heat of melting, which brings the contact temperature down, and one must solve for the contact temperature and the amount of melting simultaneously.

11.4 Cooling by convection

As shown above, magma reservoirs are so large that even minute density perturbations due to temperature heterogeneities are able to drive convective instabilities. In such conditions, cooling proceeds at different rates than those envisioned before, due to two effects. One effect is that convective motions operate over the whole body of fluid and act to cool the melt away from the boundary layers. Another effect is that convection in the melt maintains large heat fluxes into the boundary layers and affects the heat balance for solidification. Here, we address the fundamentals of convection and its consequences for the fully solidified body that is eventually generated. The temperature difference that drives convection has been a matter of debate and depends in part on the initial conditions upon emplacement.

We use the same approximations as before, namely that melt and solid have the same thermal properties and furthermore that there is no jump in the solid fraction at the melt/mush interface. We consider simple models of a magma reservoir of thickness h that is cooled from above, such that the well-mixed interior is at temperature $T_m(t)$. We assume as before that, within a mushy layer, crystals and melt do not separate so that fluid motions are limited to the molten interior. In the reservoir, magma loses heat through all boundaries, including the roof and floor regions. Thus, mushy layers grow at both the top and bottom. In order to focus on the effects of thermal convection, however, we shall first neglect cooling through the floor and shall evaluate its importance in a separate section at the end.

11.4.1 Convection in superheated melt

We consider first melt that is superheated when it gets emplaced. In this case, the roof region must evacuate sensible heat from the melt as well as latent heat released by solidification. In this simple set-up (Figure 11.1), crystallization proceeds only from the roof downwards, such that the solid and mushy layers extend to depth $l(t)$ in the fluid layer. As above, we assume that $T[l(t),t] = T_L$.

The bulk thermal balance for the melt layer, which extends over thickness $h - l(t)$, is written as follows:

$$\rho C_p [h - l(t)] \frac{dT_m}{dt} = -Q_T, \tag{11.59}$$

where Q_T is the heat flux at the top of the melt layer. There is a thin thermal boundary layer below the mush which must be accounted for in the heat budget. We lump this thin boundary layer with the mushy layer in a thermal balance. We find,

$$\rho C_p (T_L - T_m) \frac{dl}{dt} = \lambda \left(\frac{\partial T}{\partial z} \right)_{l(t)^-} - Q_T, \tag{11.60}$$

which accounts for changes of temperature in the thermal boundary layer. Within the mushy layer, heat transport proceeds by diffusion such that,

$$\rho C_p \frac{\partial T}{\partial t} = \lambda \frac{\partial^2 T}{\partial z^2} + \rho L \frac{\partial \Phi}{\partial t}, \tag{11.61}$$

where Φ is again the solid fraction.

To close this system of governing equations, the convective heat flux Q_T must be specified. We note that the Rayleigh number for the melt layer is very large, so that it is appropriate to use the local heat flux scaling law (see equation 5.82):

$$Q_T = C_Q \lambda \left(\frac{\alpha g}{\kappa \nu_i} \right)^{1/3} (T_m - T_L)^{4/3}, \tag{11.62}$$

where ν_i is the kinematic viscosity of the melt and C_Q is a proportionality constant equal to ≈ 0.16 (Table 5.3). We also use the simple binary eutectic diagram with a constant composition for the solid phase to write Φ as a function of temperature within the mushy layer (Figure 11.11). We specify that the temperature at the top of the magma body is kept at some value T_B and denote the liquidus temperature for melt at the initial composition C_o by T_o, i.e. such that $T_o = T_L(C_o)$.

The governing equations are put in dimensionless form by scaling length with the magma layer thickness h, time with the diffusive scale h^2/κ and temperature with the temperature difference across the mushy layer $\Delta T = T_o - T_B$. This leads to,

$$\theta = \frac{T - T_o}{\Delta T} \tag{11.63}$$

$$\Phi = -\frac{\theta}{Ct - \theta} \tag{11.64}$$

$$\text{where} \quad Ct = \frac{C_S - C_o}{C_o - C_B} = \frac{T_S - T_o}{T_o - T_B}, \tag{11.65}$$

for the linear liquidus boundary that we have assumed for simplicity (Figure 11.11). This introduces one new dimensionless number, Ct, which provides a measure of the extent of compositional changes due to solidification. Changing the value of Ct affects the thickness of the mushy layer because it specifies the crystallization range. Differentiating the equation for solid fraction Φ, we obtain,

$$\frac{\partial \Phi}{\partial t} = -\frac{1}{Ct} (1 - \Phi)^2 \frac{\partial \theta}{\partial t}. \tag{11.66}$$

Substituting for the dimensionless variables into the governing equations, we find,

$$\left[1 + \frac{St}{Ct} (1 - \Phi)^2 \right] \frac{\partial \theta}{\partial t} = \frac{\partial^2 \theta}{\partial z^2}, \quad \text{for } 0 \le z \le l(t) \tag{11.67}$$

$$[1 - l(t)] \dot{\theta}_m = -\mathrm{Nu}\, \theta_m^{4/3} \tag{11.68}$$

$$\theta_m \frac{dl}{dt} = \left(\frac{\partial \theta}{\partial z} \right)_{l(t)^-} - \mathrm{Nu}\, \theta_m^{4/3}, \tag{11.69}$$

where θ_m corresponds to the interior temperature T_m. Three dimensionless numbers have been introduced:

$$\mathrm{St} = \frac{L}{C_p \Delta T} \tag{11.70}$$

$$\mathrm{Nu} = C_Q \left(\frac{\mathrm{Ra}_i}{\theta_{mo}} \right)^{1/3} \tag{11.71}$$

$$\mathrm{Ra}_i = \frac{\alpha g \left(T_{mo} - T_o\right)) h^3}{\kappa \nu_i}. \tag{11.72}$$

The dimensionless composition range Ct and the Stefan number have already been discussed. Ra_i is a Rayleigh number which provides a measure of the intensity of convection. The solution and the behavior of the roof mushy layer depend strongly on a key input, the initial amount of superheat, or in the dimensionless form used here, the initial value θ_{mo} for θ_m.

The behavior of the mushy layer is best understood by comparing it to the case without convection, (Figure 11.14). With convection, solidification initially proceeds at a slower rate than in a diffusive regime but eventually overtakes diffusive growth. One may identify three different phases in the growth of the mushy layer (Figure 11.15). The first phase lasts for a very short time and is such that the conductive heat flux is larger than the convective one. This reflects the fact that convection only develops once cooling has propagated over a finite thickness. In the second phase, fully developed convection supplies a large heat flux to the mushy layer,

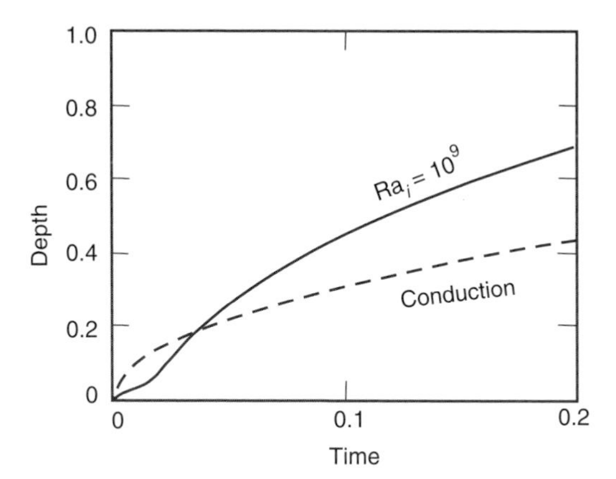

Figure 11.14. Dimensionless depth of the crystallization front $l(t)$ as a function of dimensionless time for a binary eutectic model with convection in the superheated melt. Dimensionless parameters are $Ct = 0.5$, $\mathrm{St} = 3$ and $\theta_{mo} = 1$. Solid line: solution for a Rayleigh number $\mathrm{Ra}_i = 10^9$. Dashed line: solution in a diffusive regime with no convection in the melt. Adapted from Kerr *et al.* (1990a).

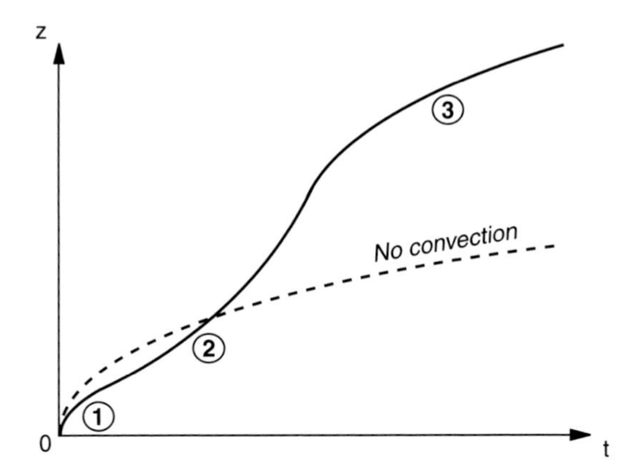

Figure 11.15. Diagram illustrating the three phases of crystallization in a superheated melt body that cools by convection. Phases 1 and 3 are controlled by diffusion. Phase 2 is controlled by convective heat transport and is limited in time due to fast cooling.

whose thickness remains small in consequence. Convection simultaneously acts to cool the melt, so that the convective heat flux decreases rapidly. The amount of superheat also decreases and so does the temperature gradient through the mush, which enhances crystallization. In the third phase, convection has died down, so that the mush grows essentially by conduction. The final phase of diffusive growth is faster than in the model without convection. This is because, with no convection, the amount of superheat ahead of the crystallization front does not change, which slows down solidification in comparison to the convective regime where the superheat is gradually suppressed. In the first and third phases, diffusive heat transport dominates implying that $l(t) \sim t^{1/2}$.

A simple argument illustrates what happens in the second, convective, phase, which is the most interesting one. We may neglect the thin mush in the bulk heat balance for the melt layer, so that

$$\dot{\theta_m} \sim -\text{Nu}\, \theta_m^{4/3}, \tag{11.73}$$

which leads to

$$\theta_m \sim (1+\gamma t)^{-3}, \tag{11.74}$$

where $\gamma = \text{Nu}/3$. This shows that a time scale for the duration of this phase, such that superheat is not negligible, is $\tau = \gamma^{-1} \sim \text{Nu}^{-1}$. In dimensional form,

$$\tau \sim \frac{h^2}{\kappa} \text{Ra}_i^{-1/3}. \tag{11.75}$$

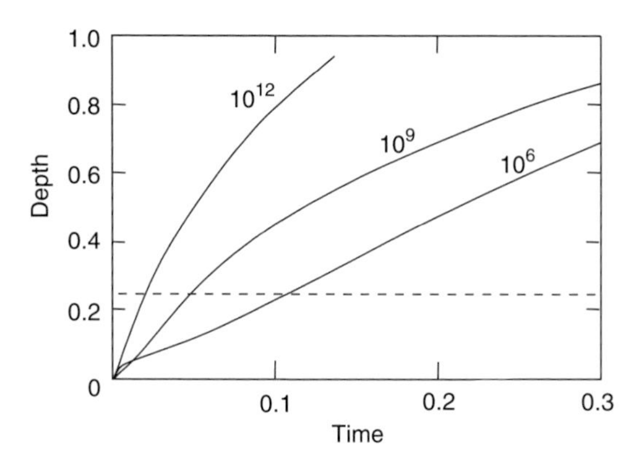

Figure 11.16. Dimensionless depth of the crystallization front $l(t)$ as a function of time for a binary eutectic model with convection in the superheated melt. Dimensionless parameters are $Ct = 0.5$, $\mathrm{St} = 3$ and $\theta_{mo} = 1$. Solutions corresponding to three values of the Rayleigh number Ra_i are shown, illustrating how convection acts to enhance the rate of solidification. The dashed line indicates approximately the end of the second, convective, phase of cooling. Adapted from Kerr *et al.* (1990a).

For increasingly vigorous convection, as measured by the value of Ra_i, Nu increases and the duration of the convective phase therefore diminishes. Writing that the convective heat flux into the mushy layer is balanced by conduction through the mushy layer, we obtain,

$$Q_T \sim \theta_m^{4/3} \sim (1 + \gamma\, t)^{-4} \sim \frac{\Delta T}{l(t)}, \tag{11.76}$$

which leads to $l(t) \sim (1 + \gamma\, t)^4$ (i.e. within a constant of proportionality). This accounts for the full numerical solutions shown in Figures 11.14 and 11.16.

As the vigor of convection is increased, the duration of the convective phase decreases, but not the amount of solid that is formed in the convective phase. This can be understood using the simple argument given above, which indicates that $l(t) \sim \theta_m^{-4/3}$. Assuming for simplicity that the convective phase ends when the dimensionless superheat θ_m reaches some threshold value (0.1, say), the amount of solid formed is therefore also the same. For the particular set of parameters of Figure 11.16, the amount of solid formed at the end of the convective phase is ≈ 0.25, regardless of the vigor of convection. This emphasizes the key effects of superheat and obviates the argument that superheat is not significant because it is rapidly suppressed by convection.

The three phases of mush evolution are characterized by different time dependence, such that the evolution of $l(t)$ is concave downwards in the first and third phases, and concave upwards in the second phase (Figure 11.15). The vigor of convection, as measured by the value of Ra_i, is a key factor in the rate of solidification. Convection ultimately acts to suppress superheat and hence to steer the cooling regime towards diffusion-controlled conditions.

11.4.2 Zero superheat: Convection in undercooled melt

The previous examples were designed to illustrate the effects of convection on the heat budget at the roof and on the thermal evolution of a melt body. One important parameter is the initial superheat, which is imposed as an initial condition. In reality, of course, the initial superheat depends on the prior history of melt extraction from the source and migration away from source. Here, we investigate the behavior of magma emplaced with no superheat and show that convection also develops in this case. As discussed above, the diffusion cooling model breaks down when there is no superheat because it implies that crystallization affects the whole melt layer. In reality, the crystallization front cannot be exactly at the liquidus temperature because the nucleation of new crystals requires a finite amount of undercooling.

The amount of undercooling

In kinetically controlled conditions, the position of the crystallization front, which is defined as the interface separating pure melt and mush, is not determined by temperature, i.e. by the condition that it is at the liquidus. This front is instead at a lower temperature noted T_c, which is unknown and must be solved for. Denoting the amount of undercooling by $\Delta T_u = T_L - T_c$, the standard kinetic formulation leads to a relationship between the rate of advance of the crystallization front, dl/dt, and ΔT_u. Because one has $dl/dt = 0$ when $(\Delta T_u) = 0$, one seeks a power-law relationship valid at small undercoolings:

$$\frac{dl}{dt} = \Gamma \left(T_L - T_c\right)^n , \tag{11.77}$$

where Γ is some constant and n an exponent. This kinetic equation replaces the requirement that temperature lies at the liquidus on the crystallization front and provides the closure equation needed for a quantitative model. The key consequence of kinetic effects is that there is always a finite temperature contrast across a crystal-free thermal boundary layer ahead of the mush, and this is sufficient to drive convection even in the absence of superheat.

Experimental data suggest that $n \approx 1$ for salt–water solutions and that $n \approx 2$ for silicate melts. We obtain estimates for the magnitude of the kinetic undercooling using the heat balance for solidification which has been introduced above (equation 11.12): $l(t) \sim \mathrm{St}^{-1/2}\sqrt{\kappa t}$, where the Stefan number is now calculated using the total temperature difference between the melt and the reservoir roof, i.e.

$\mathrm{St} = L/\left[C_p(T_L - T_B)\right]$. This balance is expected to be valid even in the presence of convection because, as shown above, the convective heat flux is small and has a minor influence on the heat budget of the roof region. Still, we shall show that convection has important consequences for the evolution of a magma reservoir. Using the expression for the crystallization front position as a function of time, we obtain an expression for the kinetic undercooling:

$$(T_L - T_c) \sim \Gamma^{-1/n}\mathrm{St}^{-1/2n}\left(\frac{\kappa}{t}\right)^{1/2n}. \tag{11.78}$$

This shows that kinetic undercooling decreases with time as solidification proceeds.

Crystallization at the roof and at the floor

Substituting for the various parameters in equation 11.78 for the amount of undercooling, Worster *et al.* (1990, 1993) found that the kinetic undercooling is in a typical range of 1–10 K for silicate melts. As discussed above, such temperature differences are sufficient to drive vigorous convective motions in the melt. The impact of convective motions is profound because they effectively remove undercooled melt from the roof region and take it to the base of the reservoir. The undercooled melt must eventually crystallize, which may happen as it descends through the melt body or as it spreads along the floor. In this case, thermal convection at the roof acts to promote solidification at the floor. We now develop full solutions.

For clarity, we again do not allow for heat loss through the floor region. The heat balance for the region below the upper thermal boundary layer is

$$Q_T = \rho L\frac{dh_f}{dt} - \rho C_p\left[h - l(t)\right]\frac{dT_m}{dt}, \tag{11.79}$$

where $h_f(t)$ is the solid fraction. For simplicity, we assume that all crystals accumulate in a solid layer with no trapped interstitial liquid, so that h_f is in fact the thickness of solid that grows at the floor (Figure 11.17). Crystallization of the undercooled melt generates residual magma with a different composition from the initial one. The key fact is that such crystallization occurs away from the upper boundary layer, i.e. away from the cooling interface. Thus, convection makes the melt body evolve chemically as well as thermally. The mass balance constraint can be written as follows:

$$(C_S - C_m)\frac{dh_f}{dt} = -\left[h - l(t) - h_f\right]\frac{dC_m}{dt}, \tag{11.80}$$

where C_m is the melt composition. This dictates how the melt evolves chemically.

One final relationship is needed to specify the crystallization rate dh_f/dt in the interior of the melt body as a function of roof heat flux Q_T. The simplest

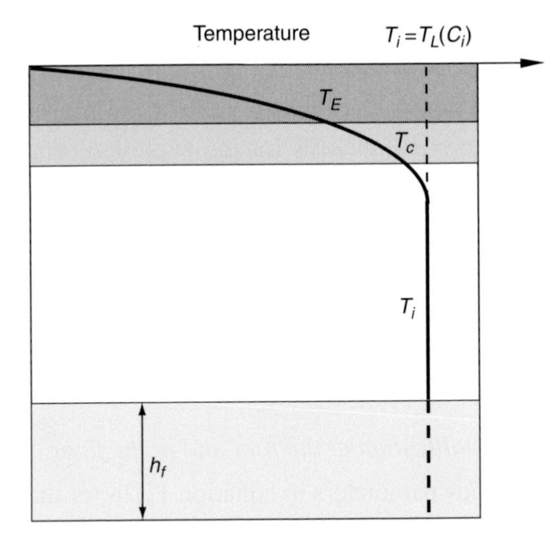

Figure 11.17. Thermal structure of a magma reservoir which is cooled from above with zero superheat. The crystallization front temperature, T_c, is lower than the liquidus due to kinetic controls on nucleation. Undercooled melt from the reservoir roof is carried to the floor by convective downwellings and crystallizes on the way down. We assume that all such crystals accumulate at the floor in a layer of thickness h_f. Another assumption is that crystals and melt are in thermodynamic equilibrium in the convective molten interior, so that the interior temperature T_i is at the liquidus of the melt with the interior composition C_i. We do not solve for the thermal structure of the floor solid layer. Adapted from Kerr *et al.* (1990b).

assumption is that such crystallization occurs in thermodynamic equilibrium, so that the interior temperature is kept at the liquidus of the evolving melt composition: i.e. $T_m = T_L(C_m)$. Thus, one has,

$$\frac{dT_m}{dt} = \frac{dT_L}{dC}\frac{dC_m}{dt}. \tag{11.81}$$

For emplacement into colder country rock, the contact temperature is initially set in a diffusive regime. Convection driven by the kinetic undercooling supplies heat to the roof region, so that the temperature at the top of the reservoir increases with time in an early phase and must be solved for. For a realistic model of magma reservoirs, one needs a large number of physical properties and thermodynamic parameters, including data on crystallization kinetics. Such information is only available for a few simple silicate systems. Worster *et al.* (1990) used the diopside–anorthite system (Figure 11.10) and adjusted the viscosity value to allow realistic predictions for the cooling and crystallization times of basaltic magma reservoirs. Their results are shown in Figure 11.18 for two different intrusion thicknesses. In both cases, solidification proceeds in a similar fashion. For this eutectic system, the solids

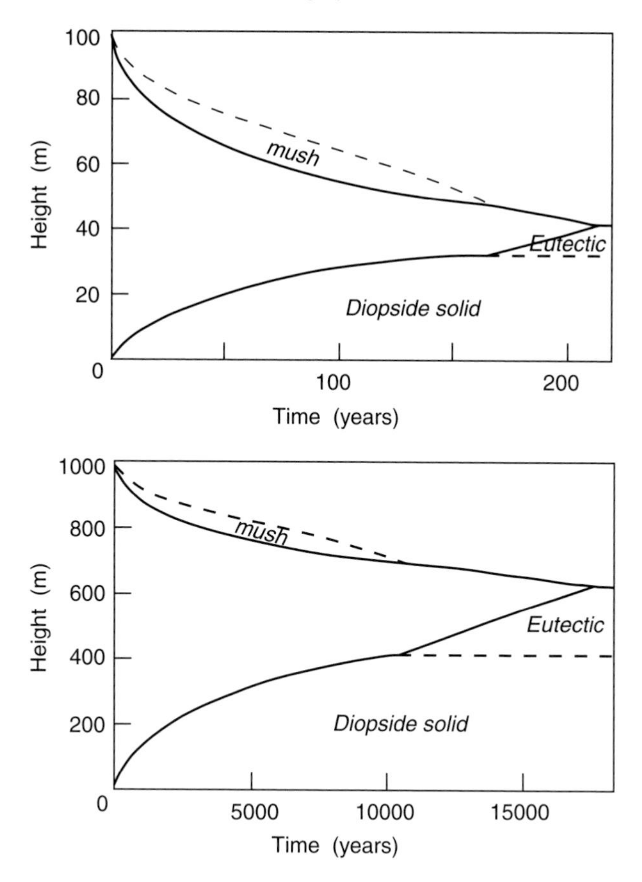

Figure 11.18. Crystallization in two reservoirs of different thicknesses filled with magma at its liquidus temperature, with thermal convection occurring in undercooled melt. Solutions are made for the diopside–anorthite system (Figure 11.10) with a $Di_{80}An_{20}$ initial composition and a viscosity appropriate for basaltic magma. Adapted from Worster *et al.* (1990).

that are formed in the mushy layer and in the reservoir interior are pure diopside crystals. Those that are formed in the interior accumulate at the floor. At the roof, the thermal boundary layer includes regions that are below the eutectic temperature and one must calculate the composition of the solid that forms (Elliott, 1977; Woods and Huppert, 1989). The composition of this solid is obtained by solving a mass balance equation and the constraints of local thermodynamic equilibrium in the mushy layer. This solid is in general not at the eutectic composition and is enriched in the high melting point end-member, diopside in this case. Thus, a composite solid with both anorthite and diopside crystals grows at the roof whilst a diopside solid grows at the floor. Convection eventually brings the interior temperature to the eutectic, so that the remaining liquid crystallizes a eutectic solid. The end result is a compositionally stratified igneous complex. The thicknesses of the floor and roof sequences depend on the intrusion size (Figure 11.18). For sizes less than about 100 m, the roof sequence is the thickest. For larger intrusions, the floor sequence is the thickest because convection maintains a large heat flux at the roof and inhibits cooling and solidification there.

In these calculations, the upper thermal boundary layer evolves in a cooling regime that does not depart markedly from a diffusive one. The effect of convection is felt mostly through its consequences on the melt composition and on the growth of crystals in the intrusion interior, away from the thermal boundary layers. The end result is that crystallization proceeds at different rates at the roof and at the floor of the melt body, leading to a crystallized roof sequence that is thinner than that at the floor in large magma reservoirs, as observed (Jaupart and Tait, 1995).

11.4.3 The floor and side-walls of an intrusion

Cooling at the side-walls

Many interesting processes are active at the side-walls of a magma reservoir, where the lateral gradients of temperature and composition that occur are unstable in all circumstances and generate local flows. Their importance for the cooling of the whole melt body depends of course on the aspect ratio of the intrusion, which is small in many cases, as indicated by Table 2.2.

The floor thermal boundary layer

Crystallization at the floor may occur due to heat absorbed by the underlying rocks. If we assume that encasing rocks at the top and bottom of the reservoir are initially at the same temperature, cooling initially proceeds in a symmetric fashion at the floor and at the roof. The thermal boundary layer at the floor, however, is stably stratified thermally and does not generate thermal convection. As shown by Jaupart and Brandeis (1986), descending plumes coming from the top do not penetrate into the stable temperature gradient. The floor boundary layer therefore grows in a diffusive regime with an upper edge that evolves thermally as the reservoir interior is cooling down. To estimate its thickness, we use the approximate solution for the interior temperature (equation 11.75) which specifies that the characteristic cooling time of the magma reservoir is $\tau \sim \mathrm{Ra}_i^{-1/3}$, where Ra_i is a Rayleigh number. For a diffusive cooling regime, thickness scales as $\sqrt{\kappa\tau}$ and we deduce that the lower thermal boundary layer extends over $\delta \sim \mathrm{Ra}_i^{-1/6}$. This is very small and can be neglected in the global heat balance.

We now estimate the importance of *in situ* crystallization at the floor. The floor sequence grows by diffusion with a thickness of solid that scales as $2\sqrt{\kappa t}$ and also receives crystals from the interior. Remote crystallization in the reservoir interior is due to the convective heat flux, Q_T (equation 5.82), and proceeds at a rate $R_{\rm int} \sim Q_T/\rho L$. If these crystals pile up at the floor, $R_{\rm int}$ is also the accumulation rate measured in thickness per unit time. In contrast, *in situ* crystallization proceeds at a rate $R_{\rm local} \sim \sqrt{\kappa/t}$. The ratio between the two crystallization rates, $R_{\rm int}/R_{\rm local}$, is $\sim t^{1/2}$. The convective heat flux is typically 100 W m^{-2} for a basaltic melt with a driving temperature difference of 10 K and generates more crystals that *in situ* cooling once the floor solid sequence is thicker than about 10 m.

The mushy layer that grows at the floor is affected by the settling of crystals from above, which is controlled by the crystal size, and by the dynamics of the density stratified interstitial liquid that develops within it. It may be the host of fascinating phenomena due to the interplay between dissolution and precipitation (Tait and Jaupart, 1992).

11.5 Kinetic controls on crystallization

In the models that have been developed above, the kinetic undercooling was estimated for a well-developed thermal boundary layer, such that the cooling rate is not large (the meaning of "large" and "not large" will be made clear below). In geological reality, magmas can intrude shallow crustal environments where temperatures are low, leading to high cooling rates. For initial temperatures T_i and T_{ro} in magma and country rock, respectively, the initial contact temperature is $T_B \approx (T_i + T_{ro})/2$, where we have assumed that magma and country rock have the same thermal properties and that there is no crystallization. For tholeiitic basalt at its liquidus temperature of 1200 °C and a regional crustal temperature of less than 200 °C, corresponding to emplacement at depths less than about 10 km, the initial contact temperature is less than 700 °C, which is much less than the solidus of about 980 °C. It is also lower than the glass transition that occurs at $\approx$ 725 °C (Ryan and Sammis, 1981). In this case, a chill zone forms and crystallization evolves at least initially in a kinetically controlled regime.

There is plenty of evidence for some kinetic control on crystallization in geological conditions. These include crystal size variations away from chilled margins, as well as quench textures and crystal morphologies that may be found even in the deep interior of plutons (Tegner *et al.*, 1983; Sisson *et al.*, 1996). In some cases, the order of appearance of certain mineral phases is kinetically controlled (Gibb, 1974). Thin intrusions contain significant amounts of glass, both in their chilled margins and in their interiors. The base of the Eskdalemuir tholeiitic dyke, in the southern uplands of Scotland, contains 42 % glass (Elliott, 1956). In the interior of basaltic tertiary Scottish dykes, the glass fraction ranges from 6 to 30% (Walker, 1930, 1935). All samples analyzed contain significant amounts of phases that can be attributed to late breakdown of the glass (Walker, 1935), so that the amount of pristine glass was probably larger than the value found today. The ~65 Ma Delakhari sill is part of the Deccan Trap intrusion sequence and exhibits upper and lower chilled zones containing $> 12\%$ glass (Sen, 1980). Glass is present (about 5 to 10%) throughout the whole intrusion, even though it is quite thick (~200 m). This large sill is noteworthy because it fed eruptive fissures and was probably active for a long time, which is not favorable to the formation of glass.

The crystallization kinetics of a Fe-rich basalt thought to be the parental liquid for the famous Skaergaard intrusion of Greenland have been documented in detail (Pupier *et al.*, 2008). For this melt, the natural order of crystallization is plagioclase (1175 °C), olivine (1165 °C), clinopyroxene (1130 °C) and Fe–Ti oxides (1100 °C). Experiments were conducted by heating powders and nucleation of the liquidus phase critically depended on the time spent above the liquidus. The longer this time was, the less efficient nucleation was. Crystallization could in fact be completely suppressed for lack of suitable nucleation sites. One expects that natural conditions favour heterogeneous nucleation due to the presence of xenoliths and perhaps antecrysts from the deep magma source,[2] but the fact remains that crystallization efficiency is sensitive to pre-existing nucleation sites. Even when the liquidus phase crystallizes easily, nucleation can be difficult for the other phases. In the experiments of Pupier *et al.* (2008), the nucleation delay for the second phase on the liquidus, olivine, is between 10 °C and 20 °C, and that for the third phase, clinopyroxene, is even larger and exceeds 30 °C. In fact, clinopyroxene was absent from almost all the kinetic experiments. The crystallization behavior is sensitive to the presence of water dissolved in the melt. In hydrous basalts, the amount of solid phases formed is smaller than in dry ones at the same cooling rate and the precipitation of plagioclase is suppressed (Gaudio *et al.*, 2010).

At the liquidus, the nucleation and growth rates of crystals are zero. These rates increase with increasing undercooling until they reach a maximum and then decrease to zero at large undercoolings. Denoting the rates of nucleation and growth by I and Y respectively, the kinetic crystallization time scale is

$$\tau_K = (IY^3)^{-1/4}. \tag{11.82}$$

One may define one characteristic time, τ_m, using I_m and Y_m, the largest crystallization rates of nucleation and growth that can be reached. One may also determine the characteristic time for crystallization at the small undercoolings which prevail in thick magma reservoirs some time after emplacement. In such conditions, the rates of nucleation and growth are necessarily smaller than the maximum values and the

[2] Antecrysts denote crystals from a previous intrusion that have been incorporated in the magma.

characteristic kinetic time is larger than τ_m. For silicate melts, very few measurements of the maximum growth rate have been attempted and we are aware of none for the maximum nucleation rate. For basaltic magmas, the growth and nucleation rates of olivine and plagioclase crystals have been determined in lava lakes, either *in situ* or through analysis of the crystal size distribution (Kirkpatrick, 1977; Mangan, 1990). For olivine, I and Y values are in the ranges 10^{-11} to 10^{-10} cm s^{-1} and 10^{-6} to 10^{-5} cm^{-3} s^{-1}, respectively. The characteristic crystallization time of olivine in natural basalts is thus $\tau \approx 4 \times 10^8$ s. For plagioclase crystals, this time is in the range 10^7 to 10^8 s. These relatively long time scales are appropriate for small undercoolings and are larger than τ_m. By comparing crystal size measurements in thin mafic intrusions and numerical crystallization calculations, Brandeis and Jaupart (1987) were able to constrain the values of I_m and Y_m to be about 1 cm^{-3} s^{-1} and 10^{-7}cm s^{-1}. For these values, $\tau_m \approx 3 \times 10^5$ s. During cooling against cold country rock, temperatures vary within a large range and the onset time of crystallization is intermediate between crystallization times at the maximum rates and at small undercoolings. It is 10^6 s or more for basaltic melts (Brandeis and Jaupart, 1987) and is even larger for silicic melts.

For chilling to occur, cooling must be faster than crystallization. For thin sills or dykes of thickness h, conduction is the dominant heat transport mechanism and the characteristic cooling time is

$$\tau_{\text{diff}} \approx \frac{1}{4}\frac{h^2}{\kappa}. \tag{11.83}$$

The ratio of the cooling time to the kinetic time is the relevant dimensionless number to assess kinetic controls on crystallization (Brandeis and Jaupart, 1987), and has been called the *Avrami* number by Spohn *et al.* (1988):

$$\text{Av} = \left(\frac{\tau_{\text{diff}}}{\tau_K}\right)^4. \tag{11.84}$$

Magma gets chilled if Av < 1. For an intrusion thickness of 1m, $\tau_{\text{diff}} = 3 \times 10^5$ s, so that Av < 1 for $\tau_k > 3 \times 10^5$ s, which is indeed appropriate for basalt. In such conditions, one predicts that little crystallization occurs, which is consistent with field observations (Walker, 1930, 1935).

One could conclude that such effects are unimportant for the large magma reservoirs that are of real importance, but this would be misleading for several reasons. We know now that such thick bodies do not form in one go and grow instead in piecemeal fashion out of a large number of intrusions of thickness in a 1–5 m range (Sisson *et al.*, 1996; Brown, 2000; Burchardt, 2008). Early stages of reservoir formation clearly deserve study and are beginning to be investigated in a systematic manner. Another reason for paying attention to crystallization kinetics is that they specify the size of crystals that precipitate in a cooling melt, a prerequisite for assessing crystal settling and internal melt circulation within a mushy layer.

For regional thermal models, chilling implies that magmas contribute a smaller amount of heat to encasing rocks. For models of magma accumulation and differentiation, one must probably consider separately an early phase with magmas that solidify rapidly after emplacement and a mature phase involving crustal rocks that have been heated by successive intrusions and that allow the growth of large magma bodies.

11.6 Conclusion

Magmas account for large inputs of heat in the crust. Latent heat released upon solidification is an important part of the heat budget. The unstable thermal stratification that develops within magma reservoirs generates convective motions that determine how and where crystallization proceeds. Thermal convection is limited to an early phase and does not modify greatly the heat flux into rocks encasing a reservoir. When calculating temperature in the surrounding country rock, one need not account for the existence of a large crystallization interval and solutions for a pure substance represent reasonable approximations. For models of magmatic differentiation, however, one cannot neglect processes in the thick partially crystallized layers that grow at the roof and floor of a reservoir.

12

Environmental problems

Objectives of this chapter

In this chapter we show how borehole temperature profiles can be used to infer past climate variations, and discuss the usefulness and limits of such methods. We also discuss the thermal conditions in ice sheets and show the importance of boundary conditions to calculate temperature profiles in the ice.

12.1 The record of past climate in temperature profiles

Time variations of the boundary condition at the Earth's surface affect subsurface temperatures with two important consequences: (1) the perturbations to steady-state temperature profiles may systematically affect the heat flux estimates, particularly in regions that were glaciated; (2) with careful measurements, these perturbations can be detected and interpreted to infer past variations in the surface boundary conditions. As early as 1923, temperature profiles from deep holes in the United States were used to infer the timing of the glacial retreat 10,000 years ago. The first heat flux estimates from Great Britain were corrected to account for the effect of the last glaciation. Birch (1948) proposed adjustments to account for the effect of the last glaciation on heat flux estimates in previously glaciated areas. The main obstacle for such corrections is that we still do not know what the temperature was at the base of the ice sheets. When the glacial retreat started, the temperature in the bedrock beneath the glacier was not in equilibrium. Although warming after the last glacial retreat 10,000 years ago is the dominating component of the temperature perturbation, the entire history of glacial retreats and advances must be included to calculate present perturbations to the temperature profiles (Figure 12.1). Following Jessop (1971), a correction has been applied to heat flux estimates from the Canadian Shield. This correction includes climatic variations for 400,000 years before present (BP) and assumes that the ground surface temperature was the same as present during interglacial and $-1\,^{\circ}$C during glacial episodes. In regions that were glaciated,

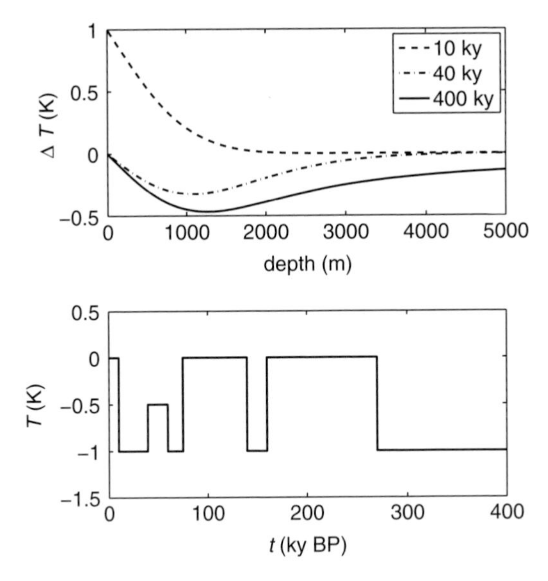

Figure 12.1. Example of glacial–interglacial surface temperature variations of the past 400 years (lower panel). Perturbations of the temperature depth profile calculated for various lengths of the climate history (upper panel). The figure demonstrates that the glacial correction must account for the entire climate history and not only the most recent warming.

where present ground surface temperatures are low, the effect of the glacial retreat on the temperature profiles remains small if temperature was near melting at the base of the glacier. A temperature profile measured in a very deep borehole (2800 m) near Flin-Flon, Manitoba, Canada, shows almost no variation in heat flux with depth which is the expected signal of post-glacial warming (Sass *et al.*, 1971). Deep (> 2000 m) boreholes from several sites in southern Canada confirm that the post-glacial warming signal is very subdued suggesting that temperatures at the base of the Laurentide ice sheet were not much colder than in the ground today (Rolandone *et al.*, 2003; Chouinard and Mareschal, 2009). In the permafrost regions of northern Canada, the adjustment for glacial retreat is negative.

The post-glacial warming affects temperature profiles down to 3000 meters, while variations in ground surface temperature of the past 200 years affect temperature profiles in the first 200 m. Borehole temperature data have thus been used to provide additional evidence for recent climate changes in several parts of the world (Cermak, 1971; Vasseur *et al.*, 1983; Lachenbruch and Marshall, 1986; Nielsen and Beck, 1989; Beltrami *et al.*, 1992; Wang, 1992; Pollack *et al.*, 1996).

The evidence for global climate change rests partly on meteorological records. These records are short; in many parts of the world, they span less than 100 years. They have been complemented by different types of proxy data to infer climate variations on a longer time scale. These proxies include historical records, such as records of the dates of harvest, advances and retreats of mountain glaciers. On continents, the main proxies come from dendrochronology, the width of the rings of annual tree growth, palynology, the distribution of pollens. Although related to climate, these proxies do not provide a direct measure of temperature. Recently, the study of the isotopic composition of air bubbles trapped in the ice of large ice sheets in Antarctica and Greenland has provided high resolution proxies of temperature variations at the surface of these glaciers. These records can not be extrapolated on a world-wide scale without supporting evidence. To a limited extent, borehole temperature data could provide some of this evidence. Many temperature profiles are available and more could be obtained if necessary. Meteorological records of climate are mostly based on surface air temperature and precipitation. The link between ground surface temperature and these climatic variables is still not very well understood and depends on many environmental factors. It appears that ground surface temperature tracks surface air temperature well, at least as long as there are no variations in the duration of the winter snow cover period (Bartlett *et al.*, 2005, 2006).

Because of the diffusive character of the heat equation, the determination of past ground temperature variations from temperature depth profiles is an ill-posed geophysical inverse problem: its solution is not unique and it is unstable, i.e. if not carefully analyzed the impact of noise on the solution may be overwhelming. Different regularization methods have been proposed and tested to invert the ground surface temperature history (GSTH) from one or several temperature depth profiles. Regardless of the method used, stability of the solution can only be achieved at the expense of resolution: the number of independent parameters that one can retrieve from such temperature profiles is small.

12.1.1 General formulation. The direct problem

The data used to infer climate variations consist of a one or several temperature profiles, with thermal conductivity, and heat production measurements. Because the temperature profiles are sparse and measured in boreholes far apart, it is common to neglect lateral variations in physical properties and in the boundary conditions. These assumptions are not always satisfied either because the surface boundary condition varies laterally (effect of lakes, vegetation cover, topography, etc.) or because heat is refracted by lateral variations in thermal conductivity. In the latter case, there is seldom sufficient information on the conductivity structure and temperature data to warrant using two- or three-dimensional models.

For a horizontally layered Earth, the steady-state temperature profile can be written as,

$$T_e(z) = T_{\text{ref}} + q_{\text{ref}} R(z) - M(z) \tag{12.1}$$

$$R(z) = \int_0^z \frac{dz'}{\lambda(z')}$$

$$M(z) = \int_0^z \frac{dz'}{\lambda(z')} \int_0^{z'} dz'' H(z''),$$

where λ is the thermal conductivity, H is the heat generation, z is depth, positive downwards. The reference surface temperature T_{ref} and heat flux q_{ref} should be understood as long-term averages on a time scale that depends on the depth of the profile. $R(z)$ is referred to as the thermal resistance and $M(z)$ includes the effect of heat production on the temperature profiles. If a temperature variation $T_0(t)$ is applied uniformly on the surface $z = 0$, it will induce a time-varying perturbation of the temperature in the half-space (equation 4.59):

$$T_t(z,t) = \frac{z}{2\sqrt{\pi\kappa}} \int_0^t \frac{T_0(t')}{(t-t')^{3/2}} \times \exp\left(\frac{-z^2}{4\kappa(t-t')}\right) dt', \qquad (12.2)$$

where the thermal diffusivity κ is assumed constant. This integral can be evaluated for different simple models of surface temperature variation.

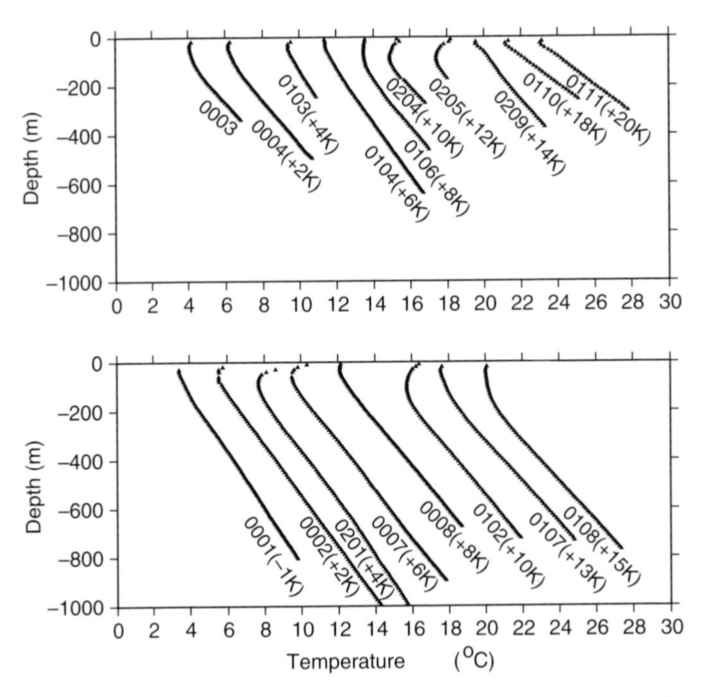

Figure 12.2. Temperature depth profiles measured in western Ontario show a decrease or an inversion of the gradient in the uppermost 200 m. Such profiles suggest recent (<200 years) warming of the ground surface. Data from Gosselin and Mareschal (2003).

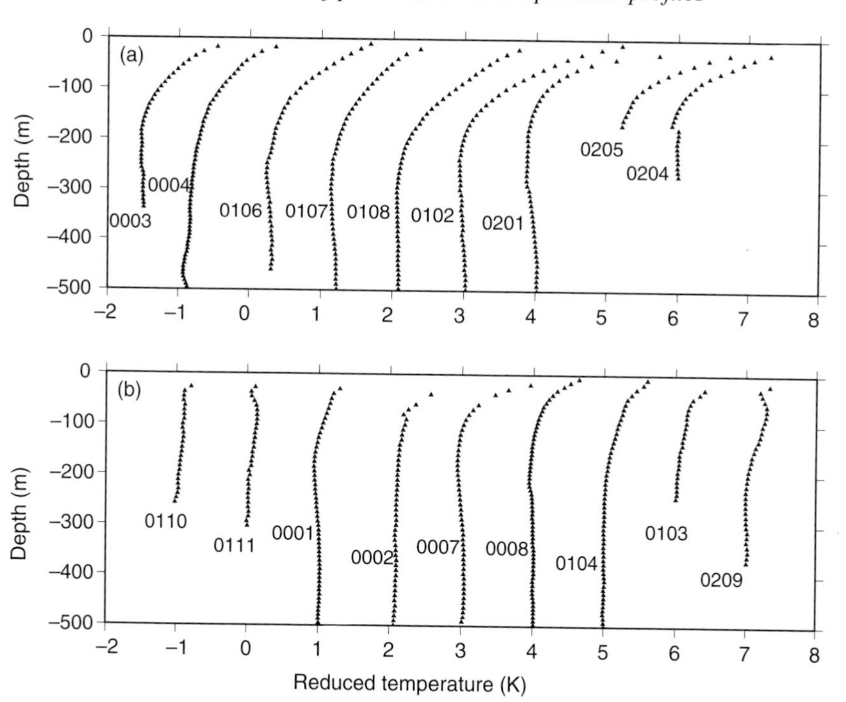

Figure 12.3. Reduced temperature profiles from western Ontario, obtained by removing the reference from the observed profiles in Figure 12.2.

For a jump ΔT in surface temperature at time t before present, the temperature perturbation at present is given by,

$$T_t(z) = \Delta T \times \operatorname{erfc}\frac{z}{2\sqrt{\kappa t}}. \tag{12.3}$$

If temperature started increasing linearly at time t before present to reach a value ΔT now, the temperature perturbation is given by,

$$T_t(z) = \Delta T\left(\left(1 + \frac{z^2}{2\kappa t}\right) \times \operatorname{erfc}\frac{z}{2\sqrt{\kappa t}} - \frac{z}{2\sqrt{\pi\kappa t}} \times \exp\frac{-z^2}{4\kappa t}\right). \tag{12.4}$$

For a constant change in surface heat flux Δq starting at time t before present, the temperature perturbation is (see Chapter 4),

$$T_t(z) = \frac{2\Delta q}{\lambda}\left(\left(\frac{\kappa t}{\pi}\right)^{1/2} \times \exp\frac{-z^2}{4\kappa t} - \frac{z}{2} \times \operatorname{erfc}\frac{z}{2\sqrt{\kappa t}}\right), \tag{12.5}$$

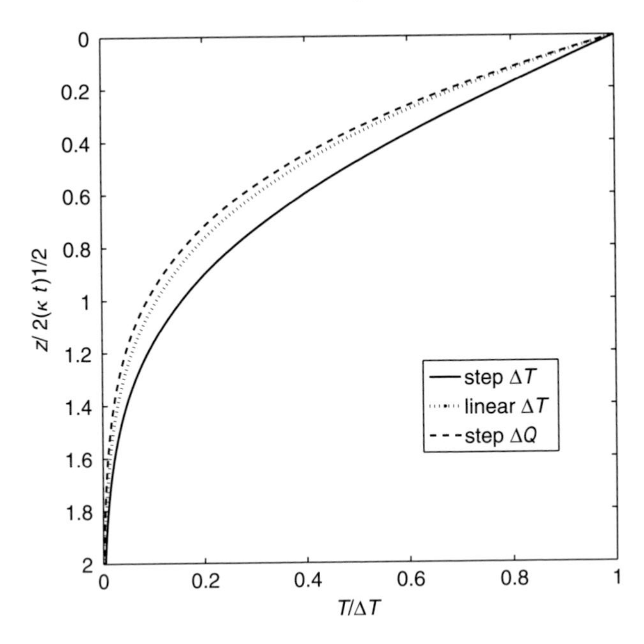

Figure 12.4. Profiles of temperature perturbations following different changes in surface boundary conditions: stepwise change in temperature, linear increase in temperature, change in heat flux. Depth is normalized to diffusion length scale.

and, in particular, the change in surface temperature is given by,

$$T(z=0) = \frac{2\Delta q}{\lambda}\left(\frac{\kappa t}{\pi}\right)^{1/2}. \tag{12.6}$$

Figure 12.4 shows the temperature profiles for three different surface boundary conditions leading the same present surface temperature. As one would expect, warming is more rapid after a jump in surface temperature than after a jump in surface heat flux. This is well known, but the point is that the temperature profile depends on the surface boundary condition, which is poorly understood, and so does the interpretation that is made of the temperature profile. For instance, using a heat flux rather than temperature boundary condition leads to underestimating the time when surface conditions changed.

It is possible to account for variations in thermal diffusivity with depth. Formal solutions for the transient temperature in a horizontally layered half-space could be obtained with the Laplace transform through rather tedious calculations. Because thermal diffusivity variations are usually small, their effect on the transient temperature profile is a second-order perturbation on the temperature profile and they can be safely neglected in view of all the other sources of error, provided that the average diffusivity is well determined. This does not hold true for the effect of conductivity variations on the steady-state temperature profile (12.1) which must be accounted for.

The loss of resolution of the record of surface temperature variations can be estimated by considering how a surface temperature pulse diffuses in the ground. A short temperature perturbation of amplitude ΔT and duration δt can be seen as a temperature pulse of strength $\Delta T \times \delta t$. At time t after the pulse, the maximum amplitude of the temperature perturbation is found at depth $z = \sqrt{2\kappa t}$ and its amplitude is

$$\Delta T_{\mathrm{max}} = \frac{\Delta T}{e\sqrt{2\pi}} \frac{\delta t}{t} \approx 0.15 \Delta T \frac{\delta t}{t}. \tag{12.7}$$

For the last glaciation with $\delta t / t > 1$, the signal is large, spread over a wide depth range, and it can be identified. Fluctuations on the centennial time scale are identifiable only if their amplitude is sufficient and if they occurred very recently. Considering for instance the medieval climatic optimum that may have occurred some 700 years ago, surface temperature being 1 degree warmer for 200 years increases temperature at depth by <0.05 K. Temperature measurements in boreholes are sufficiently precise to detect variations of <0.01 K, but the noise level (mostly due to geological noise, i.e. small-scale heterogeneities in thermal conductivity) is much higher than the signal and the very small temperature perturbation left by a possibly important climatic event is likely to be elusive. For recent Little Ice Age that occurred between 1600 and 1800, the effect is not always well identified in the borehole temperature records. The point is that one should not over-interpret the lack of identifiable climatic signals in borehole temperature profiles, even when the data are of excellent quality.

12.1.2 The inverse problem

For borehole temperature data, the inverse problem consists of determining, from the temperature–depth profile, the reference surface temperature and heat flux, and the ground surface temperature history (GSTH). Determining the reference heat flux requires knowledge of the thermal conductivity variations, usually measured on core samples. Alternatively, the thermal conductivity structure could be introduced as free model parameters through the thermal resistance versus depth in equation (12.1), but in this case the inverse problem becomes non-linear.

Formally, the inverse problem can be expressed as an integral equation:

$$T(z) = \int_{-\infty}^{0} \Delta T(t') K(z, t') dt', \tag{12.8}$$

where the kernel $K(z,t')$ is given by equation 12.2. It turns out that this type of integral equation always describes an "ill-posed" problem. If $T(z)$ is known approximately, there is no solution to the inverse problem. Furthermore, an exact solution is useless because the inverse operator is not continuous. It will thus be necessary to "regularize" the solution. The physical meaning of this instability is easy to understand. We can always add to the solution $\Delta T(t)$ a periodic function $N\sin(\omega t)$. Regardless how large N, the effect on the temperature profile $T(z)$ can be made arbitrarily small by increasing frequency ω. In other words, the difference between the exact and the approximate surface temperatures could be arbitrarily large at almost any time. This is paradoxical, but most fortunate, and we do take advantage of this property because we are mainly concerned with long period trends. In inverting temperature depth profiles several hundred meters deep, we can thus safely neglect the daily or the annual cycles although their amplitudes are at least ten times larger than those of the long-term trends that we are trying to detect.

The inverse problem discretized

Because surface temperature variations of short duration are filtered out of the temperature profile, any parameterization that reproduces the gross features of the ground surface temperature history can be used. Many different parameterizations have been proposed for the GSTH: a discontinuous function corresponding to the mean surface temperature during K time intervals Δ_k $(k = 1,\ldots,K)$, a continuous function varying linearly within K intervals Δ_k, a Fourier series, etc.

We shall assume that the GSTH is approximated by a discontinuous function corresponding to the mean surface temperature during K intervals of duration Δ_k (where Δ_k can be adjusted to the resolution decreasing with time).

For a single temperature profile, temperature Θ_j measured at depth z_j can be written as,

$$\Theta_j = A_{jl} X_l, \tag{12.9}$$

where Θ_j is the measured temperature at depth z_j (eventually corrected for the effect of heat production between the surface and that depth), X_l is a vector containing the unknowns $\{T_{\text{ref}}, q_{\text{ref}}, T_1, \ldots, T_K\}$, and A_{jl} is a matrix containing 1 in the first column and the thermal resistances to depth z_j, $R(z_j)$ in the second column. In columns 3 to $K+2$ the elements $A_{j,k+2}$ are obtained by calculating the perturbation at depth z_j due to the surface temperature difference between times t_k and t_{k-1}. It is expressed with the error function,

$$A_{j,k+2} = \operatorname{erfc}\left(\frac{z_j}{2\sqrt{\kappa t_k}}\right) - \operatorname{erfc}\left(\frac{z_j}{2\sqrt{\kappa t_{k-1}}}\right), \tag{12.10}$$

where κ is the thermal diffusivity. The other parameterizations mentioned above would yield a system of equations of similar structure.

Some authors prefer to invert reduced temperature profiles, i.e. the temperature anomaly. The reference temperature profile is obtained as a best fit to the deepest part of the temperature versus thermal resistance profile, and the temperature anomaly is the difference between the observed and reference profiles (see Figure 12.3 for an example).

Joint inversion of several profiles Temperature profiles from boreholes a few hundred meters apart often show similar but never identical transient perturbations. This is in part due to different surface boundary conditions and/or the effect of noise. Larger differences are observed on a regional scale when comparing profiles recorded at sites a few tens to hundreds of km apart, although the meteorological trends remain correlated over distances of $\approx$500 km (Jones *et al.*, 1986; Hansen and Lebedeff, 1987). One may thus hope to recover a common trend by using all temperature profiles recorded either locally or regionally. This may be achieved either by simultaneous inversion of all temperature profiles with the hope that the climatic signals are coherent and noise (i.e. any perturbation that can not be correlated with regional trends of climate variation) is not.

For I boreholes, the data are all the temperature measurements from all the boreholes. If N_i is the number of temperature measurements at borehole i, we have $N = N_1 + N_2 + \cdots N_I$ temperature data. The unknown parameters are the I reference surface temperature and reference heat flux values and the K parameters of the ground temperature history. We thus set up a system of N equations with $K + 2 \times I$ unknowns (see Clauser and Mareschal (1995) for more details).

Regularization by singular value decomposition

The system of N linear equations defined by equation 12.9 must be solved for the $M = K + 2 \times I$ unknown parameters. In general, the system is both under-determined and over-determined, and it is unstable. If the system of equations $Ax = b$ is mixed–determined, a generalized solution can be obtained by singular value decomposition (SVD) (Lanczos, 1961; Press *et al.*, 1992). This involves decomposition of the $(N \times M)$ matrix A as follows:

$$A = U \Lambda V^T, \tag{12.11}$$

where superscript T denotes the transpose of a matrix. The matrix U is an $(N \times N)$ orthonormal matrix (i.e. a rotation matrix) in data space, V is an $(M \times M)$ orthonormal matrix in parameter space, and Λ is an $(N \times M)$ diagonal matrix; the only non-zero elements are the L "singular values" λ_l on the diagonal, $L \leq \min(N, M)$.

The generalized solution is given by,

$$x = V \Lambda^{-1} U^T b, \tag{12.12}$$

where the $(M \times N)$ matrix Λ^{-1} is a diagonal matrix with the L elements $1/\lambda_l$ on the diagonal (for $\lambda_l \neq 0$) completed with zeros. The instability of the inversion results from the existence of very small singular values. In practice, this problem can be alleviated either by retaining only the $P \leq L$ singular values larger than a given "cutoff" or by damping the reciprocals of the smaller singular values. The damping is done by replacing the reciprocals of the smaller singular values λ_l by

$$\frac{1}{\lambda_l} \rightarrow \frac{\lambda_l}{\lambda_l^2 + \epsilon^2}, \tag{12.13}$$

where ϵ will be referred to as the damping or regularization parameter. The impact of noise can be reduced by selecting a higher value ϵ, which, however, decreases the resolution (see Figure 12.6).

By cutting off the small singular values, we have chosen to represent the solution in terms of just a few vectors in model space. These vectors are linear combinations of model parameters which will be best resolved by our data. This has the advantage that we do not need to search for an "optimal" parameterization; it will be selected by the method. The obvious disadvantage is that we are often unable to resolve individual parameters, with two important exceptions: the reference heat flux and surface temperature which are always well resolved. The singular values depend only on the distribution of the data and on the parameterization. For the particular problem of reconstructing ground surface temperature history, the singular value spectrum does not change much when the temperature profile is sampled more finely (Figure 12.5).

Figure 12.6 shows the result of inversion on synthetic data. The 600 m synthetic temperature profile was calculated for a temperature history that might be representative of long-term climate trends inferred for the Northern hemisphere. The ground surface temperature is initially at 5 °C , with 1 K cooling 400 years ago, followed by 2 K warming 200 years ago. The model parameters are surface temperatures during 80 intervals of 10 years. The inversion shows that the very recent past is reasonably well resolved regardless of the cutoff value. As we go further back in time, the resolution decreases even with an unrealistically low cutoff value (10^{-5}). For a cutoff value 10^{-2}, the errors on reference surface temperature and gradient are small (5.04 versus 5.00 °C and 9.94 versus 10 K km^{-1}, respectively), but the reconstructed cooling period is more diffuse than in the input. With 0.01 K Gaussian noise added to the synthetic profile, oscillations of ± 200 K appear in the GSTH for a cutoff of 10^{-5}, but the GSTH for a cutoff of 10^{-2} is almost identical to that for noise-free data.

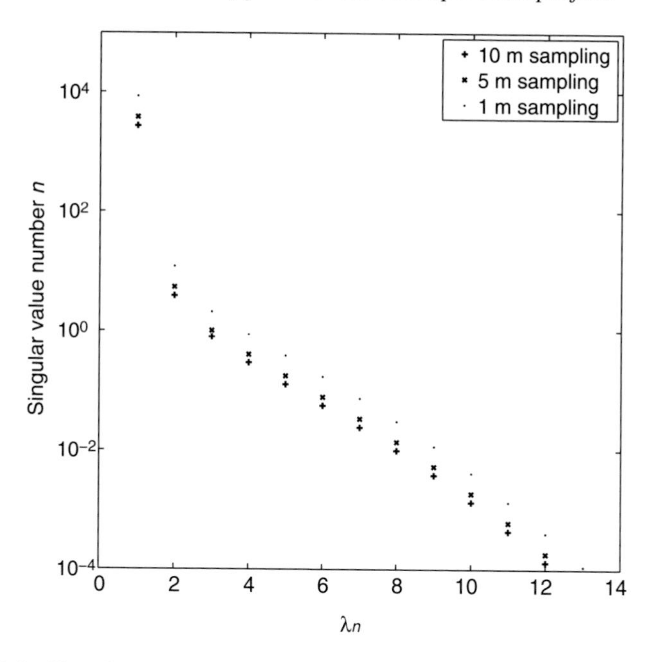

Figure 12.5. Singular value spectrum for inversion of ground temperature history when the data consist of a 600 m temperature profile sampled at different intervals and the surface temperature history covers the last 800 years with 10 year time steps. Note that finer sampling does not significantly improve resolution, as the ratio of the singular values is approximately the same.

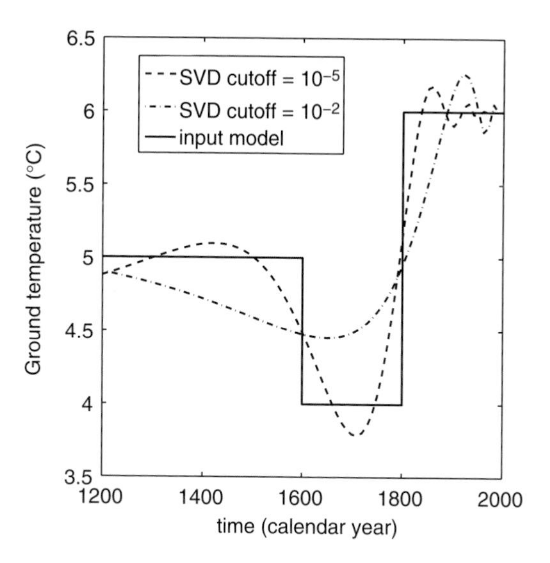

Figure 12.6. Impact of the singular value cutoff on the resolution of the inversion. A 600 m noise-free synthetic temperature profile was calculated and inverted with two different cutoffs for the singular values. The lowest cutoff value of 10^{-5} is shown only for the sake of the example; in practice, with noisy data, the cutoff must always be higher than 0.01.

Another feature of the inverted temperature history is the presence of short period oscillations in the recent past which can be reduced by damping the inverse of the small singular values rather than cutting them off. The damping parameter is usually slightly higher than the singular value cutoff ($\approx$0.02). For borehole temperature data, the value of ϵ ranges between 0.1 and 0.3. In practice, we make some assumptions on the regularity of the solution and select the damping parameter accordingly. The procedure to select the cutoff or damping parameter values consists of starting with a relatively low value, and examining the resulting GSTH; if oscillations of large amplitude appear in the solution, the value of the damping parameter is increased until these oscillations are attenuated. Oscillations in the very recent past are the result of the structure of the eigenvectors in model space and are impossible to eliminate altogether with a sharp cutoff. There is no compelling argument to prefer one method over the other, but damping usually gives smoother results than the sharp cutoff. In particular, a proper selection of the damping parameter will reduce the amplitude of the oscillations during the twentieth century to a level comparable to that found in the meteorological records, but these oscillations have a completely different cause.

12.1.3 Examples

We shall briefly present three examples to illustrate the potential and the limitations of using borehole temperature data for climate reconstructions (Figure 12.7). We shall restrict these examples to Canada because (1) the field conditions and physical properties of the rocks are well known and documented, (2) the data were collected in crystalline rocks where groundwater flow is less likely, (3) the radioactive heat production is usually small and can be neglected for shallow profiles.

The Raglan mine is located on the northernmost tip of Quebec, Canada. Except for a few Inuit communities, the region is not populated, and within a 1,000 km radius no meteorological record is available before 1945. Because climate models indicate that the northernmost latitudes might be the most affected by global warming, it is particularly important to evaluate the long-term climatic trends in northern Canada, and borehole temperature measurements could provide a substitute for the missing meteorological records. The mine is located on a barren plateau, where mean annual temperatures are low and permafrost extends deeper than 500 m. Temperature profiles measured in three exploration drill-holes have the characteristic signature of recent warming, with reversed gradient in the uppermost 60 m (Figure 12.8). Two of the profiles (0614 and 0501) cross a highly conductive layer of argillite; they are also affected by very mild topography. These two profiles were inverted, but even after accounting for conductivity variations, the inversions show large amplitude $\pm$2 K surface temperature oscillations (Figure 12.9). On the other hand,

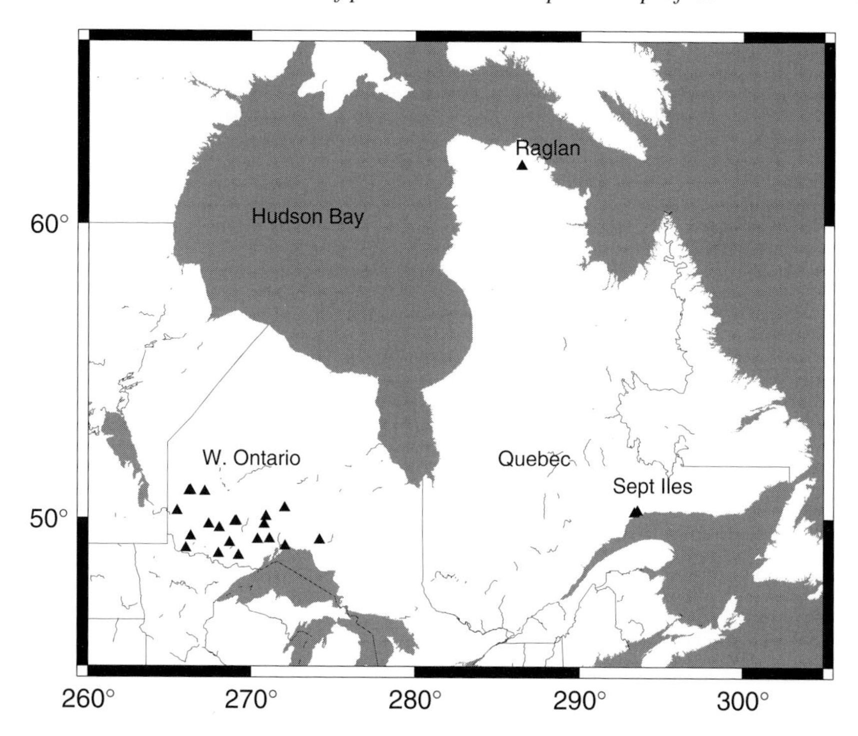

Figure 12.7. Map of eastern Canada and location of boreholes used for climate studies. The Raglan site is in thick permafrost, north of the tree line. Sept-Iles is on the north shore of the Bay of Saint Lawrence. Population density is very low both on the north shore of the Bay of Saint Lawrence and in western Ontario, north of Lake Superior.

profile 0615, which does not intersect the argillite, yields a smoother ground surface temperature history. At least for the second half of the twentieth century, this GSTH could be compared with meteorological records. The cooling indicated between 1940 and 1980 is consistent with records at two weather stations that are within 500 km of Raglan. The very strong warming (1.5 K) between 1990 and 2005 is also supported by observations at these weather stations (Figure 12.9). There is no local meteorologic record for the first half of the twentieth century but the warming is consistent with global trends.

For the two other holes, a lower singular value cutoff yields a smooth GSTH but less well resolved than the one from hole 0615. A good understanding of site characteristics facilitated interpretation of the borehole temperature profiles and selection of the least noisy of the three profiles for detailed interpretation.

Long meteorological records do not exist in central and western Canada. In north western Ontario, the oldest records start in 1880. More than 50 borehole temperatures logged between 2000 and 2005 have well-documented surface conditions.

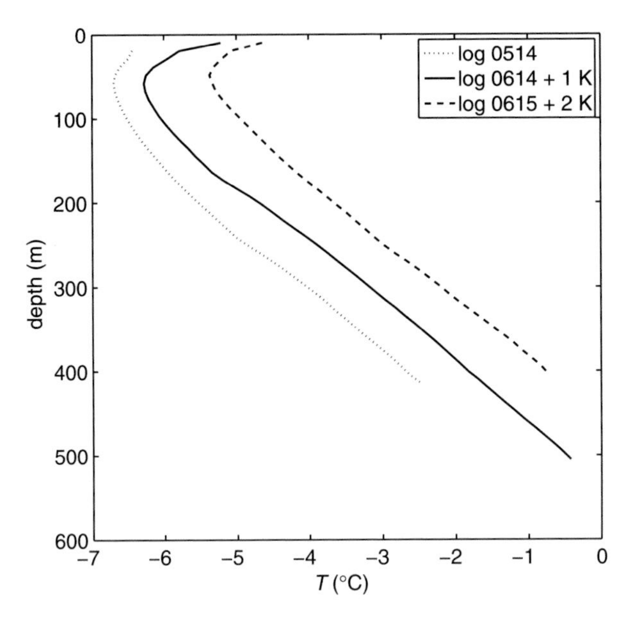

Figure 12.8. Three temperature profiles measured in the permafrost near the Raglan mine in northern Quebec, Canada, from Chouinard *et al.* (2007). Although the trends of the three profiles are similar, there are marked differences due to mild topography and thermal conductivity variations in holes 0614 and 0501. Profile 0615 is the least affected by surface conditions and does not intersect a high conductivity layer of argillite.

Many of the temperature profiles show reduced or inverted temperature gradients characteristic of recent ground surface warming, but are often also affected by other perturbations. Without eliminating the noisy profiles, the GSTH shows large oscillations during the twentieth century (Figure 12.10), which can not be reconciled with the meteorological records for the region. The amplitude of the oscillations can be reduced by selecting a high singular value cutoff, so high that the GSTH shows very little twentieth century warming. In an attempt to reconstruct the recent regional climate variations, we have selected all the holes that are not noisy, i.e. where we have not identified non-climate perturbations (see Table 12.1) and jointly inverted all these profiles (Figure 12.10). This procedure has yielded a relatively robust GSTH which is consistent with the short meteorological records in the region. This GSTH does not show any cooling before the warming that started around 1800, but we can not ascertain whether Little Ice Age cooling was absent or simply can not be resolved from the data available.

On a different time scale, deep boreholes can be used to try and reconstruct temperature conditions for the last glacial episode. During the last ice ages, between 40 and 10 ky BP, eastern Canada was covered by the Laurentide ice sheet which reached a maximum thickness $>$ 3,000 m at the last glacial maximum, 21 ky BP.

Table 12.1. *Non-climate causes of perturbations identified in the temperature profiles measured in Manitoba and Saskatchewan (Guillou-Frottier et al., 1998)*

Cause of perturbation	Expected effect
Intermittent permafrost	depends on location
Heat refraction	depends on location
Land use	usually apparent warming
Tree clearing	usually apparent warming
Slope	apparent warming uphill apparent cooling downhill
Lake	depends on dip of hole apparent cooling if dip towards lake
Forest fire	warming, stronger than tree clearing
Water flow	depends on flow direction

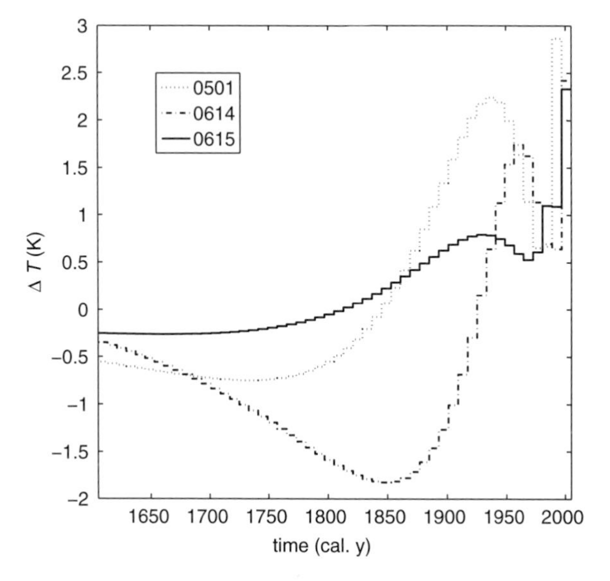

Figure 12.9. Ground surface temperature history for each of the three Raglan boreholes interpreted by Chouinard *et al.* (2007). Using the same singular value cutoff, the GSTH reconstructed from hole 0615 is much more stable than for the two other holes. The twentieth century variations including the marked warming after 1990 are consistent with proxy data and the short meteorological records for this region.

Marine sediments throughout the North Atlantic contain several layers of ice rafted debris, the Heinrich layers, supposedly carried by icebergs following sudden surges of the glacier. Models of glacial instability depend on poorly known thermal conditions at the base of the glacier. As will be seen later, the present temperature

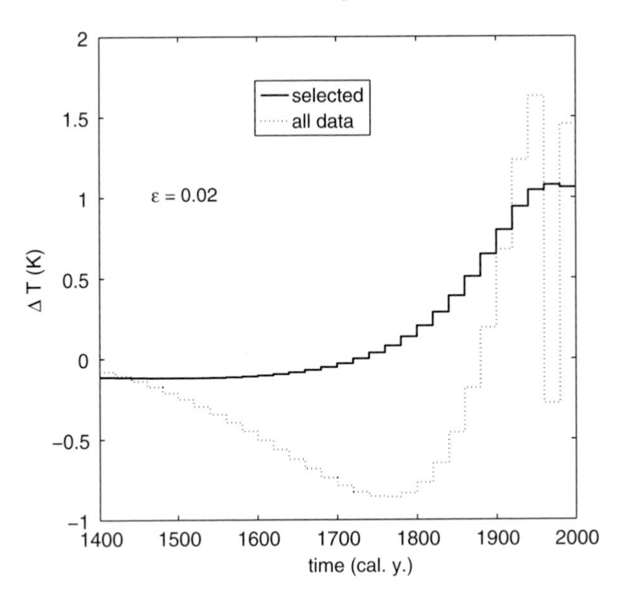

Figure 12.10. Reconstruction of the GSTH for western Ontario by joint inversion of the temperature profiles measured in the region. The inversion of selected profiles yields a smoother and presumably more robust GSTH than the joint inversion of all the profiles regardless of surface conditions (Chouinard and Mareschal, 2007).

conditions at the base of the large glaciers in Antarctica and Greenland are not well understood. The unexpected discovery of lakes at the base of the Antarctic ice sheet, as well as temperature profiles in boreholes through the ice in both Greenland and Antarctica suggest variable conditions at their base today. Temperature data from deep boreholes can be used to obtain some constraints on the conditions at the base of the Laurentide glacier.

An 1800 m deep temperature profile was measured through an anorthosite intrusion near Sept-Iles, Quebec, Canada, on the north shore of the Bay of Saint Lawrence (Figure 12.7). The hole is well suited for climate studies because: (1) with an homogeneous lithology there are no systematic thermal conductivity variations, and (2) there is no need to account for heat production which is negligible in anorthosite. The profile of heat flux shows a marked increase between 1200 and 1600 m deep, which we interpreted as the signal of the last glacial retreat (Figure 12.11). The temperature history was parameterized with a logarithmic distribution of time intervals to limit the numbers of parameters and account for the loss of resolution with time. For this site, the inversion shows that temperatures were well below freezing, ($< -4\,^\circ$C) during the last glaciation. It is noteworthy that the heat flux does not increase much in the topmost 800 m. The GSTH suggests that the ice retreat was

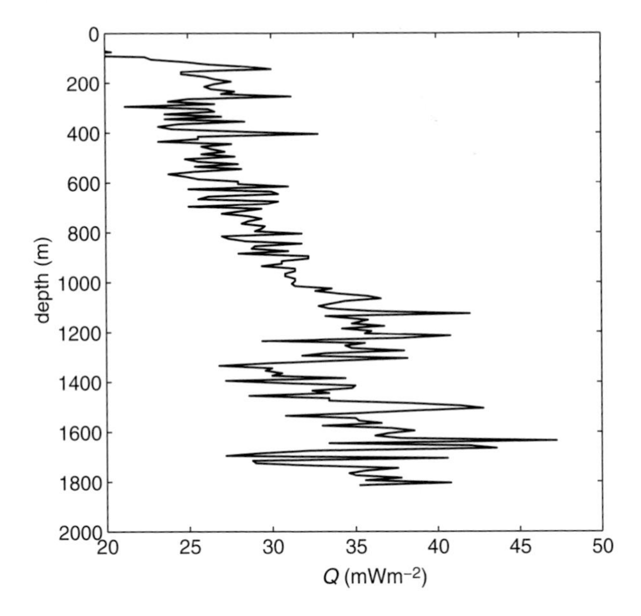

Figure 12.11. Heat flux profile measured in a deep borehole near Sept-Iles, Quebec, Canada, on the north shore of the Bay of Saint Lawrence from Mareschal *et al.* (1999). The profile was measured in an homogeneous anorthosite intrusion with few thermal conductivity variations. There is a marked increase in heat flux between 1200 and 1600 m.

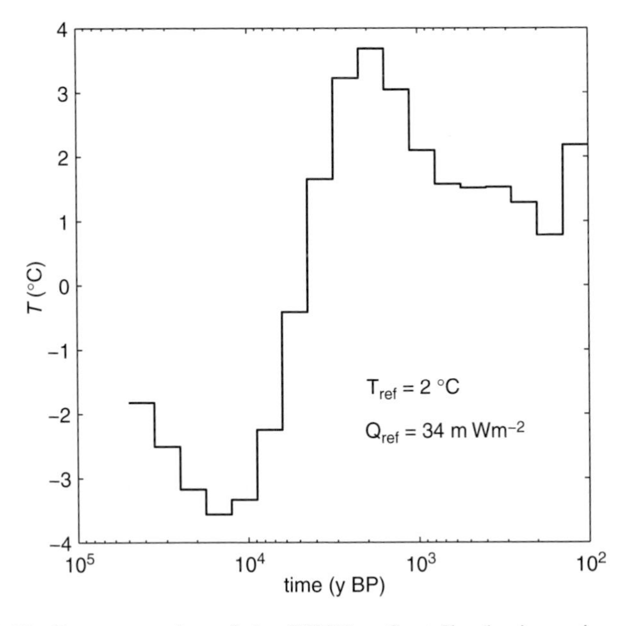

Figure 12.12. Reconstruction of the GSTH at Sept-Iles by inversion of the heat flux profile 12.11. A Monte-Carlo inversion of the heat flux profile yielded a similar GSTH (Rolandone *et al.*, 2003).

followed by a warm episode that may have ended 5000 y BP (Figure 12.12). This warm episode may be non-climatic as it coincides with the period when the entire region was under sea level, after the glacial retreat. It slowly rebounded to emerge above sea level 6000 y ago.

Monte Carlo inversion, i.e. a random search in parameter space, and direct models have confirmed that in the Sept-Iles region, the temperature was well below freezing and could have been as low as $-10\,^{\circ}\mathrm{C}$. Regional variations in temperatures at the base of the ice sheet are suggested by other data from deep holes: deep temperature profiles from western Ontario and Manitoba show almost no variation of heat flux with depth indicating that temperatures at the base of the glacier were warmer in central Canada than in the East.

12.1.4 Discussion

Temperature depth profiles are available from many regions on the continents that can be interpreted to infer on different spatial and temporal scales the climate trends of the last few hundred years and complement meteorological records. Such an approach is not as straightforward as may seem for several reasons: (1) the inherent loss of resolution of the signal due to the character of the diffusion equation; (2) the geological noise (i.e. the subsurface is not a conductive half-space); (3) the noise due to non-uniform surface boundary conditions (e.g. topography, proximity to lakes or large rivers); (4) cultural noise, i.e. man-made temporal variations in surface boundary conditions (changes in land use and vegetation cover, deforestation). Some of these perturbations cause distortions of temperature profiles similar to those due to ground surface temperature variations and they might overwhelm the real climatic signal in the profile. It is obvious that more information can be extracted from a single noise-free temperature profile than from many noisy profiles, as shown by the example in Raglan. Unfortunately, (1) there is no such thing as a perfect temperature profile in an homogeneous half-space, and (2) it is not possible to select between several profiles which ones are most suitable for climate studies, without information on site characteristics. It was hoped that simultaneous inversion of all the temperature profiles from one region, might improve the signal-to-noise ratio and yield GSTHs with good resolution if the signal were correlated between boreholes and the noise was not. This did not turn out to be true for several reasons.

(1) The number of temperature profiles from one region remains small and insufficient to produce a significant improvement in the signal-to-noise ratio which is $\propto \sqrt{N}$.
(2) The assumption that the transient perturbations are identical is almost never verified. This obviously can not hold at a global scale as there are regional differences in climate

variations. Even at the regional scale, visual inspection of the reduced temperature profiles reveals that they have not recorded the same GSTH. Thus, the joint inversion of real data seldom improves signal/noise ratio.

(3) The resolution is limited by the profile with the highest noise level which determines how much regularization is required .
(4) Beltrami *et al.* (1997) have emphasized the need to combine profiles with comparable vertical depths in order to avoid bias. The minimum depth sampled varies much between boreholes, because measurements above the water table are extremely noisy and are often discarded. This is an important bias because temperature perturbations are largest near the surface.

In practice, the examples from western Ontario and Manitoba show that the inversion yields more stable GSTHs after eliminating those profiles that are not suitable than when including all data. With good quality temperature profiles, it is possible to document regional climate variations of the past 200 y, which, in many regions, could represent a substantial complement to meteorological records. The resolution appears marginal to determine the amplitude of the Little Ice Age cooling and other episodes that may have preceded them. On a much longer time scale, deep boreholes in crystalline rocks are useful to determine average ground surface temperature during the last glaciation, and document regional variations in the thermal conditions at the base of glaciers.

12.2 Ice sheets and glaciers

Presently, there are two large continental ice sheets that cover most of Greenland and Antarctica. The temperature conditions at the base of these glaciers are extremely variable with large differences in temperature from one location to another. In Antarctica, the temperature is well below melting under most of the ice sheet, but several lakes have been found beneath the ice. In Greenland, temperatures seem to be near melting under a large part of the glacier. The stability of these glaciers is important because their melting would cause an important rise in sea level (on average 6 m for Greenland alone). Also, there is evidence that during the last glaciation, the Laurentide ice sheet that covered most of Canada was unstable, and that it surged at least five times between 45 and 10 ky BP. During these surges, enormous amounts of ice broke from the glacier and the icebergs transported rock debris across the North Atlantic. The cause of these cyclical instabilities remain poorly understood. The accumulation of snow and ice at the top of the ice sheets is balanced by the flow of ice and/or by flow of melt water near the base of the glacier. How instabilities of the glacier develop depends on the mass and energy budget of the glaciers, and particularly on the conditions near the base of the ice sheet.

The melting temperature of ice, at the base of the glacier, is about $-1\,°C$. If the base of the glacier is wet, this boundary condition must be verified, together with a Stefan boundary condition to account for the latent heat of melting (see equation 11.6). If the base of the glacier is dry, the temperature is below melting and one can not decouple the ice and ground problems. The heat equation must be solved for the ice and the rock basement with the continuity of temperature and heat flux conditions at the interface.

The mean temperature at the surface of the ice sheets is very cold ($\approx -30\,°C$ at the site of the GRIP ice core in Greenland, $-55\,°C$ at the location of the Vostok ice core in Antarctica). The present rate of accumulation of ice near the South Pole is $\approx 7\ \mathrm{cm\,y^{-1}}$, and in the order of 23 cm/y near the site of the GRIP ice core in Greenland. The thermal conductivity of ice varies from 2.22 to 2.75 $\mathrm{W\,m^{-1}\,K^{-1}}$ between 0 and $-50\,°C$. If transport of heat across the ice was by conduction only, a gradient of 20 to 30 $\mathrm{K\,km^{-1}}$ would be needed to maintain a heat flux of 40 to 60 $\mathrm{mW\,m^{-2}}$. And the temperature at the base of a 3 km thick glacier would be more than 60 K higher than at the surface. The boundary condition is thus,

$$\lambda_g \frac{\partial T_g}{\partial z} - \lambda_i \frac{\partial T_i}{\partial z} = L\rho \frac{dh_i}{dt}, \tag{12.14}$$

where the indices g and i refer to ground and ice respectively and dh_i/dt is the melting rate of ice at the base. A melting rate of 1 $\mathrm{mm\,y^{-1}}$ absorbs 10 $\mathrm{mW\,m^{-2}}$ ($L = 334\mathrm{kJ\ kg^{-1}}$). For $Q_g = 60\ \mathrm{mW\,m^{-2}}$, a rate of melting $dh_i/dt \approx 6\ \mathrm{mm\,y^{-1}}$ is sufficient to absorb all the heat flowing out of the bedrock and reduce the gradient through the ice to 0. These melting rates are one or two orders less than the rate of ice accumulation. Melting of ice by the geothermal heat flux could not maintain the glacier in steady state. For steady-state conditions, most of the ice must be removed by horizontal flow.

The energy conservation equation must thus include the transport of heat by ice flow. Near the base of the glacier, heat can be produced by shear heating in the ice and friction at the base of the glacier. Because the horizontal temperature gradient is smaller than the vertical, the energy conservation equation can be written as,

$$\rho(T)C_p(T)\left(\frac{\partial T}{\partial t} + v_z \frac{\partial T}{\partial z}\right) = \frac{\partial}{\partial z}\lambda(T)\frac{\partial T}{\partial z} + \Phi(z), \tag{12.15}$$

where Φ is the strain heating and the other terms have been defined,

$$\frac{\partial T}{\partial t} = -v_z \frac{\partial T}{\partial z} + \kappa(T)\frac{\partial^2 T}{\partial z^2} + \frac{\partial \lambda}{\partial T}\left(\frac{\partial T}{\partial z}\right)^2 + \frac{\Phi(z)}{\rho(T)C_p(T)}. \tag{12.16}$$

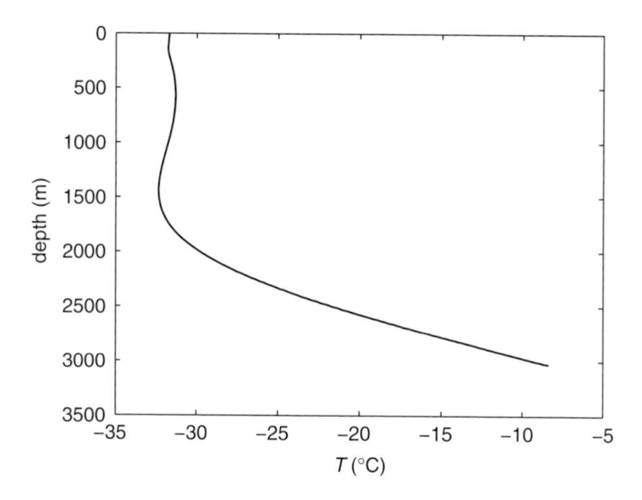

Figure 12.13. Temperature profile measured in the GRIP drill hole through the Greenland ice sheet. Data from Dahl-Jensen *et al.* (1998).

When transients are small and the time derivative can be neglected, the equation reduces to,

$$\rho C_p v_z \frac{\partial T}{\partial z} = \frac{\partial}{\partial z} \lambda \frac{\partial T}{\partial z} + \Phi(z), \tag{12.17}$$

where the shear heating rate $\Phi(z)$ depends on the rheology, which strongly depends on temperature. Over the range of temperature and stress in an ice sheet, the effective viscosity of ice varies by 6 order.

Temperature profiles measured through the ice show that the gradient increases markedly at depth, but there is very little curvature in the deepest part of the profile (Figure 12.13). This suggests that the contribution of shear heating is small, except very near the base of the glacier. As a first approximation, the heat equation can be reduced to,

$$\rho C_p v_z \frac{dT}{dz} = \lambda \frac{d^2 T}{dz^2}. \tag{12.18}$$

If the ice thickness is constant, the boundary conditions at the surface $T(z=0) = T_0$ and at the base $dT/dz(z=D) = \Gamma_i$.

For a constant ice thickness, mass conservation requires that the melting rate should be equal to the accumulation rate, which, as we have seen, is impossible to achieve because the heat flux is too low, or that there is horizontal ice transport. To account for horizontal transport of ice and satisfy mass conservation, we can assume that velocity decreases linearly from v_0 at the surface to 0 at the base of the

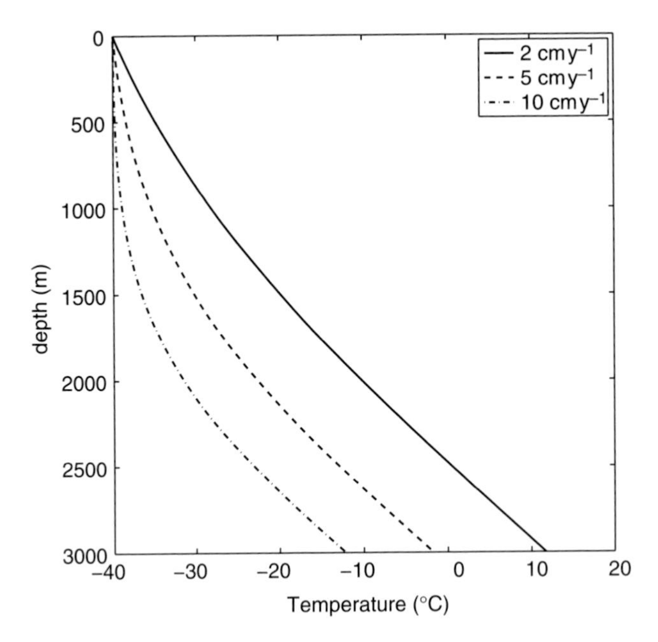

Figure 12.14. Temperature in an ice sheet when the velocity decreases linearly with depth. For this calculation the heat flux at the base of the glacier is 50 mW m^{-2}.

ice $v_z(z) = v_0(1 - z/D)$. The temperature profile is then,

$$T = T_0 + \Gamma_i \left(\frac{\pi\kappa D}{2v_0}\right)^{1/2} \left(\operatorname{erf}\left(\sqrt{\frac{v_0}{2\kappa D}}(z - D)\right) + \operatorname{erf}\left(\sqrt{\frac{v_0 D}{2\kappa}}\right)\right). \quad (12.19)$$

This solution is valid only if there is no melting at the base, but the profiles reproduce the gross features of the GRIP temperature profile (Figure 12.13). The temperature at the base depends very largely on the accumulation rate (Figure 12.14). This shows that the accumulation rate must exceed a critical value to avoid melting at the base. The model is oversimplified because it does not include coupling with the ground. It does not account for the feedbacks between ice temperature, the absorbtion of latent heat which stabilizes and shear heating which destabilizes the ice.

If the velocity decreases exponentially with depth $v_z = v_0 \exp(-z/\delta)$, the temperature is obtained in terms of the exponential integral:

$$T = A + B \times \operatorname{Ei}(-v_0\delta/\kappa) \exp(-z/\delta)), \quad (12.20)$$

where A and B are determined by the boundary conditions and the exponential integral is defined by,

$$\operatorname{Ei}(t) = \int_{-\infty}^{t} \frac{\exp(-u)}{u} du. \quad (12.21)$$

Solving for A, B gives the following temperature profile:

$$T = T_0 + \Gamma_i \delta \exp\left(\frac{-v_0\delta}{\kappa}\exp(-D/\delta)\right)\dots$$
$$\times\left(\mathrm{Ei}\left(-\frac{v_0\delta}{\kappa}\exp(-z/\delta)\right) - \mathrm{Ei}\left(-\frac{v_0\delta}{\kappa}\right)\right). \tag{12.22}$$

The purpose of this discussion was not to present detailed models of ice sheet dynamics. It was included only to show the importance of thermal boundary conditions in analyzing their stability.

Exercises

12.1 Find the temperature in the half-space $z = 0$ initially at temperature $T = 0$ when surface temperature increases linearly with time $T(z = 0, t) = \dot{T}t$.

12.2 Find the temperature in the half-space $z > 0$ initially at $T = 0$ when surface heat flux for $t > 0$ is Q_0.

12.3 Assuming a 1 K increase in temperature at the Earth's surface, determine how much energy per unit surface is absorbed by the ground in 100 years. (Assume $\lambda = 3$ W m^{-1}, and $\kappa = 10^{-6}$ m^2 s^{-1}).

12.4 Determine the total amount of energy per unit surface that the Earth has absorbed following a 1 K warming that occurred 200 years ago. ($\rho = 2700$ kg m^{-3}, $C_p = 1000$ J kg^{-1}, $\lambda = 3$ W m^{-1} K^{-1}). Assuming that on average, the continents have warmed by 0.5 K between the end of the Little Ice Age and present, how much energy has the ground absorbed?

13

New and old challenges

At the start of this book, we were reminded that the Earth's thermal evolution is an old question that was first addressed by Kelvin. Our understanding of this problem has been turned upside down several times since the time of Kelvin, but the questions that he raised have not all been fully answered. The length of this book, which does not even cover everything we know, is a clear demonstration of the enormous progress that we have made since 1862. Still, many questions remain, and understanding the Earth's thermal evolution is still a major challenge.

Thanks to decades of careful measurements, we know well the total energy loss of the Earth, and we know that the largest fraction of this energy is brought to the surface by convection in the Earth's mantle. Mantle convection is driven by secular cooling of the Earth, radiogenic heat production in mantle rocks, and heating from the core. The balance between these three sources of energy is very poorly constrained and it is likely to remain one of the most significant stumbling blocks in studies of Earth's evolution through geological time. The development of neutrino observatories opens new and exciting perspectives for the geosciences. It will require many years of observations with land detectors, and the deployment of observatories on the sea floor to narrow down our estimates of the heat production of the mantle but there is real hope that, in the end, we shall have stronger constraints on one of the sources that enters the budget.

Our best understanding of the present energy budget implies a rate of cooling of the mantle that seems high, higher than suggested by petrological studies on old Archean rocks. This raises the possibility that the energy loss has varied through time. This may be because of variation of the average rate of sea-floor spreading through time, and/or because subduction operated differently resulting in different distribution of sea-floor ages. This is an intriguing question because today we still do not understand well what triggers subduction of the oceanic plates.

We now know that continental crust started forming very early in Earth's history, more than 4 billion years ago. Much evidence suggests that the roots underlying the core of the continents formed at about the same time the crust became stable through processes that still elude us. How did they form? How did the roots remain cold and stiff enough not to be entrained in mantle convection? How did the volume and surface of continents change through time? The answers to these questions will have important implications for the Earth's thermal evolution, in part because the formation of continental crust removes some of the heat sources from the mantle, in part because very little heat from the mantle escapes through thick continental lids.

Should we solve all these questions, we might be able to calculate how much energy the core now loses, and when the inner core started to grow, and the implications for the geodynamo.

The Earth is the only internal planet which experiences plate tectonics today. Differences between the evolution of the planets have given one of the most striking demonstrations of the control of temperature and rheology on the internal dynamics of planets. These differences also demonstrate feedback between the atmosphere and hydrosphere and the planetary interiors.

In addition to these global questions, many other important geological questions involving heat transport in the Earth are challenging us today. If there is any lesson to be learned from the history of the geosciences it is that the answers may surprise us and come from where they are most unexpected.

Appendix A

A primer on Fourier and Laplace transforms

Many solutions of the heat equation can be derived by using integral transforms. Indeed, the Fourier series were introduced by Fourier (1822) in order to find solutions of the heat equation. The Fourier series and transforms are most suitable to solve boundary value problems. The Laplace transform was introduced by Heaviside as the basis for operational calculus. The Laplace transform is most useful for solving initial value problems.

This appendix and the following one only provide a short introduction to basic techniques used for solving heat conduction problems. These topics are thoroughly covered in the references.

A.1 Impulse response and Green's functions

For the sake of simplicity, we shall use time t as a variable in this section.

A.1.1 Dirac's delta function

Definition

The Dirac delta function $\delta(t)$ is defined by its properties:

$$\begin{aligned} &\delta(t-t') = 0 \; t \neq t' \\ &\int_{-\infty}^{\infty} \delta(t-t')dt = 1, \end{aligned} \tag{A.1}$$

so that for any function $f(t)$ continuous in t':

$$\int_{-\infty}^{\infty} \delta(t'-t)f(t')dt' = f(t). \tag{A.2}$$

The delta function is not a standard function but has an operational meaning when it is integrated as in equation A.2.

Some properties of the delta function

Although the Dirac function is not continuous, it is possible to give a meaning to its derivative $\delta'(t)$ in the sense:

$$\begin{aligned}\frac{d}{dt}\int_{-\infty}^{\infty}\delta(t'-t)f(t')dt' &= \frac{df(t)}{dt}\\ \int_{-\infty}^{\infty}\frac{d}{dt}\delta(t'-t)f(t')dt' &= \frac{df(t)}{dt}\\ -\int_{-\infty}^{\infty}\delta'(t'-t)f(t')dt' &= \frac{df(t)}{dt}.\end{aligned} \tag{A.3}$$

This implies

$$\frac{d}{dt}\delta(t) = -\delta(t)\frac{d}{dt}. \tag{A.4}$$

As for most operations with the delta function, the derivative of the delta function is meaningful only when the function is introduced in an integral:

$$\delta'(t)f(t) = -\delta(t)f'(t). \tag{A.5}$$

A.1.2 Impulse response of a linear time invariant system: the convolution theorem

The differential operator $\mathcal{L}_t$ defines a linear time invariant system if:

$$\begin{aligned}\mathcal{L}_t(f_1(t)+f_2(t)) &= \mathcal{L}_t(f_1(t)+\mathcal{L}_t(f_2(t)\\ \mathcal{L}_t f(t) = u(t) &\Rightarrow \mathcal{L}_t f(t-t') = u(t-t').\end{aligned} \tag{A.6}$$

The function $g(t)$, the solution of equation A.7 for a Dirac delta function is the Green's function or impulse response of the system and it is sufficient to entirely characterize a linear time invariant system:

$$\mathcal{L}_t g(t) = \delta(t), \tag{A.7}$$

with $g(t=0)=0$. The solution $f(t)$ to equation:

$$\mathcal{L}_t f(t) = u(t) \tag{A.8}$$

for any function $u(t)$ can be written as the integral:

$$f(t) = \int_{-\infty}^{\infty} g(t-t')u(t')dt'. \tag{A.9}$$

This type of integral is called a convolution. Although formally simple, the convolution (A.9) is not always straightforward to calculate, and it is time consuming to perform numerically. One can use the differential equation to check that the response of the system is given by the convolution of the impulse response with the input. The impulse response $g(t)$ satisfies the equation:

$$\mathcal{L}_t g(t) = \delta(t) \tag{A.10}$$

which implies

$$\mathcal{L}_t \int_{-\infty}^{\infty} u(t')g(t-t')dt' = \int_{-\infty}^{\infty} u(t')\mathcal{L}_t g(t-t')dt' = \int_{-\infty}^{\infty} u(t')\delta(t-t')dt' = u(t). \tag{A.11}$$

The order of the operations can be changed because of the linearity.

The convolution of two functions is often noted as "$f*g$". One can verify that the convolution is commutative $f*g=g*f$ and linear $f*(g_1+g_2)=f*g_1+f*g_2$. The time invariance results from

$$f*g(t+\tau) = \int_{-\infty}^{\infty} f(t+\tau-t')g(t')dt'. \tag{A.12}$$

A.1.3 Other useful functions

The Heaviside function

The indefinite integral of the delta function:

$$H(t) = \int_{-\infty}^{t} \delta(t')dt' = 0 \quad t<0 \\ = 1 \quad t>0 \tag{A.13}$$

is the Heaviside step function.

Boxcar (gate) function

The boxcar or gate function is defined as:

$$\begin{aligned}\Pi(t) &= 1 \ |t| < 1/2 \\ &= 0, \ |t| > 1/2.\end{aligned} \tag{A.14}$$

It is the difference of two Heaviside functions:

$$\Pi(t) = H(t+1/2) - H(t-1/2). \tag{A.15}$$

A.1.4 The Laplace transform

The Laplace transform is particularly useful to solve initial value problems for partial differential equations (PDE) or to determine solutions to ordinary differential equations (ODE) with initial conditions. The Laplace transform pair is defined by:

$$\begin{aligned}F(s) &= \int_0^\infty f(t)\exp(-st)dt \\ f(t) &= \frac{1}{2\pi i}\int_{\gamma-i\infty}^{\gamma+i\infty} F(s)\exp(st)ds,\end{aligned} \tag{A.16}$$

where s is a complex variable, and γ is a real number such that the path of integration of the inverse transform is to the right of all the poles of $F(s)$ in the complex plane.

When $F(s)$ is analytic, the integral can be evaluated by closing the contour of integration at ∞ in the complex plane. For $t > 0$, the contour must be closed to the left and the integral yields the residues at the poles (Figure A.1). For $t < 0$, the contour is closed to the right and the integral yields 0 because there are no poles. When $F(s)$ is not single valued and has a branch cut, the contour of integration must be closed along the branch cut and around the branch point. This is the case for many transforms encountered in problems with the heat equation.

If $F(s)$ is the transform of $f(t)$, the transform of the derivative of $f(t)$ is obtained by multiplying the transform by s and subtracting the initial value $f(0)$, as

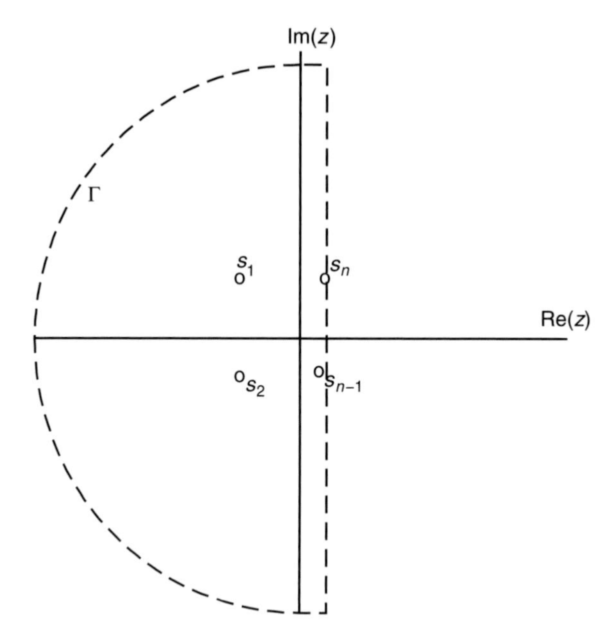

Figure A.1. Contour of integration in the complex plane for calculating an inverse Laplace transform when $F(s)$ is single valued and analytic except for n poles at $s_1,\ldots,s_n$. If the contour of integration can be closed along Γ in the left half of the complex plane, the inverse transform is obtained as the sum of residues at the poles.

follows:

$$\int_0^\infty f'(t)\exp(-st)dt = sF(s) - f(t=0)$$
$$\int_0^\infty f''(t)\exp(-st)dt = s^2F(s) - f'(0) - sf(0) \tag{A.17}$$
$$\ldots$$

These properties make the Laplace transform convenient to solve initial value problems. Ordinary differential equations can be transformed into an algebraic equation where the initial conditions are in the right-hand side. The heat equation in 1D is transformed into an ordinary differential equation with the initial conditions on the right-hand side. In 2D or 3D, the heat equation is transformed into Poisson's equation.

Useful Laplace transforms

Several useful tables of direct and inverse Laplace transforms are available (e.g., Erdélyi *et al.*, 1954a; Gradshteyn and Ryzhik, 2000). Many Laplace transforms useful for heat conduction problems can be found in the appendix of Carslaw and Jaeger (1959).

The symbol $\mathcal{L}$ is used to denote the Laplace transform.

$$
\begin{aligned}
F(s) = \mathcal{L}(f(t)) &= \int_0^\infty f(t)\exp(-st)dt \\
sF(s) - f(0) &= \mathcal{L}(f'(t)) \\
\frac{F(s)}{s} &= \mathcal{L}\int_0^t f(t')dt' \\
F(s+a) &= \mathcal{L}(f(t)\exp(-at)) \\
F(as) &= \mathcal{L}(af(t/a)) \\
1 &= \mathcal{L}(\delta(t)) \\
\frac{1}{s} &= \mathcal{L}(H(t)) \\
\frac{1}{s+a} &= \mathcal{L}(\exp(-at)) \\
\frac{\Gamma(\nu+1)}{s^{\nu+1}}\ddagger &= \mathcal{L}(t^\nu) \quad \nu > -1 \\
\exp(-\sqrt{\tau s}) &= \mathcal{L}\left(\frac{1}{2}\sqrt{\frac{\tau}{\pi t^3}}\exp\left(\frac{-\tau}{4t}\right)\right) \\
\frac{\exp(-\sqrt{\tau s})}{s} &= \mathcal{L}\left(\operatorname{erfc}\frac{\tau}{2t}\right) \\
\sqrt{\frac{\tau}{s}}\exp(-\sqrt{\tau s}) &= \mathcal{L}\left(2\left(\frac{t}{\pi\tau}\right)^{1/2}\exp\left(\frac{-\tau}{4t}\right)\right) \\
\frac{\exp(-\sqrt{\tau s})}{s\sqrt{\tau s}} &= \mathcal{L}\left(2\left(\frac{t}{\pi\tau}\right)^{1/2}\exp\left(\frac{-\tau}{4t}\right) - \operatorname{erfc}\left(\frac{1}{2}\sqrt{\frac{\tau}{t}}\right)\right) \\
\frac{Q(s)}{P(s)}\dagger &= \mathcal{L}\sum_{k=1}^{K}\frac{Q(s_k)}{P'(s_k)}\exp(-s_k t)
\end{aligned}
\tag{A.18}
$$

† P and Q are polynomials, and $P(s_k) = 0, k = 1, \ldots K$.
‡ where $\Gamma(\nu)$ is the gamma function:

$$
\Gamma(\nu) = \int_0^\infty t^{\nu-1}\exp(-t)dt \tag{A.19}
$$

and $\Gamma(\nu+1) = \nu\Gamma(\nu)$, $\Gamma(1) = 1$, $\Gamma(1/2) = \sqrt{\pi}$.

One can recognize in A.18 some of solutions to the heat equation that have been derived in Chapter 4.

Operational calculus

A linear ordinary differential equation of order n can be written as a polynomial of order n $P_n(\frac{d}{dt})$. In transform domain, it becomes a polynomial with s replacing the differential operator. The equation becomes $P_n(s)y(s) = Q_{n-1}(s)$, where $Q_{n-1}(s)$ contains the initial conditions. The Laplace transform of the solution is $Q_{n-1}(s)/P_n(s)$ and it can be inverted directly. This method to solve initial value problems for ODEs was introduced by Heaviside and is often referred to as "operational calculus".

Operational calculus can also be used to obtain the solution of a non-homogeneous ODE. It can be seen that the solution is obtained as a product of the transform of the input function with the impulse response. The Laplace transform of the convolution of two functions is equal to the product of the transforms of these functions:

$$F(s) = \mathcal{L}(f(t)) \quad G(s) = \mathcal{L}(g(t)) \tag{A.20}$$

$$F(s)G(s) = \mathcal{L} \int_0^t g(t')f(t-t')dt'. \tag{A.21}$$

This property is useful to calculate the response of a system to any input because it is sometimes easier to calculate an inverse Laplace transform than to integrate the convolution.

As stated before, the Laplace transform method can be used to solve initial value problems with PDEs, in particular for the heat equation (Carslaw and Jaeger, 1959).

A.1.5 Evaluating inverse Laplace transforms

Applications of Laplace transforms to the heat equation can be found in standard textbooks (e.g., Churchill, 1960; Carslaw and Jaeger, 1959). Many of the Laplace transforms of solutions of the heat equation contain terms in $\sqrt{s}$, i.e. the contour of integration in the complex plane must be closed along a branch cut. For example, in order to evaluate the inverse Laplace transform of $1/\sqrt{s}$,

$$\frac{1}{2\pi i} \int_{-i\infty}^{i\infty} \frac{\exp(st)ds}{\sqrt{s}}, \tag{A.22}$$

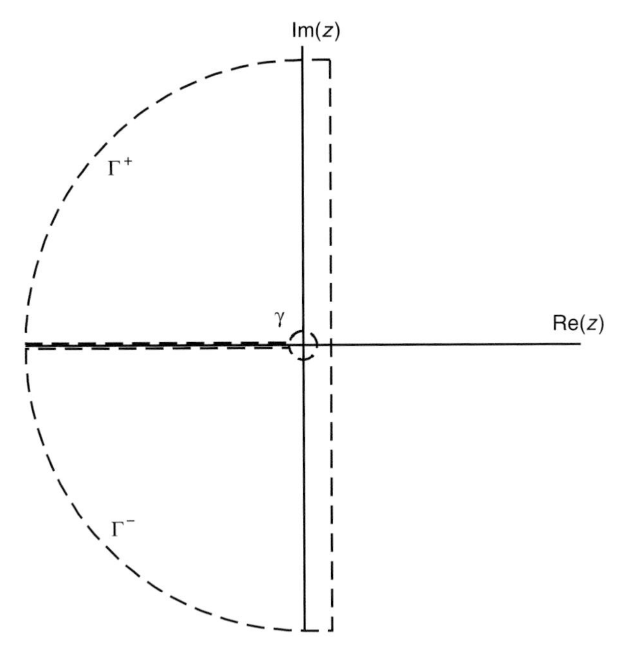

Figure A.2. Contour of integration in the complex plane for calculating an inverse Laplace transform when the branch cut follows the negative real axis.

the contour of integration must be closed along the branch cut following the negative real axis (Figure A.2),

$$\left\{\int_{-i\infty}^{i\infty}+\int_{\Gamma_+}+\int_{-\infty+i\epsilon}^{0}+\int_{\gamma}+\int_{0}^{-\infty-i\epsilon}+\int_{\Gamma_-}\right\}\frac{\exp(st)ds}{\sqrt{s}}=0. \tag{A.23}$$

As the integrals along Γ_+, Γ_-, and γ give 0, we have

$$\frac{1}{2\pi i}\int_{-i\infty}^{i\infty}\frac{\exp(st)ds}{\sqrt{s}}=\frac{1}{\pi}\int_{0}^{\infty}\frac{\exp(-ut)du}{\sqrt{u}}=\frac{\Gamma(1/2)}{\pi\sqrt{t}}=\frac{1}{\sqrt{\pi t}} \tag{A.24}$$

Example: The cooling half-space

For a temperature pulse at time $t=0$ on the surface of the half-space $z>0$ with initial temperature 0, the Laplace transforms of the equation and boundary conditions are:

$$sT(s,z)-T_0=\kappa\frac{d^2T}{dz^2}, \tag{A.25}$$

with the boundary conditions $T(s,z=0)=T_0$ and $T(s,z\to\infty)$ finite. The Laplace transform of the solution is:

$$\frac{T(s,z)}{T_0}=\exp\left(-\sqrt{\frac{s}{\kappa}}z\right). \tag{A.26}$$

For $t>0$, we can then evaluate the inverse Laplace transform:

$$\frac{1}{2\pi}\int_{c-i\infty}^{c+i\infty}\exp\left(-\sqrt{\frac{s}{\kappa}}z\right)\exp(st)ds \tag{A.27}$$

by closing the contour of integration along a branch cut following the negative real axis (Figure A.2) in such a way that:

$$\begin{aligned}\left(\int_{-i\infty}^{i\infty}+\int_{\Gamma_+}+\int_{-\infty+i\epsilon}^{0}+\int_{\gamma}+\int_{0}^{-\infty-i\epsilon}+\int_{\Gamma_-}\right)\cdots\\ \ldots\exp\left(-\sqrt{\frac{s}{\kappa}}z\right)\exp(st)ds=0.\end{aligned} \tag{A.28}$$

The integration along Γ_+ and Γ_- and γ gives 0, and the integration on the small circle gives 1. Therefore,

$$\frac{1}{2\pi i}\int_{-i\infty}^{i\infty}\exp\left(-\sqrt{\frac{s}{\kappa}}z\right)\exp(st)ds=\frac{1}{\pi}\int_{0}^{\infty}\sin\left(\sqrt{\frac{u}{\kappa}}z\right)\exp(-ut)du, \tag{A.29}$$

which has the form of a direct Laplace transform. It gives:

$$\frac{z}{2\sqrt{\pi\kappa t^3}}\exp\left(\frac{-z^2}{4\kappa t}\right), \tag{A.30}$$

which is the Green's function for a pulse at the surface of the half-space that we obtained in Chapter 4.

For a cooling half-space initially at temperature $T(z,t=0)=T_0$ and with surface temperature $T(z=0,t>0)=0$, the solution to the heat equation can easily be obtained by the Laplace transform. The transform of the heat equation with its initial condition is:

$$sT(s,z)-T_0=\kappa\frac{d^2T}{dz^2}, \tag{A.31}$$

with the boundary conditions $T(s,z=0)=0$ and $T(s,z\to\infty)$ finite. The solution in Laplace transform domain is:

$$\frac{T(s,z)}{T_0}=\frac{1}{s}\left(1-\exp\left(-\sqrt{\frac{s}{\kappa}}z\right)\right). \tag{A.32}$$

The first term is the Laplace transform of the Heaviside step function. To calculate:

$$\frac{1}{2\pi}\int_{c-i\infty}^{c+i\infty}\frac{1}{s}\exp\left(-\sqrt{\frac{s}{\kappa}}z\right)\exp(st)ds. \tag{A.33}$$

We could calculate the integral by closing the contour in the complex plane as before. It is simpler to observe that the Laplace transform is $1/s$ times the transform (A.27). As dividing the Laplace transform by s is equivalent to integrating in time, we obtain the solution by integrating the Green's function:

$$\int_0^t dt'\frac{z}{2\sqrt{\pi\kappa t'^3}}\exp\left(\frac{-z^2}{4\kappa t'}\right)=\frac{2}{\sqrt{\pi}}\int_{z/(2\sqrt{\kappa t})}^{\infty}\exp(-v^2)dv=\operatorname{erfc}\frac{z}{2\sqrt{\kappa t}}. \tag{A.34}$$

Removing this term from the Heaviside function gives the complete solution:

$$T(z,t)=T_0\operatorname{erf}\frac{z}{2\sqrt{\kappa t}}. \tag{A.35}$$

A.1.6 Asymptotic expansions

The behavior of a function for very small or very large t can often be determined directly from the series expansion of a Laplace transform from very large or very small values of s. The form of the direct Laplace transform suggests that the behavior of the function for $t\to\infty$ is related to the behavior of the transform for $s\to 0$. For instance, if a function tends to a constant value for $t\to\infty$, this value can be determined as:

$$\lim_{t\to\infty}f(t)=\lim_{s\to 0}sF(s). \tag{A.36}$$

A more general result is that if the transform can be expanded as:

$$F(s)=\frac{1}{s}\left(a_o+\sum_{n=0}^{\infty}b_n s^{n+1/2}\right)\quad |s|<1, \tag{A.37}$$

then the function is

$$f(t)=a_0+\sum_{n=0}^{\infty}b_n\frac{(-)^n\Gamma(n+1/2)t^{-n-1/2}}{\pi}\quad t\to\infty. \tag{A.38}$$

Also, the initial value of the function is related to the behavior of the transform for $s \to \infty$:

$$\lim_{t \to 0} f(t) = \lim_{s \to \infty} sF(s). \tag{A.39}$$

A.2 Fourier series and transforms

Fourier series and transforms are useful to solve linear partial differential equations, when any linear combination of solutions is also a solution. The Fourier series (and transforms) decompose any periodic (non-periodic) function into sines and cosines or complex exponentials. This is convenient because these functions form a complete orthogonal basis to the space, which implies that: (1) any function sufficiently regular can be decomposed in such functions; (2) the functions are orthogonal. Therefore, each coefficient can be determined independently of all the others.

A.2.1 Periodic function: Fourier series

A function $f(x)$ is periodic of period λ if

$$f(x+\lambda) = f(x) = f(x+n\lambda). \tag{A.40}$$

The wave-number $k = 2\pi/\lambda$.

Any periodic function that is sufficiently "regular" (in general this means piece-wise continuous) can be expressed as an infinite series of trigonometric functions of period λ/k:

$$f(x) = \frac{a_0}{2} + \sum_{n=1}^{\infty} \left(a_n \cos\left(\frac{2n\pi x}{\lambda}\right) + b_n \sin\left(\frac{2n\pi x}{\lambda}\right) \right). \tag{A.41}$$

The orthogonality of the trigonometric functions implies:

$$\int_{-\lambda/2}^{\lambda/2} \cos\left(\frac{2m\pi x}{\lambda}\right) \sin\left(\frac{2n\pi x}{\lambda}\right) dx = 0 \tag{A.42}$$

$$\int_{-\lambda/2}^{\lambda/2} \cos\left(\frac{2m\pi x}{\lambda}\right) \cos\left(\frac{2n\pi x}{\lambda}\right) dx = \delta_{mn} \frac{\lambda}{2} \tag{A.43}$$

$$\int_{-\lambda/2}^{\lambda/2} \sin\left(\frac{2m\pi x}{\lambda}\right) \sin\left(\frac{2n\pi x}{\lambda}\right) dx = \delta_{mn} \frac{\lambda}{2}, \tag{A.44}$$

with $\delta_{mn} = 1$ if $m = n$ and $\delta_{mn} = 0$ if $m \neq n$. The symbol δ_{mn} is known as the Kronecker delta. With the orthogonality conditions, the coefficients of the series are obtained as:

$$a_0 = \frac{2}{\lambda} \int_{-\lambda/2}^{\lambda/2} f(x) dx \tag{A.45}$$

$$a_n = \frac{2}{\lambda} \int_{-\lambda/2}^{\lambda/2} f(x) \cos\left(\frac{2n\pi x}{\lambda}\right) dx \tag{A.46}$$

$$b_n = \frac{2}{\lambda} \int_{-\lambda/2}^{\lambda/2} f(x) \sin\left(\frac{2n\pi x}{\lambda}\right) dx. \tag{A.47}$$

There are several standard tables of Fourier series (Gradshteyn and Ryzhik, 2000).

It is often more convenient to replace the trigonometric functions by complex exponentials. For a real function:

$$f(x) = \sum_{n=-\infty}^{\infty} c_n \exp\left(\frac{-i2n\pi x}{\lambda}\right) \tag{A.48}$$

with $i = \sqrt{-1}$. The orthogonality condition is:

$$\int_{-\lambda/2}^{\lambda/2} \exp\left(\frac{-i2m\pi x}{\lambda}\right) \exp\left(\frac{+i2n\pi x}{\lambda}\right) dx = \lambda \delta_{mn}. \tag{A.49}$$

Note the signs. The coefficients c_n are given by:

$$c_n = \frac{1}{\lambda} \int_{-\lambda/2}^{\lambda/2} f(x) \exp\left(\frac{+i2n\pi x}{\lambda}\right) dx \tag{A.50}$$

or

$$f(x) = \sum_{n=-\infty}^{\infty} c_n \exp(-ik_n x) \tag{A.51}$$

$$c_n = \frac{1}{\lambda} \int_{-\lambda/2}^{\lambda/2} f(x) \exp(+ik_n x) dx \tag{A.52}$$

with $k_n = 2\pi n/\lambda$ are the discrete wave-numbers. Note the sign convention for the complex exponentials.

In general, the numbers c_n are complex, i.e. $c_n = a'_n + ib'_n$. If the funtion $f(x)$ is real, then $f(x) = f^*(x)$ and:

$$c_{-n} = c_n^* \quad a'_{-n} = a'_n \quad b'_{-n} = -b'_n, \tag{A.53}$$

where * indicates the complex conjugation. Note that the a'_n and b'_n in the above equation differ from the coefficients of the trigonometric series expansion.

A.2.2 Useful Fourier series

We consider functions of period $2L$.
The square wave: $f(x) = -1$ for $-L < x < 0$, $f(x) = 1$ for $0 < x < L$

$$f(x) = \frac{4}{\pi} \sum_{n=0}^{\infty} \frac{\sin((2n+1)\pi x/L)}{(2n+1)}. \tag{A.54}$$

The triangular wave: $f(x) = x/L$ for $0 < x < L$, and $f(x) = 2 - x/L$ for $L < x < 2L$

$$f(x) = \frac{1}{2} - \frac{4}{\pi^2} \sum_{n=0}^{\infty} \frac{\cos((2n+1)\pi x/L)}{(2n+1)^2}, \tag{A.55}$$

or the function $f(x) = 2x/L$ for $0 < x < L/2$ and $f(x) = 2 - 2x/L$ for $L/2 < x < L$; $f(x+L) = -f(x)$,

$$f(x) = \frac{8}{\pi^2} \sum_{n=0}^{\infty} \frac{(-)^n}{(2n+1)^2} \sin\left(\frac{(2n+1)\pi x}{L}\right). \tag{A.56}$$

The "saw tooth" function, $f(x) = x/L$ for $-L < x < L$

$$f(x) = \frac{2}{\pi} \sum_{n=0}^{\infty} \frac{(-)^{n+1}}{n} \sin\left(\frac{n\pi x}{L}\right). \tag{A.57}$$

A.2.3 Non-periodic functions: Fourier transforms

Note that as the wavelength λ increases, the distance between the wave-numbers k_n in the discrete Fourier spectrum decreases as $\Delta k = 2\pi/\lambda$. One can see that as $\lambda \to \infty$, the distance between spectral lines, $\Delta k_n \to 0$, and the Fourier spectrum becomes continuous.

$$f(x) = \sum_{n=-\infty}^{\infty} \exp(-ik_n x) \Delta k \frac{1}{2\pi} \int_{-\lambda/2}^{\lambda/2} f(x) \exp(+ik_n x) dx \tag{A.58}$$

as $\Delta k \to 0$, the sum becomes an integral. We thus have:

$$F(k) = \int_{-\infty}^{\infty} f(x) \exp(ikx) dx \tag{A.59}$$

$$f(x) = \frac{1}{2\pi} \int_{-\infty}^{\infty} \exp(-ikx) F(k) dk. \tag{A.60}$$

The equations above define the Fourier transform pair $f(x)$, $F(k)$. The first equation defines the (direct) Fourier transform of $f(x)$. The second equation is the inverse Fourier transform. Note the perfect symmetry of the relationships above (the sign of the complex exponent changes, but the sign convention is arbitrary). The Fourier transform is often used to solve boundary value problems for partial differential equations (Sneddon, 1950). Standard tables are available to avoid calculating the integrals (Erdélyi *et al.*, 1954a,b).

The Fourier transform is often referred to as the spectrum of the function $f(x)$. If a function is real, then its spectrum must be symmetric to its complex conjugate $F(-k) = F^*(k)$. Reciprocally, if a function is symmetric, its spectrum must be real. If a function is antisymmetric $f(-k) = -f(k)$, its spectrum is imaginary.

Scaling in space and wave-number

If $f(x) \leftrightarrow F(k)$, one can calculate the Fourier transform of $f(\alpha x)$:

$$\int_{-\infty}^{\infty} f(\alpha x) \exp(ikx) dx = \frac{1}{\alpha} F(k/\alpha). \tag{A.61}$$

If $\alpha > 1$, the effect of the scaling $x \to \alpha x$ is to compress the function towards the origin. The transform will be wider and its amplitude will be smaller. As a general rule, the more concentrated the space function, the wider the transform (i.e. the transform must contain higher wave-numbers) and vice versa.

Translation

If $f(x) \leftrightarrow F(k)$, the Fourier transform of $f(x - b)$ is,

$$\int_{-\infty}^{\infty} f(x - b) \exp(ikx) dx = \exp(ikb) F(k). \tag{A.62}$$

Note that because $|\exp(i\phi)| = 1$, the amplitude of the transform is unchanged, only the phase is modified by the translation.

Fourier transform of the δ function

The Fourier transform of the δ function is,

$$\int_{-\infty}^{\infty} \exp(ikx)\delta(x)dx = 1. \tag{A.63}$$

The inverse Fourier transform must give the δ function, therefore,

$$\int_{-\infty}^{\infty} 1 \exp(-ikx)dk = 2\pi\,\delta(x). \tag{A.64}$$

The orthogonality conditions follow:

$$\int_{-\infty}^{\infty} \exp(ikx)\exp(-ik'x)dx = 2\pi\,\delta(k-k') \tag{A.65}$$

This equation is the equivalent to the orthogonality conditions (A.49) for the discrete set of functions for the Fourier series.

Fourier transform of the Heaviside function

The Heaviside function was obtained by integrating the δ function. We thus have the relationship,

$$\frac{d}{dx}(H(x)+C) = \delta(x), \tag{A.66}$$

where the arbitrary constant $C = -1/2$ so that

$$\int_{-A}^{A} (H(x) - 1/2)dx = 0. \tag{A.67}$$

Its Fourier transform is therefore

$$-ik(H(k) - \pi\,\delta(k)) = 1. \tag{A.68}$$

We therefore have that

$$H(x) \leftrightarrow \frac{-1}{ik} + \pi\,\delta(k). \tag{A.69}$$

Note that the inverse Fourier transform of $-1/ik$ does not converge in the classical sense but it has a Cauchy principal value:

$$\int_{-\infty}^{\infty} \frac{\exp(-ikx)}{-ik} dk = \pi\,\mathrm{sgn}(x). \tag{A.70}$$

where $\mathrm{sgn}(x) = x/|x|$.

Fourier transform of the boxcar (gate) function

Let the gate function be defined as:

$$\begin{cases} \Pi(x) = 0 & |x| > 1/2 \\ \Pi(x) = 1/2 & |x| = 1/2 \\ \Pi(x) = 1 & |x| < 1/2. \end{cases} \tag{A.71}$$

The Fourier transform is given by

$$\int_{-1/2}^{1/2} \exp(ikx)dx = 2\frac{\sin(k/2)}{k}. \tag{A.72}$$

Fourier transform of the Gaussian distribution function

It can be shown that

$$\exp(-x^2/a^2) \leftrightarrow a\sqrt{\pi}\exp(-k^2a^2/4). \tag{A.73}$$

This pair of transforms shows again that if $a \to \infty$, the time function is 1, and its transform $\to \delta(k)$. When the time function is concentrated about the origin ($a \to 0$), the spectrum becomes wider.

Fourier transform of the trigonometric functions

Expressing the sin and cos functions as complex exponentials, it is easy to see that:

$$\int_{-\infty}^{\infty} \exp(ikx)\cos(\alpha x)dx = \pi\,(\delta(k-\alpha)+\delta(k+\alpha)) \tag{A.74}$$

$$\int_{-\infty}^{\infty} \exp(ikx)\sin(\alpha x)dx = \frac{\pi}{i}\,(\delta(k+\alpha)-\delta(k-\alpha)) \tag{A.75}$$

i.e. their spectrum consists of the two discrete lines corresponding to $\pm$ the frequency of the function. More generally, the Fourier series can be seen as a special case of Fourier transform. The transform of the series:

$$\sum_{n=-\infty}^{\infty} c_n \exp(-ik_n x) \leftrightarrow 2\pi \sum_{n=-\infty}^{\infty} c_n \delta(k-k_n). \tag{A.76}$$

A.2.4 Space and wave-number domains

The function of position $f(x)$ or its transform $F(k)$ contain exactly the same information and one can be substituted for the other. The Fourier transform variable k is the wave-number. One can speak of space (time) or wave-number (frequency) domains. Using the wave-number domain may often prove more convenient or show information that is not apparent in the space domain. It may also result (although this is seldom used) in compressing the information. For instance, the function $f(x) = \cos(\alpha x) + \sin(\beta x)$ is more succinctly described in wave-number than in space domain.

A.2.5 Differential operators in frequency domain

Derivatives and integrals in Fourier domain

Taking the time derivative of the inverse Fourier transform yields:

$$\frac{df}{dx} = \frac{1}{2\pi} \int_{-\infty}^{\infty} F(k)(-ik)\exp(-ikx)dx. \tag{A.77}$$

In other words, the Fourier transform of the differential operator $\frac{d}{dx}$ is $-ik$. The nth derivative is obtained in a similar manner by multiplying the transform by $(-ik)^n$.

The inverse operator is the integral. It follows that the transform of the integral is obtained by dividing the transform of the function by $-ik$.

A.2.6 The impulse response in frequency domain. The convolution theorem

Consider first a periodic function with a discrete Fourier series. If the response of the system to each complex exponential $\exp(-ik_l x)$ is known and is $g_l \exp(-ik_l x)$, then, the response of the system to $f(x)$ is obtained by superposition as follows:

$$\sum_{l=-\infty}^{\infty} g_l c_l \exp(-ik_l x) \tag{A.78}$$

i.e. it is the product of each Fourier component of the function and the response of the system.

For non-periodic functions, we have a continuous spectrum; the response of the system for each wave-number is $G(k)$. The response of the system to the function $f(x)$ is obtained by superposition:

$$\frac{1}{2\pi} \int_{-\infty}^{\infty} \exp(-ikx)F(k)G(k)dk. \tag{A.79}$$

The function $G(k)$ is the Fourier transform of the Green's function. Indeed, it is the spectrum of the response when the input function is a delta function (whose spectrum is uniformly 1). In other words, the convolution in the space domain (equation A.9) has been replaced by a product in the wave-number domain.

Alternatively, one can calculate the Fourier transform of the convolution $f(x) * g(x)$:

$$\int_{-\infty}^{\infty} \exp(ikx)dx \int_{-\infty}^{\infty} f(x-x')g(x')dx' = G(k)F(k). \tag{A.80}$$

This correspondence between convolution in the space domain and product in the wave-number domain is important for practical reasons. Indeed it is in general easier to calculate a product than the convolution. To perform numerically a convolution it is convenient to calculate the Fourier transforms, perform their product in wave-number domain, and then back-transform to space domain.

A.2.7 Solution of boundary value problems. Steady-state solutions for the heat equation

The Fourier transformation is very useful to replace a partial differential equation by an ordinary differential equation and to determine the solution to simple boundary value problems. As an example, we shall show how the downward continuation of potential fields is performed in Fourier transform domain. The steady-state temperature field $T(x,z)$ satisfies the 2D Laplace equation $\nabla^2 T = 0$ for $z > 0$ and, on the surface $z = 0$, $T(x,0) = T_0(x)$. The Fourier transform of the Laplace equation is obtained as,

$$\frac{\partial^2 g}{\partial z^2} - k^2 g = 0. \tag{A.81}$$

Its general solution in the transform domain is,

$$T(k,z) = A(k)\exp(-kz) + B(k)\exp(kz), \tag{A.82}$$

where $A(k)$ and $B(k)$ are two constants to be determined by the boundary conditions. Because the solution must remain finite for $z \to \infty, A(k) = 0$ for $k < 0$ and $B(k) = 0$ for $k > 0$. We can thus write the solution as,

$$T(k,z) = C(k)\exp -|k|z. \tag{A.83}$$

Using the boundary condition at $z = 0$, we obtain,

$$T(k,z) = T_0(k)\exp(-|k|z). \tag{A.84}$$

This downward continuation operator can be generalized in 3D.

The inverse Fourier transform of the downward continuation operator gives in space domain,

$$\frac{1}{2\pi}\int_{-\infty}^{\infty} \exp(-|k|z)\exp(-ikx)dk$$
$$= \frac{1}{\pi}\int_{0}^{\infty} \cos(kx)\exp(-kz)dk = \frac{z}{\pi(x^2+z^2)}. \tag{A.85}$$

Note that the inversion integral has the form of a Laplace transform.

Note that we have assumed that x and k are real variables. However, it is common to use the theory of analytical functions to calculate Fourier transforms by using the analytical continuation of the functions and closing the contour of integration in the complex plane. For example, the (direct) Fourier transform of $1/(x^2+a^2)$ is,

$$\int_{-\infty}^{\infty} \frac{\exp(ikx)}{(x^2+a^2)}dx. \tag{A.86}$$

To evaluate this transform, we consider that x is the real part of a complex variable z. For $k > 0$ and $k < 0$ we can close the contour of integration in the upper and in the lower half of the complex z plane respectively. We obtain the integral as the residue at the pole. For $k > 0$,

$$\oint \frac{\exp(ikz)}{z^2+a^2}dz = 2\pi i(z-ia) \times \left.\frac{\exp(ikz)}{z^2+a^2}\right|_{z=ia} = \frac{\pi}{a}\exp(-ka), \tag{A.87}$$

and, for $k < 0$, we obtain

$$-\oint \frac{\exp(ikz)}{z^2+a^2}dz = \frac{\pi}{a}\exp(ka), \tag{A.88}$$

which is the transform of the downward continuation operator that we calculated before (see equation 4.85 in Chapter 4).

A.2.8 Applications to initial value problems for the heat equation

The infinite rod, periodic initial conditions

For an infinite rod initially at temperature

$$T(x,t=0) = \frac{1}{2\pi}\int_{-\infty}^{\infty} T_0(k)\exp(-ikx)dk. \tag{A.89}$$

The Fourier transform of the heat equation is,

$$\frac{d}{dt}T(k,t) = -\kappa k^2 T(k,t). \tag{A.90}$$

The solution to the heat equation will be written in the form:

$$T(x,t) = \frac{1}{2\pi}\int_{-\infty}^{\infty} T_0(k)\exp(-\kappa k^2 t)\exp(-ikx)dk. \tag{A.91}$$

Each wave-number relaxes with a time constant $\lambda^2/(4\pi^2\kappa)$.

Cooling of an infinite layer. Uniform temperature at both surfaces

For a layer of thickness L, initially at temperature T_0 and such that $T(z=0,t) = T(z=L,t) = 0$, the initial temperature can be written in Fourier series as,

$$T(z,t=0) = \frac{4T_0}{\pi}\sum_{n=1}^{\infty}\frac{1}{2n-1}\sin\left(\frac{(2n-1)\pi z}{L}\right). \tag{A.92}$$

Each Fourier component will be damped exponentially with time as

$$\exp\left(\frac{-\kappa(2n-1)^2\pi^2 t}{L^2}\right). \tag{A.93}$$

A.2.9 Cooling of an infinite layer. Fixed temperature on upper surface, fixed heat flux on lower boundary

For a layer initially at $T(z,t=0) = T_0$ and such that $T(z=0,t) = 0$ and $\frac{\partial T}{\partial z}(z = L,t) = 0$, it is convenient to write $T(z,t=0)$ as

$$\frac{2T_0}{\pi}\sum_{n=1}^{\infty}\frac{1}{(2n-1)}\left(1-\cos(\frac{(2n-1)\pi}{2}\right)\sin\left(\frac{(2n-1)\pi z}{2L}\right). \tag{A.94}$$

Each Fourier component will be damped exponentially with time as

$$\exp\left(\frac{-\kappa(2n-1)^2\pi^2 t}{4L^2}\right). \tag{A.95}$$

A.3 Cylindrical symmetry. Hankel transform

In a cylindrical coordinate system r,θ,z, Laplace's equation is written as

$$\frac{1}{r}\frac{\partial}{\partial r}r\frac{\partial T}{\partial r} + \frac{1}{r\sin\theta}\frac{\partial}{\partial\theta}\sin\theta\frac{\partial T}{\partial\theta} + \frac{\partial^2 T}{\partial z^2} = 0. \tag{A.96}$$

For axially symmetric problems:

$$\frac{\partial^2 T}{\partial r^2} + \frac{1}{r}\frac{\partial T}{\partial r} + \frac{\partial^2 T}{\partial z^2} = 0. \tag{A.97}$$

Separation of variables $T(r,z) = R(r)Z(z)$ yields

$$\begin{aligned} \frac{d^2 R}{dr^2} + \frac{1}{r}\frac{dR}{dr} &= \alpha^2 R \\ \frac{d^2 Z}{dz^2} &= -\alpha^2 Z. \end{aligned} \tag{A.98}$$

The solutions of the equation in z are the exponentials $\exp(\pm\alpha z)$ The solutions of the radial equation are obtained in terms of Bessel functions of the first and second kind $J_0(\alpha r)$ and $Y_0(\alpha r)$.

The Bessel functions J_ν and Y_ν are solutions of the equation

$$\frac{d^2 J_\nu}{dr^2} + \frac{1}{r}\frac{dJ_\nu}{dr} - \left(1 - \frac{\nu^2}{r^2}\right) J_\nu = 0. \tag{A.99}$$

If ν is not an integer, the equation has two independent solutions $J_{\pm\nu}$ which are finite for $r = 0$. If $\nu = n$ is an integer, $J_{-n} = (-)^n J_n$ is not independent, and the solution distinct of J_n is the Bessel function of the second kind Y_n which is singular at the origin.

The Bessel functions satisfy orthogonality conditions:

$$\int_0^\infty J_\nu(\alpha r) J_\nu(\beta r) r dr = \frac{1}{\sqrt{\alpha\beta}} \delta(\alpha - \beta). \tag{A.100}$$

The Hankel transform is defined by:

$$\begin{aligned} \bar{F}_\nu(v) &= \int_0^\infty f(r) J_\nu(vr) r dr \\ f(r) &= \int_0^\infty \bar{F}_\nu(v) J_\nu(vr) v dv. \end{aligned} \tag{A.101}$$

Some useful Hankel transforms:

$$\int_0^\infty \exp(-a^2 r^2) J_0(vr) r dr = \frac{v}{2a^2} \exp\left(\frac{-v^2}{4a^2}\right) \tag{A.102}$$

$$\int_0^\infty \frac{1}{\sqrt{(r^2 + a^2)}} J_0(vr) r dr = \frac{\exp(-av)}{v} \tag{A.103}$$

$$\int_0^\infty \exp(-vz) J_0(vr) v dv = \frac{z}{(z^2 + r^2)^{3/2}}. \tag{A.104}$$

The modified Bessel functions I_ν and K_ν are solutions of the equation

$$\frac{d^2 I_\nu}{dr^2} + \frac{1}{r}\frac{dI_\nu}{dr} - \left(1 + \frac{\nu^2}{r^2}\right) I_\nu = 0. \tag{A.105}$$

If ν is not an integer, the equation has two independent solutions $I_{\pm\nu}$ that are finite. If $\nu = n$ is an integer, $I_{-n} = (-)^n I_n$ is not independent of I_n for $r = 0$, and the solution distinct of I_n is the Bessel function of the second kind K_n which is singular at the origin. The two independent solutions $I_\nu(r)$ and $K_\nu(r)$ remain finite for $r \to 0$ and $r \to \infty$ respectively.

The spherical Bessel functions of the first kind which appear in the solution of Helmholtz equation in spherical geometry are defined as:

$$j_{n+1/2}(r) = \sqrt{\frac{\pi}{2r}} J_{m+1/2}(r). \tag{A.106}$$

Appendix B

Green's functions

This short introduction follows from the discussion on integral transforms and the convolution theorem.

B.1 Steady-state heat equation

B.1.1 Full-space 3D

For an isotropic and homogeneous body, the solution to Poisson's equation for a point heat source $H\delta(x)\delta(y)\delta(z)$ is obtained by integrating Poisson's equation over a sphere centered on the source:

$$\int_V -\nabla \cdot \lambda \nabla T dV = \nabla \cdot \vec{q} dV = \oint_S q_R dS = 4\pi R^2 q_R = H. \tag{B.1}$$

It follows that

$$q_R = \frac{H}{4\pi R^2} \tag{B.2}$$

and

$$T(R) = \int_\infty^R + \frac{\partial T}{\partial R'} dR' = -\int_\infty^R \frac{H}{4\pi \lambda R'^2} dR' = \frac{H}{4\lambda\pi R}. \tag{B.3}$$

B.1.2 Half-space 3D

The solution to Poisson's equation for a point source at depth $z = a$ below the surface $z = 0$ with boundary condition $T = 0$ is obtained by adding a sink -H at the image $x = 0$, $y = 0$, $z = -a$ of the source:

$$T(x,y,z) = \frac{H}{4\pi\lambda} \left(\frac{1}{(x^2+y^2+(z-a)^2)} - \frac{1}{(x^2+y^2+(z+a)^2)} \right). \tag{B.4}$$

B.1.3 2D Green's functions

For an infinitely long line source perpendicular to the plane (x,y), the Green's function is obtained as above by integrating the flux across a cylinder of radius r whose axis coincides with the line. It gives:

$$q_r = \frac{H}{2\pi r} \tag{B.5}$$

and

$$T(r) = \frac{H \log r - \log(\infty)}{2\pi \lambda}. \tag{B.6}$$

B.1.4 2D problems. Application of complex variables theory

For potential fields in 2D, there are many direct applications of the theory of functions of a complex variable. If $f(z)$ is an analytical function of the complex variable $z = x + iy, f(z) = u(x,y) + iv(x,y)$, the (real) functions $u(x,y)$ and $v(x,y)$ are harmonic (i.e. they satisfy Laplace's equation). It also follows from the Cauchy–Riemann relationships that $\nabla u \cdot \nabla v = 0$.

If a function is analytical within any domain in the complex plane, it can be continued outside that domain. Analytical continuation can be used to solve the boundary value problems for Laplace's equation. The solutions can be derived from Cauchy's integral formula:

$$f(z) = \frac{1}{2\pi i} \oint \frac{f(\zeta)}{\zeta - z} d\zeta. \tag{B.7}$$

If $\mathrm{Re}(f(z)) = u(x,0)$ or $\mathrm{Im}(f(z)) = v(x,0)$ is known on the line $y = 0$, it can be shown that (e.g. Morse and Feshbach, 1953)

$$u(x,y) = \frac{y}{\pi} \int_{-\infty}^{\infty} \frac{u(\xi,0)}{(x-\xi)^2 + y^2} d\xi = \frac{1}{\pi} \int_{-\infty}^{\infty} \frac{(x-\xi)v(\xi,0)}{(x-\xi)^2 + y^2} d\xi. \tag{B.8}$$

On the real line $x = 0$, the two conjugate functions $u(x,y)$ and $v(x,y)$ are Hilbert transforms of each other:

$$\begin{aligned} u(x,0) &= \frac{1}{\pi} \wp \int_{-\infty}^{\infty} \frac{v(\xi,0)}{\xi - x} d\xi \\ v(x,0) &= \frac{-1}{\pi} \wp \int_{-\infty}^{\infty} \frac{u(\xi,0)}{\xi - x} d\xi. \end{aligned} \tag{B.9}$$

The symbol $\wp$ indicates a Cauchy principal value.

If the functions $u(x,y)$ or $v(x,y)$ are known on a circle $x = a\cos\theta$, $y = a\sin\theta$, they can be calculated within the circle from their values on the circle with Poisson's formula:

$$u(r,\theta) = \frac{1}{2\pi}\int_0^{2\pi} \frac{(a^2 - r^2)u(a,\phi)}{a^2 - 2ar\cos(\phi - \theta) + r^2} d\phi, \tag{B.10}$$

or they be calculated from the value of the conjugate:

$$\begin{aligned} u(r,\theta) &= u(0) + \frac{ar}{\pi}\int_0^{2\pi} \frac{\sin(\phi - \theta)v(a,\phi)}{a^2 + r^2 - 2ar\cos(\phi - \theta)} d\phi \\ v(r,\theta) &= v(0) - \frac{ar}{\pi}\int_0^{2\pi} \frac{\sin(\phi - \theta)u(a,\phi)}{a^2 + r^2 - 2ar\cos(\phi - \theta)} d\phi. \end{aligned} \tag{B.11}$$

B.2 Transient heat equation

B.2.1 1D Green's function

The Green's function is derived using the combination of Fourier and Laplace transforms. The Fourier and Laplace transforms of the heat equation with a delta function source are

$$\begin{aligned} \frac{\partial T}{\partial t} &= \kappa \frac{\partial^2}{\partial z^2} + \delta(t)\delta(z) \\ sT(s,k) &= -k^2\kappa T(s,k) + 1. \end{aligned} \tag{B.12}$$

We obtain directly the double transform,

$$G(s,k) = \frac{1}{s - \kappa k^2}. \tag{B.13}$$

The Laplace transform has a simple pole at $s = -\kappa k^2$ and its inversion is straightforward:

$$G(t,k) = \exp(-\kappa k^2 t). \tag{B.14}$$

This is also a standard Fourier transform which is directly inverted:

$$G(z,t) = \frac{1}{2\sqrt{\pi\kappa t}} \exp(-z^2/(4\kappa t)). \tag{B.15}$$

One can check that

$$\int_{-\infty}^{\infty} G(z,t)dx = 1. \tag{B.16}$$

In n dimensions, the solution is the product of the solutions for each direction:

$$G^n(\vec{r},t) = \frac{1}{(4\pi\kappa t)^{n/2}} \exp\left(\frac{-r^2}{4\kappa t}\right) \tag{B.17}$$

$$T(r,t) = \frac{H}{8(\pi\kappa t)^{3/2}} \exp\left(\frac{-r^2}{4\kappa t}\right). \tag{B.18}$$

For a temperature pulse at the surface of a half-space, the initial temperature is $T(z > 0, t = 0) = 0$, and the boundary condition at the surface is $T(z = 0, t) = \delta(t)$. This gives,

$$T(z,t) = \frac{z}{2\sqrt{\pi\kappa t^3}} \exp\left(\frac{-z^2}{4\kappa t}\right). \tag{B.19}$$

Appendix C

About measurements

Measuring heat flux

Heat flux is never measured directly but its vertical component is obtained by Fourier's law:

$$q = -\lambda \frac{\partial T}{\partial z}. \tag{C.1}$$

Measurements thus require determination of the vertical temperature gradient and of the thermal conductivity. Heat flux is a vector quantity and, in principle, measurement requires determination of three components. This vector is normal to an isothermal surface and, if the Earth's surface is truly an isotherm, the heat flux is vertical. We shall see that, in some cases, this assumption is violated and thus requires a correction.

C.1 Land heat flow measurements

C.1.1 Conventional land heat flow measurements

On land, measurements are obtained in drill-holes (usually holes of opportunity, mostly from mining exploration). During drilling, the injection and circulation of mud and fluids induces a temperature perturbation that can be approximated as that due to an infinite line source of heat activated between the start and the end of drilling (Bullard, 1947). In practice, the temperature perturbation induced by drilling becomes sufficiently small after a time at least six times the duration of the drilling (see Exercise 4.15). In permafrost areas, the perturbation persists for a much longer time because freezing liberates latent heat. Monitoring of temperature in drill-holes in permafrost shows that return to equilibrium usually requires several years.

Temperature is measured with a calibrated thermistor at the end of an electrical cable. Temperature can be measured with an accuracy better than 0.005 K. The probe is lowered into the hole and measurements are made at regular intervals, typically 10 m. Thermal equilibration of the probe in the water-filled hole requires a few minutes, and logging a 1000 m deep hole requires about 8 hours. Continuous measurements could be done provided that the probe is lowered at very slow speed and the movement is mechanically controlled to insure precise recording of the depth. With present technology, self-supporting electrical cables are sufficiently light (20 kg for 1000 m) and the equipment is portable. The thermistors are designed to insure maximum sensitivity in the relevant temperature range.

Continuous core samples are routinely recovered in mining exploration allowing measurement of thermal conductivity. Such natural rock samples contain micro-cracks and pores which get closed by the confining pressure at depth. Thus, conductivity measurements at room pressure may be artificially biased because they correspond to larger porosity values than those of *in situ* conditions (Simmons and Nur, 1968). For this reason, conductivity must be measured on water-saturated samples under an applied pressure. The divided bar method (Misener and Beck, 1960) provides the most accurate measurements because it involves relatively large samples and is insensitive to small-scale variations in lithology, but it is time consuming and only a limited number of samples can be processed. Continuous measurement of conductivity on an entire core can be made with an optical scanning device (Popov *et al.*, 1999) but its accuracy is less than for the divided bar apparatus.

The heat flux may be obtained as the slope of the best fitting line to the "Bullard plot" of temperature versus thermal resistivity $R(z)$:

$$R(z) = \int_0^z \frac{dz'}{\lambda(z')}. \tag{C.2}$$

Alternatively heat flux can be obtained from Fourier's law over depth intervals where the conductivity is constant. Both methods give comparable results when the conductivity sampling is adequate. Because the temperature field in the upper 200 m is often perturbed by surface effects that we shall discuss later (see Figure C.1), reliable measurements require deep boreholes (at least 200 m).

Temperature in a borehole can be continuously monitored by sending electromagenetic signals along a fiber optic cable, but the accuracy is typically ≈ 1 K over a distance interval of ≈ 1 m, which is not sufficient for the purpose of heat flow measurements.

C.1.2 Bottom hole temperature (BHT) data

Temperature measurements are also routinely available from oil exploration wells, either as bottom hole temperature or drill-stem tests, with a precision never better

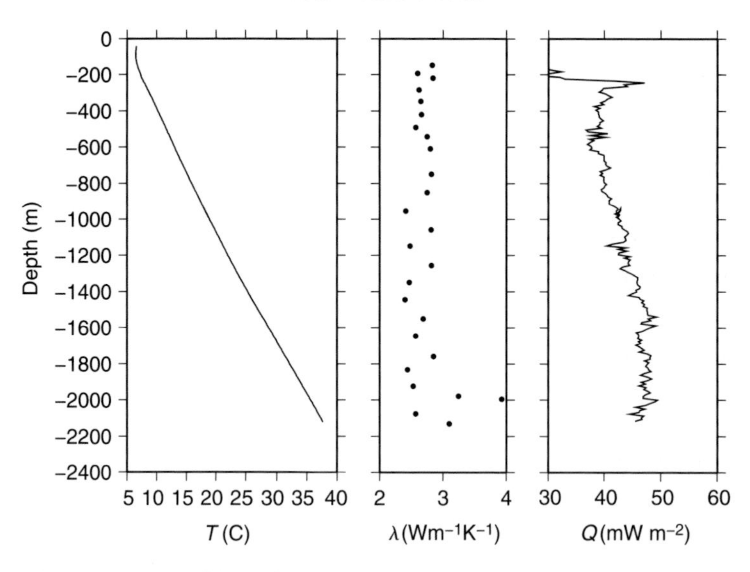

Figure C.1. Example of heat flux measurements in a deep borehole in Sudbury (Ontario). Temperature profile, thermal conductivity and heat flux as a function of depth. Note the marked perturbations of the gradient in the topmost 300 m that are due to a combination of topographic and climatic effects. Also note the steady increase in gradient to $\approx$ 1800 m. This increase is probably due to post-glacial warming. Calculations of the perturbation due to post-glacial warming near Sudbury predict almost exactly the same heat flux increase with depth.

than 5–10 K after corrections for thermal transients. In these deep wells, the gradient can thus be estimated with a precision of 10–15%. The other difficulty of these measurements is the lack of core samples for thermal conductivity, which has to be estimated from the lithology or from other physical properties (density, porosity, etc.) that are routinely logged. Although less precise than conventional methods, these data have provided most of the estimates of heat flux in sedimentary basins and continental margins.

C.1.3 Corrections

Heat flux determinations rely on the assumption that heat is transported vertically in steady state, and thus require no lateral variations in surface boundary conditions or physical properties. Changes in vegetation, the proximity to a lake, topography can distort the temperature field and affect the heat flux estimate (Jeffreys, 1938). Rapid erosion (or sedimentation) also affect the temperature field (Benfield, 1949). These effects are largest near the surface and the error on the heat flux is usually small in sufficiently deep boreholes (more than 200 m). The effect of a lake or change in vegetation cannot be estimated without extensive data coverage in the horizontal direction.

Effect of topography

Topography distorts the variation of temperature with depth and the pattern of heat flow near the Earth's surface. In a first approximation, the isotherms are stretched to follow the topographic relief, resulting in steeper gradient and higher heat flux in valleys than below ridges. Corrections can be made for such effects. One method consists of projecting the temperature perturbation on a plane horizontal surface and downward continuing the horizontal temperature distribution below that surface (Jeffreys, 1938; Bullard, 1938; Birch, 1950). Alternatively, it is possible to approximate the topography using a surface with simple geometry and to calculate its effect on the temperature gradient (Jaeger and Sass, 1963; Lachenbruch, 1968).

In first order, the temperature perturbation at any point on a horizontal surface will be proportional to the height of the topography times the regional vertical gradient, $\Delta T(x,y) = \Gamma h(x,y)$. In 2D, the perturbation of the temperature field in the region below is

$$\Delta T(x,z) = \frac{1}{\pi}\Gamma \int_{-\infty}^{\infty} \frac{zh(x')}{(x-x')^2+z^2} dx'. \tag{C.3}$$

The perturbation of the gradient on the plane $z = 0$ is thus obtained as,

$$\frac{\Delta\Gamma}{\Gamma} = \frac{1}{\pi} \int_{-\infty}^{\infty} \frac{h(x')}{(x-x')^2} dx'. \tag{C.4}$$

For example, a vertical cliff of height δh will induce a perturbation $\Delta\Gamma/\Gamma = \delta h/x$ where x is the distance to the edge of the cliff.

One can use the 2D Fourier transform to estimate these perturbations. If $\Delta h(k_x, k_y)$ is the transform of the topography, the perturbation to the gradient at depth z can be written as,

$$\frac{\Delta\Gamma(\vec{k},z)}{\Gamma} = -|\vec{k}|\Delta h(\vec{k})\exp(-|\vec{k}|z) = \frac{2\pi}{\lambda}\Delta h(\vec{k})\exp(-2\pi z/\lambda), \tag{C.5}$$

where $\lambda = 2\pi/\sqrt{k_x^2+k_y^2}$ here denotes the wavelength. It shows that the correction becomes small if the amplitude of the topography is much less than the wavelength or if the depth is in the order of the wavelength. Except in mountainous regions, one of these two conditions will be fulfilled for continental heat flux measurements. Topography corrections are often necessary for oceanic heat flux measurements because they are made at very small depths (typically less than 10 m).

Lakes. Horizontal changes in surface boundary conditions

Lakes or change in vegetation cover induce lateral variations in surface temperature than can reach a few degrees. Such effects can be accounted for using the same methods as for topography.

Convective heat transport

Perturbation of the temperature profile by a vertical flow of water has been calculated (see equation 5.151). In this case, the temperature profile depends on two parameters, the surface temperature gradient Γ_0 and Pe, the Péclet number. If the temperature profile is deep enough, both parameters can be determined from the measurements and the conductive heat flow component can in principle be calculated.

Erosion sedimentation

If the erosion rate is known, a correction can be calculated (from equation 4.122). The effect of erosion on the shallow temperature gradient increases with time and is negligible for $t \ll \kappa v^{-2}$, where v is the erosion rate.

Climatic effects

Temperatures near the Earth's surface keep a memory of past surface boundary conditions (Chapter 12). For periodic variations of the surface temperature, the temperature wave is attenuated exponentially as it propagates downward with a skin depth $\delta = \sqrt{\kappa \tau / \pi}$, where τ is the period and κ the thermal diffusivity. The daily and annual temperature cycles are damped over less than 0.5 or 10 m. They do not affect temperature at the depth of land heat flux measurements. Long-term variations in surface temperature could potentially significantly affect the temperature gradient. Following the last glacial episode that ended *c.* 10,000 years BP, surface temperature warming could affect the temperature gradient down to 2,000 m and, if not accounted for, lead to underestimating the heat flow (Birch, 1948). If the time-varying surface boundary condition was known, it would be easy to make a correction. For the part of Canada covered by the Laurentide ice sheet, Jessop (1971) proposed a correction for a detailed climate history of the past 400,000 years with temperature equal to present during the interglacials and to $-1\,°\mathrm{C}$ during the glacial episodes. Because the present temperature of the ground surface in Canada is quite low, the correction is usually small ($< 10\%$ the heat flux). Measurements in very deep boreholes in Canada have indeed shown that heat flow does not increase much at depth and that the effect of the last glaciation is small (Sass *et al.*, 1971; Rolandone *et al.*, 2003). These measurements also show that the thermal boundary condition at the base of the glacier might have been quite variable (possibly because it depends on the heat flow). Similar corrections have been applied to data from Finland and Siberia.

During the past 200 years, there has been a general warming trend following the "Little Ice Age" with an acceleration since 1960. This recent warming affects temperature profiles down to 200 m. In regions where heat flux is low and the

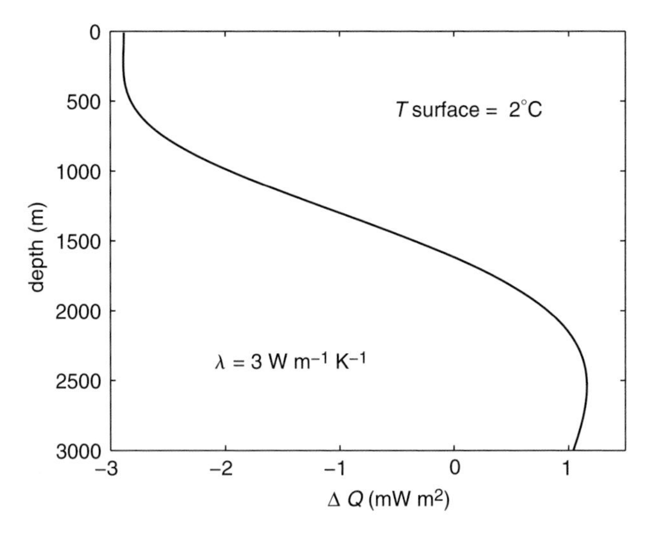

Figure C.2. Perturbation of the heat flux profile due to post-glacial warming 10,000 years ago. The present surface temperature (1 °C) is 2 K higher than at the base of the glacier and thermal conductivity 3 W m^{-1} K^{-1}. Note that the correction varies with depth.

warming has been particularly strong, the temperature gradients are inverted down to 50–80 m. Borehole temperature profiles have been used to reconstitute the surface temperature evolution of the past centuries (see Chapter 12). Accurate heat flux measurements require relatively deep boreholes (>200 m) to filter out the effect of recent warming and measure a stable temperature gradient.

C.2 Oceanic heat flux measurements

The difficulty of making reliable oceanic heat flux measurements was summarized by this statement known as Bullard's law: *Never duplicate an oceanic heat flow measurement for fear that it differs from the first by two orders of magnitude*. Heat flow measurements on the sea floor are usually done by dropping a probe that penetrates the soft sediment of sea floor and measuring the temperature (Bullard, 1954; Revelle and Maxwell, 1952). The Bullard probe consists of a 15 m long shaft fitted with several thermistors that measure the temperature in the sediment. The Ewing probe is similar to the the Bullard probe, but the sensors are mounted on fins attached to the side of the shaft. One very significant improvement to these probes was the "violin bow" thermal probe of C. R. B. Lister. In this probe the sensors are mounted on a thin tube 5 cm away from the main shaft. The introduction of digital recording and computer chips has permitted the design of probes that do not need to be brought back to the surface after each measurement. With such POGO-type probes, the measurement rate is about one per hour, i.e. many times better than with the older probes.

Probe penetration causes friction and a thermal perturbation that requires about half an hour to decay. Penetration depths are in the order of 10 m and great precision is required to reduce the error on measurements. Thermal conductivity can be determined by continuously recording the temperature and determining the decay of the perturbation due to probe insertion (Lee and von Herzen, 1994). It can also be determined *in situ* by a controlled thermal perturbation or on sediment cores brought back to the ship. In some cases, it is simply estimated from the density and porosity of the sediment, using a relationship of the type:

$$\lambda_e = \lambda_s^{1-\phi} \lambda_w^{\phi}, \tag{C.6}$$

where ϕ is the porosity and λ_s and λ_w are the conductivities of the sediment and water respectively (Von Herzen and Maxwell, 1959). For more than 40% of reported marine heat flux measurements, the conductivity was estimated in this way.

Measurements in lakes

Oceanic heat flux probes have also been used in lakes in the continents. Because the probe penetrates only a few meters, it is important to insure that the temperature of the water at the lake bottom is stable. Measurements in deep lakes (Lake Malawi, Lake Baikal, Lake Superior) have indeed provided reliable heat flux values (Von Herzen and Vacquier, 1967; Poort and Klerkx, 2004). On the other hand, some heat flux measurements in shallow lakes were found to be extremely noisy and should not be considered reliable.

Appendix D

Physical properties

D.1 Thermal conductivity

The thermal conductivities of the main silicate minerals are well known and vary within a rather large range of about 1.6–7.7 W m^{-1} K^{-1} (Horai and Simmons, 1969; Diment and Pratt, 1988). The lowest and largest conductivity values are those of plagioclase and quartz, respectively. In principle, it is possible to calculate the thermal conductivity of a rock from knowledge of its mineral phases and their proportions. The procedure is discussed in a separate section below but is rarely implemented. Silicate minerals are anisotropic and belong to solid solutions with end-members that may have very different conductivities. An accurate prediction would thus require determination of the composition and orientation of each mineral phase. For heat flux measurements, furthermore, such determinations would need to be done over a representative rock volume and not at the scale of a petrological thin section. For this reason, thermal conductivity must be measured on each and every rock type encountered in a borehole. For regional thermal models, one may choose representative values for the dominant rock types, but one must pay attention to the level of approximation that is entailed.

Table D.1 lists values for thermal properties of the main rock types and emphasizes the large ranges that exist for some of them. An excellent compilation of thermal conductivity measurements was made by Robertson (1988). Most of the available measurements have been made on upper crustal rocks and ignore deeper crustal lithologies. Thermal properties of rocks from the middle and lower crust are listed in Table D.2.

Variation with temperature

Thermal conductivity can be broken into lattice and radiative components, which have different behaviors as temperature and pressure change.

Table D.1. *Thermal properties of some rocks and water at room temperature*

Material	λ $\mathrm{W\,m^{-1}\,K^{-1}}$	κ $\mathrm{m^2\,s^{-1}}(\times 10^{-6})$	C_p $\mathrm{J\,kg^{-1}\,K^{-1}}(\times 10^3)$
Water	0.6	0.14	4.18
Soil	0.17–1.1	0.2–0.6	0.8
Snow	0.11	0.5	
Granite	1.7–3.9	1	0.8
Rocksalt	5.5–6.5		
Basalt glass	1.3	0.9	
Gabbro	1.9–2.3	1.2	
Gneiss $\perp$ †	1.5–3.2		
Gneiss $\parallel$ †	2.5–4.8		
Ice	1.6–2.2	1.2	2.14

† Conductivity values measured in directions perpendicular and parallel to the foliation.

Table D.2. *Thermal properties of deep crustal rocks at room temperature*

Material	λ $\mathrm{W\,m^{-1}\,K^{-1}}$	κ $\mathrm{m^2\,s^{-1}}(\times 10^{-6})$	Reference
Amphibolite facies gneisses			
Orthogneiss (India)	2.20	1.57	Ray *et al.*, 2006
Paragneiss (India)	2.43–2.94	2.10–1.96 †	Ray *et al.*, 2006
Granulite-facies rocks			
Charnockite (India)	2.33–2.81	1.31–1.81	Ray *et al.*, 2006
Enderbite (India)	2.53–3.50	1.53–2.22	Ray *et al.*, 2006
Mafic granulites (India)	2.27–2.60	2.24–1.36	Ray *et al.*, 2006
Mafic granulites (Finland)	2.34–2.41	0.74–0.79	Kukkonen *et al.*, 1999
Pyroxene norite (Finland)	2.22	0.73	Kukkonen *et al.*, 1999

† Values of conductivity and diffusivity on two different rocks, listed in the same order.

The lattice thermal conductivity decreases with temperature. For individual crystals, theoretical arguments suggest that, below the Debye temperature, the temperature dependence of conductivity takes the following form:

$$\lambda(T) = \lambda_{298}\sqrt{\frac{298}{T}}, \qquad \text{(D.1)}$$

where T is absolute temperature. For mantle olivine, Xu *et al.* (2004) have shown that this equation is consistent with their experimental measurements with $\lambda_{298} = 4.49\ \mathrm{W\,m^{-1}\,K^{-1}}$. For polycrystalline assemblages, this equation is no longer valid and one relies on empirical fits to the data. Reasonable agreement can be obtained

Table D.3. *Constants for calculating the lattice thermal conductivity of different rock types, from Clauser and Huenges (1995)*

Rock type	T range (°C)	A	B
Rock salt	−20, 40	−2.11	2960
Limestones	0, 500	0.13	1073
Metamorphic rocks	0, 1200	0.75	705
Felsic rocks	0, 1400	0.64	807
Mafic rocks	50, 1100	1.18	474
Ultra-mafic rocks	20-1400	0.73	1293

with a relationship of the type:

$$\lambda(T) = A + \frac{B}{350 + T}, \tag{D.2}$$

where T is in °C and where constants A and B depend on the rock type (Table D.3).

It is sufficient to use the lattice conductivity component over the temperature range of the crust. In the mantle, one must also account for radiative heat transport. For a pure substance, the radiative component of conductivity is given by the grey body law (see Chapter 3):

$$\lambda_r = \frac{16}{3}\frac{n^2}{\epsilon}\sigma T^3, \tag{D.3}$$

where n is the refractive index, ϵ is the opacity, σ is the Stefan constant of black-body radiation ($\sigma = 5.67 \times 10^{-8}$ W m^{-2} K^{-4}). For olivine, several independent laboratory measurements suggest that (Schatz and Simmons, 1972; Beck *et al.*, 1978; Schärmeli, 1979)

$$\lambda_r = 0.368 \times 10^{-9}\, T^3. \tag{D.4}$$

This equation leads to $0.13 < \lambda_r < 1.8$ W m^{-1} K^{-1} for $700 < T < 1700$ K. It is only valid in a single crystal if the mean free path of photons is independent of temperature. For mantle rocks, one must account for scattering and for the effect of interfaces in a mineral assemblage. Such complications led Marton *et al.* (2005) to use a constant radiative conductivity component $\lambda_r = 1$ W m^{-1} K^{-1} for temperatures higher than 700 K. According to Gibert *et al.* (2005), however, the conductivity of poly-crystalline dunite samples conforms to relation (D.4) and is close to that of single olivine crystals. McKenzie *et al.* (2005) have proposed a more complicated equation:

$$\lambda_r(T) = \sum_{m=0}^{3} d_m (T + 273)^m, \tag{D.5}$$

with $d_0 = 1.753 \times 10^{-2}$, $d_1 = -1.0365 \times 10^{-4}$, $d_2 = 2.2451 \times 10^{-7}$ and $d_3 = -3.4071 \times 10^{-11}$.

The upper mantle of the Earth and the lithospheric roots of continents are made of slightly different rocks. Differences are essentially in the clinopyroxene, orthopyroxene and olivine contents. The thermal conductivities of olivine and orthopyroxene differ by about 30% (Schatz and Simmons, 1972). Changes in the amounts of these two minerals are complementary to one another, so that the net effect on the bulk rock conductivity is small (≈ 0.1 W m^{-1}K^{-1}). In fact, the slight differences of olivine composition that exist between the different types of peridotite (between about Fo_{92} and Fo_{90}) have an opposite effect on the bulk conductivity. Lack of data on clinopyroxene prevents comprehensive calculation, but the resulting uncertainty is negligible for such a minor constituent.

Variation with pressure

Lattice thermal conductivity strongly depends on density:

$$\frac{\delta\lambda}{\lambda} \approx 7\frac{\delta\rho}{\rho}, \tag{D.6}$$

and hence depends on pressure. The effect of pressure on conductivity can be treated independently of temperature, so that one can write

$$\lambda(T,P) = \lambda_o(T)(1 + \beta P), \tag{D.7}$$

where β is a constant coefficient (Table D.4). The effect of pressure increases conductivity by about 20% between the surface and 200 km. In contrast, the effect of pressure on the radiative conductivity component is negligible.

Thermal properties of magmas

There are very few measurements of the physical properties of magma. At the high temperatures that are required, large amounts of heat are transported by radiation through any apparatus, and it is difficult to properly account for the energy that is required to maintain a temperature difference across a sample. One may use transient techniques and monitor the propagation of a thermal front due to a sudden change of conditions at a boundary, but the procedure is error prone. An alternative method is to use glass, because it is solid and has the structure of a liquid, and to

Table D.4. *Pressure dependence of lattice conductivity*

Material	β † $\times 10^{-2}$ GPa^{-1}	Reference
Mafic granulites (Finland)	2.4 ± 0.8	Kukkonen *et al.*, 1999
Olivine	3.2	Xu *et al.*, 2004

† Coefficient in equation D.7 for thermal conductivity.

Table D.5. *Thermal conductivity of silicate melts and glasses, from Spera (2000)*

Material	Temperature (K)	λ (W m^{-1} K^{-1})
SiO_2 (glass)	1170	2.57
Obsidian (glass)	770	1.89
$CaNa_4Si_3O_9$ (melt)	1520	0.22
Na_2SiO_3 (melt)	1460	0.22
$CaMgSi_2O_6$ (melt)	1670	0.31

extrapolate the results to high temperatures. In spite of the small data set and the lack of inter-laboratory comparisons and calibrations, it is clear that the thermal conductivity of a silicate melt is much smaller than that of the fully crystallized solid. In experiments on olivine-melilitite over a large temperature range ($\approx 300-1500$ K), Buttner *et al.* (1998) found that conductivity drops significantly above the solidus temperature. They were unfortunately not able to work on a pure melt phase at high temperature but their measurements on partially molten samples provide an upper bound of 1 W m^{-1} K^{-1} for the thermal conductivity of the silicate liquid. Table D.5 lists the very few measurements available for silicate melts and glasses.

Mixing laws for thermal conductivity

The thermal conductivity of a mixture depends on the volume fractions, the shapes and the geometrical arrangement of the different constituents, and cannot be calculated with a single equation valid for all rocks. Bounds on the conductivity can be obtained with the Reuss and Voigt averaging methods. For N components each occupying a fraction ϕ_i of the total volume, the *Reuss* average is:

$$\lambda_{\text{Reuss}} = \left(\sum_{i=1}^{N} \frac{\phi_i}{\lambda_i} \right)^{-1} \tag{D.8}$$

and the *Voigt* average:

$$\lambda_{\text{Voigt}} = \sum_{i=1}^{N} \phi_i \lambda_i. \tag{D.9}$$

These two values correspond to the parallel and series arrangements discussed in Chapter 4 and provide absolute bounds for the conductivity of the mixture, λ_e:

$$\lambda_{\text{Reuss}} < \lambda_e < \lambda_{\text{Voigt}}. \tag{D.10}$$

The average of these two bounds, called the *Voigt–Reuss–Hill* average, is often used to estimate conductivity. Another frequent model is the geometric mean:

$$\lambda_e = \sum_{i=1}^{N} \lambda_i^{\phi_i}, \tag{D.11}$$

which allows accurate predictions for marine sediments (equation C.6).

More restrictive bounds on the average conductivity can be obtained by the variational method of Hashin and Sthrikhman (1962). For a mixture of N components with concentration x_i and thermal conductivity λ_i, the upper and lower bounds for thermal conductivity are given by (Berryman, 1995),

$$\lambda^{u/l} = \left(\sum_{i=1}^{N} \frac{x_i}{\lambda_i + 2\lambda_{\max/\min}} \right)^{-1} - 2\lambda_{\max/\min}. \tag{D.12}$$

For two phases occupying volume fraction ϕ and $1-\phi$ with $\lambda_1 < \lambda_2$, we have,

$$\lambda_1 + \frac{\phi\lambda_1}{\lambda_1/(\lambda_2-\lambda_1)+(1-\phi)/3} < \lambda_e < \lambda_2 + \frac{(1-\phi)\lambda_2}{\lambda_2/(\lambda_1-\lambda_2)+\phi/3} \tag{D.13}$$

In practice, it is found that the geometric mean conductivity is very close to the measured value in crystalline rocks (Troschke and Burkhardt, 1998). A review of mixing laws and their validity limits can be found in Berryman (1995).

Tensor of thermal conductivity

Most silicate minerals are anisotropic and their thermal conductivity depends on direction (Table D.6). For anisotropic materials, the linear relationship between the heat flux and the temperature gradient involves a second-order thermal conductivity tensor:

$$q_i = -\lambda_{ij} \frac{\partial T}{\partial x_j}, \tag{D.14}$$

Table D.6. *Components of the thermal diffusivity tensor in San Carlos Olivine (Fo91) at $T = 300$ K, from Gibert et al. (2003)*

Crystallographic orientation	κ ($\times 10^{-6}$ m^2 s^{-1})
[100]	2.73 ± 0.04
[010]	1.70 ± 0.12
[010]	2.49 ± 0.14

with the standard convention that summation over repeated indices is implied. In many rocks, however, anisotropy is subdued because of the extremely variable orientations of their mineral constituents. For heat flow measurements, the shallow temperature gradient is vertical so that one need only measure conductivity in the vertical direction.

D.1.1 Heat capacity

The heat capacity of rocks and minerals depends on temperature. The average heat capacity varies from $\approx$700 J kg^{-1} K^{-1} at 300 K to 1000 J kg^{-1} K^{-1} at 600 K. At higher temperatures the variations of C_P level off.

For olivine, an equation of the following form fits experimental data (Berman and Aranovich, 1996):

$$C_p = c_0 + c_1 T^{-1/2} + c_3 T^{-3}. \tag{D.15}$$

For forsterite, $c_0 = 233.18$, $c_1 = -1801.6$ and $c_3 = -26.794 \times 10^7$; for fayalite, $c_0 = 252$, $c_1 = -2013.7$ and $c_3 = -6.219 \times 10^7$ when C_p is in kJ mol^{-1}, and T is in Kelvin.

D.1.2 Thermal diffusivity

The thermal diffusivity of most rocks ranges between 0.8 and 2.5×10^{-6} m^2 s^{-1}. Average values for some rocks are given in Table D.7.

D.1.3 Latent heat

The latent heat of fusion is the amount of energy, and more specifically the change of enthalpy, that must be supplied to melt a substance. For multicomponent materials such as rocks, heat must be supplied over a finite temperature range and the amount may not depend linearly on either temperature or crystal fraction. Table D.8 lists the latent heat of fusion of water and some pure silicate end-members. For order-of-magnitude arguments, a reasonable value is $L \approx 3 \times 10^5$ J kg^{-1}.

Table D.7. *Thermal diffusivity of water, ice and some rock types*

Rock	κ ($\times 10^{-6}$ m^2 s^{-1})
Basalt	0.9
Gabbro	1.2
Granite	1.6
Peridotite	1.7
Sandstone	1.3
Quartzite	2.6
Limestone	1.2
Dolomite	2.6
Marble	1.0
Shale	0.8
Water	0.15
Ice	1.2

Table D.8. *Latent heat of fusion for some pure silicate phases, from Spera (2000)*

Formula	T_L (K)†	L (kJ kg^{-1})
Water	273	333
SiO_2 (quartz)	1700	157
$CaMgSi_2O_6$	1665	636
$Ca_2MgSi_2O_7$	1727	454
Fe_2SiO_4	1490	438
Mg_2SiO_4	2174	1010
$NaAlSi_3O_8$	1393	246
$Na_2Si_2O_5$	1147	196
$CaAl_2SiO_8$	1830	478

† Melting temperature.

D.1.4 Porosity–Permeability

In shallow crustal environments, rocks are permeable when they are fractured and porous. Fractures get closed as the confining pressure increases, but pores can remain open at large depths. Porosity, measured in %, refers to the volume of open space to the total rock volume and can be deduced from density measurements. Some of the pore space, however, may not be connected and one should determine the effective porosity that is available to fluid flow. This may be measured directly or through its relationship to electrical conductivity. Effective porosity can be one order of magnitude less than porosity.

Table D.9. *Permeability of rocks*

Rock	k (m^2)
Fractured rocks	$10^{-7} - 10^{-10}$
Fresh granite	$10^{-17} - 10^{-18}$
Sandstone	10^{-14}
Limestone	10^{-16}

Permeability k depends on the dimensions of the open network of pores and fractures. For pores of average diameter a, $k = Ca^2$, where C is a geometric factor depending on pore configuration and on porosity. It is often reported in units of 1 Darcy $= 10^{-12}$ m^2.

D.1.5 Thermal expansion

The volumetric thermal expansion coefficient α is defined as,

$$\alpha = \frac{1}{V}\left(\frac{\partial V}{\partial T}\right)_P. \quad \text{(D.16)}$$

At atmospheric pressure, the vast majority of substances expand as temperature is increased. One notable exception is water, for which density is a maximum at 4°C. For water, the thermal expansion coefficient changes sign at 4°C and goes from negative to positive as T increases. Its value is $\approx 2 \times 10^{-4}$ K^{-1} at 20°C and increases with increasing temperature.

More generally, the thermal expansion properties of a material are defined by a relationship between the strain tensor u_{ik} and temperature defining the thermal expansion tensor:

$$u_{ik} = \alpha_{ik} \Delta T. \quad \text{(D.17)}$$

For an isotropic material, the thermal expansion tensor is diagonal,

$$\alpha_{ik} = \frac{\alpha}{3}\delta_{ik}. \quad \text{(D.18)}$$

Thermal expansion depends on temperature and pressure and is anisotropic for many minerals (Fei, 1995). Values for olivine and mantle minerals are in the order of $\alpha = 4 \times 10^{-5}$ K^{-1}.

D.1.6 Viscosity

Natural magmas behave as perfect Newtonian fluids when they are fully liquid. Their rheological behavior is affected by the presence of crystals and bubbles, and

Table D.10. *Viscosities of common volatile-free magma types at their liquidus temperatures.*†

Magma	Liquidus temperature (K)	Viscosity (Pa s)
Komatiite	1600	1
Basalt	1500	10^2
Andesite	1300	10^4
Dacite	1100	10^8
Rhyolite	1000	10^{11}

† Values have been rounded off for clarity.

they become non-Newtonian above certain threshold values of crystal content and bubble content.

The viscosity of a magma depends strongly on composition, dissolved water content and temperature. For volatile-free melts, the liquidus temperature depends on composition and it is useful to consider the viscosity at the liquidus, which varies within a very large range of about $1 - 10^{12}$ Pa s for common magma types including basalts and high-silica rhyolites. At a given temperature above the highest liquidus, the viscosity range for the same magma population is significantly smaller, $\approx 1 - 10^5$ Pa s at 1500 K. The range decreases with increasing temperature, so that all viscosities seem to converge to the same limit value. The temperature dependence of viscosity does not conform to an Arrhenius law and can be accounted for by the following equation (Giordano *et al.*, 2008):

$$\mu = \mu_\infty \exp\left(\frac{B}{T-C}\right), \tag{D.19}$$

where $\mu_\infty = 10^{-4.55}$ Pa s is the high-T limit for silicate viscosity and where coefficients B and C depend on composition. Giordano *et al.* (2008) give explicit relationships between these two coefficients and magma composition, including water and fluor contents. For convenience, Table D.10 lists values of viscosity for a few volatile-free magmas at their liquidus temperature. Each magma type involves in fact a range of melt compositions, and hence a range of viscosities, so that one should use the true viscosity value when dealing with a specific natural magma.

Water has a very strong effect on viscosity because it acts to break the long molecular chains that characterize silicate melts. A rough rule of thumb is that the addition of 1 wt% water lowers the viscosity by one order of magnitude, but the magnitude of this effect depends on melt composition.

Appendix E

Heat production

Rocks contain long-lived radioactive isotopes that release heat as they decay. Such heat generation is a key energy source for convection in the Earth's mantle and for metamorphic reactions in the continental crust. Its main characteristics are that (1) it decays with time, by definition, (2) it is strongly concentrated in continental crustal rocks and (3) it is distributed very unevenly amongst crustal rock types and cannot be predicted from physical properties such as seismic velocity and density. In many cases, the amount of radiogenic heat production is poorly known and makes thermal calculations speculative.

E.1 Heat production rate due to uranium, thorium and potassium

Table E.1 gives the decay constant and the heat produced for the radioactive isotopes important to the energy budget. Using present-day isotopic ratios, the bulk heat production of a rock sample (in W kg^{-1}) is calculated by summing the contributions of each element as follows:

$$H = 10^{-11}(9.52[U] + 2.56[Th] + 3.48[K]), \tag{E.1}$$

where [U] and [Th] are the uranium and thorium concentrations in ppm and [K] the potassium concentration in %. The heat production per unit volume, A, is such that $A = \rho H$, where ρ is the sample density. For an average crustal density of 2700 kg m^{-3}, one has,

$$A = 0.257[U] + 0.069[Th] + 0.094[K], \tag{E.2}$$

where A is given in μW m^{-3}.

We have examined in detail the uranium, thorium and potassium contents of the Bulk Earth and of the Earth's upper mantle in Chapter 8. Here we focus on crustal rocks because they are markedly enriched compared to the mantle. We shall not

Table E.1. *Heat production constants*

Isotope / element	Natural abundance (%)	Half-life (y)	Energy per atom ($\times 10^{-12}$ J)	Heat production per unit mass of isotope/element (W kg^{-1})
^{238}U	99.27	4.46×10^9	7.41	9.17×10^{-5}
^{235}U	0.72	7.04×10^8	7.24	5.75×10^{-4}
U				9.52×10^{-5}
^{232}Th	100	1.40×10^{10}	6.24	2.56×10^{-5}
Th				2.56×10^{-5}
^{40}K	0.0117	1.26×10^9	0.114	2.97×10^{-5}
K				3.48×10^{-9}

From Rybach (1988).

Table E.2. *Element concentration and heat production for some common rocks. Element concentrations are given in ppm. A density of* ≈ 3.0 *Mg m*$^{-3}$ *is assumed for all rocks*

Rock type	U	Th	K	K/U	pW kg^{-1}	μW m^{-3}
Granite	4.6	18	33,000	7,000	1,050	3
Alkali basalt	0.75	2.5	12,000	16,000	180	0.5
Tholeitic basalt	0.11	0.4	1,500	13,600	27	0.08
Peridotite	0.006	0.02	100	17,000	1.5	0.005

discuss in detail mafic rocks because they are strongly depleted (Table E.2). Heat production in these rocks is so small that it can be safely neglected in both regional energy budgets in geological provinces and thermal calculations. Heat production is high in granite and felsic rock samples, but it shows a lot of variability both between and within intrusions, as demonstrated by systematic measurements.

E.1.1 Heat production in the continental crust

The continental crust is the most important repository of uranium, thorium and potassium. Exactly how much it contains is a key issue for the thermal regime of continents and for understanding how the Earth's mantle has evolved through geological time due to crust extraction. Uranium and thorium tend to be located in accessory minerals and grain boundaries, which are not related simply to bulk chemical composition. Thus, their concentrations vary on the scale of a petrological thin section, a hand sample, an outcrop and a whole massif. Estimates of the average U, Th and K concentrations in the continental crust require large extrapolations in

Table E.3. *Element concentration and heat production in the continental crust, from Rudnick and Gao (2003). Element concentrations are given in ppm*

	U	Th	K	μW m^{-3}
Upper crust	2.7	10.5	23,300	1.65
Middle crust	1.3	6.5	19,200	1.00
Lower crust	0.2	1.2	5,080	0.19
Total crust	1.3	5.6	15,080	0.89

scale from tiny specimens to the whole crust of a geological province. Several factors limit the accuracy of the end result. Abundant rocks such as gneisses and metasedimentary rocks are usually under-studied because of their complex origin and metamorphic history. The composition of intermediate and lower crustal levels, which are as heterogeneous as shallow ones, can only be determined in a few scattered outcrops. One consequence is that crustal heat production has in fact been determined from heat flow data (Table E.4). Table E.3 lists estimates derived from geochemical models of the crust. These data have been obtained from a very large number of measurements and have been separated in three different crustal levels corresponding to regional average seismic models. The middle crust lies in the 12–23 km depth range, and the lower crust is taken to be 17 km thick on average, so that the total crust is 40 km thick. The two global crustal models derived from geochemical and heat flow data are consistent with one another. There are considerable variations amongst provinces of different ages and even amongst provinces of the same age and one should not use the average values from Tables 8.4 and E.4 without care.

E.1.2 Large-scale averages

Large data sets are available for several provinces and allow determination of reliable average values of heat production (Table E.5). One should note that the standard deviation of the distribution is large, and in almost all cases as large as the mean, which emphasizes the large spread of values that exist. As a consequence, a single thermal model for the crust in a geological province would be meaningless because it would gloss over the important lateral and vertical temperature variations that probably occur and which may be recorded in metamorphic assemblages, for example. For a reliable thermal calculation, one should also avoid selecting the local heat production values of a few rock types because temperatures are sensitive to heat that is generated over rather large volumes. In short, the average heat production values must be determined on a scale that relates to the problem.

Table E.4. *Estimates of bulk continental crust heat production from heat flow data for 40 km crust, from Jaupart and Mareschal (2003)*

Age group	A^a	Q_C^b	% Area[c]
Archean	0.56–0.73	23–30	9
Proterozoic	0.73–0.90	30–37	56
Phanerozoic	0.95–1.21	37–47	35
Total continents	0.79–0.99	32–40	

[a] range of heat production in μW m^{-3}.
[b] range of crustal heat flow component in mW m^{-2}.
[c] Fraction of total continental surface, from Model 2 in Rudnick and Fountain (1995).

Table E.5. *Average surface heat production for different geological provinces obtained by systematic regional sampling over large areas*

	$\langle A \rangle^a$	σ_A^b	N^c	References
	μW m^{-3}			
Yilgarn (Archean, western Australia)	3.3	3.3	540	Heier and Lambert, 1978
Superior (Archean, Canada)				
New Québec region	1.22	[d]	3085	Eade and Fahrig, 1971
Wawa Gneiss Terrane	1.01	[e]	56	Shaw *et al.*, 1994
Sachigo subprovince	1.04	[e]	20	Fountain *et al.*, 1987
"Churchill" (Archean to Proterozoic, Canada)	2.0	[d]	1510	Eade and Fahrig, 1971
Gawler (Proterozoic, central Australia)	3.6	3.4	90	Heier and Lambert, 1978
Baltic Shield (Proterozoic, Finland + Russia)	1.2	1.2	284	[f]
Central East China (Archean to Neogene, China)	1.22[g]	[d]	11451	Gao *et al.*, (1998)

[a] Mean heat production.
[b] Standard deviation on the heat production distribution.
[c] Number of sites.
[d] Analyses were made on mixed powders, implying that the standard deviation of the analyses underestimates the true spread of values for individual rock samples.
[e] Average calculated by weighting according to the abundances of the different rock types.
[f] Baltic Shield data compiled from Hanski (1992); Eilu (1994); Salonsaari (1995); Lahtinen (1996).
[g] Heat production value recalculated for a bulk density of 2700 kg m^{-3}.

Table E.6. *Element concentration and heat production for sedimentary rocks. Element concentrations are given in ppm. A density of ≈ 2.5 Mg m^{-3} is assumed for all rocks*

Rock type	U	Th	K	K/U	pW kg^{-1}	μW m^{-3}
Shales	3.7 ± 0.5	12 ± 1	27,000	7,300	775	1.94
Carbonate rocks	2.2 ± 0.1	1.7 ± 0.7	2,700	270	180	0.7

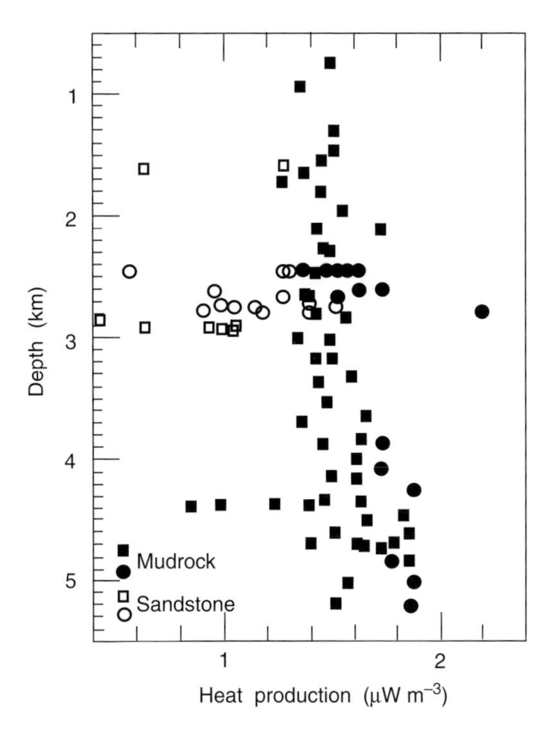

Figure E.1. Heat production profile through sedimentary rocks of the Gulf of Mexico, (from McKenna and Sharp, 1998). Square and circular symbols refer to two different sedimentary formations. Filled symbols: mudrocks, open symbols: sandstones. Note that mudrock samples are on average enriched compared to sandstone samples.

E.1.3 Heat production in sediments and sedimentary rocks

Sediments usually do not contain large amounts of uranium, thorium and potassium (Table E.6). They can accumulate over large thicknesses, however, and may account for a significant fraction of the surface heat flux. Such heat production must be included in thermal calculations. Figure E.1 shows a vertical profile of heat production in sedimentary packages of the Gulf of Mexico. Over about 5 km

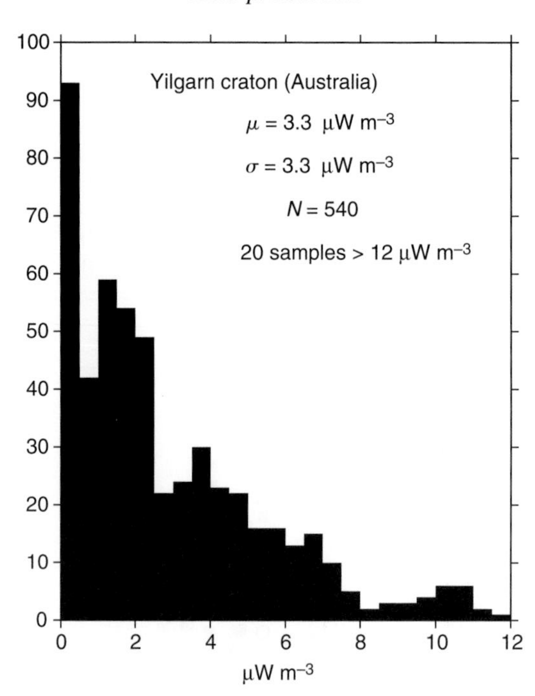

Figure E.2. Histogram of heat production in all the samples from the Archean Yilgarn province, in western Australia. Data from Heier and Lambert (1978).

thickness, this sedimentary sequence adds about 8 mW m^{-2} to the heat flux from the crystalline basement.

E.1.4 Heat production variations

In the continental crust, heat production varies on all sorts of scales. Between all the samples collected in the Archean Yilgarn province (Australia), heat production varies by a factor > 10 (Figure E.2). Even within a single rock type, the gneiss of the Yilgarn, heat production varies from < 0.5 to 6 μW m^{-3}, excluding the outliers in the distribution (Figure E.3). Within a single pluton, radioelement concentrations can change by large amounts in both vertical and horizontal directions. These changes may be due to a host of different causes, such as facies changes, intrinsic heterogeneity of the source material, fluid migration and late-stage alteration. In the Bohus granite, Sweden, for example, thorium concentrations vary by a factor of 5 over horizontal distances as small as a few tens of meters and as large as a few kilometers (Landstrom *et al.*, 1980).

Radioelement concentrations change according to lithology and a geological map allows a rough idea of the spatial variation of heat production. Within Cordilleran

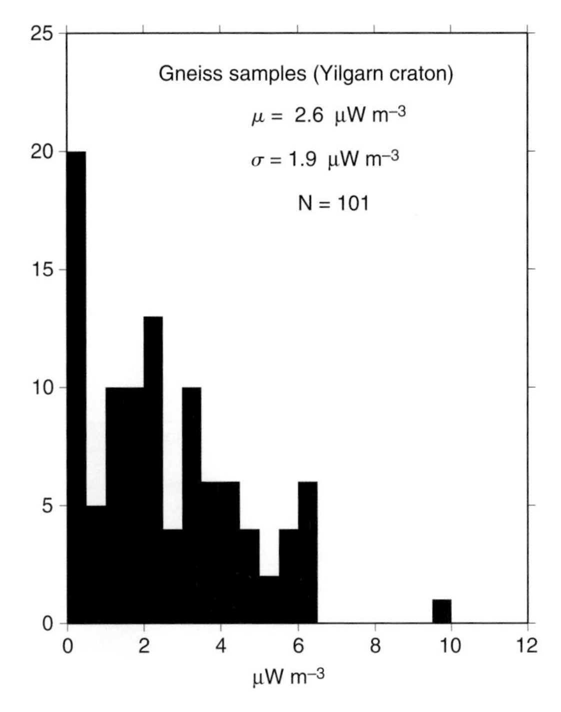

Figure E.3. Histogram of heat production in the gneiss of the Archean Yilgarn province, in western Australia. Data from Heier and Lambert (1978).

core complexes, significant variations occur over distances of a few kilometers (Ketcham, 1996). Heat production may vary significantly even for the same rock type in a single geological province, which does not allow extrapolation from one unit to another. In New Hampshire, USA, which forms part of the Appalachians province, for example, there are large differences in Th abundance amongst plutons of the same White Mountain magma series (Rogers and Adams, 1969). Precambrian Shield regions exhibit remarkable heat production variability. For example, the exposed part of the Proterozoic Trans-Hudson Orogen in Saskatchewan and Manitoba ($\approx 500 \times 500$ km) is a mosaic of different belts of different origins and compositions with contrasting heat production values. Another example is provided by the Abitibi province of eastern Canada, which stands out as a low heat production region within the Archean Superior Province.

E.1.5 Crustal stratification

On a vertical scale of ≈ 10 km, measurements on samples from the deep boreholes at Kola, Russia, (Kremenentsky *et al.*, 1989) and the KTB, Germany, (Clauser *et al.*, 1997), indicate no systematic variation of heat production with depth. At KTB,

Table E.7. *Mean heat production rates of different rock types in the main hole of China Continental Scientific Drilling (He et al., 2008)*

Rock type	N	ρ,kg m^{-3}	Th, ppm	U, ppm	K,%	Th/U	A, μW m^{-3}
Eclogite	37	3272	2.58 ± 3.88	0.41 ± 0.40	0.77 ± 0.71	6.3	0.42 ± 0.46
Amphibolite	21	2965	3.26 ± 3.81	0.75 ± 1.11	1.55 ± 0.67	4.3	0.61 ± 0.55
Peridotite	3	2981	0.02 ± 0.01	0.01 ± 0.01	0.01 ± 0.01	1.4	0.01 ± 0.13
Paragneiss	88	2727	10.93 ± 13.9	1.40 ± 0.93	2.76 ± 1.12	7.8	1.37 ± 1.13
Orthogneiss	100	2665	13.06 ± 6.19	1.71 ± 1.30	3.88 ± 0.99	7.6	1.65 ± 0.70

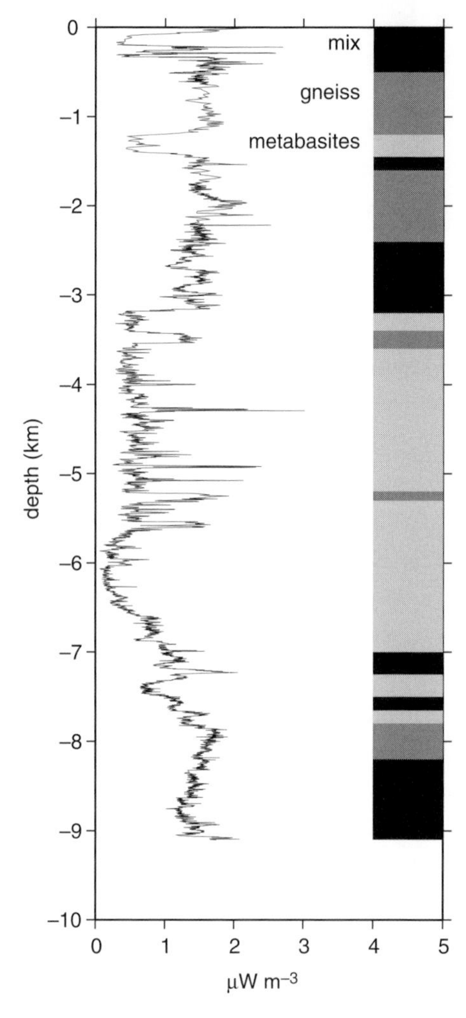

Figure E.4. Heat production profile in the KTB deep borehole in Germany. Measurements of radioactivity were made with a sampling interval of 2 m.

Table E.8. *Heat production estimates for the subcontinental lithospheric mantle*

Sample	Heat production rate (μW m^{-3})	Reference
Peridotite samples		
Off-craton, massif peridotite	0.018	(1)
Off-craton, spinel peridotite	0.033	(1)
Average values for xenolith suites worldwide		
On craton, all	0.093	(1)
On craton, kimberlite hosted	0.104	(1)
On craton, non-kimberlite hosted	0.028	(1)
Jericho kimberlite, Slave Province, Canada		
Low T xenoliths	0.15–0.27	(2)
High T xenoliths	0.096–0.461	(2)
Inversion of (P,T) array		
Slave province	<0.025	(2), (3)
Kaapvaal craton	<0.03	(3)

References: (1) Rudnick *et al.* (1998), (2) Russell *et al.* (2001), (3) Michaut *et al.* (2007).

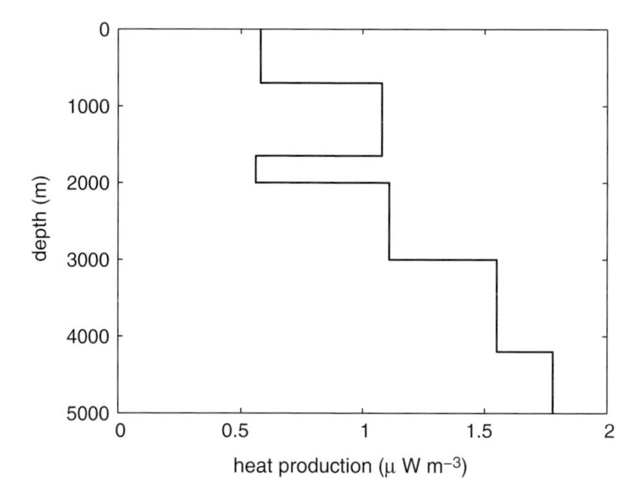

Figure E.5. Average heat production profile in the the main hole of the Chinese Continental Scientific Drilling program. The 5 km hole intersects the Dabie-Sulu ultra-high metamorphic belt. Data from He *et al.* (2008).

heat production between 8 and 9 km depth is the same as between 1 and 2 km, and higher than above 1 km (Figure E.4). The heat production in the 5 km deep hole of the Chinese continental scientific program increases with depth (Figure E.5). The average heat production varies between the main rock types (Table E.7). At Kola, in the Russian part of the Baltic Shield, heat generation in the Archean rocks between 8 and 12 km is higher (1.47 μW m^{-3}) than in the shallower Proterozoic section (0.4 μW m^{-3}) (Kremenentsky *et al.*, 1989). Over a crustal thickness scale, heat production is lower in mid-crustal assemblages than in the upper crust (Fountain *et al.*, 1987). This vertical variation is not monotonous by any means and cannot be described by a simple function valid everywhere, as shown by exposed crustal sections such as the Vredefort in South Africa (Nicolaysen *et al.*, 1981), the Cordilleran core complexes of Arizona (Ketcham, 1996), or the Pikwitonei-Sachigo and Kapuskasing-Wawa areas of Canada (Fountain *et al.*, 1987; Ashwal *et al.*, 1987).

On a large scale, the crust is stratified, such that the lower crust is strongly depleted compared to the upper and middle crust (Table E.3, Figure 7.7).

E.1.6 Subcontinental lithospheric mantle

This is probably the least documented part of the continental lithosphere, despite its importance for the thermal evolution and rheology of continents. Some estimates of the heat production in the lithospheric mantle are listed in Table E.8. Xenoliths in kimberlite samples have been infiltrated by enriched metasomatic melts and are not representative of the undisturbed lithospheric mantle. Peridotites allow large-scale sampling of lithosphere that has not been affected by kimberlite magmatism.

List of symbols

Many symbols are used throughout the book. It is almost unavoidable that the same symbol may appear with different meanings in different places. For instance λ is the symbol for thermal conductivity throughout the book, but sometimes it is also used for wavelength; thermal expansion coefficient is always α, but α is sometimes used as a decay constant. Pressure is always denoted by p or P, but P is sometimes used for power. We trust that the context always makes the meaning of the symbols clear.

Table E.9. *Symbols used*

Symbol	Definition	Units (commonly used units/ or value)
C_p	Heat capacity	J kg^{-1} K^{-1}
C_Q	Heat flux / age$^{1/2}$	470–510 mW m^{-2}/(Myr)$^{1/2}$
C_Q	dimensionless constant in scaling laws	
C_A	Sea-floor accretion rate	3.23 km^2 Myr^{-1}
C	also used as composition	
E_c	Kinetic energy	J
E_g	Gravitational potential energy	J
E_{rot}	Rotational energy	J
E_D	Strain energy	J
F	Buoyancy flux	m^4 s^{-3}
g	Acceleration of gravity	m s^{-1}
G	Gravitational constant	6.67×10^{-11} kg m^3 s^{-2}
h	Thickness of convecting layer	m
H (sometimes A)	Heat production (volumetric)	W m^{-3} (μW m^{-3})
H	Enthalpy	J
I	Moment of inertia	kg m^2
K (K_S, K_T)	Bulk modulus	Pa
k	Permeability also used for wavenumber	m^2

Table E.9. *(Cont'd) Symbols used*

Symbol	Definition	Units (commonly used units/ or value)
L	Latent heat	J kg^{-1}
L	Length of oceanic plate (also length-scale)	m
L	Lithospheric thickness	m
M	Mass of Earth	kg
Nu	Nusselt number	
p	Pressure also Power	Pa (MPa, GPa)
Pr	Prandlt number	
Q, q	Heat flux	W m^{-2} (mW m^{-2})
R	Radius of Earth	km (6,378)
Ra	Rayleigh number	/
Re	Reynolds number	
s	Entropy per unit mass	J kg^{-1} K^{-1}
St	Stefan number	
T	Temperature	K (°C)
u	Internal energy per unit mass (also used for horizontal velocity component)	J kg^{-1}
U	Internal energy (also used for horizontal velocity component)	J
Ur	Urey number	
V	Volume	m^3
V_ϕ^2	Equivalent sound velocity	km^2 s^{-2}
w	Radial velocity	
α	Volumetric expansion coefficient	K^{-1}
γ	Grüneisen parameter	
δ	Thickness of boundary layer	
κ	Thermal diffusivity	m^2 s^{-1}
λ	Thermal conductivity Occasionally used for wavelength	W m^{-1} K^{-1}
μ	Viscosity	Pa s
ν	Dynamic viscosity	m^2 s^{-1}
Ω	Angular velocity	rad s^{-1}
Π	Stress tensor	
Γ	Temperature gradient	
Ψ	Heat dissipated by friction	J
φ	External energy sources	J
ρ	Density	kg m^{-3}
ΔS_{cond}	Entropy production	J K^{-1}
σ	also Stefan constant	W m^{-2} K^{-4}
σ	Deviatoric stress tensor	Pa (MPa)
τ	Relaxation time constant	s

References

Abbott, D. 1991. The case for accretion of the tectosphere by buoyant subduction. *Geophys. Res. Lett.*, **18**, 585–588.

Abbott, D., Burgess, L., and Longhi, J. 1994. An empirical thermal history of the Earth's upper mantle. *J. Geophys. Res.*, **99**, 13835–13850.

Abe, Y. 1993. Physical state of the very early Earth. *Lithos*, **30**, 223–235.

Abe, Y. 1997. Thermal and chemical evolution of the terrestrial magma ocean. *Phys. Earth Planet. Inter.*, **100**, 27–39.

Ahern, J. L., and Mrkvicka, S. R. 1984. A mechanical and thermal model for the evolution of the Williston Basin. *Tectonics*, **3**, 79–102.

Albarède, F. 2003. The thermal history of leaky chronometers above their closure temperature. *Geophys. Res. Lett.*, **30**, 1015.

Alfé, D., Gillan, M. J., and Price, G. D. 2002. Ab initio chemical potentials of solid and liquid solutions and the chemistry of the Earth's core. *J. Chem. Phys.*, **116**, 7127–7136.

Allegre, C. J., and Jaupart, C. 1985. Continental tectonics and continental kinetics. *Earth Planet. Sci. Lett.*, **74**, 171–186.

Anderson, D. L., and Sammis, C. 1970. Partial melting in the upper mantle. *Phys. Earth Planet. Inter.*, **3**, 41–50.

Anderson, E. M. 1934. Earth contraction and mountain building. *Beitr. Z. Geophys.*, **42**, 133–159.

Ashwal, L. D., Morgan, P, Kelley, S. A., and Percival, J. 1987. Heat production in an Archean crustal profile and implications for heat flow and mobilization of heat producing elements. *Earth Planet. Sci. Lett.*, **85**, 439–450.

Audet, P., and Mareschal, J.-C. 2007. Wavelet analysis of the coherence between Bouguer gravity and topography: application to the elastic thickness anisotropy in the Canadian Shield. *Geophys. J. Int.*, **168**, 287–298.

Ballard, S., Pollack, H. N., and Skinner, N. J. 1987. Terrestrial heat flow in Botswana and Namibia. *J. Geophys. Res.*, **92**, 6291–6300.

Balmino, G. 2003. Gravity field recovery from GRACE: Unique aspects of the high precision inter-satellite data and analysis methods. *Space Sci. Rev.*, **108**, 47–54.

Bank, C.-G., Bostock, M. G., Ellis, R. M., Hajnal, Z., and VanDecar, J. C. 1998. Lithospheric mantle structure beneath the Trans-Hudson Orogen and the origin of diamondiferous kimberlites. *J. Geophys. Res.*, **103**, 10103–10114.

Bartlett, M. G., Chapman, D. S., and Harris, R. N. 2005. Snow effect on North American ground temperatures, 1950-2002. *J. Geophys. Res.*, **110**, 3008.

Bartlett, M. G., Chapman, D. S., and Harris, R. N. 2006. A decade of ground air temperature tracking at Emigrant Pass Observatory, Utah. *Journal of Climate*, **19**, 3722–3731.

Batchelor, G. K. 1954. Heat convection and buoyancy effects in fluids. *Quart. J. R. Met. Soc.*, **80**, 339–358.

Beaumont, C., Jamieson, R. A., Nguyen, M. H., and Medvedev, S. 2004. Crustal channel flows: 1. Numerical models with applications to the tectonics of the Himalayan-Tibetan orogen. *J. Geophys. Res.*, **109**, B06406.

Beck, A. E., Dharba, D. M., and Schloessin, H. H. 1978. Lattice conductivities of single crystal and polycrystalline materials at mantle pressures and temperatures. *Phys. Earth Planet. Inter.*, **17**, 35–53.

Becker, T. W., Conrad, C. P., Buffett, B., and Muller, R. D. 2009. Past and present seafloor age distributions and the temporal evolution of plate tectonic transport. *Earth Planet. Sci. Lett.* **278**, 233–242.

Bedini, R.-M., Blichert-Toft, J., Boyet, M., and Albarède, F. 2004. Isotopic constraints on the cooling of the continental lithosphere. *Earth Planet. Sci. Lett.*, **223**, 99–111.

Beltrami, H., and Mareschal, J.-C. 1995. Resolution of ground temperature histories inverted from borehole temperature data. *Global Planet. Change*, **11**, 57–70.

Beltrami, H., Jessop, A. M., and Mareschal, J.-C. 1992. Ground temperature histories in eastern and central Canada from geothermal measurements: evidence of climatic change. *Global Planet. Change*, **19**, 167–184.

Beltrami, H., Cheng, L., and Mareschal, J.-C. 1997. Simultaneous inversion of borehole temperature data for determination of ground surface temperature history. *Geophys. J. Int.*, **129**, 311–318.

Benfield, A. E. 1939. Terrestrial heat flow in Great Britain. *Proc. Roy. Soc. London, Ser. A*, **173**, 428–450.

Benfield, A. E. 1949. The effect of uplift and denudation on underground temperatures. *J. App. Phys.*, **20**, 66–70.

Berman, R. G., and Aranovich, L. Y. 1996. Optimized standard state and solution properties of minerals: 1. Model calibration for olivine, orthopyroxene, cordierite, garnet, and ilmenite in the system FeO–MgO–CaO–Al2O3–TiO2–SiO2. *Contrib. Mineral. Petrol.*, **126**, 1–24.

Bernal, J. D. 1936. Hypothesis on the 20° discontinuity. *Observatory*, **59**, 268.

Berryman, J. 1995. Mixture Theories for Rock Properties. Pages 205–228 of: Ahrens, T. J. (ed), *Rock physics and phase relationship. A Handbook of Physical constants*, vol. 3. AGU.

Birch, F. 1948. The effects of Pleistocene climatic variations upon geothermal gradients. *Am. J. Sci.*, **246**, 729–760.

Birch, F. 1950. Flow of heat in the Front Range, Colorado. *Bull. Geol. Soc. Amer.*, **61**, 567–630.

Birch, F. 1952. Elasticity and constitution of the Earth's interior. *J. Geophys. Res.*, **57**, 227–285.

Birch, F. 1964. Density and composition of mantle and core. *J. Geophys. Res.*, **69**, 4377–4388.

Birch, F. 1965. Speculations on the thermal history of the Earth. *Bull. Geol. Soc. Amer.*, **76**, 133–154.

Birch, F., Roy, R. F., and Decker, E. R. 1968. Heat flow and thermal history in New England and New York. Pages 437–451 of: An-Zen, E. (ed), *Studies of Appalachian Geology*. New York: Wiley-Interscience.

Bird, P. 1979. Continental delamination and the Colorado Plateau. *J. Geophys. Res.*, **84**, 7561–7571.

Bird, R. B., Stewart, W. E., and Lightfoot, E. N. 1960. *Transport Phenomena*. John Wiley & Sons.

Blundy, J., and Cashman, K. 2001. Ascent-driven crystallization of dacite magmas at Mount St Helens, 1980-1986. *Contrib. Mineral. Petrol.*, **140**, 631–650.

Bodell, J. M., and Chapman, D. S. 1982. Heat flow in the north central Colorado Plateau. *J. Geophys. Res.*, **87**, 2869–2884.

Bodvarsson, G., and Lowell, R. P. 1972. Ocean-floor heat flow and the circulation of interstitial waters. *J. Geophys. Res.*, **77**, 4472–4475.

Bond, G. C., and Kominz, M. A. 1991. Disentangling middle Paleozoic sea level and tectonic events in cratonic margins and cratonic basins of North America. *J. Geophys. Res.*, **94**, 6619–6639.

Bonneville, A., Von Herzen, R. P., and Lucazeau, F. 1997. Heat flow over Reunion hot spot track: Additional evidence for thermal rejuvenation of oceanic lithosphere. *J. Geophys. Res.*, **102**, 22731–22748.

Brady, R. J., Ducea, M. N., Kidder, S. B., and Saleeby, J. B. 2006. The distribution of radiogenic heat production as a function of depth in the Sierra Nevada batholith, California. *Lithos*, **86**, 229–244.

Braginsky, S. I., and Roberts, P. H. 1995. Equations governing convection in Earth's core and the geodynamo. *Geophys. Astrophys. Fluid Dyn*, **79**, 1–97.

Brandeis, G., and Jaupart, C. 1987. The kinetics of nucleation and growth and scaling laws for magmatic crystallization. *Contrib. Mineral. Petrol.*, **96**, 24–34.

Brown, M. 2000. Pluton emplacement by sheeting and vertical ballooning in part of the Southeast Coast plutonic complex, British Columbia. *Geol. Soc Am. Bull.*, **112**, 708–719.

Buffett, B. A. 2007. Core-Mantle Interactions. Chap. 12, pages 345–358 of: Olson, P. (ed), *Treatise of Geophysics*, vol. 8. Elsevier.

Bullard, E. C. 1938. The disturbance of the temperature gradient in the Earth's crust by inequalities of height. *Mon. Not. R. Astron. Soc., Geophys. Suppl.*, **4**, 360–362.

Bullard, E. C. 1939. Terrestrial heat flow in South Africa. *Proc. Roy. Soc. London, Ser. A*, **173**, 474–502.

Bullard, E. C. 1947. The time necessary for a bore hole to attain temperature equilibrium. *Mon. Not. R. Astron. Soc., Geophys. Suppl.*, **5**, 127–130.

Bullard, E. C. 1954. The flow of heat through the floor of the Atlantic ocean. *Proc. Roy. Soc. London, Ser. A*, **222**, 408–422.

Bullard, E. C. 1962. The deeper structure of the ocean floor. *Proc. Roy. Soc. London, ser. A*, **265**, 386–395.

Bullard, E. C., Maxwell, A. E., and Revelle, R. 1956. Heat flow through the deep sea floor. *Advances in Geophysics*, **3**, 153–181.

Bullen, K. E. 1952. Cores of Terrestrial Planets. *Nature*, **170**, 363–364.

Bullen, K. E. 1963. An index of degree of chemical inhomogeneity in the earth. *Geophysical Journal International*, **7**, 584–592.

Burchardt, S. 2008. New insights into the mechanics of sill emplacement provided by field observations of the Njardvik sill, Northeast Iceland. *J. Volcanol. Geotherm. Res.*, **173**, 280–288.

Buttner, R., Zimanowski, B., Blumm, J., and Hageman, L. 1998. Thermal conductivity of a volcanic rock material (olivine-melilitite) in the temperature range between 288 and 1470 K. *J. Volcan. Geotherm. Res.*, **80**, 293–302.

Byerlee, J. 1978. Friction of rocks. *Pure and Applied Geophysics*, **116**, 615–626.
Cameron, A. G. W. 2001. From interstellar gas to the Earth-Moon system. *Meteorit. Planet. Sci.*, **36**, 9–22.
Canup, R. M. 2004. Simulations of a late lunar-forming impact. *Icarus*, **168**, 433–456.
Canup, R. M., and Righter, K. (eds). 2000. *Origin of the Earth and Moon*. Tucson: The University of Arizona Press.
Carlson, R. L., and Johnson, H. P. 1994. On modeling the thermal evolution of the oceanic upper mantle: an assessment of the cooling plate model. *J. Geophys. Res.*, **99**, 3201–3214.
Carlson, R. W., Pearson, D. G., and James, D. E. 2005. Physical, chemical, and chronological characteristics of continental mantle. *Rev. Geophys.*, **43**, RG1001.
Caro, G., Bourdon, B., Birck, J.-L., and Moorbath, S. 2003. ^{146}Sm-^{142}Nd evidence from Isua metamorphosed sediments for early differentiation of the Earth's mantle. *Nature*, **423**, 428–432.
Carslaw, H. S., and Jaeger, J. C. 1959. *Conduction of Heat in Solids*. 2nd edn. Oxford.
Cermak, V. 1971. Underground temperature and inferred climatic temperature of the past millenium. *Palaeogeogr. Palaeoclimatol., Palaeoecol.*, **98**, 167–182.
Chambers, K., Woodhouse, J. H., and Deuss, A. 2005. Topography of the 410-km discontinuity from PP and SS precursors [rapid communication]. *Earth Planet. Sci. Lett.*, **235**, 610–622.
Chandrasekhar, S. 1961. *Hydrodynamic and Hydromagnetic Stability*. Oxford University Press.
Cherkaoui, A. S. M., and Wilcock, W. S. D. 1999. Characteristics of high Rayleigh number two-dimensional convection in an open-top porous layer heated from below. *J. Fluid Mech.*, **394**, 241–260.
Cherkaoui, A. S. M., and Wilcock, W. S. D. 2001. Laboratory studies of high Rayleigh number circulation in an open-top Hele-Shaw cell: An analog to mid-ocean ridge hydrothermal systems. *J. Geophys. Res.*, **106**, 10983–11000.
Chouinard, C., and Mareschal, J.-C. 2007. Selection of borehole temperature depth profiles for regional climate reconstructions. *Clim. Past*, **3**, 297–313.
Chouinard, C., and Mareschal, J.-C. 2009. Ground surface temperature history in southern Canada: Temperatures at the base of the Laurentide ice sheet and during the Holocene. *Earth Planet. Sci. Lett.*, **277**, 280–289.
Chouinard, C., Fortier, R., and Mareschal, J.-C. 2007. Recent climate variations in the subarctic inferred from three borehole temperature profiles in northern Quebec, Canada. *Earth Planet. Sci. Lett.*, **263**, 355–369.
Christensen, U. R. 1985a. Heat transport by variable viscosity convection II: Pressure influence, non-Newtonian rheology and decaying heat sources. *Phys. Earth Planet. Inter.*, **37**, 183–205.
Christensen, U. R. 1985b. Thermal evolution models for the earth. *J. Geophys. Res.*, **90**, 2995–3007.
Christensen, Ulrich R., and Tilgner, Andreas. 2004. Power requirement of the geodynamo from ohmic losses in numerical and laboratory dynamos. *Nature*, **429**, 169–171.
Churchill, R. V. 1960. *Complex Variables and Applications*. Second edn. New York: McGraw-Hill.
Clark, S. P. 1966. Thermal conductivity. Pages 459–482 of: Clark, S. P. (ed), *Handbook of Physical Constants*. Boulder (CO): GSA.
Clarke, G. K. C. 2005. Subglacial Processes. *Annual Review of Earth and Planetary Sciences*, **33**, 247–276.

Clauser, C., and Huenges, E. 1995. Thermal conductivity of rocks and minerals. Pages 105–126 of: Ahrens, T. J. (ed), *Rock Physics and Phase Relations: A Handbook of Physical Constants. AGU Reference Shelf 3*. AGU.

Clauser, C., and Mareschal, J.-C. 1995. Ground temperature history in central Europe from borehole temperature data. *Geophys. J. Int.*, **121**, 805–817.

Clauser, C., Gieses, P., Huenges, E., Kohl, T., Lehmann, H., Rybach, L., Safanda, J., Wilhelm, H., Windlow, K., and Zoth, G. 1997. The thermal regime of the crystalline continental crust: implications from the KTB. *J. Geophys. Res.*, **102**, 18417–18441.

Cogley, J. G. 1984. Continental margins and the extent and number of continents. *Rev. Geophys. Space Phys.*, **22**, 101–122.

Cogné, J.-P., and Humler, E. 2004. Temporal variation of oceanic spreading and crustal production rates during the last 180 My. *Earth Planet. Sci. Lett.*, **227**, 427–439.

Conrad, C. P., and Hager, B. H. 1999. The thermal evolution of an earth with strong subduction zones. *Geophys. Res. Lett.*, **26**, 3041–3044.

Courtney, R. C., and White, R. S. 1986. Anomalous heat flow and geoid across the Cape Verde Rise: evidence of dynamic support from a thermal plume in the mantle. *Geophys. J. R. Astr. Soc.*, **87**, 815–868.

Crane, K., and O'Connell, S. 1983. The distribution and implications of the heat flow from the Gregory rift in Kenya. *Tectonophys.*, **94**, 253–272.

Crough, S. T., and Thompson, G. A. 1976. Thermal model of continental lithosphere. *J. Geophys. Res.*, **81**, 4857–4862.

Cull, J. P. 1991. Heat flow and regional geophysics in Australia. Pages 486–500 of: Cermak, V., and Rybach, L. (eds), *Terrestrial Heat Flow and the Lithosphere Structure*. Berlin: Springer Verlag.

Dahl-Jensen, D., Mosegaard, K., Gundestrup, N., Clow, G. D., Johnsen, S. J., Hansen, A. W., and Balling, N. 1998. Past temperatures directly from the Greenland ice sheet. *Science*, **282**, 268–271.

Davaille, A., and Jaupart, C. 1993a. Thermal convection in lava lakes. *Geophys. Res. Lett.*, **20**, 1827–1830.

Davaille, A., and Jaupart, C. 1993b. Transient high-Rayleigh-number thermal convection with large viscosity variations. *J. Fluid Mech.*, **253**, 141–166.

Davaille, A., Limare, A., Touitou, F. Kumagai, I., and Valteville, J. 2010. Anatomy of a laminar starting plume in high-Prandtl number. *Experiments in Fluids*, **49**.

Davies, G. F. 1980. Review of oceanic and global heat flow estimates. *Rev. Geophys.*, **18**, 718–722.

Davies, G. F. 1980. Thermal histories of convective Earth models and constraints on radiogenic heat production in the Earth. *J. Geophys. Res.*, **85**, 2517–2530.

Davies, G. F. 1988. Ocean bathymetry and mantle convection, 1. Large-scale flow and hotspots. *J. Geophys. Res.*, **93**, 10467–10480.

Davies, G. F. 1999. *Dynamic Earth: Plates, Plumes, and Mantle Convection*. Cambridge, (U.K.): Cambridge University Press.

Davis, E. E., and Elderfield, H. 2004. *Hydrogeology of the Oceanic Lithosphere*. Cambridge Univ. Press.

Davis, E. E., Chapman, D. S., Wand, K., Villinger, H., Fisher, A. T., Robinson, S. W., Grigel, J., Pribnow, D., Stein, J., and Becker, K. 1999. Regional heat flow variations across the sedimented Juan de Fuca ridge eastern flank: Constraints on lithospheric cooling and lateral hydrothermal heat transport. *J. Geophys. Res.*, **104**, 17,675–17,688.

Debayle, E., Kennett, B., and Priestley, K. 2005. Global azimuthal seismic anisotropy and the unique plate-motion deformation of Australia. *Nature*, **433**, 509–512.

Decker, E. R., Baker, K. R., Bucher, G. J., and Heasler, H. P. 1980. Preliminary heat flow and radioactivity studies in Wyoming. *J. Geophys. Res.*, **85**, 311–321.

Detrick, R. S., and Crough, T. 1978. Island subsidence, hot spots, and lithospheric thinning. *J. Geophys. Res.*, **83**, 1236–1244.

Dietsche, C., and Muller, U. 1985. Influence of Benard convection on solid-liquid interfaces. *J. Fluid Mech.*, **161**, 249–268.

Diment, W. H., and Pratt, H. R. 1988. *Thermal conductivity of some rock-forming minerals: a tabulation.* Open file report 88-690. U.S. Geological Survey.

Doin, M.-P., Fleitout, L., and McKenzie, D. 1996. Geoid anomalies and the structure of continental and oceanic lithospheres. *J. Geophys. Res.*, **101**, 16119–16136.

Doin, M. P., Fleitout, L., and Christensen, U. 1997. Mantle convection and stability of depleted and undepleted continental lithosphere. *J. Geophys. Res.*, **102**, 2771–2787.

Doin, M. P., and Fleitout, L. 1996. Thermal evolution of the oceanic lithosphere: An alternative view. *Earth Planet. Sci. Lett.*, **142**, 121–136.

Domenico, P. A., and Palciauskas, V. V. 1973. Theoretical analysis of forced convective heat transfer in regional ground water flow. *Bull. Geol. Soc. Amer.*, **84**, 3803–3819.

Donnelly, K. E., Goldstein, S. L., Langmuir, C. H., and Spiegelman, M. 2004. Origin of enriched ocean ridge basalts and implications for mantle dynamics. *Earth Planet. Sci. Lett.*, **226**, 347–366.

Duchkov, A. D. 1991. Review of Siberian heat flow data. Pages 426–443 of: Cermak, V., and Rybach, L. (eds), *Terrestrial Heat Flow and the Lithosphere Structure.* Berlin: Springer-Verlag.

Dziewonski, A. M., and Anderson, D. L. 1981. Preliminary reference Earth model. *Phys. Earth Planet. Inter.*, **25**, 297–356.

Eade, K. E., and Fahrig, W. F. 1971. Geochemical evolutionary trends of continental plates, A preliminary study of the Canadian Shield. *Geol. Surv. Canada Bull.*, **179**, 59pp.

Elder, J. W. 1967. Steady free convection in a porous medium heated from below. *J. Fluid Mech.*, **27**, 29–48.

Elkins-Tanton, L. T., Parmentier, E. M., and Hess, P. C. 2003. Magma ocean fractional crystallization and cumulate overturn in terrestrial planets: Implications for Mars. *Meteorit. Planet. Sci.*, **38**, 1753–1771.

Elliott, R. 1977. Eutectic solidification. *Int. Metal. Rev.*, **219**, 161–186.

Elliott, R. B. 1956. The Eskdalemuir tholeiite and its contribution to an understanding of tholeiite genesis. *Min. Mag.*, **31**, 245–254.

England, P., Molnar, P., and Richter, F. 2007. John Perry's neglected critique of Kelvin's age for the Earth: A missed opportunity in geodynamics. *GSA Today*, **17**, 4–9.

England, P. C., and Thompson, A. B. 1984. Pressure-temperature-time paths of regional metamorphism. I Heat transfer during the evolution of regions of thickened continental crust. *J. Petrol.*, **25**, 894–928.

Enomoto, S., Ohtani, E., Inoue, K., and Suzuki, A. 2007. Neutrino geophysics with KamLAND and future prospects. *Earth Planet. Sci. Lett.*, **258**, 147–159.

Erdélyi, A., Magnus, W., Oberhettinger, F., and Tricomi, F. G. 1954a. *Tables of Integral Transforms. Vol. I.* New-York: McGraw-Hill Book Co.

Erdélyi, A., Magnus, W., Oberhettinger, F., and Tricomi, F. G. 1954b. *Tables of Integral Transforms. Vol. II.* New-York: McGraw-Hill Book Co.

Erickson, A. J., Von Herzen, R. P., Sclater, J. G., Girdler, R. W., Marshall, B. V., and Hyndman, R. 1975. Geothermal measurements in Deep-Sea drill holes. *J. Geophys. Res.*, **80**, 2515–2528.

Fei, Y. 1995. Thermal Expansion. Pages 29–44 of: Ahrens, T. (ed), *Mineral Physics and Crystallography. A Handbook of Physical Constants. AGU Reference Shelf 2*. Washington (DC): AGU.

Fiorentini, G., Lissia, M., Mantovani, F., and Vannucci, R. 2005. Geo-neutrinos: A new probe of Earth's interior. *Earth Planet. Sci. Lett.*, **238**, 235–247.

Flasar, F. M., and Birch, F. 1973. Energetics of Core Formation: A Correction. *J. Geophys. Res.*, **78**, 6101–6103.

Fountain, D. M., Salisbury, M. H., and Furlong, K. P. 1987. Heat production and thermal conductivity of rocks from the Pikwitonei-Sachigo continental cross-section, central Manitoba: Implications for the thermal structure of Archean crust. *Can. J. Earth Sci.*, **24**, 1583–1594.

Fourier, J. B. J. 1820. Extrait d'un mémoire sur le refroidissement du globe terrestre. *Bull. Sci. par la Société philomatique de Paris*.

Fourier, J. B. J. 1822. *Theorie Analytique de la Chaleur*. 1988 reprint by Editions Jacques Gabay, Sceaux, France: Firmin Didot.

Fourier, J. B. J. 1827. Mémoire sur les températures du globe terrestre. *Mémoires de l'académie royale des sciences de l'Institut de France*, **VII**, 570–604.

Fournier, R. O. 1989. Geochemistry and dynamics of the Yellowstone National Park hydrothermal system. *Ann. Rev. Earth Planet. Sci.*, **17**, 13–53.

Francheteau, J., Jaupart, C., Jie, Shen Xian, Wen-Huaà, Kang, De-Luà, Lee, Jia-Chià, Bai, Hung-Pinà, Wei, and Hsia-Yeu, Deng. 1984. High heat flow in southern Tibet. **307**, 32–36.

Funfschilling, D., Brown, E., Nikolaenko, A., and Ahlers, G. 2005. Heat transport by turbulent Rayleigh-Benard convection in cylindrical cells with aspect ratio one and larger. *J. Fluid Mech.*, **536**, 145–154.

Galsa, A., and Lenkey, L. 2007. Quantitative investigation of physical properties of mantle plumes in three-dimensional numerical models. *Phys. Fluids*, **19**, 116601.

Gass, I. G., Thorpe, R. S., Pollack, H. N., and Chapman, D. S. 1978. Geological and geophysical parameters for mid-plate volcanism. *Phil. Trans. Roy. Soc. (London), Ser. A*, **301**, 581–597.

Gaudemer, Y., Jaupart, C., and Tapponier, P. 1988. Thermal control on post-orogenic extension in collision belts. *Earth Planet. Sci. Lett.*, **89**, 48–62.

Gaudio, P. Del, Mollo, S., Ventuyra, G., Iezzi, G., Taddeucci, J., and Cavallo, A. 2010. Cooling rate-induced differentiation in anhydrous and hydrous basalts at 500 MPa: Implications for the storage and transport of magmas in dikes. *Chem. Geol.*, **270**, 164–178.

Gibb, F. G. F. 1974. Supercooling and crystallization of plagioclase from a basaltic magma. *Min. Mag.*, **39**, 641–653.

Gibert, B., Schilling, F. R., Tommasi, A., and Mainprice, D. 2003. Thermal diffusivity of olivine single-crystals and polycrystalline aggregates at ambient conditions: a comparison. *Geophys. Res. Lett.*, **30**, 2172.

Gibert, B., Schilling, F. R., Gratz, K., and Tommasi, A. 2005. Thermal diffusivity of olivine single crystals and a dunite at high temperature: Evidence for heat transfer by radiation in the upper mantle. *Phys. Earth Planet. Inter.*, **151**, 129–141.

Giordano, D., Russell, J. K., and Dingwell, D. B. 2008. Viscosity of magmatic liquids: A model. *Earth Planet. Sci. Lett.*, **271**, 123–134.

Gliko, A. O., and Mareschal, J.-C. 1989. Non-linear asymptotic solution to Stefan-like problems and the validity of the linear approximation. *Geophys. J. Int.*, **99**, 801–809.

Goes, S., Govers, R., and Vacher, P. 2000. Shallow mantle temperatures under Europe from *P* and *S* wave tomography. *J. Geophys. Res.*, **105**, 11153–11169.

Goldstein, R. J., Chiang, H. D., and See, D. L. 1990. High Rayleigh number convection in a horizontal enclosure. *J. Fluid Mech.*, **213**, 111–126.

Gordon, R. G. 1998. The plate tectonic approximation: Plate nonrigidity, diffuse plate boundaries, and global plate reconstructions. *Ann. Rev. Earth Planet. Sci.*, **26**, 615–642.

Gosselin, C., and Mareschal, J.-C. 2003. Recent warming in northwestern Ontario inferred from borehole temperature profiles. *J. Geophys. Res.*, **108**, B02542.

Gradshteyn, I. S., and Ryzhik, I. M. 2000. *Table of Integrals, Series, and Products. 6th ed.* Academic Press. San Diego, CA.

Grand, S. P. 1997. Global seismic tomography: a snapshot of convection in the Earth. *GSA Today*, **7**, 1–7.

Green, D. H. 1982. Anatexis of mafic crust and high pressure crystallization of andesite. Pages 241–258 of: Thorpe, R. S. (ed), *Andesites: Orogenic Andesites and Related Rocks*. New York: J. Wiley.

Griffin, W. L., O'Reilly, S. Y., Abe, N., Aulbach, S., Davies, R. M., Pearson, N. J., Doyle, B. J., and Kivi, K. 2003. The origin and evolution of Archean lithospheric mantle. *Precam. Res.*, **127**, 19–41.

Griffiths, R. W. 1986. Thermal in extremely viscous fluids, including the effects of temperature-dependent viscosity. *J. Fluid Mech.*, **166**, 115–138.

Grigné, C., Labrosse, S., and Tackley, P. 2005. Convective heat transfer as a function of wavelength: Implications for the cooling of the Earth. *J. Geophys. Res.*, **110**, B03409.

Grigné, C., Labrosse, S., and Tackley, P. J. 2007. Convection under a lid of finite conductivity: Heat flux scaling and application to continents. *J. Geophys. Res.*, **112**, B08402.

Grossmann, S., and Lohse, D. 2000. Scaling in thermal convection: a unifying theory. *J. Fluid Mech.*, **407**, 27–56.

Grossmann, S., and Lohse, D. 2001. Thermal convection for large Prandtl numbers. *Phys. Rev. Lett.*, **86**, 3316–3319.

Grove, T., and Parman, S. 2004. Thermal evolution of the Earth as recorded by komatiites. *Earth Planet. Sci. Lett.*, **219**, 173–187.

Guillou, L., Mareschal, J.-C., Jaupart, C., Gariépy, C., Bienfait, G., and Lapointe, R. 1994. Heat flow and gravity structure of the Abitibi belt, Superior Province, Canada. *Earth Planet. Sci. Lett.*, **122**, 447–460.

Guillou, L., and Jaupart, C. 1995. On the effect of continents on mantle convection. *J. Geophys. Res.*, **100**, 24217–24238.

Guillou-Frottier, L., Mareschal, J.-C., Jaupart, C., Gariépy, C., Lapointe, R., and Bienfait, G. 1995. Heat flow variations in the Grenville Province, Canada. *Earth Planet. Sci. Lett.*, **136**, 447–460.

Guillou-Frottier, L., Mareschal, J.-C., and Musset, J. 1998. Ground surface temperature history in central Canada inferred from 10 selected borehole temperature profiles. *J. Geophys. Res.*, **103**, 7385–7398.

Gung, Y., Panning, M., and Romanowiz, B. 2003. Global anisotropy and the thickness af the continents. *Nature*, **422**, 707–710.

Gupta, M. L., Sharma, S. R., and Sundar, A. 1991. Heat flow and heat generation in the Archean Dharwar craton and implication for the southern Indian Shield geotherm. *Tectonophys.*, **194**, 107–122.

Hales, A. L. 1935. Convection currents in the Earth. *Mon. Notices Roy. astr. Soc. Geophys. Supp.*, **3**, 372.

Halliday, A. N., Mahood, G. A., Holden, P., Metz, J. M., Dempster, T. J., and Davidson, J. P. 1989. Evidence for long residence times of rhyolitic magma in the Long Valley magmatic 1245 system: the isotopic record in precaldera lavas of Glass Mountain. *Earth Planet. Sci. Lett.*, **94**, 274–290.

Hamdani, Y., Mareschal, J.-C., and Arkani-Hamed, J. 1991. Phase change and thermal subsidence in intracontinental sedimentary basins. *Geophys. J. Int.*, **106**, 657–665.

Hamdani, Y., Mareschal, J.-C., and Arkani-Hamed, J. 1994. Phase change and thermal subsidence of the Williston basin. *Geophys. J. Int.*, **116**, 585–597.

Hamoudi, M., Cohen, Y., and Achache, J. 1998. Can the thermal thickness of the continental lithosphere be estimated from Magsat data. *Tectonophys.*, **284**, 19–29.

Hansen, J., and Lebedeff, S. 1987. Global trends of measured surface air temperature. *J. Geophys. Res.*, **92**, 13345–13372.

Hansen, U., Yuen, D. A., and Malevsky, A. V. 1992. Comparison of steady-state and strongly chaotic thermal convection at high Rayleigh number. *Phys. Rev. A*, **46**, 4742–4754.

Harris, R. N., and Chapman, D. S. 2004. Deep-seated oceanic heat flow, heat deficits, and hydrothermal circulation. Pages 311–336 of: Elderfield, H. and Davis, E. E. (eds), *Hydrology of the Oceanic Lithosphere*. Cambridge University Press.

Harris, R. N., Garven, G., Georgen, J., McNutt, M. K., Christiansen, L., and Von Herzen, R. P. 2000. Submarine hydrogeology of the Hawaiian archipelagic apron 2. Numerical simulations of coupled heat transport and fluid flow. *J. Geophys. Res.*, **105**, 21353–21370.

Hart, S. R., and Zindler, A. 1986. In search of a bulk Earth composition. *Chem. Geol.*, **57**, 247–267.

Hashin, Z., and Sthrikhman, S. 1962. A variational approach to the theory of effective thermal conductivity of multiphase materials. *J. App. Phys.*, **33**, 3125–3133.

Haskell, N. A. 1935. The motion of a viscous fluid under a surface load. *J. App. Phys.*, **6**, 265–269.

Haskell, N. A. 1936. The motion of a viscous fluid under a surface load. Part II. *J. App. Phys.*, **7**, 56–61.

Hasterok, D., and Chapman, D. S. 2007a. Continental thermal isostasy: 1. Methods and sensitivity. *J. Geophys. Res.*, **112**, B06414.

Hasterok, D., and Chapman, D. S. 2007b. Continental thermal isostasy: 2. Application to North America. *J. Geophys. Res.*, **112**, B06415.

Haxby, W. F., and Turcotte, D. L. 1978. On isostatic geoid anomalies. *J. Geophys. Res.*, **83**, 5473–5478.

Haxby, W. F., Turcotte, D. L., and Bird, J. M. 1976. Thermal and mechanical evolution of the Michigan basin. *Tectonophys.*, **36**, 57–75.

He, L., Hu, S., Huang, S., Yang, W., Wang, J., Yuan, Y., and Yang, S. 2008. Heat flow study at the Chinese Continental Scientific Drilling site: Borehole temperature, thermal conductivity, and radiogenic heat production. *J. Geophys. Res.*, **113**, B02404.

Heier, K. S., and Lambert, I. B. 1978. *A compilation of potassium, uranium and thorium abundances and heat production of Australian rocks*. Research School of Earth Science, Australian National University, Canberra.

Helmstaedt, H. H., and Schulze, D. J. 1989. Southern African kimberlites and their mantle sample: implications for Archean tectonics and lithosphere evolution. Pages 358–368 of: Ross, J. (ed), *Kimberlites and Related Rocks: Volume 1. Their Composition, Occurrence, Origin, and Emplacement. Geol. Soc. Australia Spec. Publ.*, vol. 14. Geol. Soc. Australia.

Helz, R. T. 1976. Differentiation behavior of Kilauea volcano, Hawaii: an overview of past and current work. Pages 241–258 of: Mysen, B. O. (ed), *Magmatic Processes: Physicochemical Principles, Special Publ. 1*. Geochem. Soc.

Helz, R. T. 1980. Crystallization history of Kilauea Iki lava lake as seen in drill core recovered in 1967 1979. *Bull. Volc.*, **43**, 675–701.

Helz, R. T., and Thornber, C. R. 1987. Geothermometry of Kilauea Iki lava lake, Hawaii. *Bull. Volc.*, **49**, 651–668.

Henry, S. G., and Pollack, H. N. 1988. Terrestrial heat flow above the Andean subduction zone in Bolivia and Peru. *J. Geophys. Res.*, **93**, 15153–15162.

Hernlund, J. W., Thomas, C., and Tackley, P. J. 2005. Phase Boundary Double Crossing and the Structure of Earth's Deep Mantle. *Nature*, **434**, 882–886.

Herzberg, C., and Zhang, J. 1996. Melting experiments on anhydrous peridotite KLB-1: Compositions of magmas in the upper mantle and transition zone. *J. Geophys. Res.*, **101**, 8271–8296.

Hildreth, W., and Lanphere, M. A. 1994. Potassium-argon geochronology of a basalt-andesite-dacite arc system: The Mount Adams volcanic field, Cascade Range of southern Washington. *Geol. Soc. Am. Bull.*, **106**, 1,413–1,429.

Hildreth, W., and Wilson, C. J. N. 2007. Compositional zoning of the Bishop tuff. *J. Petrol.*, **48**, 951–999.

Hills, R. N., Loper, D. E., and Roberts, P. H. 1983. A thermodynamically consistent model of a slushy zone. *Quart. J. Mech. Appl. Math.*, **36**, 505–539.

Hirose, K. 2006. Postperovskite phase transition and its geophysical implications. *Rev. Geophys.*, **44**, RG3001.

Hirose, K., Sinmyo, R., Sata, N., and Ohishi, Y. 2006. Determination of post-perovskite phase transition boundary in $MgSiO_3$ using Au and MgO pressure standards. *Geophys. Res. Lett.*, **33**, L01310.

Hirth, G., Evans, R. L., and Chave, A. D. 2000. Comparison of continental and oceanic mantle electrical conductivity: Is the Archean lithosphere dry? *Geochem. Geophys. Geosyst.*, **1**, 1030.

Hoffman, P. F. 1997. Tectonic Geology of North America. Pages 459–464 of: van der Pluijm, BA, and Marshak, S. (eds), *Earth Structure: An Introduction to Structural Geology and Tectonics*. New York: McGraw Hill.

Holmes, A. 1915a. Radioactivity and the Earth's thermal history: Part 1. The concentration of radioactive elements in the Earth's crust. *Geol. Mag.*, **2**, 60–71.

Holmes, A. 1915b. Radioactivity and the Earth's thermal history: Part 2. Radioactivity and the Earth as a cooling body. *Geol. Mag.*, **2**, 102–112.

Holmes, A. 1931. Radioactivity and Earth's movements. *Proc. Geol. Soc. Glasgow*, **18**, 559–606.

Holness, M. B., and Humphreys, M. C. S. 2003. The Traigh Bhan Na Sgurra sill, Isle of Mull: 1251 flow localization in a major magma conduit. *J. Petrol.*, **44**, 1961–1976.

Horai, K., and Simmons, G. 1969. Thermal conductivity of rock forming minerals. *Earth Planet. Sci. Lett.*, **6**, 359–368.

Horton, C. W., and Rogers, Jr., F. T. 1945. Convection Currents in a Porous Medium. *J. App. Phys.*, **16**, 367–370.

Hotchkiss, W. O., and Ingersoll, L. R. 1934. Post-glacial time calculations from recent measurements in the Calumet Copper Mine. *J. Geol.*, **42**, 113–142.

Howard, L. N. 1966. Convection at high Rayleigh numbers. Pages 1109–1115 of: Görtler, H. (ed), *Proc. 11th Intl Congress of Applied Mech.* Springer.

Hu, S., He, L., and Wang, J. 2000. Heat flow in the continental area of China: a new data set. *Earth Planet. Sci. Lett.*, **179**, 407–419.

Huang, S., Pollack, H. N., and Shen, P. Y. 2000. Temperature trends over the past five centuries reconstructed from borehole temperatures. *Nature*, **403**, 756–758.

Huerta, A. D., Royden, L. H., and Hodges, K. V. 1998. The thermal structure of collisional orogens as a response to accretion, erosion, and radiogenic heating. *J. Geophys. Res.*, **103**, 15287–15302.

Hulot, G., Eymin, C., Langlais, B., Mandea, M., and Olsen, N. 2002. Small-scale structure of the geodynamo inferred from Oersted and Magsat satellite data. *Nature*, **416**, 620–623.

Humler, E., Langmuir, C. H., and Daux, V. 1999. Depth versus age: New perspectives from the chemical compositions of ancient crust. *Earth Planet. Sci. Lett.*, **173**, 7–23.

Huppert, H. E., and Worster, M. G. 1985. Dynamic solidification of a binary melt. *Nature*, **314**, 703–707.

Hurter, S. J., and Pollack, H. N. 1996. Terrestrial heat flow in the Paraná Basin, southern Brazil. *J. Geophys. Res.*, **101**, 8659–8672.

Hutko, A. R., Lay, T., Garnero, E. J., and Revenaugh, J. 2006. Seismic detection of folded, subducted lithosphere at the core-mantle boundary. *Nature* **441**, 333–336.

Jackson, D. D. 1972. Interpretation of inaccurate, insufficient, and inconsistent data. *Geophys. J. R. Astron. Soc.*, **28**, 97–110.

Jacobs, J., and Allan, D. W. 1956. The thermal history of the Earth. *Nature*, **177**, 155–157.

Jaeger, J. C., and Sass, J. H. 1963. Lees's topographic correction in heat flow and the geothermal flux in Tasmania. *Geofisica Pura e Applicata*, **54**, 53–63.

James, D. E., Fouch, M. J., VanDecar, J. C., and van der Lee, S. 2001. Tectospheric structure beneath southern Africa. *Geophys. Res. Lett.*, **28**, 2485–2488.

Jamieson, R. A., Beaumont, C., Medvedev, S., and Nguyen, M. H. 2004. Crustal channel flows: 2. Numerical models with implications for metamorphism in the Himalayan-Tibetan orogen. *J. Geophys. Res.*, **109**, B06407.

Jarvis, G. T., and McKenzie, D. P. 1980. Convection in a compressible fluid with infinite Prandtl number. *J. Fluid Mech.*, **96**, 515–583.

Jaupart, C. 1983a. The effects of alteration and the interpretation of heat flow and radioactivity data - a reply to R.U.M. Rao. *Earth Planet. Sci. Lett.*, **62**, 430–438.

Jaupart, C. 1983b. Horizontal heat transfer due to radioactivity contrasts: Causes and consequences of the linear heat flow-heat production relationship. *Geophys. J. R. Astron. Soc.*, **75**, 411–435.

Jaupart, C., and Brandeis, G. 1986. The stagnant bottom layer of convecting magma chambers. *Earth Planet. Sci. Lett.*, **80**, 183–199.

Jaupart, C., and Mareschal, J.-C. 1999. The thermal structure and thickness of continental roots. *Lithos*, **48**, 93–114.

Jaupart, C., and Mareschal, J. C. 2003. Constraints on crustal heat production from heat flow data. Pages 65–84 of: Rudnick, R. L. (ed), *Treatise on Geochemistry, The Crust*, vol. 3. New York: Permagon.

Jaupart, C., and Tait, S. R. 1995. Dynamics of differentiation in magma reservoirs. *J. Geophys. Res.*, **100**, 17615–17636.

Jaupart, C., Francheteau, J., and Shen, X. J. 1985. On the thermal structure of the southern Tibetan crust. *Geophys. J. R. Astr. Soc.*, **81**, 131–155.

Jaupart, C., Mareschal, J. C., Guillou-Frottier, L., and Davaille, A. 1998. Heat flow and thickness of the lithosphere in the Canadian Shield. *J. Geophys. Res., 103*, **103**, 15269–15286.

Jaupart, C., Labrosse, S., and Mareschal, J. C. 2007. Temperatures, heat and energy in the mantle of the Earth. Pages 253–303 of: Bercovici, D. (ed), *Treatise on Geophysics, The Mantle*, vol. 7. New York: Elsevier.

Javoy, M. 1999. Chemical Earth models. *C. R. Acad. Sci. Paris*, **329**, 537–555.

Jeffreys, H. 1921. On certain geological effects of the cooling of the Earth. *Proc. R. Soc. London, ser A*, **100**, 122–149.

Jeffreys, H. 1938. The disturbance of the temperature gradient in the Earth's crust by inequalities of height. *Mon. Not. R. Astron. Soc., Geophys. Suppl.*, **4**, 309–312.

Jeffreys, H. 1942. On the radioactivities of rocks. *Mon. Notices Roy. astr. Soc. Geophys. Supp.*, **5**, 37–40.

Jeffreys, H. 1959. *The Earth. Its Origin, History and Physical Constitution*. Fourth edn. Cambridge: Cambridge University Press.

Jessop, A. M. 1971. The distribution of glacial perturbation of heat flow in Canada. *Can. J. Earth Sci.*, **8**, 162–166.

Joeleht, T. H., and Kukkonen, I. T. 1998. Thermal properties of granulite facies rocks in the Precambrian basement of Finland and Estonia. *Tectonophys.*, **291**, 195–203.

Johnson, H. P., and Carlson, R. L. 1992. Variation of sea floor depth with age: A test of models based on drilling results. *Geophys. Res. Lett.*, **19**, 1971–1974.

Joly, J. 1909. *Radioactivity and Geology: An Account of the Influence of Radioactive Energy on the Earth's History*. London, U.K.: Archibald Constable and Co. (London, U.K.).

Jones, A. G., Evans, R. L., and Eaton, D. W. 2009. Velocity-conductivity relationships for mantle mineral assemblages in Archean cratonic lithosphere based on a review of laboratory data and Hashin-Shtrikman extremal bounds. *Lithos*, **109**, 131–143.

Jones, M. Q. W. 1987. Heat flow and heat production in the Namaqua mobile belt, South Africa. *J. Geophys. Res.*, **92**, 6273–6289.

Jones, M. Q. W. 1988. Heat flow in the Witwatersrand Basin and environs and its significance for the South African Shield geotherm and lithosphere thickness. *J. Geophys. Res.*, **93**, 3243–3260.

Jones, P. D., Wigley, T. M. L., and Wright, P. B. 1986. Global variations between 1861 and 1984. *Nature*, **322**, 430–434.

Jordan, T. H. 1981. Continents as a chemical boundary layer. *Phil. Trans. R. Soc. Lond. A*, **301**, 359–373.

Jurine, D., Jaupart, C., Brandeis, G., and Tackley, P. J. 2005. Penetration of mantle plumes through depleted lithosphere. *J. Geophys. Res.*, **110**, B10104.

Kaminski, E. 1997. *Deux aspects de la dynamique des fluides geologiques*. Unpub. Ph.D. thesis Université Paris Diderot.

Kaminski, E., and Jaupart, C. 2000. Lithospheric structure beneath the Phanerozoic intracratonic basins of North America. *Earth Planet. Sci. Lett.*, **178**, 139–149.

Kaminski, E., and Jaupart, C. 2003. Laminar starting plumes in high-Prandtl-number fluids. *J. Fluid Mech.*, **478**, 287–298.

Karner, G. D., Steckler, M. S., and Thorne, J. 1983. Long-term mechanical properties of the continental lithosphere. *Nature*, **304**, 250–253.

Katsaros, K. B., Liu, W. T., Businger, J. A., and Tillman, J. E. 1977. Heat transport and thermal structure in the interfacial layer in an open tank in turbulent free convection. *J. Fluid Mech.*, **83**, 311–335.

Katsura, T., Yamada, H., Shinmei, T., Kubo, A., Ono, S., Kanzaki, M., Yoneda, A., Walter, M. J., Ito, E., and Urakawa, S. 2003. Post-spinel transition in Mg_2SiO_4 determined by in-situ X-ray diffractometry. *Phys. Earth Planet. Inter.*, **136**, 11–24.

Katsura, T., Yamada, H., Nishikawa, O., Song, M., Kubo, A., Shinmei, T., Yokoshi, S., Aizawa, Y., Yoshino, T., Walter, M. J., and Ito, E. 2004. Olivine-Wadsleyite transition in the system $(Mg,Fe)Si0_4$. *J. Geophys. Res.*, **109**, B02209.

Kennett, B. L. N., and Engdahl, E. R. 1991. Traveltimes for global earthquake location and phase identification. *Geophys. J. Int.*, **105**, 429–465.

Kerr, R. C., Woods, A. W., Worster, M. G., and Huppert, H. E. 1990a. Solidification of an alloy cooled from above Part 1. Equilibrium growth. *J. Fluid Mech.*, **216**, 323–342.

Kerr, R. C., Woods, A. W., Worster, M. G., and Huppert, H. E. 1990b. Solidification of an alloy cooled from above Part 2. Non-equilibrium interfacial kinetics. *J. Fluid Mech.*, **217**, 331–348.

Ketcham, R. A. 1996. Distribution of heat-producing elements in the upper and middle crust of southern and west central Arizona: Evidence from the core complexes. *J. Geophys. Res.*, **101**, 13,611–13,632.

Kimura, S., Schubert, G., and Straus, J. M. 1986. Route to chaos in porous-medium thermal convection. *J. Fluid Mech.*, **166**, 305–324.

Kinzler, R. J., and Grove, T. L. 1992. Primary Magmas of Mid-Ocean Ridge Basalts 2. Applications. *J. Geophys. Res.*, **97**, 6907–6926.

Kirkpatrick, R. J. 1977. Nucleation and growth of plagioclase, Makaopuhi lava lakes, Kilauea volcano, Hawaii. *Geol. Soc Am. Bull.*, **81**, 799–800.

Klein, E. M., and Langmuir, C. H. 1987. Global correlations of ocean ridge basalt chemistry with axial depth and crustal thickness. *J. Geophys. Res.*, **92**, 8089–8115.

Kleine, T., Münker, C., Mezger, K., and Palme, H. 2002. Rapid accretion and early core formation on asteroids and the terrestrial planets from Hf-W chronometry. *Nature*, **418**, 952–955.

Kohlstedt, D. L., and Goetze, C. 1974. Low-stress high-temperature creep in olivine single crystals. *J. Geophys. Res.*, **79**, 2045–2051.

Korenaga, J. 2003. Energetics of mantle convection and the fate of fossil heat. *Geophys. Res. Lett.*, **30**, 1437.

Korenaga, J., and Karato, S. 2008. A new analysis of experimental data on olivine rheology. *J. Geophys. Res.*, **113**, B02403.

Kremenentsky, A. A., Milanovsky, S. Y., and Ovchinnikov, L. N. 1989. A heat generation model for the continental crust based on deep drilling in the Baltic Shield. *Tectonophys.*, **159**, 231–246.

Kukkonen, I. T., and Peltonen, P. 1999. Xenolith-controlled geotherm for the central Fennoscandian Shield: implications for lithosphere-asthenosphere relations. *Tectonophys.*, **304**, 301–315.

Kukkonen, I. T., Golovanova, Y. V., Druzhinin, V. S., Kosarev, A. M., and Schapov, V. A. 1997. Low geothermal heat flow of the Urals fold belt: Implication of low heat production, fluid circulation or paleoclimate. *Tectonophys.*, **276**, 63–85.

Kukkonen, I. T., Jokinen, J., and Seipold, U. 1999. Temperature and pressure dependencies of thermal transport properties of rocks: Implications for uncertainties in thermal lithosphere models and new laboratory measurements of high-grade rocks in the central fennoscandian shield. *Surveys in Geophysics*, **20**, 33–59.

Kutas, R. I. 1984. Heat flow, radiogenic heat production, and crustal thickness in southwest USSR. *Tectonophys*, **103**, 167–174.

Labrosse, S. 2002. Hotspots, Mantle plumes and core heat loss. *Earth Planet. Sci. Lett.*, **199**, 147–156.

Labrosse, S. 2003. Thermal and magnetic evolution of the Earth's core. *Phys. Earth Planet. Inter.*, **140**, 127–143.

Labrosse, S., and Jaupart, C. 2007. Thermal evolution of the Earth: secular changes and fluctuations of plate characteristics. *Earth Planet. Sci. Lett.*, **260**, 465–481.

Lachenbruch, A. H. 1968. Rapid estimation of the topographic disturbance to superficial thermal gradients. *Rev. Geophys. Space Phys.*, **6**, 347–363.

Lachenbruch, A. H. 1970. Crustal temperature and heat production: implications of the linear heat flow heat production relationship. *J. Geophys. Res.*, **73**, 3292–3300.

Lachenbruch, A. H., and Marshall, B. V. 1986. Changing climate: Geothermal evidence from permafrost in the Alaskan Arctic. *Science*, **234**, 689–696.

Lachenbruch, A. H., and Morgan, P. 1990. Continental extension, magmatism and elevation; formal relations and rules of thumb. *Tectonophys.*, **174**, 39–62.

Lachenbruch, A. H., and Sass, J. H. 1978. Models of an extending lithosphere and heat flow in the Basin and Range Province. *Mem. Geol. Soc. Amer.*, **152**, 209–258.

Lachenbruch, A. H., Sass, J. H., and Morgan, P. 1994. Thermal regime of the southern Basin and Range Province: 2. Implications of heat flow for regional extension and metamorphic core complexes. *J. Geophys. Res.*, **99**, 22,121–22,133.

Lago, B., Cazenave, A., and Marty, J.-C. 1990. Regional variations in subsidence rate of lithospheric plates: implication for thermal cooling models. *Phys. Earth Planet. Inter.*, **61**, 253–259.

Lambeck, K. 1977. Tidal disspation in the oceans: Astronomical, geophysical, and oceanographic consequences. *Phil. Trans. R. Soc. London, ser. A*, **287**, 545–594.

Lanczos, C. 1961. *Linear Differential Operators*. Princeton, N.J.: D. Van Nostrand.

Landstrom, O., Larson, S. A., Lind, G., and Malmqvist, D. 1980. Geothermal investigations in the Bohus granite area in southwestern Sweden. *Tectonophys.*, **64**, 131–162.

Lane, A. C. 1923. Geotherms from the Lake Superior Copper country. *Bull. Geol. Soc. Am*, **34**, 703–720.

Langmuir, C. H., Goldstein, S. L., Donnelly, K., and Su, Y. K. 2005. Origins of Enriched and Depleted Mantle Reservoirs. Page 1 of: Suppl., Fall Meet. (ed), *Eos Trans.*, vol. 86. Abstract V23D-02: American Geophysical Union.

Langseth, M. G., Grim, P. J., and Ewing, M. 1965. Heat-Flow Measurements in the East Pacific Ocean. *J. Geophys. Res.*, **70**, 367–380.

Lapwood, E. R. 1948. Convection of a fluid in a porous medium. *Math. Proc. Cambridge Phil. Soc.*, **44**, 508–521.

Lapwood, E. R. 1952. The effect of contraction in the cooling of a gravitating sphere, with special reference to the Earth. *Mon. Not. Roy. Astron. Soc., Geophys. Suppl.*, **6**, 402–407.

Lay, T., Williams, Q., and Garnero, E. J. 1998. The core-mantle boundary layer and deep Earth dynamics. *Nature*, **392**, 461–468.

Le Pichon, X., and Sibuet, J.-C. 1981. Passive margins: A model of formation. *J. Geophys. Res.*, **86**, 3708–3720.

Lee, M. K., Brown, G. C., Webb, P. C., Wheildon, J., and Rollin, K. E. 1987. Heat flow, heat production and thermo-tectonic setting in mainland UK. *J. Geol. Soc.*, **144**, 35–42.

Lee, T.-C., and von Herzen, R. P. 1994. In-situ determination of thermal properties of sediments using a friction-heated probe source. *J. Geophys. Res.*, **99**, 12121–12132.

Lewis, T. J., Hyndman, R. D., and Fluck, P. 2003. Heat flow, heat generation and crustal temperatures in the northern Canadian Cordillera: Thermal control on tectonics. *J. Geophys. Res.*, **108**, 2316.

Li, C., van der Hilst, R. D., Engdahl, E. R., and Burdick, S. 2008. A new global model for P wave speed variations in Earth's mantle. *Geochemistry, Geophysics, Geosystems*, **9**, 5018.

Li, X., Kind, R., Yuan, X., Wölbern, and Hanka, W. 2004. Rejuvenation of the lithosphere by the Hawaiian plume. *Nature*, **427**, 827–829.

Lister, C. R. B. 1977. Estimators for heat flow and deep rock properties based on boundary layer theory. *Tectonophysics*, **41**, 157–171.

Lister, C. R. B. 1982. Geoid anomalies over cooling lithosphere: source for a third kernel of upper mantle thermal parameters and thus an inversion. *Geophys. J. Int.*, **68**, 219–240.

Lister, C. R. B. 1990. An explanation for the multivalued heat transport found experimentally for convection in a porous medium. *J. Fluid Mech.*, **214**, 287–320.

Lister, C. R. B., Sclater, J. G., Nagihara, S., Davis, E. E., and Villinger, H. 1990. Heat flow maintained in ocean basins of great age - Investigations in the north-equatorial West Pacific. *Geophys. J. Int.*, **102**, 603–630.

Lister, J. R. 2003. Expressions for the dissipation driven by convection in the Earth's core. *Phys. Earth Planet. Inter.*, **140**, 145–158.

Litasov, K., and Ohtani, E. 2002. Phase relations and melt compositions in CMAS-pyrolite-H_2O system up to 25 GPa. *Phys. Earth Planet. Inter.*, **134**, 105–127.

Lovett, D. R. 1999. *Tensor Properties of Crystal.* 2nd edn. CRC Press.

Lowry, A. R., and Smith, R. B. 1995. Strength and rheology of the western US Cordillera. *J. Geophys. Res.*, **100**, 17,947–17–963.

Lyubetskaya, T., and Korenaga, J. 2007. Chemical composition of Earth's primitive mantle and its variance, 1, Method and results. *J. Geophys. Res.*, **112**, B03211.

MacDonald, G. J. F. 1959. Calculations on the thermal history of the Earth. *J. Geophys. Res.*, **64**, 1967–2000.

MacDonald, G. J. F. 1964. Dependence of the surface heat flow on the radioactivity of the Earth. *J. Geophys. Res.*, **69**, 2933–2946.

Maggi, A., Jackson, J. A., McKenzie, D. P., and Priestley, K. 2000. Earthquake focal depths, effective elastic thickness, and the strength of the continental lithosphere. *Geology*, **28**, 495–498.

Mangan, M. 1990. Crystal size distribution systematics and the determination of magma storage times: the 1959 eruption of Kilauea volcano, Hawaii. *J. Volcanol. Geotherm. Res.*, **44**, 295–302.

Mareschal, J. C. 1983. Mechanisms of uplift preceding rifting. *Tectonophys.*, **94**, 51–66.

Mareschal, J.-C. 1983. Uplift and heat flow following the injection of magmas into the lithosphere. *Geophys. J. Int.*, **73**, 109–127.

Mareschal, J.-C., and Jaupart, C. 2004. Variations of surface heat flow and lithospheric thermal structure beneath the North American craton. *Earth Planet. Sci. Lett.*, **223**, 65–77.

Mareschal, J.-C., and Jaupart, C. 2005. Archean thermal regime and stabilization of the cratons. Pages 61–73 of: Benn, K., Condie, K., and Mareschal, J. C. (eds), *Archean Geodynamic Processes*. Washington (DC): AGU.

Mareschal, J.-C., Jaupart, C., Cheng, L. Z., Rolandone, F., Gariépy, C., Bienfait, G., Guillou-Frottier, L., and Lapointe, R. 1999. Heat flow in the Trans-Hudson Orogen of the Canadian Shield: implications for Proterozoic continental growth. *J. Geophys. Res.*, **104**, 29007–29024.

Mareschal, J.-C., Jaupart, C., Gariépy, C., Cheng, L. Z., Guillou-Frottier, L., Bienfait, G., and Lapointe, R. 2000. Heat flow and deep thermal structure near the southeastern edge of the Canadian Shield. *Can. J. Earth Sci.*, **37**, 399–414.

Mareschal, J.-C., Nyblade, A., Perry, H. K. C., Jaupart, C., and Bienfait, G. 2004. Heat flow and deep lithospheric thermal structure at Lac de Gras, Slave Province, Canada. *Geophys. Res. Lett.*, **31**, L12611.

Marshall, S. J. 2005. Recent advances in understanding ice sheet dynamics. *Earth and Planetary Science Letters*, **240**, 191–204.

Marton, F. C., Shankland, T. J., Rubie, D. C., and Xu, Y. 2005. Effects of variable thermal conductivity on the mineralogy of subducting slabs and implications for mechanisms of deep earthquakes. *Phys. Earth Planet. Inter.*, **149**, 53–64.

Marty, J. C., and Cazenave, A. 1989. Regional variations in subsidence rate of oceanic plates: a global analysis. *Earth Planet. Sci. Lett.*, **94**, 301–315.

Maus, S., Rother, M., Hemant, K., Stolle, C., Lühr, H., Kuvshinov, A., and Olsen, N. 2006. Earth's lithospheric magnetic field determined to spherical harmonic degree 90 from CHAMP satellite measurements. *Geophys. J. Int.*, **164**, 319–330.

Mayhew, M. A. 1982. Application of satellite magnetic anomaly to Curie isotherm mapping. *J. Geophys. Res.*, **87**, 4846–4854.

McConnell, Jr., R. K. 1965. Isostatic Adjustment in a Layered Earth. *J. Geophys. Res.*, **70**, 5171–5188.

McDonough, W. F., and Rudnick, R. L. 1998. Mineralogy and composition of the upper mantle. Pages 139–164 of: Hemley, R. J. (ed), *Ultrahigh-Pressure Mineralogy: Physics and Chemistry of the Earth's Deep Interior*. Washington DC: Mineralogical Society of America.

McDonough, W. F., and Sun, S. S. 1995. The composition of the Earth. *Chem. Geol.*, **120**, 223–253.

McKenna, T. E., and Sharp, J. M. 1998. Radiogenic heat production in sedimentary rocks of the Gulf of Mexico basin, South Texas. *AAPG Bull.*, **82**, 484–496.

McKenzie, D. 1967. Some remarks on heat flow and gravity anomalies. *J. Geophys. Res.*, **72**, 6261–6273.

McKenzie, D. 1978. Some remarks on the development of sedimentary basins. *Earth Planet. Sci. Lett.*, **40**, 25–32.

McKenzie, D., and Bickle, M. J. 1988. The volume and composition of melt generated by extension of the lithosphere. *J. Petrol.*, **29**, 625–679.

McKenzie, D., and Jarvis, G. 1980. The conversion of heat into mechanical work by mantle convection. *J. Geophys. Res.*, **85**, 6093–6096.

McKenzie, D., Jackson, J., and Priestley, K. 2005. Thermal structure of oceanic and continental lithosphere. *Earth Planet. Sci. Lett.*, **233**, 337–349.

Melosh, H. J., and Ivanov, B. A. 1999. Impact crater collapse. *Ann. Rev. Earth Planet. Sci.*, **27**, 385–415.

Menke, W. 1989. *Geophysical Data Analysis: Discrete Inverse Theory*. International Geophysical Series, no. 45. San Diego: Academic Press.

Michaut, C., and Jaupart, C. 2004. Nonequilibrium temperatures and cooling rates in thick continental lithosphere. *Geophys. Res. Lett.*, **31**, L24602.

Michaut, C., Jaupart, C., and Bell, D. R. 2007. Transient geotherms in Archean continental lithosphere: New constraints on thickness and heat production of the subcontinental lithospheric mantle. *J. Geophys. Res.*, **112**, B04408.

Misener, A. D., and Beck, A. E. 1960. The measurement of heat flow over land. Pages 11–61 of: Runcorn, S. K. (ed), *Methods and Techniques in Geophysics*. New-York: Wiley-Interscience.

Mitrovica, J. X., and Forte, A. M. 1997. Radial profile of mantle viscosity: Results from the joint inversion of convection and postglacial rebound observables. *J. Geophys. Res.*, **102**, 2751–2770.

Mittelstaedt, Eric, and Tackley, Paul J. 2006. Plume heat flow is much lower than CMB heat flow. *Earth Planet. Sci. Lett.*, **241**, 202–210.

Mojzsis, S. J., Harrison, T. M., and Pidgeon, R. T. 2001. Oxygen-isotope evidence from ancient zircons for liquid water at the Earth's surface 4,300 Myr ago. *Nature*, **409**, 178–181.

Mooney, W. D., Laske, G., and Guy Masters, T. 1998. CRUST 5.1: A global crustal model at $5° \times 5°$. *J. Geophys. Res.*, **103**, 727–748.

Moore, W. B. 2008. Heat transport in a convecting layer heated from within and below. *J. Geophys. Res.*, **113**, B11407.

Moore, W. B., Schubert, G., and Tackley, P. J. 1999. The role of rheology in lithospheric thinning by mantle plumes. *Geophys. Res. Lett.*, **26**, 1073–1076.

Morgan, P. 1983. Constraints on rift thermal processes from heat flow and uplift. *Tectonophys.*, **94**, 277–298.

Morgan, P. 1985. Crustal radiogenic heat production and the selective survival of ancient continental crust. *J. Geophys. Res.*, **90**, C561–C570.

Morse, P. M., and Feshbach, H. 1953. *Methods of Theoretical Physics, Part I*. McGraw-Hill.

Morse, S. A. 1980. *Basalts and Phase Diagrams*. New York: Springer-Verlag.

Moses, E., Zocchi, G., and Libchaber, A. 1993. An experimental study of laminar plumes. *J. Fluid Mech.*, **251**, 581–601.

Müller, R. D., Sdrolias, M., Gaina, C., and Roest, W. R. 2008. Age, spreading rates, and spreading asymmetry of the world's ocean crust. *Geochemistry, Geophysics, Geosystems*, **9**, 4006.

Murakami, M., Hirose, K., Kawamura, K., Sata, N., and Ohishi, Y. 2004. Post-Perovskite Phase Transition in $MgSiO_3$. *Science*, **304**, 855–858.

Newsom, Horton E., and Jones, John H. (eds). 1990. *Origin of the Earth*. Oxford University Press.

Nicolaysen, L. O., Hart, R. J., and Gale, N. H. 1981. The Vredefort radioelement profile extended to supracrustal strata at Carletonville, with implications for continental heat flow. *J. Geophys. Res.*, **86**, 10653–10662.

Nielsen, S. B. 1987. Steady state heat flow in a random medium and the linear heat flow-heat production relationship. *Geophys. Res. Lett.*, **14**, 318–322.

Nielsen, S. B., and Beck, A. E. 1989. Heat flow density values and paleoclimate determined from stochastic inversion of four temperature-depth profiles from the Superior Province of the Canadian Shield. *Tectonophys.*, **164**, 345–359.

Nisbet, E. G., Cheadle, M. J., Arndt, N. T., and Bickle, M. J. 1995. Constraining the potential temperature of the Archean mantle: a review of the evidence from komatiites. *Lithos*, **30**, 291–307.

Nunn, Jeffrey A., and Sleep, Norman H. 1984. Thermal contraction and flexure of intracratonic basins: a three dimensional study of the Michigan Basin. *Geophys. J. R. Astr. Soc.*, **79**, 587–635.

Nyblade, A. A. 1997. Heat flow across the East African Plateau. *Geophys. Res. Lett.*, **24**, 2083–2086.

Nye, J. F. 1985. *Physical Properties of Crystals: Their Representations by Tensors and Matrices*. Oxford University Press (Oxford, U.K.).

Oganov, A. R., and Ono, S. 2004. Theoretical and Experimental Evidence for a Post-Perovskite phase of $MgSiO_3$ in Earth's D'' Layer. *Nature*, **430**, 445–448.

Oldenburg, D. W. 1975. A Physical Model for the Creation of the Lithosphere. *Geophys. J. R. Astr. Soc.*, **43**, 425–451.

Palme, H., and O'Neill, H. St. C. 2003. Cosmochemical estimates of mantle composition. In: Carlson, R. W. (ed), *Treatise on Geochemistry, vol. 2, Mantle and Core*. Elsevier.

Parker, R. L. 1994. *Geophysical Inverse Theory*. Princeton, New Jersey: Princeton University Press.

Parmentier, E. M., and Sotin, C. 2000. Three-dimensional numerical experiments on thermal convection in a very viscous fluid: Implications for the dynamics of a thermal boundary layer at high Rayleigh number. *Phys. Fluids*, **12**, 609–617.

Parmentier, E. M., Sotin, C., and Travis, B. J. 1994. Turbulent 3D thermal convection in an infinite Prandtl number, volumetrically heated fluid: implications for mantle dynamics. *Geophys. J. Int.*, **116**, 241–251.

Parsons, B. 1982. Causes and consequences of the relation between area and age of the ocean floor. *J. Geophys. Res.*, **87**, 289–302.

Parsons, B., and Sclater, J. G. 1977. An analysis of the variation of ocean floor bathymetry and heat flow with age. *J. Geophys. Res.*, **82**, 803–827.

Patterson, C. 1956. Age of meteorites and the earth. *Geochim. Cosmochim. Acta*, **10**, 230–237.

Patterson, C., Tilton, G., and Inghram, M. 1955. Age of the Earth. *Science*, **121**, 69–75.

Pekeris, C. L. 1935. Thermal convection in the interior of the earth. *Mon. Notices Roy. astr. Soc. Geophys. Supp.*, **3**, 346–367.

Peltier, W. R. 1974. The impulse response of a Maxwell earth. *Rev. Geophys. Space Phys.*, **12**, 649–669.

Peltier, W. R. 1998. Postglacial variations in the level of the sea: Implications for climate dynamics and solid-Earth geophysics. *Rev. Geophys.*, **36**, 603–689.

Pérez-Gussinyé, M., Metois, M., Fernández, M., Vergés, J., Fullea, J., and Lowry, A. R. 2009. Effective elastic thickness of Africa and its relationship to other proxies for lithospheric structure and surface tectonics. *Earth Planet. Sci. Lett.*, **287**, 152–167.

Perry, C., Mareschal, J. C., and Jaupart, C. 2008. Enhanced crustal geo-neutrino production near the Sudbury neutrino observatory, Ontario, Canada. *Earth Planet. Sci. Lett.*, **288**, 301–308.

Perry, H. K. C., Jaupart, C., Mareschal, J. C., Rolandone, F., and Bienfait, G. 2004. Heat flow in the Nipigon arm of the Keweenawan Rift, northwestern Ontario, Canada. *Geophys. Res. Lett.*, **31**, L15607.

Perry, H. K. C., Jaupart, C., Mareschal, J.-C., and Bienfait, G. 2006. Crustal heat production in the Superior Province, Canadian Shield, and in North America inferred from Heat flow data. *J. Geophys. Res.*, **111**, B04401.

Perry, J. 1895a. On the age of the Earth. *Nature*, **51**, 224–227.

Perry, J. 1895b. On the age of the Earth. *Nature*, **51**, 341–342.

Perry, J. 1895c. On the age of the Earth. *Nature*, **51**, 582–585.

Petterson, H. 1949. Exploring the bed of the ocean. *Nature*, **164**, 468–470.

Pilidou, S., Priestley, K., Debayle, E., and Gudmundsson, Ó. 2005. Rayleigh wave tomography in the North Atlantic: high resolution images of the Iceland, Azores and Eifel mantle plumes. *Lithos*, **79**, 453–474.

Pinet, C., and Jaupart, C. 1987. The vertical distribution of radiogenic heat production in the Precambrian crust of Norway and Sweden: Geothermal implications. *Geophys. Res. Lett.*, **14**, 260–263.

Pinet, C., Jaupart, C., Mareschal, J.-C., Gariepy, C., Bienfait, G., and Lapointe, R. 1991. Heat flow and structure of the lithosphere in the eastern Canadian shield. *J. Geophys. Res.*, **96**, 19941–19963.

Pollack, H. N., Hurter, S. J., and Johnston, J. R. 1993. Heat flow from the Earth's interior: analysis of the global data set. *Rev. Geophys.*, **31**, 267–280.

Pollack, H. N., Shen, P. Y., and Huang, S. 1996. Inference of ground surface temperature history from subsurface temperature data: Interpreting ensembles of temperature logs. *Pure Appl. Geophys.*, **147**, 537–550.

Poort, J., and Klerkx, J. 2004. Absence of a regional surface thermal high in the Baikal rift; new insights from detailed contouring of heat flow anomalies. *Tectonophysics*, **383**, 217–241.

Popov, Y. A., Pribnow, D. F. C., Sass, J. H., Williams, C. F., and Burkhardt, H. 1999. Characterization of rock thermal conductivity by high resolution optical scanning. *Geothermics*, **28**, 253–276.

Poupinet, G., Arndt, N., and Vacher, P. 2003. Seismic tomography beneath stable tectonic regions and the origin and composition of the continental lithospheric mantle. *Earth Planet. Sci. Lett.*, **212**, 89–101.

Press, W. H., Teukolsky, W. T., Vetterling, W. T., and Flannery, B. P. 1992. *Numerical Recipes in Fortran. The Art of Scientific Computing*. Cambridge, UK: Cambridge University Press.

Pupier, E., Duchene, S., and Toplis, M. J. 2008. Experimental quantification of plagioclase crystal size distribution during cooling of a basaltic liquid. *Contrib. Mineral. Petrol.*, **155**.

Purucker, M., Langlais, B., Olsen, N., Hulot, G., and Mandea, M. 2002. The southern edge of cratonic North America: Evidence from new satellite magnetometer observations. *Geophys. Res. Lett.*, **29**, 56–1.

Ranalli, G. 1995. *Rheology of the Earth*. 2nd ed. edn. Springer.

Ray, L., Förster, H.-J., Schilling, F. R., and Förster, A. 2006. Thermal diffusivity of felsic to mafic granulites at elevated temperatures. *Earth Planet. Sci. Lett.*, **251**, 241–253.

Ray, R. D., Eanes, R. J., and Chao, B. F. 1996. Detection of tidal dissipation in the solid earth by satellite tracking and altimetry. *Nature*, **381**, 595–597.

Revelle, R., and Maxwell, A. E. 1952. Heat flow through the floor of the eastern North Pacific ocean. *Nature*, **170**, 199–200.

Ribando, R. J., Torrance, K. E., and Turcotte, D. L. 1976. Numerical models for hydrothermal circulation in the oceanic crust. *J. Geophys. Res.*, **81**, 3007–3012.

Richter, F. M., Nataf, H. C., and Daly, S. F. 1983. Heat-transfer and horizontally averaged temperature of convection with large viscosity variations. *J. Fluid Mech.*, **129**, 173–192.

Ringwood, A. E. 1958. The constitution of the mantle–III Consequences of the olivine-spinel transition. *Geochim. Cosmochim. Acta*, **15**, 195–212.

Ringwood, A. E. 1962. A model for the upper mantle. *J. Geophys. Res.*, **67**, 857–867.

Ringwood, A. E. 1969. Phase transformations in the mantle. *Earth Planet. Sci. Lett.*, **5**, 401–412.

Ringwood, A. E. 1979. *Origin of the Earth and Moon*. Berlin: Springer-Verlag.

Roberts, G. O. 1979. Fast viscous Bénard convection. *Geophys. Astrophys. Fluid Dyn*, **12**, 235–272.

Roberts, P. H., and Loper, D. E. 1983. Towards a theory of the structure and evolution of a dendrite layer. Pages 329–349 of: Soward, A. (ed), *Stella and Planetary Magnetism*. New York: Gordon and Breach.

Roberts, P. H., Jones, C. A., and Calderwood, A. R. 2003. Energy Fluxes and Ohmic Dissipation in the Earth's Core. Pages 100–129 of: Jones, C. A., Soward, A. M., and Zhang, K. (eds), *Earth's Core and Lower Mantle*. London: Taylor & Francis.

Robertson, E. C. 1988. *Thermal properties of rocks*. Open file report 88-441. U.S. Geological Survey.

Rogers, J. J. W., and Adams, J. A. S. 1969. Thorium. Pages 90B–90O of: Wedepohl, K. H. (ed), *Handbook of Geochemistry*. Berlin: Springer Verlag.

Rolandone, F., Jaupart, C., Mareschal, J.-C., Gariépy, C., Bienfait, G., Carbonne, C., and Lapointe, R. 2002. Surface heat flow, crustal temperatures and mantle heat flow in the Proterozoic Trans-Hudson Orogen, Canadian Shield. *J. Geophys. Res.*, **107**, BO2314.

Rolandone, F., Mareschal, J.-C., and Jaupart, C. 2003. Temperatures at the base of the Laurentide Ice Sheet inferred from borehole temperature data. *Geophys. Res. Lett.*, **30**, 1944.

Rondenay, S., Bostock, M. G., Hearn, T. M., White, D. J., and Ellis, R. M. 2000. Lithospheric assembly and modification of the SE Canadian Shield: Abitibi-Grenville teleseismic experiment. *J. Geophys. Res.*, **105**, 13735–13754.

Rowley, D. B. 2002. Rate of plate creation and destruction; 180 Ma to present. *GSA Bull.*, **114**, 927–933.

Roy, S., and Rao, R. U. M. 2000. Heat flow in the Indian Shield. *J. Geophys. Res.*, **105**, 25587–25604.

Roy, S., and Rao, R. U. M. 2003. Towards a crustal thermal model for the Archean Dharwar craton, southern India. *Physics and Chemistry of the Earth*, **28**, 361–373.

Roy, S., Ray, L., Bhattacharya, A., and Srinivasan, R. 2008. Heat flow and crustal thermal structure in the Late Archean Closepet Granite batholith, South India. *Int. J. Earth Sci.*, **97**, 245–256.

Royden, L., and Keen, C. E. 1980. Rifting process and thermal evolution of the continental margin of Eastern Canada determined from subsidence curves. *Earth Planet. Sci. Lett.*, **51**, 343–361.

Rubie, D. C., Melosh, H. J., Reid, J. E., Liebske, C., and Righter, K. 2003. Mechanisms of metal-silicate equilibration in the terrestrial magma ocean. *Earth Planet. Sci. Lett.*, **205**, 239–255.

Rudnick, R. L., and Fountain, D. M. 1995. Nature and composition of the continental crust: A lower crustal perspective. *Rev. Geophys.*, **33**, 267–310.

Rudnick, R. L., and Gao, S. 2003. Composition of the continental crust. Pages 1–64 of: Rudnick, R. L. (ed), *Treatise on Geochemistry, The Crust*, vol. 3. New York: Permagon.

Rudnick, R. L., and Nyblade, A. A. 1999. The thickness of Archean lithosphere: constraints from xenolith thermobarometry and surface heat flow. Pages 3–11 of: Fei, Y., Bertka, C. M., and Mysen, B. O. (eds), *Mantle Petrology; Field Observations and High Pressure Experimentation: A Tribute to Francis R. (Joe) Boyd*. Geochemical Society.

Rudnick, R. L., McDonough, W. F., and O'Connell, R. J. 1998. Thermal structure, thickness and composition of continental lithosphere. *Chem. Geol.*, **145**, 395–411.

Rushmer, T., Minarik, W. G., and Taylor, G. J. 2000. Physical Processes of Core Formation. Pages 227–243 of: Canup, R. M., and Drake, K. (eds), *Origin of the Earth and Moon*. The University of Arizona Press.

Russell, J. K., Dipple, G. M., and Kopylova, M. G. 2001. Heat production and heat flow in the mantle lithosphere, Slave craton, Canada. *Phys. Earth Planet. Int.*, **123**, 27–44.

Ryan, M. P., and Sammis, C. G. 1981. The glass transition in basalt. *J. Geophys. Res.*, **86**, 9519–9535.

Rybach, L. 1988. Determination of heat production rate. Pages 125–142 of: Haenel, R., Rybach, L., and Stegena, L. (eds), *Handbook of Terrestrial Heat-Flow Density Determination*. Dordrecht(The Netherlands): Kluwer.

Sandiford, M., and McLaren, S. 2002. Tectonic feedback and the ordering of heat producing elements within the continental lithosphere. *Earth Planet. Sci. Lett.*, **204**, 133–150.

Sandwell, D., and Schubert, G. 1980. Geoid height versus age for symmetric spreading ridges. *J. Geophys. Res.*, **85**, 7235–7241.

Sass, J. H., Lachenbruch, A. H., and Jessop, A. M. 1971. Uniform heat flow in a deep hole in the Canadian Shield and its paleoclimatic implications. *J. Geophys. Res.*, **76**, 8586–8596.

Sass, J. H., Lachenbruch, A. H., Jr, S. P. Galanis, Morgan, P., Priest, S. S., Jr, T. H. Moses, and Munroe, R. J. 1994. Thermal regime of the southern Basin and Range: 1. Heat flow data from Arizona and the Mojave desert of California and Nevada. *J. Geophys. Res.*, **99**, 22,093–22,120.

Schärmeli, G. 1979. Identification of radiactive thermal conductivity in olivine up to 25 kbar and 1500K. Pages 60–74 of: Timmerhauf, K. D., and Barber, M. S. (eds), *Proceedings of the 6th AIRAPT Conference*. New York: Plenum Press.

Schatz, J. F., and Simmons, G. 1972. Thermal conductivity of Earth materials. *J. Geophys. Res.*, **77**, 6966–6983.

Scholz, Christopher H. 2002. *The Mechanics of Earthquakes and Faulting*. 2nd edn. Cambridge University Press.

Schubert, G., Turcotte, D. L., and Olson, P. 2001. *Mantle Convection in the Earth and Planets*. Cambridge University Press.

Sclater, J. G., and Christie, P. A. F. 1980. Continental stretching: An explanation of the post-mid-Cretaceous subsidence of the central North Sea Basin. J. Geophys. Res., **85**, 3711–3739.

Sclater, J. G., Crowe, J., and Anderson, R. N. 1976. On the reliability of oceanic heat flow averages. *J. Geophys. Res.*, **81**, 2997–3006.

Sclater, J. G., Jaupart, C., and Galson, D. 1980a. The heat flow through oceanic and continental crust and the heat loss from the Earth. *Rev. Geophys.*, **18**, 269–311.

Sclater, J. G., Royden, L., Horvath, F., Burchfiel, B. C., Semken, S., and Stegena, L. 1980b. The formation of the intra-Carpathian basins as determined from subsidence data. *Earth Planet. Sci. Lett.*, **51**, 139–162.

Sclater, J. G., Parsons, B., and Jaupart, C. 1981. Oceans and continents - Similarities and differences in the mechanisms of heat loss. *J. Geophys. Res.*, **86**, 11535–11552.

Sclater, John G., and Francheteau, Jean. 1970. The implications of terrestrial heat flow observations on current tectonic and geochemical models of the crust and upper mantle of the Earth. *Geophys. J. R. Astr. Soc.*, **20**, 509–542.

Scorer, R. S. 1957. Experiments on convection of isolated masses of buoyant fluid. *J. Fluid Mech.*, **2**, 583–594.

Sen, G. 1980. Mineralogical variations in the Delakhari sill, Deccan trapp intrusion, central india. *Contrib. Mineral. Petrol.*, **75**, 71–78.

Shapiro, N. M., and Ritzwoller, M. H. 2002. Monte-Carlo inversion for a global shear-velocity model of the crust and upper mantle. *Geophys. J. Int.*, **151**, 88–105.

Shapiro, N. M., and Ritzwoller, M. H. 2004. Thermodynamic constraints on seismic inversions. *Geophys. J. Int.*, **157**, 1175–1188.

Shaw, D. M., Dickin, A. P., Li, H., McNutt, R. H., Schwarcz, H. P., and Truscott, M. G. 1994. Crustal geochemistry in the Wawa-Foleyet region, Ontario. *Can. J. Earth Sci.*, **31**, 1104–1121.

Siggia, E. D. 1994. High Rayleigh number convection. *Ann. Rev. Fluid Mech.*, **26**, 137–168.

Silver, P. G. 1996. Seismic anisotropy beneath the continents: Probing the depths of geology. *Ann. Rev. Earth Planet. Sci.*, **24**, 385–432.

Simmons, G., and Nur, A. 1968. Granites: relation of properties in situ to laboratory measurements. *Science*, **162**, 789–791.

Sisson, T., Grove, T. L., and Coleman, D. S. 1996. Hornblende gabbro sill complex at Onion Valley, California, and a mixing origin for the Sierra Nevada batholith. *Contrib. Mineral. Petrol.*, **126**, 81–108.

Sleep, N. H. 1971. Thermal effects of the formation of atlantic continental margins by continental break up. *Geophys. J. R. Astr. Soc.*, **24**, 325–350.

Sleep, N. H. 1990. Hotspots and mantle plumes: some phenomenology. *J. Geophys. Res.*, **95**, 6715–6736.

Sleep, N. H. 2000. Evolution of the mode of convection within terrestrial planets. *J. Geophys. Res.*, **105**, 17563–17578.

Sleep, N. H., and Snell, N. S. 1976. Thermal contraction and flexure of mid-continent and Atlantic marginal basins. *Geophys. J. R. Astr. Soc.*, **45**, 125–154.

Sneddon, I. N. 1950. *Fourier transforms.* New York: McGraw-Hill Book Company, Inc. .

Solomatov, V. S. 1995. Scaling of temperature- and stress-dependent viscosity convection. *Phys. fluids*, **7**, 266–274.

Solomatov, V. S. 2000. *Fluid Dynamics of a Terrestrial Magma Ocean.* Origin of the Earth and Moon, edited by R. M. Canup and K. Righter and 69 collaborating authors. Tucson: University of Arizona Press., p.323-338. Pages 323–338.

Solomatov, V. S., and Moresi, L. N. 2000. Scaling of time-dependent stagnant lid convection: Application to small-scale convection on Earth and other terrestrial planets. *J. Geophys. Res.*, **105**, 21795–21817.

Sommerfeld, A. 1949. *Partial Differential Equations in Physics (Lectures on Theoretical Physics).* Vol. 6. New York: Academic Press.

Sotin, C., and Labrosse, S. 1999. Three-dimensional thermal convection in an iso-viscous, infinite Prandtl number fluid heated from within and from below: applications to the transfer of heat through planetary mantles. *Phys. Earth Planet. Inter.*, **112**, 171–190.

Spera, F. 2000. Physical Properties of Magmas. Pages 171–190 of: Sigurdsson, H. (ed), *Encyclopedia of Volcanoes*. New York: Academic Press.

Spohn, T., Hort, M., and Fischer, H. 1988. Numerical simulation of the crystallization of multicomponent melts in thin dikes or sills, 1, The liquidus phase. *J. Geophys. Res.*, **93**, 4880–4894.

Stacey, F. D., and Davis, P. M. 2008. *Physics of the Earth.* 4th edn. Cambridge University Press.

Stein, C. A. 1995. Heat Flow of the Earth. Pages 144–158 of: Ahrens, T. J. (ed), *Global Earth Physics: A Handbook of Physical Constants.* Washington: AGU.

Stein, C. A., and Stein, S. 1992. A model for the global variations in oceanic depth and heat flow with lithospheric age. *Nature*, **359**, 123–129.

Stevenson, D. J. 1989. Formation and early evolution of the Earth. Pages 817–873 of: Peltier, W. R. (ed), *Mantle Convection.* New York: Gordon and Breach.

Stevenson, D. J. 1990. Fluid Dynamics of Core Formation. Pages 231–249 of: Newsom, H. E., and Jones, J. H. (eds), *Origin of the Earth.* New York: Oxford University Press.

Strutt, R. J. 1906. On the distribution of radium in the Earth's crust and on internal heat. *Proc. Roy. Soc. Ser. A*, **77**, 472–485.

Su, Y. J. 2000. Mid-ocean ridge basalt trace element systematics: Constraints from database management, ICPMS analysis, global data compilation and petrologic modeling. *Unpub. Ph.D. thesis, Columbia University, New York*, 569 pp.

Swanberg, C. A., Chessman, M. D., Simmons, G., Smithson, S. B., Groenlie, G., and Heier, K. S. 1974. Heat-flow — heat-generation studies in Norway. *Tectonophys.*, **23**, 31–48.

Tait, S. R., and Jaupart, C. 1992. Compositional convection in a reactive crystalline mush and melt differentiation. *J. Geophys. Res.*, **97**, 6735–6756.

Talwani, M., Worzel, J. L., and Landisman, M. 1959. Rapid gravity computations for two-dimensional bodies with application to the Mendocino submarine fracture zone. *J. Geophys. Res.* **64**, 49–59.

Tapley, B., Ries, J., Bettadpur, S., Chambers, D., Cheng, M., Condi, F., Gunter, B., Kang, Z., Nagel, P., Pastor, R., Pekker, T., Poole, S., and Wang, F. 2005. GGM02 An improved Earth gravity field model from GRACE. *J. Geodesy*, **79**, 467–478.

Tegner, C., Wilson, J. R., and Brooks, C. K. 1983. Intraplutonic quench zones in the Kap Edvard Holm layered gabbro complex, East Greenland. *J. Petrol.*, **34**, 681–710.

Thompson, W. 1862. On the secular cooling of the Earth. *Trans. Royal Soc. Edinburgh*, **XXIII**, 295–311.

Thordarson, T., and Self, S. 1993. The Laki (Skaftar Fires) and Grimsvotn eruptions in 1783-1785. *Bull. Volc.*, **55**, 233–263.

Tikhonov, A. N., and Arsenin, V. Y. 1977. *Solution of ill posed problems*. New-York: Wiley.

Townsend, A. A. 1959. Temperature fluctuations over a heated horizontal surface. *J. Fluid Mech.*, **5**, 209–241.

Townsend, A. A. 1964. Natural convection in water over an ice surface. *Q.J.R. Met. Soc.*, **90**, 248–259.

Troschke, B., and Burkhardt, H. 1998. Thermal conductivity models for two-phase systems. *Phys. Chem. Earth*, **23**, 351–355.

Turcotte, D. L., and Ahern, J. L. 1977. On the thermal and subsidence history of sedimentary basins. *J. Geophys. Res.*, **82**, 3762–3766.

Turcotte, D. L., and McAdoo, D. C. 1979. Geoid anomalies and the thickness of the lithosphere. *J. Geophys. Res.*, **84**, 2381–2387.

Turcotte, D. L., and Oxburgh, E. R. 1967. Finite amplitude convective cells and continental drift. *J. Fluid Mech.*, **28**, 29–42.

Turner, J. S. 1973. *Buoyancy Effects in Fluids*. Cambridge University Press.

Urey, H. C. 1951. The origin and development of the earth and other terrestrial planets. *Geochim. Cosmochim. Acta*, **1**, 209–277.

Urey, H. C. 1964. A review of atomic abundances in chondrites and the origin of meteorites. *Rev. Geophys.*, **2**, 1–34.

van der Lee, S., and A., Frederiksen. 2005. Surface wave tomography applied to the North America upper mantle. In: Nolet, G., and Levander, A. (eds), *Seismic Data Analysis and Imaging With Global and Local Arrays*. AGU.

van der Lee, S., and Nolet, G. 1997. Upper mantle S velocity structure of North America. *J. Geophys. Res.*, **102**, 22815–22838.

van der Velden, A. J., Cook, F., Drummond, B. J., and Goleby, B. R. 2005. Reflections on the Neoarchean: A global prespective. Pages 255–265 of: Benn, K., Mareschal, J. C., and Condie, K. (eds), *Archean Geodynamics and Environments*. Geophys. Mono., vol. 164. AGU.

Vasco, D. W., and Johnson, L. R. 1998. Whole earth structure estimated from seismic arrival times. *J. Geophys. Res.*, **103**, 2633–2672.

Vasseur, G., and Singh, R. N. 1986. Effects of random horizontal variations in radiogenic heat source distribution on its relationship with heat flow. *J. Geophys. Res.*, **91**, 10397–10404.

Vasseur, G., Bernard, P., Van de Meulebrouck, J., Kast, Y., and Jolivet, J. 1983. Holocene paleotemperatures deduced from geothermal measurements. *Palaeogeogr., Palaeocl., Palaeoecol.*, **43**, 237–259.

Vitorello, I., Hamza, V. M., and Pollack, H. N. 1980. Terrestrial heat flow in the Brazilian highlands. *J. Geophys. Res.*, **85**, 3778–3788.

Vogt, P., and Ostenso, N. 1967. Steady state crustal spreading. *Nature*, **215**, 810–817.

Von Herzen, R. P., and Maxwell, A. E. 1959. The measurement of thermal conductivity of deep-sea sediments by a needleprobe method. *J. Geophys. Res.*, **64**, 1557–1563.

Von Herzen, R. P., and Vacquier, V. 1967. Terrestrial heat flow in Lake Malawi, Africa. *J. Geophys. Res.*, **72**, 4221–4226.

Von Herzen, R. P., Cordery, M. J., Detrick, R. S., and Fang, C. 1989. Heat flow and the thermal origin of hot spot swells: The Hawaiian Swell revisited. *J. Geophys. Res.*, **94**, 13783–13799.

Von Herzen, R. P., Detrick, R. S., Crough, S. T., Epp, D., and Fehn, U. 1982. Thermal origin of the Hawaiian swell: Heat flow evidence and thermal models. *J. Geophys. Res.*, **87**, 6711–6724.

Vosteen, H. D., Rath, V., Clauser, C., and Lammerer, B. 2003. The thermal regime of the Eastern Alps from inverse analyses along the TRANSALP profile. *Phys Chem Earth*, **28**, 393–405.

Wager, L. R., and Brown, G. M. 1968. *Layered Igneous Rocks*. Edinburgh: Oliver and Boyd.

Walker, F. A. 1930. A tholeiitic phase of the quartz-dolerite magma of central Scotland. *Min. Mag.* **22**, 368–376.

Walker, F. A. 1935. The late palaeozoic quartz-dolerites and tholeiites of Scotland. *Min. Mag.* **24**, 131–159.

Wang, K. 1992. Estimation of ground surface temperatures from borehole temperature data. *J. Geophys. Res.*, **97**, 2095–2106.

Wasserburg, G. J., MacDonald, G. J. F., Hoyle, F., and Fowler, W. A. 1964. Relative contributions of Uranium, Thorium, and Potassium to heat production in the Earth. *Science*, **143**, 465–467.

Watts, A. B. 2001. *Isostasy and Flexure of the Lithosphere*. Cambridge, U.K.: Cambridge University Press.

Watts, A. B. 2007. Crust and Lithosphere Dynamics. An overview. In: Watts, A. B. (ed), *Treatise on Geophysics, vol 6, Crust and Lithosphere Dynamics*. Elsevier.

Watts, A. B., and Daly, S. F. 1981. Long wavelength gravity and topography anomalies. *Ann. Rev. Earth Planet. Sci.*, **9**, 415–448.

Wegener, A. 1928. *The Origin of Oceans and Continents*. 4 edn. Dover (Reprinted 1966).

Wessel, P., and Smith, W. H. F. 1995. New version of the generic mapping tools released. *EOS Transactions*, **76**, 329–329.

Wiens, D. A., and Stein, S. 1983. Age dependence of oceanic intraplate seismicity and implications for lithospheric evolution. *J. Geophys. Res.*, **88**, 6455–6468.

Wiens, D. A., and Stein, S. 1984. Intraplate seismicity and stresses in young oceanic lithosphere. *J. Geophys. Res.*, **89**, 11442–11464.

Williams, D. L., and von Herzen, R. P. 1974. Heat Loss from the Earth: New Estimate. *Geology*, **2**, 327–330.

Wilson, J. T. 1963. A possible origin of the Hawaiian Islands. *Can. J. Phys.*, **1**, 863–868.

Woodhouse, J. H., and Dziewonski, A. M. 1989. Seismic modelling of the Earth's large-scale three-dimensional structure. *Roy. Soc. London, Phil. Trans Ser. A*, **328**, 291–308.

Woods, A. W., and Huppert, H. E. 1989. The growth of a compositionally stratified solid above a horizontal boundary. *J. Fluid Mech.*, **199**, 29–53.

Workman, R. K., and Hart, S. R. 2005. Major and trace element composition of the depleted MORB mantle (DMM). *Earth and Planet. Sci. Lett.*, **231**, 53–72.

Worster, M. G. 1986. The axisymmetric laminar plume: asymptotic solution for large Prandtl number. *Studies App. Math.*, **75**, 139–152.

Worster, M. G. 1986b. Solidification of an alloy from a cooled boundary, *J. Fluid Mech.*, **167**, 481–501.

Worster, M. G., Huppert, H. E., Stephen, R., and Sparks, J. 1990. Convection and crystallization in magma cooled from above. *Earth Planet. Sci. Lett.*, **101**, 75–89.

Worster, M. G., Huppert, H. E., Stephen, R., and Sparks, J. 1993. The crystallization of lava lakes. *J. Geophys. Res.*, **98**, 15891–15901.

Wright, T. L., and Okamura, R. T. 1977. *Cooling and crystallization of tholeiitic basalt - 1965 Makaopuhi lava lake, Hawaii*. U.S. Geol. Surv. Prof. Pap. 1004.

Wright, T. L., Peck, D. L., and Shaw, H. R. 1976. Kilauea lava lakes: Natural laboratories for study of cooling, crystallization and differentiation of basaltic magma. Pages 375–390 of: Sutton, G. H., Manghnani, M. H., and Moberly, R. (eds), *The Geophysics of the Pacific Ocean Basins and its Margin, Geophys. Monograph. Ser. vol. 19*. Washington, D.C.: Am. Geophys. Union.

Xu, Y., Shankland, T. J., Linhardt, S., Rubie, D. C., Langenhorst, F., and Klasinski, K. 2004. Thermal diffusivity and conductivity of olivine, wadsleyite and ringwoodite to 20 GPa and 1373 K. *Phys. Earth Planet. Inter.*, **143**, 321–336.

Zhong, S. J. 2005. Dynamics of thermal plumes in three-dimensional isoviscous thermal convection. *Geophys. J. Int.*, **162**, 289–300.

Ziagos, J. P., Blackwell, D. D., and Mooser, F. 1985. Heat flow in southern Mexico and the thermal effects of subduction. *J. Geophys. Res.*, **90**, 5410–5419.

Zschau, J. 1986. Constraints from the Chandler wobble period. Pages 315–344 of: Anderson, A. J., and Cazenave, A. (eds), *Space geodesy and geodynamics*. New York: Academic Press.

Index

Accretion, 10, 300
Activation energy, 284
Activation volume, 284, 299
Adams–Williamson equation, 11, 46
Adiabat, 44
Avrami number, 355

Basin and Range, 201, 209, 228, 249
Basin subsidence
  Stretching model, 221
Bessel functions, 401
  Modified, 401
Biot number, 267
Boundary
  Free, 122, 134
  Rigid, 122, 291
Boundary layer, 117, 120, 127, 135, 147, 275, 279, 281, 282, 291, 294
  Thermal, 289
  Thickness, 283
Boussinesq approximation, 45
Bulk modulus, 38
Bulk silicate Earth, 254, 260
Bullen parameter, 22
Buoyancy flux, 107, 125, 273, 280

Climate
  Correction, 358, 412
Climate change, 413
Conductivity
  Layered system, 59
  Mixture, 60
Constitutive equation, 42, 47
Continental growth, 315
Continents
  Conductive lid, 267
  Lithospheric roots, 266
Convection
  Cell width, 266
  Cells, 262, 268
  Heat flux, 118
  Internal heating, 271
  Mantle, 134, 261
  Onset, 112
  Rayleigh–Benard, 111, 121, 123, 134, 262, 270, 292
  Stress scale, 286
  Velocity scale, 292
Convolution, 66, 384
Cooling
  Dike, 86
  Sill, 86
Cooling half-space, 149
Cooling plate, 159
Core, 257
Core cooling, 296
Core formation, 10, 302
Core–mantle boundary, 19
Craton, 176, 194
Crystal
  Growth, 318, 354
  Nucleation, 318, 354
Curie isotherm, 228

Darcy's law, 138
Deccan traps, 250
Delamination, 214, 215
Deviatoric stress, 42
Differentiation index, 185
Diffusion creep, 285
Diffusivity, 51, 52, 421
Dislocation creep, 285
Dissipation, 123, 125, 131, 133, 274, 279
Downward continuation, 73, 400

Effective elastic thickness, 13, 173, 227
Entrainment rate, 106
Erosion, 85, 205
Error function, 62
Extension, 215, 249

Fourier series, 72, 392
Fourier transform, 73, 394
Fourier's law, 36

Geoneutrinos, 256
Geodynamo, 257

Geoid, 12, 153
Geotherm, 55
Giant impact, 10, 302
Glacier, 85, 375
Gravitational energy, 11, 234
Gravity
  Free air, 12, 13
Green's function, 66, 404, 407
  Half-space, 404
Gruneisen parameter, 39
GSTH, 363

Hankel transform, 74, 91, 401
Heat equation, 51
Heat flow
  Basin and Range, 249
  Compressional orogens, 249
  Continental margins, 250
  Continents, 245, 248
  Oceans, 239
  Rifts, 250
  Yellowstone, 250
Heat flux, 36
  Core, 257, 260, 313
  Hotspot, 168, 170
  Measurements, 408
  Moho, 187, 189
  Sea floor, 151, 153
Heat production, 52, 425
  Bulk silicate Earth, 255
  Chondrites, 255
  Crust, 182, 426
  Dimensionless, 276
  KTB, 431
  Mantle, 253
Heat refraction, 77
Heinrich events, 372
Hot-spots, 165, 169, 259
Hydrodynamic velocity, 38
Hydrothermal circulation, 143, 153, 164, 239, 240
Hydrothermal convection, 137

Igneous complexes, 318
Intra-cratonic basins, 218
Intrusions, 33, 137
Isentrope, 20, 21, 43, 306
Isostasy, 12, 13, 228

Komatiites, 305

Laplace transform, 82, 385
Laplace's equation, 54
Latent heat, 39, 326, 328, 376, 421
Laurentide ice sheet, 358, 372, 375, 412
Lid, 297
Liquidus, 323, 334
Lithosphere, 9, 29, 147
Lithospheric thickening, 163
Little Ice Age, 363

Magma
  Conductive cooling, 339
  Convection, 123, 133, 321, 343, 347, 348
  Nusselt number
    Magma, 321
  Rayleigh number, 321
  Superheated, 321, 323, 326, 332, 333, 343, 346
  Thermal properties, 418
  Undercooling, 319, 348
Magma ocean, 302
Magma reservoir, 127
Mantle
  Depleted, 299
Mantle convection, 123
Mantle plumes
  Heat flux, 296
Mantle plumes, 291
Material derivative, 41
Medieval climate optimum, 363
Metamorphism, 205
Mixing laws
  Hashin–Sthrikhman bounds, 420
  Reuss, 420
  Voigt, 420
  Voigt–Reuss–Hill, 420
Moment of inertia, 11
Momentum boundary layer, 104
Mush, 319, 334, 339

Navier–Stokes equation, 48
Newtonian fluid, 47
Nusselt number, 113, 119, 123, 126, 131, 136, 142, 264, 266, 271, 276, 277, 287, 345

Oceanic plates, 296
Overthrust
  Crustal scale, 203

Péclet number, 50, 139, 143
Parameterized cooling model, 308
Peridotite, 18
Permeability, 138
Perovskite, 18, 19, 24
Plate tectonics, 8, 15
Plume, 99
Plumes, 121
  Heat flux, 295
  Number, 294
Poisson's equation, 54
Post-glacial rebound, 14
Post-perovskite, 19, 24
Prandtl number, 50, 105, 110, 114, 119, 122, 127, 129, 131, 132, 136, 262
PREM, 9
Pyrolite, 18, 19, 254

Radiation, 37
Rayleigh number, 112, 114, 116, 119, 130, 140, 142, 271, 276, 287, 344, 352
  Critical, 115, 141
Refraction, 78, 80
Reynolds number, 49, 105, 108, 111, 114, 129, 131
Rheological temperature difference, 282

Rheological temperature scale, 279, 284, 286, 290
Rifts, 208
Rossby number, 50

Sea floor
  Ages, 32, 310
  Bathymetry, 30, 33, 152, 157, 168
Sea-floor spreading, 8, 149
Sedimentation, 85
Sedimentary basins, 218
Seismic tomography, 26, 27, 29, 170, 224
Seismicity, 15
  Focal depth, 173
Seismogenic zone, 227
Shear thinning, 289
Singular Value Decomposition (SVD), 365
Skin depth, 71
Slurry, 334
Solidus, 334
Specific heat, 38
Stagnant lid, 282, 298
Stefan constant, 37, 64
Stefan number, 326, 330
Stefan problem, 163, 328, 376
Stress, 41
Subduction, 8, 202
Supercontinent cycle, 313
Superposition, 53

Taylor number, 50
Thermal, 99
  Rayleigh number, 109
Thermal conductivity, 36, 415
Thermal expansion, 38
Thermal wave
  Damping, 71

Underplating, 211
Uplift
  Colorado Plateau, 214
Urey number, 307, 309

Viscosity, 47
  Arrhenius, 284
  Interior, 289
  Magmas, 423
  Mantle, 15
  Non-Newtonian, 285
  Temperature dependent, 278
Viscosity contrast, 279, 286

Xenoliths, 189, 191